# Sprinkler and Drip Irrigation

V. Ravikumar

# Sprinkler and Drip Irrigation

## Theory and Practice

 Springer

V. Ravikumar
Water Technology Centre
Tamil Nadu Agricultural University
Coimbatore, India

ISBN 978-981-19-2777-5          ISBN 978-981-19-2775-1   (eBook)
https://doi.org/10.1007/978-981-19-2775-1

This Springer imprint is published by the registered company Springer Nature Singapore Pte Ltd.
The registered company address is: 152 Beach Road, #21-01/04 Gateway East, Singapore 189721,
Singapore

*Two teachers who taught me undergraduate classes at Agricultural Engineering College and Research Institute, Tamil Nadu Agricultural University, Coimbatore, India, were motivating forces behind many of my endeavours. They did not guide me about what to do. They lived their life for teaching and are inspiration for many students like me. They were fondly called by my classmates as Tandem Plough. I do not know whether I have grown to the level of dedicating my works in memory of them.*

*1. Prof. C. Divaker Durairaj*
*2. Prof. D. Manohar Jesudas*

# Foreword

Sprinkler and drip irrigation has become an indispensable way of irrigation around the world due to the continuously shrinking availability of water. Many books on sprinkler and drip irrigation are already available in India and as well as in the world. But there is a need for bringing out a comprehensive treatise like this book at this point of time.

Sprinkler and drip Irrigation is being increasingly researched and innovations are continuously emerging. Tamil Nadu Agricultural University has provided the Author a congenial ambience to amass a rich experience on this subject and has let him continuously work in this field for 27 years. He has taught numerous students, working professionals and the farmers. He has researched on this subject and published in many acclaimed journals in the world and has given a good visibility to the university on this subject. He has developed and is continuously in the process of developing user friendly softwares for the design of drip irrigation systems.

He has endeavoured to document his experiences on this subject through this book, and I am happy to note that the renowned publisher, namely Springer Nature, also has accepted to bring out this book.

The prime objective of any scientific book is to provide the reader with the knowledge and skill on the subject. While the reader reads the book, he must be able to comfortably understand with his available capabilities. The book must be lucid, and the reader must be able to understand all the complexities of the subject in a step-by-step approach.

The author has achieved all these objectives in this book. Information on basic components to optimal design methodologies of irrigation systems is provided in the book. The book contains two sections, namely a basic section and a main section. Hence, all the students who pass out from schools can easily understand this book. The basic section would also be very helpful for the working professionals who want to refresh the fundamentals.

The author has exclusively written a chapter on a niche area, namely rhizosphere modelling, which I hope will be very useful for the researchers.

I hope this book will become a good reference book for all the professionals in this field, and I am happy to congratulate the author for having reached this feat in his career.

Dr. V. Geethalakshmi
Vice-Chancellor
Tamil Nadu Agricultural University
Coimbatore, India

# Preface

Sprinkler and Drip irrigation engineers need to have skills in both theory and practice for a successful career. I moulded myself all along my career to train myself so that theory and practice go together. I have tried to give practical information for all the questions those are raised by the professionals in this field.

I have been teaching this subject for fresh students as well as practicing engineers. Fresh students are able to grasp and appreciate the applied subjects like this when theory is built up in the class room in gradual steps with demonstrated examples. This presentation style is my hall mark and this worked well with all my students. The material in the book has been broadly classified into 2 sections. One is named as Main section containing sprinkler Irrigation and drip Irrigation and two is named as Basic section containing essential requisites needed to understand the sprinkler and drip irrigation.

The equations used for design of pipes laid on slope are implicit equations. Many research advancements have happened to convert this implicit equations into explicit equations. But the theory behind all these developments are complex. Therefore, only when the designers are able to comprehend these developments, they would get the confidence of using these equations. Mostly, practising engineers resort to trial and error solution methods to circumvent this issue. In this book, I have attempted to bridge this issue by harnessing the potential of MS Excel. An add-on tool known as Solver in MS-Excel has been extensively demonstrated in the book for handling implicit equations with case. Not much training in MS-Excel is needed to start using this tool.

Sprinkler irrigation system design is first demonstrated with conventional back step method. Relative advantages and disadvantages of explicit method are discussed by comparing with back step method. Pump selection is normally done to cater to the most difficult situations. Procedures for operation of pump under normal situations is also presented so that optimal utilization of the installed pump is achieved.

In drip irrigation design, using 'optimal wetted volume of soil' as design objective is demonstrated with a numerical example. There are many criteria in use for judging the goodness of design of sprinkler and lateral pipes. While Keller and Karmeli uniformity criterion and statistical variation criterion are much reported in research,

Relative Flow Difference criterion continues to be used by the field engineers because it is easily comprehensible. Therefore in this book, I have promoted all my design equations based on Relative Flow Difference criterion. Additional advantage of this criterion is that, both sprinkler and drip pipes can be designed with this single common criterion. Procedures for conversion from Relative Flow Difference to all the other criteria are also presented. Poly-plot method for designing the telescopic drip pipes in non-uniform slopes is often used by the field engineers. One disadvantage of poly-plot method is that it is a manual method and in this book, simple procedures for adapting this method in MS-Excel is done.

When designing main pipes, giving consideration to water hammer and also economy improves the quality of design and installation. Hence procedures for the same are presented and equations are derived from the scratch. A rudimentary knowledge on fertigation is presented because the ultimate success of the investment could be achieved only when the engineer makes the farmer win economically.

Hydrus software is really a boon for all the people involved in variably saturated flow and nutrient transport studies. Due to the phenomenal advancements in computing technology and versatility of Hydrus, application of Hydrus to improve the water and nutrient management would continue to be an active area of research in the future. Since the Hydrus software solves the Richards equation and reactive transport convection-dispersion equations, the users of the software can concentrate on optimizing the parameters of water and nutrient flow in soil by doing field research. A chapter namely, Rhizosphere modelling has found place in the book which will be very helpful for the researchers who want to initiate research work on this subject.

Irrigation planning by doing water budgeting and economic analysis also would help engineers improve the quality of their work. Hence this subject is discussed here.

The subject presented in this book is built upon the basic subjects like fluid mechanics, pumps and soil science. Therefore a humble introduction of all these subjects have been provided in a separate basic section which is sufficient for complete understanding of the material presented here. Those who feel comfortable in the basic can skip the basic section and the reader is prompted in the main text to refer the basic section wherever required. The basic section would also be useful for agricultural, civil and mechanical engineers who want to learn this subject by self-reading.

The basic section has been so oriented that the discussed examples are directly related to Sprinkler and Drip irrigation systems. I have experienced with learners that the basic section creates enthusiasm and excites them as they are able to relate with many simple real life encounters. The basic section is proven to be very useful for the practicing engineers as they would have mostly forgotten the theory.

The subject in this book is the result of my long years of experience and definitely is prone for mistakes. I will be very happy to correct them if you could inform me to my address: veeravikumar1@gmail.com or veeravikumar@hotmail.com.

Coimbatore, India                                                                    V. Ravikumar

# Acknowledgements

I am very much obliged to thank Dr. K. Shanmugasundaram, retired professor in soil and water conservation engineering of agricultural engineering college and research institute, Kumulur, Trichy, Tamil Nadu Agricultural University (TNAU). He rendered a sincere and self-less help in correcting the manuscript.

I am very much grateful to my present boss, Prof. S. Panneerselvam, Director, Water Technology Centre, TNAU, Coimbatore, for his constant encouragement.

The help rendered by my earlier bosses, Prof. S. Santhana Bosu, Prof. M. V. Ranghaswami and Prof. S. Chellamuthu, throughout my career could not be forgotten.

Prof. K. Appavu, Dr. P. Balsubramaniam, Dr. T. Sherene Genita Rajammal and Dr. K. M. Chellamuthu helped me acquire basic knowledge in soil chemistry.

The help rendered by my present colleagues Dr. G. Thiyagarajan and Ms. S. Nithya at various levels was very useful. I am happy to thank all my students whom I taught and simultaneously learnt from them back.

The works of my Ph.D. students, namely Dr. M. Angaleeswari, Dr. R. Sharmiladevi, Dr. S. Vanitha and Ms. K. K. Shaheemath Suhara, are the building blocks of this book.

The work of my M.Tech. student, Ms. M. R. Namitha, and my undergraduate students, Mr. M. Krishnan and Mr. Mohanraj, is also significant in this book.

Dr. G. Vijayakumar who worked with me in Hydrus provided a significant contribution.

I would like to immensely thank Prof. Simunek J., University of California, Riverside, for making me learn Hydrus software.

I would like to place on record of my thanks to my teacher Prof. V. Kumar, who spared the famous book, namely *Sprinkle and Trickle Irrigation*, way back in year 1996. He is also Alumni and a former student of Keller. Getting any book nowadays in India is easy and possible. But in 1996, knowing that such a book is available on earth itself was a great knowledge. The hard copy he gave me during 1996 is the starting point for all my work on this subject. I am still exploring the magnificence of this book.

The anonymous reviewers and the publisher Springer who reposed their faith on me in bringing out this book is a corner stone for everything related to this book.

# Contents

Contents

# About the Author

V. Ravikumar is currently the chief Scientist in an All India coordinated Indian council of Agricultural Research (ICAR) project namely Irrigation Water Management at Water Technology Centre, Tamil Nadu Agricultural University (TNAU), Coimbatore, India. He obtained Bachelor of Engineering degree from College of Agricultural Engineering, TNAU, Coimbatore; and masters and doctorate degrees in Irrigation Water Management in College of Engineering, Guindy, Chennai. He has more than 30 research publications which include two in American Society of Civil Engineers, three in Agricultural Water Management and one in Irrigation Science. He has taught this subject of Sprinkler and Drip irrigation to students of agricultural engineering, working professionals and farmers for a period of 27 years. He is a Co-developer of a popular free software for the design of drip irrigation systems namely, Drip System Simulation Programme (DSSP). He was awarded with Shrimati Saroma Sanyal Memorial prize by Institution of Engineers (India) in 2011.

# Main Section

# Sprinkler Irrigation Types and Components

## Introduction

In sprinkler irrigation artificial rainfall is provided for irrigating crops. In a Sprinkler Irrigation System (SIS), pipes are used for conveyance and sprinkling of water is brought about by sprinkler heads. Relatively, sprinkler irrigation needs more pressure than drip irrigation. In this chapter, advantages, types of sprinkler irrigation, types of sprinkler heads, flow control valves and air valves used in SIS are discussed. The need of booster pump in sprinkler irrigation and water filtration methods for sprinkler irrigation are also discussed.

## Advantages

1.  When a land has a rolling topography with shallow soil, the land levelling cannot be done (Fig. 1). In such cases, the land need not be graded and sprinkler irrigation can be adopted, without producing runoff.
2.  Labour requirement is less.
3.  Modification of weather extremes by increasing humidity and cooling crops is possible (Fig. 2). Sometimes water would be available in the soil for the roots to absorb. But if the atmospheric temperature is too high, plants may not be able to extract water and show wilt symptoms. In such situations, external cooling of the crop by sprinkling water on the foliage would remove the wilt. For groundnut crop, sprinklers are very useful in external cooling during very hot weather conditions.
4.  In areas of frost problem, sprinkling water removes frost deposited over the foliage (Fig. 3).
5.  High application efficiency.

**Fig. 1**  Undulating topography and shallow soil [1]

**Fig. 2**  Wilt of leaves due to excess sunlight (Gerald Holmes, Strawberry Center, Cal Poly San Luis Obispo, Bugwood.org)

**Fig. 3** Frost on leaves (Credit: Stephan Herb, North Western Univ.)

# Disadvantages

1.  High investment cost. Many a times, pump has to be changed because the existing pumps may not be able to provide sufficient pressure.
2.  The operating cost for pressurizing water is more.
3.  Saline water may cause problems because salt water is absorbed by the leaves of some crops and high concentrations of bicarbonates in irrigation water may affect the quality of the fruit.

If sodium level is more than 70 ppm or chloride level is above 105 ppm, then leaf burning problems may occur (Fig. 4). Even with the good quality water, the crops like citrus and grapes suffer. During sprinkler irrigation, the foliage gets wetted. Before next irrigation, the water in the foliage evaporates leaving sodium chloride on the leaves. Even low concentration of sodium chloride causes leaf burning in these crops.

**Fig. 4** Burn due to salty water spray (www.lcsun-news.com)

4.    If soils have very less infiltration rate (<3 mm/h), then runoff occurs.
5.    Windy conditions are very difficult to handle (Fig. 5).
6.    For irregular shaped fields, spacing of sprinklers is very inconvenient (Fig. 6).

Note that the wind is blowing from left to right, and hence, the jet is unable to pierce through the force of wind.

**Fig. 5** Wind effect on sprinkler irrigation (https://www.senninger.com/node/66146)

**Fig. 6** Irregular shaped fields—difficult for sprinkler spacing

# Classification of Sprinkler Irrigation Systems

Primarily sprinkler irrigation systems are divided into three types. They are as follows:

(1)  Solid set system
(2)  Continuous move system
(3)  Periodic move system.

In a solid set system the sprinkler system remains at a fixed position while sprinkling is being done (Fig. 7). Pipes are buried under ground, and sprinklers are set on the ground. This system is relatively costly.

In continuous move system, sprinkler moves while sprinkling is done. There are three types in continuous move system.

1.  Travelling sprinkler
2.  Centre pivot
3.  Boom type.

The travelling sprinkler system has a flexible hose wound on a reel (Fig. 8). Usually one reel serves one large sprinkler placed on small wheels. When irrigation is done, the reel is slowly rotated to pull the sprinkler towards the reel. The energy for rotation of wheels and the reels is obtained from the pressurized water used for sprinkling.

In centre-pivot (Figs. 9 and 10) and boom-type sprinklers (Fig. 11), large number of sprinklers are fixed on a truss and the truss is mounted on wheels. The entire truss

**Fig. 7**  Solid set system

**Fig.8** Travelling sprinkler

**Fig.9** Centre pivot (China
Yulin Irrig. Equip.)

along with the sprinklers rotates about a centre while sprinkling. The boom sprinklers
are also similar to centre pivot with only one major difference. In the boom type,
the truss moves linearly and does not rotate about a centre. The centre-pivot and
boom-type sprinklers are not used in India owing to the reason that they are costly
and cover very large areal extent at a time.

In periodic move system, sprinkler pipes and fittings are moved from one location
to the next location after a place is irrigated. During sprinkling is done, the pipes and
the sprinklers will remain in a place. The components are designed in such a way
that they can be easily connected, disconnected, lifted up and fitted at another place
[2] (Fig. 12).

**Fig.10** Areal units irrigated by centre pivots (Science photo library)

**Fig.11** Boom sprinklers

## Components of a Periodic Move Sprinkler Irrigation System

Following are the components of a periodic move sprinkler irrigation system (Fig. 12)

- Pump
- Pipes and fittings
- Sprinkler heads
- Valves
- Booster pumps
- Desilting basins and debris screens.

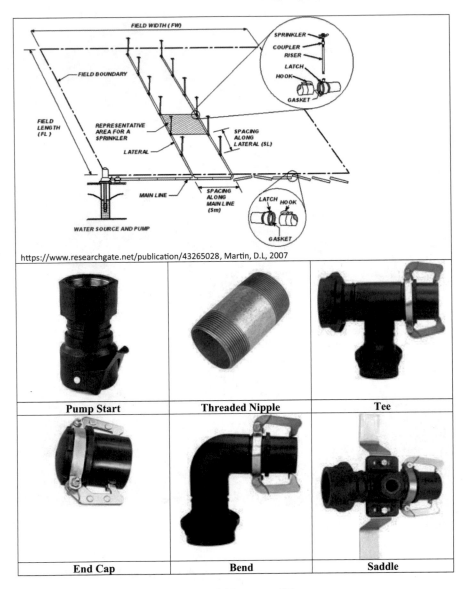

**Fig. 12** Components of a periodic move sprinkler system [3]

## *Pump*

For sprinkler irrigation, pump is a very important accessory. Pump is used for pressurizing water and to release as a jet through sprinklers. All types of pumps except airlift pumps can be used for sprinkler. At the exit of the sprinkler around 20–70 m pressure head is needed to be provided by the pump.

## Pipes

The pipes may be aluminium or high-density polyethylene (HDPE) pipes. Since HDPE pipes are cheaper than the aluminium pipes, HDPE pipes are increasingly being used nowadays. The pipes are available in lengths of 3 and 6 m. Pipe strengths are specified using maximum pressure which the pipe can withstand. The unit for pressure specification is Mega Pascals (MPa), but in India mostly the conventional pressure specification used is kg/cm$^2$ (written as ksc). There are four pressure ratings in which the pipes are available in market. They are 0.25 MPa (2.5 ksc), 0.32 MPa (3.2 ksc), 0.4 MPa (4 ksc) and 0.6 MPa (6 ksc). Table 1 provides the physical dimensions of different pipe diameters with respect to pipe strengths. Mostly 75 mm diameter pipes are used by farmers in India.

The front end of the pipes has a rubber washer for quicker connection. The clips of the rear end can be moved on to the front end of the joining pipe. When the pressure increases the washer will expand and arrest the water leakages.

Main line is the pipe section leading to the field from the pump. Lateral lines are the pipe section installed in the field. In the lateral line, sprinklers are installed in between the pipes (Fig. 12).

## Pipe Fittings

Figure 13 shows a typical pump installed in a well. At the delivery end of the pump, normally a delivery valve is provided. Downstream of the valve, a GI pipe connection fitting namely threaded nipple (shown in Fig. 12) is fitted. Then another fitting, namely pump start, is fitted using threaded connection. This connection is useful to connect GI pipe on one side and easy-to-connect HDPE pipes on the other side.

Different pipe fittings used are shown in Fig. 12. End caps are used for closing the end of pipes. Tee is used for dividing a supply line into two branches. The use of the bend is obvious. The sprinkler heads are fixed in a riser pipe usually with 1 m height. When crops like sugarcane are grown under sprinklers, the length of riser pipe has to be sufficiently higher. Sprinkler with riser pipe is fixed on a fitting called as saddle. The saddle has a threaded provision for connecting to the riser pipes.

## Sprinkler Heads

The sprinkler heads may have single nozzle or twin nozzles. Single-nozzle sprinklers wet smaller diameters (3.5–14.0 m). Twin-nozzle sprinklers wet larger diameters (Fig. 14). In these sprinklers, one nozzle is smaller and another nozzle is larger. Smaller nozzle would wet areas nearer the centre, and larger nozzle would wet peripheral areas. Rain guns are capable of wetting very large diameters as much as

**Table 1** Dimensions of sprinkler pipes for sprinkler irrigation (IS 14151: Part-1-1999) All dimensions in millimetres

| Nominal dia. | Outside dia. | Nominal tol. on outside dia. | Ovality | Wall thickness ($e$) | | | | | | | |
| | | | | Class 1 (0.25 MPa) | | Class 2 (0.32 MPa) | | Class 3 (0.4 MPa) | | Class 4 (0.6 MPa) | |
| | | | | Min | Max | Min | Max | Min | Max | Min | Max |
| (1) | (2) | (3) | (4) | (5) | (6) | (7) | (8) | (9) | (10) | (11) | (12) |
| 40 | 40.0 | +0.4 | 1.4 | – | – | – | – | – | – | 2.3 | 2.8 |
| 50 | 50.0 | +0.5 | 1.4 | – | – | – | – | 2.0 | 2.4 | 2.9 | 3.4 |
| 63 | 63.0 | +0.6 | 1.5 | – | – | 2.0 | 2.4 | 2.5 | 2.9 | 3.8 | 4.4 |
| 75 | 75.0 | +0.7 | 1.6 | 2.0 | 2.4 | 2.5 | 2.9 | 3.0 | 3.4 | 4.5 | 5.2 |
| 90 | 90.0 | +0.8 | 1.8 | 2.2 | 2.6 | 2.9 | 3.4 | 3.5 | 4.1 | 5.3 | 6.1 |
| 110 | 110.0 | +1.0 | 2.2 | 2.7 | 3.2 | 3.4 | 3.9 | 4.2 | 4.8 | 6.5 | 7.4 |
| 125 | 125.0 | +1.2 | 2.5 | 3.1 | 3.6 | 3.8 | 4.5 | 4.8 | 5.5 | 7.4 | 8.3 |
| 140 | 140.0 | +1.3 | 2.8 | 3.5 | 4.1 | 4.3 | 5.0 | 5.4 | 6.1 | 8.3 | 9.3 |
| 160 | 160.0 | +1.5 | 3.2 | 3.9 | 4.5 | 4.9 | 5.6 | 6.2 | 7.0 | 9.4 | 10.6 |
| 180 | 180.0 | +1.7 | 3.6 | 4.4 | 5.0 | 5.5 | 6.3 | 6.9 | 7.8 | 10.6 | 11.9 |
| 200 | 200.0 | +1.8 | 4.0 | 4.9 | 5.6 | 6.1 | 7.0 | 7.7 | 8.7 | 11.8 | 13.2 |

**Fig. 13** A typical pump installation in a well

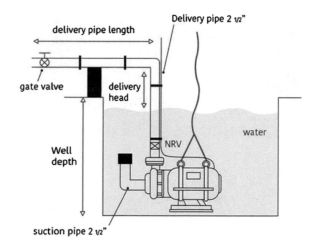

**Fig. 14** Twin-nozzle sprinkler blown-up view

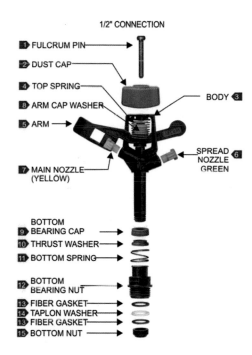

100 m (Fig. 15). Sprinklers are designed to rotate without any external energy. Water flowing through the larger nozzle hits on a rocker arm, and this rocker arm provides an impact and rotates the sprinklers. After every impact, the rocker arm returns to its original position by a spring (part 4. in Fig. 14). These types of sprinklers are called as impact sprinklers. Rain guns are large sprinklers which usually have a gear inside the sprinkler head for rotating the sprinkler. They are called as gear type, and the

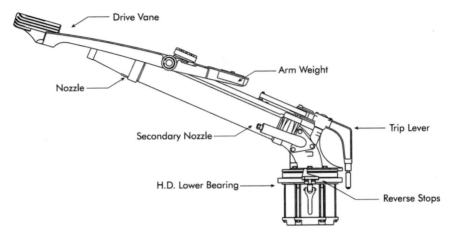

**Fig.15**  Rain gun (Top image: www.chiraharit.com & Bottom image: www.nelsonirrigation.com)

rotation of the gear type sprinkler is very smooth compared to impact-type sprinkler (Fig. 15). Impact type rain guns are also available in market.

## *Valves*

There are many types of valves used in the sprinkler and drip irrigation system. Some of the valve types are listed below:

- Control valve
- Air/vacuum valve
- Non-return valve
- Pressure relief valve.

**Fig. 16**  Ball valve

## *Control Valves*

Control valves are widely used in sprinkler and drip irrigation. They are used for allowing or shutting off water along any pipeline. There are three types of control valves.

1.   Ball valve
2.   Gate valve
3.   Butterfly valve.

Ball valve made of plastic is normally used in SIS (Fig. 16). A rotating ball inside the valve controls the flow in ball valve. In gate valve, a gate controls the flow (Fig. 17). Gate valves may be made of galvanized iron or brass. Butterfly valve (Fig. 18) uses a rotating disc to control the water flow. A true butterfly valve has two half-discs, hinged together in the centre. When the discs or "wings" are folded together, the water flows freely past them. When folded out into the water stream, the wings block the way. Most "butterfly valves" are really "rotating disc" valves. They have a single round disc that rotates on the axle. When fully open, the disc is rotated so that it is aligned along the direction of water flow. To close, the disc is rotated at a right angle so that it fully blocks the flow. Butterfly valves tend to be very reliable and trouble free. They are mostly used on larger size pipes, seldom less than 75 mm size.

## *Air Release/Vacuum Breaker Valve*

Air valves have a moving float inside (Fig. 19). When the water is flowing with pressure, the float is raised in such a way that the water does not go out. But when

**Fig. 17**  Gate valve

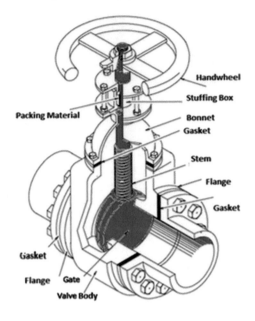

**Fig. 18**  Butterfly valve

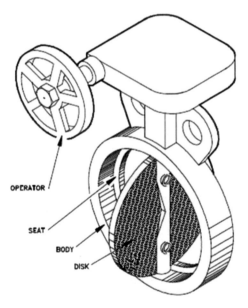

pressure drops, the float comes down and a passage for air flow is created. When the pump is shut off, after the water from pipes drains out and vacuum is created in pipes. To fill this vacuum, air from atmosphere enters into the pipe through the sprinklers. Because of formation of vacuum inside the pipe, breakage of pipe would also occur. Therefore, if an air valve is provided, this problem can be reduced to a significant

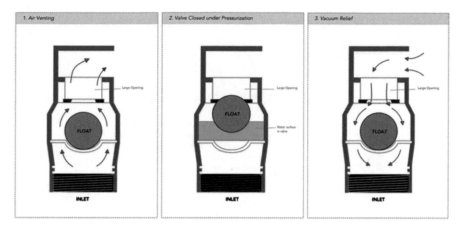

**Fig. 19** Air valve

level. The air valve also releases air from the pipes during the start of irrigation. This valve is to be provided at the location of the highest elevation of the pipe system. When a main pipe is laid on undulating slopes, where ever there is an abrupt up slope followed by a down slope, at the junctions of the up slope and the down slope, air pockets get formed as shown in Fig. 20. They tend to obstruct the flow. At those places air valves are to be provided.

## Non-return Valve

Non-return valves allow the flow to pass through in only one direction. When flow direction is reversed, the non-return valve closes. This valve is very important when fertigation and chemical application through irrigation water are done. Otherwise back flow may cause pollution of water source (Fig. 21). Non-return valves fixed close to pump on delivery line also help prevent water hammer on pump components.

## Pressure Relief Valve

Since sprinkler irrigation works with a higher pressure, there are many possibilities of pressure development beyond the limits in the pipe system. Hence it is advisable to provide a pressure relief valve downstream of non-return valve. Pressure relief valve is also provided at the upstream of the last outlet/hydrant. Figure 22 provides a line sketch showing components of a pressure relief valve. A spring-loaded adjusting screw is fixed over a plunger and a rubber diaphragm. When pressure inside the pipe increases beyond the pressure caused due to spring load, the spring is lifted up. Then

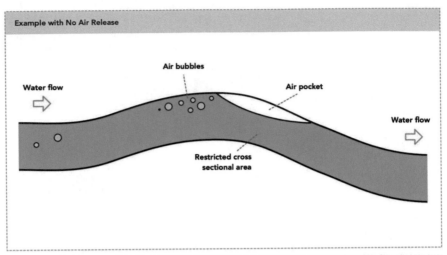

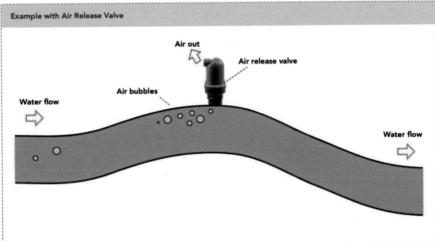

**Fig. 20**  Locating air release/vacuum breaker valve

**Fig. 21**  Non-return valve
(Credit: Jenis Jenis Valve
Kapal Tanker)

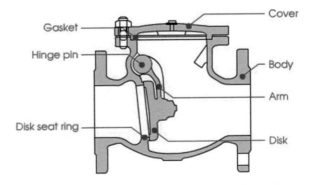

**Fig. 22** Pressure relief valve

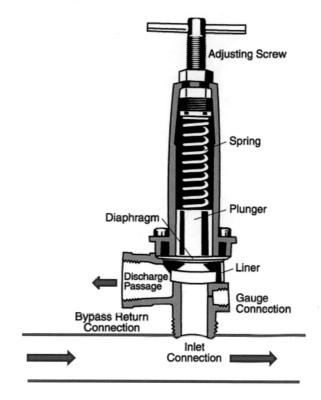

water from the pipe is released through the discharge passage. Note that, there is a provision for connection of pressure gauge in the valve. The adjusting screw is used to adjust the pressure at which the valve is to open and relieve the excess pressure. More the adjusting screw is tightened; more will be the pressure at which the valve will open up.

## Booster Pumps

Booster pumps are also basically pumps only. When water is pumped to a larger distance, along the way pressure is lost due to friction and sometimes due to elevation increase. In such a situation if we use pumps which develop very high pressure at the source of the water itself, the pipes used to carry water from the source to fields would have to be with a higher pressure rating. This is not economical. Therefore, booster pumps are used to increase the pressure at the field level where sprinklers are operated. If booster pumps are used, suction side of booster pumps need not have a pipe with higher pressure rating.

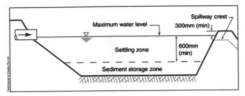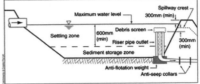

**Fig. 23**   Desilting basin with debris screen

## *Desilting Basins and Debris Screens*

Sprinkler nozzles have relatively larger diameter of flow passage and hence coarse filtration is sufficient. When water is used from surface sources such as canals, rivers and lakes water would contain silt, debris and trashes. Desilting basins are used for settling the silt by allowing water in a surface tank wherein velocity of flow is sluggish. Debris screens are simple screens at the inlet of suction pipe of pumps which prevents debris from going into the pipe Fig. 23.

Sophisticated debris screens are also available nowadays (Fig. 24). These screens clean the surface of the screen at specified time intervals automatically by redirecting a part of delivery water from the pump back to the debris screen.

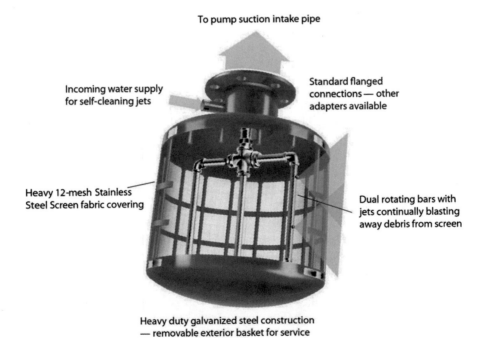

**Fig.24**   Advanced debris screen

# Summary

Sprinkler irrigation needs around 20–70 m pressure head. In India, periodic move lateral system is widely used. Hence the pipe systems and accessories which can be easily assembled and disassembled are used. Mostly 75 mm diameter sprinkler laterals are used in the periodic move system. Sprinkler heads typically have two nozzles, and single-nozzle heads are also available. Air valves are to be used when the pipes are laid on undulating grounds. Pressure relief valves are essential as the sprinkler irrigation is operated with a relatively higher pressure. Booster pumps are also sometimes needed to increase the pressure of operation. Since sprinkler nozzles have relatively larger diameter of passage, coarse filtration is sufficient.

# Assessment

### Part A. Say True or False

1. When a land has a rolling topography with shallow soil, sprinkler irrigation is very much useful—True
2. Modification of weather extremes by increasing humidity and cooling crops can be done with sprinkler irrigation—True
3. In areas of frost problem, sprinkling water removes frost deposited over the foliage—True
4. Saline water causes leaf burning problem in sprinkler irrigation—True
5. Windy conditions can be easily handled in sprinkler irrigation—False
6. Airlift pumps are more commonly used in sprinkler irrigation systems—False
7. At the exit of the sprinkler, pressure head needed is around 20–70 m in sprinkler irrigation—True
8. Saddle is the component on which riser pipe with sprinkler head is connected—True
9. High-density polyethylene pipes are used more than the aluminium pipes in sprinkler irrigation nowadays—True
10. The energy needed for rotation of sprinkler heads comes from the energy of irrigation water—True
11. In the boom-type sprinklers, truss moves linearly and does not rotate about a centre—True
12. Centre-pivot and boom-type sprinklers are not predominantly used in India—True
13. Smaller sprinklers are called as rain guns—False
14. Choose the incorrect statement

    a. When a land has a rolling topography with shallow soil, sprinkler irrigation is very much useful.
    b. Modification of weather extremes by increasing humidity and cooling of crops can be done with sprinkler irrigation.

    c.    In areas of frost problem, sprinkling water removes frost deposited over the foliage.

    d.    **Saline water does not cause leaf burning problem in sprinkler irrigation**

15.    Choose the incorrect statement

    a.    Airlift pumps cannot be used in sprinkler irrigation systems

    b.    Saddle is the component on which riser pipe with sprinkler head is connected

    c.    High-density polyethylene pipes are used more than the aluminium pipes in sprinkler irrigation nowadays

    d.    **Airlift pumps are commonly used in sprinkler irrigation**

16.    Choose the incorrect statement

    a.    The energy needed for rotation of sprinkler heads comes from the energy of irrigation water

    b.    In the boom-type sprinklers, truss moves linearly and does not rotate about a centre

    c.    Centre-pivot and boom-type sprinklers are not predominantly used in India

    d.    **Smaller sprinklers are called as rain guns**

17.    In _____, sprinklers are not moved from one place to the other after each irrigation is done

    a.    **Set System**

    b.    Continuous move system

    c.    Periodic move system

    d.    Linear move

    e.    Travelling sprinkler.

## Part B. Descriptive Answers

1.    What are the broad classifications of sprinkler irrigation systems?

2.    Write a brief note on travelling sprinkler system?

3.    What are impact sprinklers?

4.    What is the use of non-return valve in sprinkler irrigation?

5.    What is the use of a pressure relief valve?

6.    Write a short note on butterfly valves?

7.    What is the use of a booster pump in SIS?

8.    What are the effects of salinity in irrigation water when sprinkler irrigation is used?

9.    List out the components of a periodic move sprinkler irrigation system?

10.    Differentiate between impact sprinklers and gear-type sprinklers?

11. Explain with sketches about desilting basins and debris screens in sprinkler irrigation?
12. What are the advantages of sprinkler irrigation over other irrigation methods?
13. What are the disadvantages of sprinkler irrigation over other irrigation methods?
14. Explain the working principle of air/vacuum release valves and state where these valves are used?
15. Explain different components of a periodic move sprinkler irrigation system with a neat sketch.

## References and Further Readings

1. Badenoch, N. (2009). The politics of PVC: Technology and institutions in upland water management in northern Thailand. *Water Alternatives, 2*(2), 269–288.
2. Keller, J., & Bliesner, R. D. (1990). *Sprinkle and trickle irrigation.* Van Nostrand Reinhold.
3. Martin, D. L., Kincaid, C. D., & Lyle, W. M. (2007). Design and operation of sprinkler irrigation systems. https://doi.org/10.13031/2013.23699

# Uniform Irrigation with Sprinklers

## Introduction

A sprinkler is characterized by many factors such as discharge rate, number of nozzles, diameter of each nozzle, optimal operating pressure, wetting diameter, rotational speed, droplet size and mechanism of break-up of jet. The performance of a sprinkler while being in operation in the field is also influenced greatly by direction and magnitude of wind. In order to select sprinklers according to field situations, knowledge of interplay of these factors is imparted in this chapter.

All the water issued from sprinkler is not used for evapotranspiration from the root zone. Water loss occurs due to evaporation from droplet and canopy, drift loss, surface runoff and deep percolation loss (Fig. 1). Application of water cannot also be done uniformly to all the area due to the variation of pressure available at each sprinkler head along a lateral, nature of precipitation pattern of each sprinkler, wetted circle overlap provided and wind effects. In this chapter, measures of uniformity, evaluation methods for finding uniformity of application rate both in the fields as well as a theoretical method are discussed with suitable worked out examples.

## Sprinkler Discharge Rate

Sprinkler discharge rate depends on the pressure head at the nozzle. This can be found out using the following simple Torricelli orifice formula:

$$q = C_d A \sqrt{2gh}$$

where

$q$   —Discharge rate from nozzle of cross-sectional area '$A$'. Usually nozzle diameter is written near the nozzle outlet. (Fig. 2);

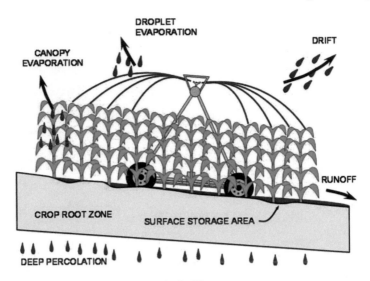

**Fig. 1** Components of loss in sprinkling. Martin [2]

$C_d$     —Coefficient of discharge is usually taken as 0.95. For special non-circular shapes, equivalent circular diameter is provided by the manufacturers;

$h$       —Pressure head (m).

Equation 1 is conventionally expressed as follows:

$$q = k\sqrt{h} \tag{1}$$

where $k$ is called as sprinkler head discharge rate coefficient.

The pressure head measurement at the exit of sprinkler nozzle can be measured directly with a simple pitot tube accessory as shown in Fig. 2. It is rather difficult and so alternatively a pressure gauge can also be fixed on the riser pipe just below

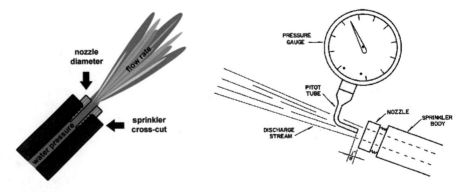

**Fig. 2** A sprinkler nozzle and pressure measurement at the exit of nozzle (www.irrigation.wsu.edu)

**Fig. 3** Pressure
measurement by fixing a
pressure gauge in the riser
pipe

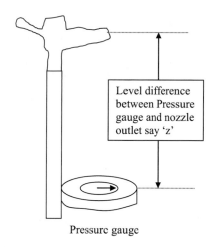

Level difference
between Pressure
gauge and nozzle
outlet say 'z'

Pressure gauge

the sprinkler (Fig. 3). Kinetic head can be estimated at the connection point of the
pressure gauge by finding the discharge rate of the sprinkler and inside diameter of
the pipe. The sum of pressure head shown by pressure gauge and the kinetic head
would provide total head at the pressure gauge. The pressure head expected at the
nozzle is the total head minus the level difference between pressure gauge and the
outlet of the nozzle, and the friction loss between the two locations (pressure gauge
location & nozzle location) can be assumed as negligible. Following problem is used
to illustrate this method.

**Problem** A riser pipe fitted with a pressure gauge at the bottom of the riser pipe
shows a pressure head of 25 m. The inside diameter of the riser pipe is 20 mm. The
discharge rate from the sprinkler is 0.5 l/s. Level difference between pressure gauge,
and the nozzle outlet is 1 m. Find out the expected pressure head at the nozzle outlet
(assume friction loss between pressure gauge and nozzle outlet as zero).

Velocity of flow = Discharge rate/area of cross sectional area of flow

$$= 0.0005 \, \text{m}^3/\text{s}/3.14 * 0.001 \, \text{m}^2 = 1.592 \, \text{m/s}$$

Kinetic head $(v^2/2\,g) = 0.129 \, \text{m}$

Total head at the pressure gauge = Pressure head + kinetic head = 25 + 0.129
= 25.129 m.
Pressure head expected at the nozzle outlet.
= 25.129 – Level difference between pressure gauge and nozzle outlet (1 m).
= 24.129 m.

## Wetted Diameter

Wetted diameter depends on the angle of throw, pressure at the nozzle and riser pipe height and prevailing wind speed and direction. An angle of throw of 30° is generally optimal. However for specific applications like under tree sprinkling less than 30° is also used.

After the jet is issued from the nozzle, the water droplets travel in air and during the process, the horizontal component of its velocity is very much reduced due to air resistance. Hence the angle subtended with respect to the horizontal during downward flight of the droplet is much more than the angle subtended with the horizontal during upward flight. Due to this fact, by increasing the height of the riser pipe, we do not get appreciable increase in diameter of wetting (Fig. 4). Therefore, increase of the height of riser pipe is not recommended for increasing the diameter of wetting.

As pressure increases, until a certain critical level, the wetted diameter continues to increase. After the critical level, the wetted diameter does not increase because the jet breaks down at the exit of nozzle and the droplets do not have sufficient energy to overcome the resistance of the atmospheric air. Normally, manufacturers provide a range of operating pressure for operation of sprinklers. In still air condition, it is advised to operate near the higher side of recommended operating pressure. In the windy conditions, it is advised to operate at the lower side of the recommended operating pressure. When a sprinkler is operated at the lower recommended pressure in windy conditions, the jet issued from nozzle does not break down fully and has the ability to pass through air for some distance and the jet gradually breaks down along the flight due to air resistance.

The depth of water applied by one sprinkler at any location can be measured by placing catch collectors as shown in Fig. 5 and operating the sprinklers for a period of time. Placing the collectors for one radial leg itself would give an idea of precipitation profile over the radial distance. However if the catch cans are placed in many radial directions for a sprinkler head as shown in Fig. 5, average depth and standard deviation of the depth for each radial distance from the sprinkler can be obtained. Figure 6 shows shapes of typical precipitation patterns obtained when a sprinkler is operated.

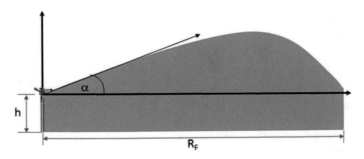

**Fig. 4**  Trajectory of sprinkler jet

**Fig. 5** Catch can
measurement for a sprinkler
head [2]

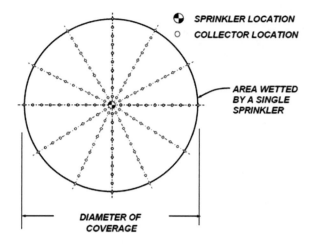

**Fig. 6** Typical precipitation
pattern for one leg [2]

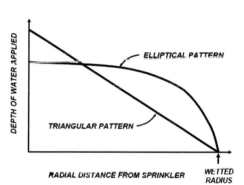

Figure 7 shows the variation of depth of application with respect to operating pressure for a twin-nozzle sprinkler. When pressure is at a lower level, the smaller nozzle throws the jet very near to the sprinkler and the other bigger nozzle throws the jet without breaking of jet at the nozzle. That is the reason why when the pressure is too low, we find two peaks in the depth of application. One peak is near the centre, and another peak is at the periphery.

It is to be noted that even when the pressure of operation is at the correct level, the depth of application is not uniform with respect to the distance from the sprinkler. The depth of application is the maximum at the centre and the precipitation pattern appears as a triangle. In case of excess pressure, the jet breaks down into very small droplets and does not have sufficient energy to overcome the air resistance, and hence, most of the droplets fall very near the sprinkler.

**Fig. 7** Variation of depth of
application with respect to
operating pressure for a twin
nozzle [1]

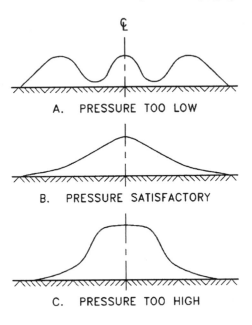

A. PRESSURE TOO LOW

B. PRESSURE SATISFACTORY

C. PRESSURE TOO HIGH

## Spacing of Sprinkler Heads

Since individual sprinkler heads do not provide uniform precipitation over an area,
the sprinkler heads should be spaced in such a way that uniform precipitation is
obtained. There are three types of spacing normally adopted. One is square, two is
rectangle, and the three is triangle.

Table 1 provides a guideline for the selection of sprinkler heads based on the
wetted diameter. Table 2 provides pressure head range for different nozzle sizes.

In the square spacing, the distance between the sprinkler heads along the lateral
line and also between the lateral lines is same. Distance between the adjacent sprinklers will be equal to 50% of wetted diameter. In rectangular spacing, the spacing
between sprinkler heads along the lateral line is smaller than the spacing between the
laterals. Smaller length is 40% of wetted diameter (s), and longer length is 67% of
wetted diameter (L) (Fig. 7). Number of lateral take off for rectangular spacing would
be less than the square spacing and hence rectangular spacing is relatively economical. In a triangular spacing, the sprinkler heads of alternate rows are as shown
in Fig. 8. The side of each triangle will be 55–60% of wetted diameter. In terms of
uniformity of application, triangular spacing is the best and the square spacing yields
lower uniformity and the rectangular spacing provides the least uniformity.

Depending on the precipitation profile of a selected sprinkler, the spacing may be
adopted using the information in Fig. 9.

### Spacing of Sprinkler during Wind
The wetted diameter provided in the manufacturers catalogue are normally for still
air conditions. Wind greatly affects the wetted diameter. In windy conditions, the
effective wetted diameter would get affected.

**Table 1** Sprinkler types and operating characteristics [1]

| Sprinkler characteristics | Impact type—low pressure (3.5–14.0 m) | Impact or gear type—medium pressure (10.5–21.0 m) | Impact or gear type—high pressure (21.0–69.0 m) | Impact or gear type—very high pressure (34.0–69.0 m) | Gear type—giant models (55.0–83.0 m) |
|---|---|---|---|---|---|
| General characteristics | Special pressure spring–single nozzle | Single or double nozzle | Single or double nozzle | Single or double nozzle | Double nozzle |
| Wetted diameter (m) | 6–15 | 16–24 | 23–37 | 34–90 | 61–122 |
| Minimum precipitation (mm/h) | 10 | 3 | 2.5 | 10 | 15 |
| Breaking of jet | Large drops | Medium drops | Medium drops | Medium drops | Small drops |
| Precipitation uniformity | Fair | Good | Very good | Good. If wind velocity is more than 6.4 km/h, uniformity is affected very much | Fair. Even for very small winds, distortion of jet is very high |

**Table 2** Pressure head range for different nozzle sizes

| No. | Nozzle diameter (mm) | Pressure range (m) |
|---|---|---|
| 1 | 2.0–2.4 | 14–31 |
| 2 | 2.8–3.6 | 17–34 |
| 3 | 4.0–4.4 | 21–38 |
| 4 | 4.8–5.6 | 24–41 |

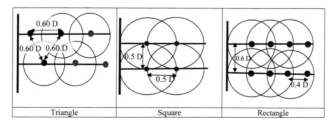

| Triangle | Square | Rectangle |

**Fig. 8** Sprinkler spacing options

| SPRINKLER PROFILE | | RECOMMENDED SPACING AS A PERCENTAGE OF DIAMETER | | |
|---|---|---|---|---|
| TYPE | SHAPE | SQUARE | TRIANGULAR EQUILATERAL | RECTANGULAR SHORT x LONG |
| A |  | 50 | 50 | 40 x 60 to 65 |
| B | 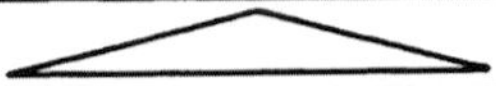 | 55 | 66 | 40 x 60 |
| C | | 60 | 65 | 40 x 60 to 65 |
| D |  | 40 70 (FAIR) | 70 to 75 | 40 x 70 to 75 |
| E |  | 40 80 (FAIR) | 80 | 40 x 80 |

**Fig. 9**  Spacing of sprinklers based on precipitation profile [1]

For a wind speed of 0–5 kms per hour (kmph), the diameter may get shortened by 10%. A reduction of 2.5% for each 1.6 kmph over 5 kmph may be done (Fig. 10).

Laterals should be at the right angle to the prevailing wind direction. Movement of laterals should be done in the direction of wind if the water is salty. By doing so, the salt deposits of previous irrigation would be washed off in the subsequent lateral settings.

**Fig. 10**  Sprinkler spacing during wind

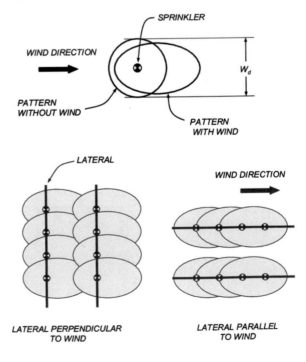

**Fig. 11** Pins for jet break-up in sprinklers

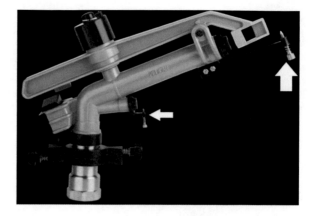

Sprinkler pipes available in the market are in lengths of 6 and 3 m. Hence, while deciding the spacing of sprinklers, it should be in multiples of 3 m for periodic move lateral systems.

## Breaking of Jet

Break-up of jet is necessary in sprinkler irrigation. The break-up is made possible by increasing the pressure. Increasing pressure causes use of more power and subsequently more operating cost. In case of large nozzles, the break-up of jet can be achieved by provision of slots in the nozzle and by introducing a pin in the path of the jet (Fig. 11). This kind of sprinkler design is called as low-energy precision application (LEPA).

Larger droplets have larger kinetic energy and tend to cause more soil erosion. But they are less interfered by wind. If the soil is sandy, the soil erosion due to bigger droplets is the minimum. Smaller droplets tend to get lost by drift due to wind.

## Application Rate

Application rate is the depth of water applied on a land per unit time. Conventionally, at least a minimum application rate of 3 mm/h is to be ensured because the application rates lower than this cause too much loss. The maximum application rate is the rate at which no appreciable runoff from the field occurs. When the rotating sprinkler is in operation, that water falling on any location should go into the soil before water is applied at the same location due to the subsequent rotation. Table 4 can be used for deciding the maximum application rates for different soil textures in different slopes.

**Table 3** Minimum riser height

| No. | Sprinkler discharge (l/s) | Minimum riser height (cm) |
|---|---|---|
| 1 | <0.6 | 15 |
| 2 | 0.6–1.6 | 23 |
| 3 | 1.6–3.2 | 30 |
| 4 | 3.2–7.6 | 45 |
| 5 | >7.6 | 90 |

**Table 4** Maximum application rates for different soils [1]

| Soil texture and profile | Slope | | | |
|---|---|---|---|---|
| | 0–5% | 5–8% | 8–12% | 12–16% |
| | Maximum application rate | | | |
| | mm/h (in./hr) | mm/h (in./hr) | mm/h (in./hr) | mm/h (in./hr) |
| Coarse sandy soil to 1.8 m (6 ft) | 50 (2.0) | 38 (1.5) | 25 (1.0) | 13 (0.50) |
| Coarse sandy soils over more compact soils | 38 (1.5) | 25 (1.0) | 19 (0.75) | 10 (0.40) |
| Light sandy loams to 1.8 m (6 ft) | 25 (1.0) | 20 (0.80) | 15 (0.60) | 10 (0.40) |
| Light sandy loams over more compact soils | 19 (0.75) | 13 (0.50) | 10 (0.40) | 8 (0.30) |
| Silt loams to 1.8 m (6 ft) | 13 (0.50) | 10 (0.40) | 8 (0.30) | 5 (0.20) |
| Silt loams over more compact soils | 8 (0.30) | 6 (0.25) | 4 (0.15) | 2.5 (0.10) |
| Heave textured clays or clay loams | 4 (0.40) | 2.5 (0.10) | 2 (0.08) | 1.5 (0.06) |

The proper break-up of jet is to occur at the exit of the nozzle. When riser pipe lengths are shorter, too much turbulence occurs and premature break-up of jet occurs before the water reaches the nozzle. This is not a conducive condition. This kind of difficulties can be solved by adopting the minimum riser pipe heights as recommended in Table 3.

## Application Efficiency

All the water applied in sprinkler irrigation will not replenish the root zone. Certain inevitable losses such as drift of water droplets due to wind, evaporation of water from the wet canopy and deep percolation occur. Application efficiency is the ratio of gross depth of water applied to the water stored in the root zone to meet evapotranspiration. Normally 60–80% application efficiency can be expected in sprinkler irrigation.

## Sprinkler Head Selection

Based on the capacity of the available pump and wind velocity, Table 1 can be used for the selection of sprinkler heads. When small sprinkler heads are used, the number of sprinkler heads, length of pipes needed are higher and so the total initial cost are very high. But, the pressure needed, drift due to wind and cost towards the power will be less. When larger sprinkler heads are used, the number of sprinkler heads, length of pipes needed are less and so initial cost will be less. But, the pressure needed drift due to wind, and cost of power will be more. In case of clay soils and also undulated lands, the sprinkler head selection should be done so that the design precipitation rate should not cause runoff.

## Uniformity Parameters

Christiansen's uniformity coefficient (CU) is the most popularly used parameter for evaluating uniformity of water application in a SIS. It is based on finding the mean of absolute deviations of precipitation depth divided by the mean precipitation depth.

$$CU = 100\left(1 - \frac{\sum_{i=1}^{i=n}|Z_i - \overline{Z}|}{n\overline{Z}}\right) \tag{2}$$

where $Z_i$ is the depth of water applied at $i$th observation; $\overline{Z}$ is the average depth of water applied in all the observations; and $n$ is the number of observations. CU range of 75–85% is acceptable.

Approximately 50 ml catch cans (collectors) are placed in an area which is representative of the total area irrigated by SIS. The method of demarcating representative areas for square, rectangular and triangular spacing of sprinklers is shown in Figs. 12, 13 and 14. We have to operate all the sprinklers which contribute to the representative area for a specified amount of time. Find out the volume of water collected in each can. Find out the area of opening of the can. Divide the volume of water collected by the area of opening of the can to get the depth of water applied at each can. Using these observations, Christiansen's uniformity coefficient can be found out.

It is not always possible to operate all the sprinklers simultaneously which contribute to a representative area. When only one lateral could be operated at a time, the data collected from only one lateral can be adjusted to get the data as though all

**Fig. 12** Representative area
for square spacing

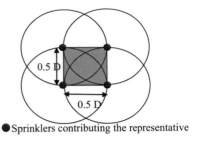

● Sprinklers contributing the representative

**Fig. 13** Representative area
for rectangular spacing

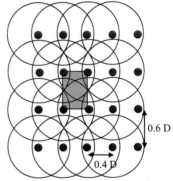

● Sprinklers contributing the representative

**Fig. 14** Representative area
for triangular spacing

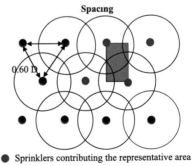

● Sprinklers contributing the representative area

the sprinklers contributing to the representative area are operated. A problem in the succeeding sections demonstrates the method of doing such a kind of data adjustment.

Distribution uniformity (DU) is also another parameter which is used to evaluate uniformity.

$$DU = \frac{\text{Average Low Quarter Depth Received}}{m} \times 100 \qquad (3)$$

where '$m$' is the mean depth received.

It has been found that the catch can observations are normally distributed when CU is greater than 70%. Therefore, based on the statistical principles, a following relationship has been found out:

$$DU = 100 - 1.59(100 - CU) \qquad (4)$$

The accuracy of the uniformity coefficient values results in a deviation of plus or minus 2%. That means if the estimated CU is 82%, the result is reported as 82±2%.

**Example-A** Volumes of water in cm$^3$ collected in catch cans from a sprinkler catch can test by an operation of four sprinklers for 2 h as shown in Fig. 15 are provided in Table 5. The surface area of opening of each catch can is 20 cm$^2$. Find out the average precipitation rate. Find out the value of CU and say whether the value is satisfactory. Find also DU based on the observations as well as by the relationship between CU and DU.

Number of Observations ($n$) = 49

**Fig. 15** Configuration of catch cans for the example-A

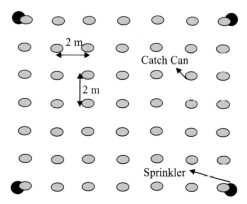

**Table 5** Catch can observations for the example-A

| | | | | | | |
|------|------|------|------|------|------|------|
| 6.15 | 6.40 | 6.64 | 6.64 | 6.64 | 6.40 | 6.15 |
| 6.89 | 6.86 | 6.83 | 6.83 | 6.83 | 6.86 | 6.89 |
| 7.63 | 7.33 | 7.02 | 7.02 | 7.02 | 7.33 | 7.63 |
| 7.56 | 7.29 | 7.02 | 7.02 | 7.02 | 7.29 | 7.56 |
| 7.49 | 7.26 | 7.02 | 7.02 | 7.02 | 7.26 | 7.49 |
| 6.82 | 6.70 | 6.58 | 6.71 | 6.83 | 6.83 | 6.82 |
| 6.15 | 6.15 | 6.14 | 6.39 | 6.64 | 6.40 | 6.15 |

**Table 6** Deviations from the mean

| | | | | | | |
|------|------|------|------|------|------|------|
| 0.72 | 0.48 | 0.23 | 0.23 | 0.23 | 0.48 | 0.72 |
| 0.02 | 0.01 | 0.04 | 0.04 | 0.04 | 0.01 | 0.02 |
| 0.76 | 0.45 | 0.15 | 0.15 | 0.15 | 0.45 | 0.76 |
| 0.69 | 0.42 | 0.15 | 0.15 | 0.15 | 0.42 | 0.69 |
| 0.62 | 0.39 | 0.15 | 0.15 | 0.15 | 0.39 | 0.62 |
| 0.05 | 0.17 | 0.29 | 0.17 | 0.04 | 0.05 | 0.05 |
| 0.72 | 0.73 | 0.73 | 0.48 | 0.23 | 0.48 | 0.72 |

Mean of the observations $(m) = 6.87\,\text{cm}^3$

Table 6 provides deviation of all the observations from the mean.
Sum of deviations $= 16.16$

$$CU = 100\left(1 - \frac{16.16}{49 \times 6.87}\right) = 95.20\%$$

$$\text{Mean depth of application} = \frac{m}{\text{Area of Can} \times \text{time}} = \frac{6.87}{20 \times 2} = 0.17\,\text{cm}\,\text{h}^{-1}$$

The set of low quarter depth of water received is obtained by writing all the data in ascending order and taking out the first top quarter data. The low quarter data obtained is as follows:

| 6.14 | 6.15 | 6.15 | 6.15 | 6.15 | 6.15 | 6.39 | 6.40 | 6.40 | 6.40 | 6.58 | 6.64 |
|------|------|------|------|------|------|------|------|------|------|------|------|

Average of low quarter depth of water received $= 6.30$ m

$$DU = \frac{6.30}{6.87} \times 100 = 91.70\%$$

DU based on the approximate formula (Eq. 4):$DU = 100 - 1.59(100 - 95.2) = 92.37\%$

It can be seen that the DU value actually calculated using the observations directly and also using the approximate formula based on the assumption of normal distribution are fairly equal. The CU value must be reported as 95.20% plus or minus 2%.

**Example-B** Mild wind blows from the east to west direction at a SIS when in operation. The representative area in which observations were collected is shown in Fig. 16. The representative area is surrounded by 6 sprinklers. At a time, only two sprinklers are operated across the direction of wind and observations are taken with only two sprinklers in operation (Fig. 17). The sprinklers operated for observations are shown with green color in Figs.16 and 17. The depths of water collected in catch

**Fig. 16** Configuration of sprinklers for example-B

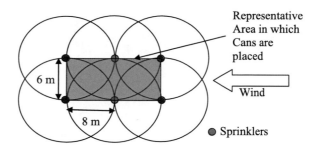

**Fig. 17** Configuration of catch cans for the example-B

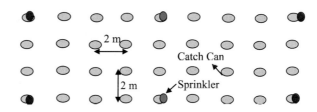

**Table 7** Observations from the catch can test (mm)

| | | | | | | | | |
|---|---|---|---|---|---|---|---|---|
| 0.2 | 1.8 | 2.9 | 4.0 | 5.3 | 3.9 | 2.8 | 1.7 | 0.0 |
| 0.1 | 1.5 | 2.1 | 2.9 | 5.1 | 2.8 | 2.0 | 1.4 | 0.0 |
| 0.1 | 1.5 | 2.2 | 2.9 | 5.1 | 2.9 | 2.1 | 1.4 | 0.0 |
| 0.2 | 1.7 | 2.9 | 3.9 | 5.3 | 3.7 | 2.7 | 1.6 | 0.0 |

**Table 8** Adjusted observations when sprinklers are operated with overlap

| Left 4 | Left 3 | Left 2 | Left 1 | Centre | Right 1 | Right 2 | Right 3 | Right 4 |
|---|---|---|---|---|---|---|---|---|
| 0.2<br>5.3-L<br>0.0 | 1.8<br>3.9-L | 2.9<br>2.8-L | 4.0<br>1.7-L | 5.3<br>0.0-L<br>0.2-R | 3.9<br>1.8-R | 2.8<br>2.9-R | 1.7<br>4.0-R | 0.0<br>5.3-R<br>0.2 |
| 0.1<br>5.1-L<br>0.0 | 1.5<br>2.8-L | 2.1<br>2.0-L | 2.9<br>1.4-L | 5.1<br>0.0-L<br>0.1-R | 2.8<br>1.5-R | 2.0<br>2.1-R | 1.4<br>2.9-R | 0.0<br>5.1-R<br>0.1 |
| 0.1<br>5.1-L<br>0.0 | 1.5<br>2.9-L | 2.2<br>2.1-L | 2.9<br>1.4-L | 5.1<br>0.0-L<br>0.1 | 2.9<br>1.5-R | 2.1<br>2.2-R | 1.4<br>2.9-R | 0.0<br>5.1-R<br>0.1 |
| 0.2<br>5.3-L<br>0.0 | 1.7<br>3.7-L | 2.9<br>2.7-L | 3.9<br>1.6-L | 5.3<br>0.0-L<br>0.2-R | 3.7<br>1.7-R | 2.7<br>2.9-R | 1.6<br>3.9-R | 0.0<br>5.3-R<br>0.2 |

cans in mm by operation of two sprinklers for 2 h are presented in Table 7. Find out the depth of precipitation in the representative area if irrigation is done by overlapping the sprinklers and also find out the Christiansen's coefficient of uniformity for the representative area.

**Table 9** Total precipitation depth in each can when sprinklers are operated with overlap

| 5.5 | 5.7 | 5.7 | 5.7 | 5.5 | 5.7 | 5.7 | 5.7 | 5.5 |
|-----|-----|-----|-----|-----|-----|-----|-----|-----|
| 5.2 | 4.3 | 4.1 | 4.3 | 5.2 | 4.3 | 4.1 | 4.3 | 5.2 |
| 5.2 | 4.4 | 4.3 | 4.3 | 5.2 | 4.4 | 4.3 | 4.3 | 5.2 |
| 5.5 | 5.4 | 5.4 | 5.3 | 5.5 | 5.4 | 5.6 | 5.5 | 5.5 |

The adjusted observations are shown in Table 8. All the upper values of each cell in Table 8 correspond to copying the value of the corresponding cell in Table 7.

In Table 8, in the second row of each cell, you can see notations L and R are used. L denotes the precipitation depth if the two left-hand-side sprinklers are operated. R denotes the precipitation depth if the two right-hand-side sprinklers are operated.

Note that columns Left 4 and Right 4 have three values for each cell. The columns Left 4, Centre and Right 4 get the same depth of precipitation. Table 9 shows the total precipitation depth in each can when all the sprinklers are operated.

Mean depth of precipitation = 5.08 mm.

Mean deviation = 0.53 mm

$$CU = 100\left(1 - \frac{0.53}{5.08}\right) = 89.6\%$$

## Design of Sprinklers Spacing

Sometimes, the precipitation rate versus radial distance relationship can be developed for still air conditions assuming no wind is blowing at the site. If such a relationship is available, we can examine different spacing configurations and find out how the precipitation is distributed within the area of our interest. This kind of analysis is very useful in the spacing of landscaping sprinklers where the geometry of the area irrigated is complex. Following example illustrates this theoretical method.

**Example** A selected sprinkler of wetting radius of 6 m has a triangular precipitation pattern with a following relationship:

$$p = 6 - r$$

where $p$ is the precipitation rate in mm/h and $r$ is the radial distance in $m$.

A square field of 20 m side is to be laid with four sprinklers in square pattern as shown in Fig. 18. The wetting pattern of the four sprinklers is as shown in Fig. 19. Find out the Christiansen's uniformity coefficient for the entire field and also for the square part by the lines joining the sprinklers A, B, C, D.

**Fig. 18** Field with sprinkler locations

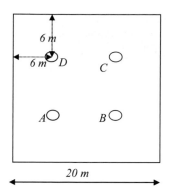

**Fig. 19** Wetting patterns of four sprinklers

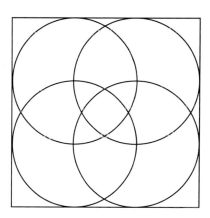

Let us assume that, only one sprinkler alone is operated and the area around the sprinkler is divided into a grid with squares of 2 m. For this situation, the precipitation rate in mm/h for one quarter is calculated and presented in Fig. 20. One sample calculation for finding precipitation rate onto the nearest cell from the sprinkler cell is as follows:

The centroid of the nearest square from the sprinkler is 1 m distance in horizontal direction and 1 m distance in vertical direction.

The radial distance of the centroid of the cell $(r) = ((x^2 + y^2))^{0.5} = (1 + 1)^{0.5} = 1.414$ m.

Precipitation rate in the cell $(d) = 6 - 1.414 = 4.59$ mm/h.

Let us assume that the total area is divided into a grid of square cells of size 2 m and in each cell, a catch can (collector) is placed (Fig. 21). If all the four sprinklers are operated, the precipitation rate in each cell due to each sprinkler is filled as in Table 10. The sum of all the values in each cell gives the total precipitation rate at each cell (Table 11). Using this data, Christiansen's uniformity can be found out (Table 12).

CU for the total area = 41.5%

CU for the area enclosed by the four sprinklers ABCD = 79.64%.

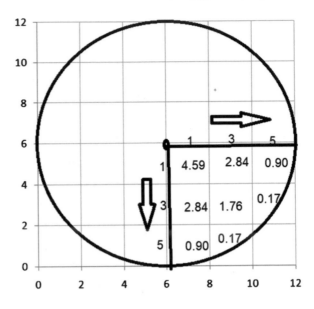

**Fig. 20** Precipitation rate for one sprinkler in one quarter of the wetted diameter

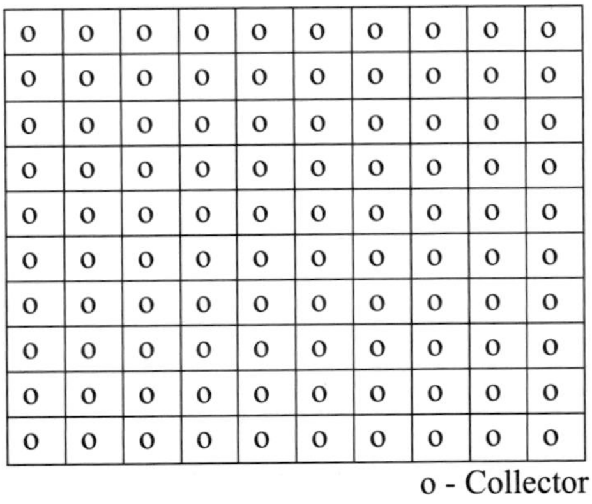

**Fig. 21** Grid division of the area

o - Collector

## Summary

For uniform application of irrigation over an area, sprinklers have to be placed with a recommended overlap. During wind conditions, the extent of overlap must be increased. Optimal pressure of operation of sprinkler heads is very important otherwise uniformity of application would get tremendously affected. The precipitation rate must be selected for each soil type so that excess water does not cause runoff. Coefficient of uniformity and distribution uniformity are the two measures used for measuring the uniformity of application.

**Table 10** Precipitation rate in each cell due to individual sprinklers (Bold-A, Italic-B, BoldItalic-C, Underline-D)

| | <u>0.17</u> | <u>0.9</u> | <u>0.9</u> | <u>0.17</u> | *0.17* | *0.90* | *0.90* | *0.17* | |
|---|---|---|---|---|---|---|---|---|---|
| <u>0.17</u> | <u>1.76</u> | <u>2.84</u> | <u>2.84</u> | *0.17*<br><u>1.76</u> | *1.76*<br>0.17 | *2.84* | *2.84* | *1.76* | *0.17* |
| <u>0.9</u> | <u>2.84</u> | <u>4.59</u> | <u>4.59</u> | *0.9*<br>2.84 | *2.84*<br><u>0.9</u> | *4.59* | *4.59* | *2.84* | *0.9* |
| <u>0.9</u> | <u>2.84</u> | <u>4.59</u> | <u>4.59</u> | *0.9*<br>2.84 | *2.84*<br><u>0.9</u> | *4.59* | *4.59* | *2.84* | *0.9* |
| <u>0.17</u> | **0.17**<br>1.76 | **0.9**<br>2.84 | **0.9**<br>2.84 | 0.17<br><u>1.76</u><br>***0.17*** | <u>0.17</u><br>***1.76***<br>*0.17* | *0.90*<br>2.84 | 0.9<br>2.84 | 0.17<br>1.76 | 0.17 |
| **0.17** | **1.76**<br><u>0.17</u> | **2.84**<br><u>0.9</u> | **2.84**<br><u>0.9</u> | **1.76**<br><u>0.17</u><br>*0.17* | **0.17**<br>*1.76*<br>***0.17*** | *2.84*<br>***0.90*** | *2.84*<br>***0.90*** | *1.76*<br>***0.17*** | *0.17* |
| **0.9** | **2.84** | **4.59** | **4.59** | **2.84**<br>*0.9* | **0.9**<br>*2.84* | *4.59* | *4.59* | *2.84* | *0.9* |
| **0.9** | **2.84** | **4.59** | **4.59** | **2.84**<br>*0.90* | **0.9**<br>*2.84* | *4.59* | *4.59* | *2.84* | *0.9* |
| **0.17** | **1.76** | **2.84** | **2.84** | **1.76**<br>*0.17* | **0.17**<br>*1.76* | *2.84* | *2.84* | *1.76* | *0.17* |
| | **0.17** | **0.9** | **0.9** | **0.17** | *0.17* | *0.9* | *0.9* | *0.17* | |

**Table 11** Cumulative precipitation rate at each cell

| | 0.17 | 0.9 | 0.9 | 0.17 | 0.17 | 0.90 | 0.90 | 0.17 | |
|---|---|---|---|---|---|---|---|---|---|
| 0.17 | 1.76 | 2.84 | 2.84 | 1.93 | 1.93 | 2.84 | 2.84 | 1.76 | 0.17 |
| 0.9 | 2.84 | 4.59 | 4.59 | 3.74 | 3.74 | 4.59 | 4.59 | 2.84 | 0.9 |
| 0.9 | 2.84 | 4.59 | **4.59** | **3.74** | **3.74** | **4.59** | 4.59 | 2.84 | 0.9 |
| 0.17 | 1.93 | 3.74 | **3.74** | **2.1** | **2.1** | **3.74** | 3.74 | 1.93 | 0.17 |
| 0.17 | 1.93 | 3.74 | **3.74** | **2.1** | **2.1** | **3.74** | 3.74 | 1.93 | 0.17 |
| 0.9 | 2.84 | 4.59 | **4.59** | **3.74** | **3.74** | **4.59** | 4.59 | 2.84 | 0.9 |
| 0.9 | 2.84 | 4.59 | 4.59 | 3.74 | 3.74 | 4.59 | 4.59 | 2.84 | 0.9 |
| 0.17 | 1.76 | 2.84 | 2.84 | 1.93 | 1.93 | 2.84 | 2.84 | 1.76 | 0.17 |
| | 0.17 | 0.9 | 0.9 | 0.17 | 0.17 | 0.9 | 0.9 | 0.17 | |

**Table 12** Precipitation depths of cells enclosed by the sprinklers ABCD

| 4.59 | 3.74 | 3.74 | 4.59 |
|---|---|---|---|
| 3.74 | 2.1 | 2.1 | 3.74 |
| 3.74 | 2.1 | 2.1 | 3.74 |
| 4.59 | 3.74 | 3.74 | 4.59 |

## Assessment

### Part A. Say True or False

1.  Larger droplets tend to get drifted more than the smaller droplets due to wind -False
2.  Breaking of jet in sprinklers by pins at the exit of nozzle is called as low-energy precision application (LEPA).True
3.  Conventionally, a minimum application rate of 3 mm/h is adopted in sprinkler irrigation because the application rates lower than this cause too much loss. True
4.  The maximum application rate in sprinkler irrigation is one above which appreciable runoff from the field occurs. True
5.  When riser pipe lengths are shorter, too much turbulence occurs and premature break-up of jet occurs before the water reaches the nozzle. True
6.  Normally 60–80% application efficiency can be expected in sprinkler irrigation. True
7.  Christiansen's uniformity coefficient is based on finding the mean of absolute deviations of precipitation depth divided by the mean precipitation depth. True
8.  Distribution coefficient of uniformity is defined as the ratio of average low quarter depth received divided by the mean depth. True
9.  When Christiansen's uniformity is greater than 70%, observed depths have been found to follow normal distribution. True
10. In a catch can test, if a collector has a surface area of 10 cm$^2$ and the volume collected is 10 cm$^3$. Then the depth of precipitation is 1 cm. True
11. Christiansen's uniformity coefficient range of 75–85% is acceptable. True

### Part B. Descriptive Answers

1.  Why increasing pressure after a critical level does not increase the wetted diameter in sprinkler irrigation?
2.  Why is it advised to operate at the lower side of the recommended operating pressure in windy conditions?
3.  Define application efficiency in sprinkler irrigation?
4.  Draw a typical trajectory of sprinkler jet?
5.  Define Christiansen's uniformity coefficient?
6.  Define Distribution coefficient of uniformity?
7.  What are the factors affecting sprinkler head operating performance?
8.  Why increasing the riser height is not a solution for increasing the wetted diameter in sprinkler irrigation?
9.  Write down the relative advantages and disadvantages in selecting smaller and larger sprinkler heads?
10. Show with a sketch, the representative area in which observations have to be made for estimation of Christiansen's uniformity coefficient for rectangular spacing?

11. Show with a sketch, the representative area in which observations have to be made for estimation of distribution uniformity coefficient for triangular spacing?
12. Explain how depth of precipitation varies with respect to operating pressure for a twin-nozzle sprinkler with line sketches?
13. Explain about three commonly used sprinkler head spacing methods?

**Problems**

1. Find out the sprinkler discharge rate coefficient if the sprinkler operating pressure is 25 m and the discharge rate is 0.5 l/s?
2. A riser pipe fitted with a pressure gauge at the bottom of the riser pipe shows a pressure head of 30 m. The inside diameter of the riser pipe is 20 mm. The discharge rate from the sprinkler is 0.5 l/s. Level difference between pressure gauge and the nozzle outlet is 1 m. Find out the expected pressure head at the nozzle outlet (assume friction loss between pressure gauge and nozzle outlet as zero).
3. Volumes of water in cm$^3$ collected in catch cans from a sprinkler catch can test by an operation of four sprinklers for 2 h are provided. The surface area of opening of each catch can is 20 cm$^2$. Find out the average precipitation rate. Find out the Christiansen's coefficient of uniformity.

| | | | |
|------|------|------|------|
| 6.64 | 6.64 | 6.40 | 6.15 |
| 6.83 | 6.83 | 6.86 | 6.89 |
| 7.02 | 7.02 | 7.26 | 7.49 |
| 6.58 | 6.83 | 6.83 | 6.82 |
| 6.14 | 6.64 | 6.40 | 6.15 |

4. Volumes of water in cm$^3$ collected in catch cans from a sprinkler catch can test by an operation of four sprinklers for 2 h are provided. The surface area of opening of each catch can is 20 cm$^2$. Find out the average precipitation rate. Find out the distribution uniformity coefficient.

| | | | |
|------|------|------|------|
| 6.64 | 6.64 | 6.40 | 6.15 |
| 6.83 | 6.83 | 6.86 | 6.89 |
| 7.02 | 7.02 | 7.26 | 7.49 |
| 6.58 | 6.83 | 6.83 | 6.82 |
| 6.14 | 6.64 | 6.40 | 6.15 |

5.  Volumes of water in $cm^3$ collected in catch cans from a sprinkler catch can test by an operation of four sprinklers for 2 h are provided. The surface area of opening of each catch can is 20 $cm^2$. Find out the average precipitation rate. Find out the Christiansen's coefficient of uniformity and distribution uniformity coefficient.

| | | | |
|------|------|------|------|
| 6.83 | 6.83 | 6.86 | 6.89 |
| 7.02 | 7.02 | 7.33 | 7.63 |
| 7.02 | 7.02 | 7.29 | 7.56 |
| 7.02 | 7.02 | 7.26 | 7.49 |

## References and Further Readings

1.  Keller, J., & Bliesner, R. D. (1990). *Sprinkle and trickle irrigation*. Van Nostrand Reinhold.
2.  Martin, D. L., Kincaid, C. D., & Lyle, W. M. (2007). Design and operation of sprinkler irrigation systems. https://doi.org/10.13031/2013.23699

# Layout of Sprinkler Systems

## Introduction

Layout of sprinkler systems is unique to every field, and it depends on the shape, topography of the field, pressure and available discharge rate. Figure 1 shows a very simple lay out. Figure 2 shows a complex layout and the configuration of main, submain and laterals at different locations of the field and also the plan of direction of shifting laterals at every location. Some general guidelines presented in this chapter are very useful in deciding the layout of sprinkler irrigation systems.

## Maximum Area of Land Irrigated from a Water Source

The maximum area that can be irrigated is derived based on the assumption that the maximum water that can be supplied from a source per day must be equal to the amount of water needed by the plants per day.

$$ET_o K_c A = \frac{QT\eta_a}{100}$$

where

$Q$ is the pump discharge rate;
$T$ is time available for irrigation per day;
$ET_o$ is the reference crop evapotranspiration;
$\eta_a$ is the application efficiency (90% for drip irrigation and 80% for sprinkler irrigation);
$K_c$ is crop coefficient depending on the type and the stage of the crop.

If $ET_o$ is expressed in mm; $A$ is expressed in m$^2$; $Q$ is expressed in l/s; $T$ is expressed in hours, the preceding equation must be expressed as follows:

© The Author(s), under exclusive license to Springer Nature Singapore Pte Ltd. 2023   47
V. Ravikumar, *Sprinkler and Drip Irrigation*,
https://doi.org/10.1007/978-981-19-2775-1_3

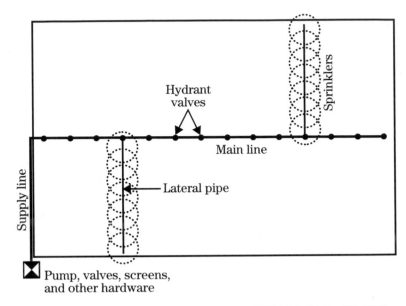

**Fig. 1** Typical simple layout of a sprinkler system. *Source* USDA-NRCS-Part623-NEH

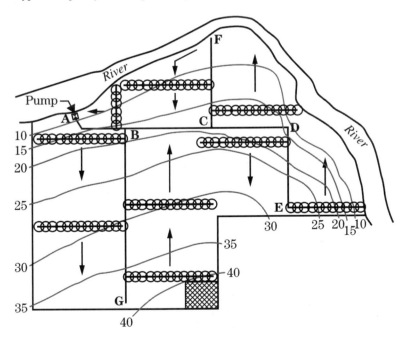

**Fig. 2** Complex layout of sprinkler irrigation system. *Source* USDA-NRCS-Part623-NEH

$$ET_o K_c A = 36 Q T_a \tag{1}$$

**Example**

**Data**: $Q = 1$ l/s;   $T = 8$ h;   $ET_o K_c = ET_c = 8$ mm;   $K_c = 1$;   $\eta_a = 90\%$

**Solution**
By applying Eq. 1, the maximum area that can be irrigated is found out as follows:
$A = 3240$ m$^2$.

## Number of Sprinklers Operated at a Time ($N_{max}$)

The maximum number of sprinklers that can be operated at a time ($N_{max}$) is found out using an equation which is as follows:

$$N_{max} = \frac{Q_d}{q_a} \tag{2}$$

where $Q_d$ is the available discharge rate at the water source and $q_a$ is the average sprinkler discharge rate.

During the design, actual number of sprinklers decided must be less than the maximum number of sprinklers obtained by using the preceding equation. Also the actual number of sprinklers used during every irrigation must not also have too much variation. Otherwise, operation would be very difficult.

In case, much variation exists in the number of sprinklers operated and if the pump is driven by electric motor, throttling of the delivery valve of the pump must be resorted for every irrigation. If the pump is driven by diesel engine, then the speed of the engine must be adjusted.

## Time of Irrigation

The representative area for one sprinkler is the product of spacing between sprinklers within a lateral row ($S_s$) and sprinkler spacing between successive laterals ($S_L$). Field length/width determines the number of sprinklers per lateral. In Fig. 1, the number of sprinklers per lateral can be seen as 7. There are 28 numbers of lateral settings in Fig. 1.

The set time per one lateral setting is the time of operation of one lateral at a location plus the time needed to shift the laterals from one location to the adjacent location. The total time of operation per one irrigation for the entire area must be less than the allowable irrigation interval for the crop irrigated.

**Fig. 3** Sprinkler lateral on contour in steep slopes (www.circlebirrigation.com/k-line-irrigation.
html)

## Pressure Variation in a Lateral

Pressure variation along a lateral is inevitable. But the variation of pressure along
a lateral is normally maintained below 20%. Generally, laying the lateral along a
mild downslope so that pressure loss due to friction and pressure gain due to fall
in elevation are approximately equal. A naive and simple method of layout is placing
a lateral across the prevailing slope (may be along a contour (Fig. 3)). Running a
lateral upslope must be avoided as far as possible. Paired laterals means laying out
of laterals on either side of the submain or main pipe. In such a situation upslope
laying of laterals becomes inevitable. In such situations, care should be taken to
design the laterals to maintain the pressure variation within 20% on both up slope
and down slope lateral.

## Typical Layouts

Figure 4 provides different layouts for different topography. When water supply is
at the centre as in Fig. 4-A, if field slope is steeper, it is better to orient the laterals
along the contours. Instead of this orientation, if the orientation of laterals is along
the slope, two cases arise. Fifty per cent area must be irrigated with upslope laterals,
and another 50% area must be irrigated with downslope laterals. Upslope laterals are
bound to have more variation of pressure.

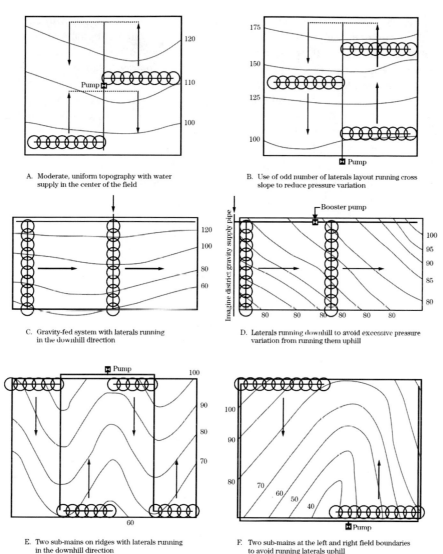

**Fig. 4** Lateral layouts for different topographical situations. *Source* USDA-NRCS-Part623-NEH

Also it must be noted that there are two laterals operated simultaneously and both the laterals are not operated one adjacent to the other but operated on both the directions of main pipe. This kind of operation is called split line operation. The advantages of this method are as follows:

The main line feeding the lateral conveys only 50% of water on either side, and hence the pressure reduction due to friction would be less in main pipe and so smaller diameter pipe could be used. The load on the prime mover also remains approximately the same always. This advantage would be perceptibly felt when the prime mover is diesel engine because when both the laterals are operated simultaneously at the highest elevation from the pump the diesel engine needs to develop more power.

In Fig. 4-B, the laterals are running down the slope in gravity fed system, and this kind of configuration is mostly used in hilly areas. In Fig. 4-D, it can be seen that the left side is level land and the right side is steep slope. So in this case, only during the operation for the right side, a booster pump is used. In Fig. 4-E&F, the main pipe is oriented on ridges and the laterals are oriented downsloping. When downslope laterals are laid, design is made in such a way that the gain in pressure due to fall in elevation is compensated by the loss of pressure head due to friction.

In Fig. 5, different typical main line and pumping plant configurations adopted are presented.

## Summary

Laying out of sprinkler irrigation pipes must start from the data regarding the capacity of the available pump. The maximum number of sprinklers operated simultaneously must not be greater than the pump capacity when optimal pressure is developed in the pump. The maximum area irrigated by the system must also be calculated based on the capacity of pump, crop water requirements and duration of power available in the area. Layout of lateral must be planned in such a way that the pressure variation in the lateral is within 20%. The lateral must be preferably laid on a downslope so that the pressure lost along the way due to friction is compensated by the gain of pressure due to fall in elevation. Upslope lateral must be avoided as far as possible. The topography of the land must be carefully studied, and alternate layouts must be examine before finalizing the final layout (Fig. 5).

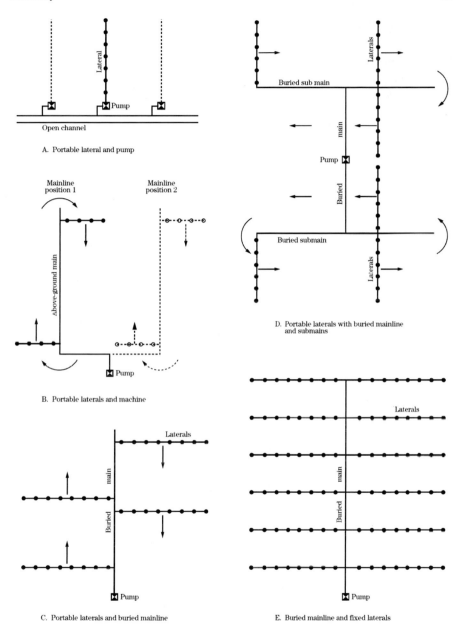

**Fig. 5** Main line and pumping plant layouts. *Source* USDA-NRCS-Part623-NEH

# Appendix

### Interpretation of Contours

From a topographical map, one can visualize the nature of topography of that area. Figure 6 shows the nature of typical contours for hillocks. The contours are closed. Inner contours have higher values than outer contours. Figure 7 shows typical nature of contours for ponds. For ponds, inner-contours have lower elevation than outer contours.

Figure 8 shows the nature of contour lines for ridges and valleys. The contours for ridges and valleys when they are not closed, one must be able to ascertain the correct feature by observing the shape of the contours. In case of ridge line, if the contours are closed, the inner contour would have larger elevation similar to hillocks. In case of valley line, if the contours are closed, the inner contour would have lower elevation similar to ponds.

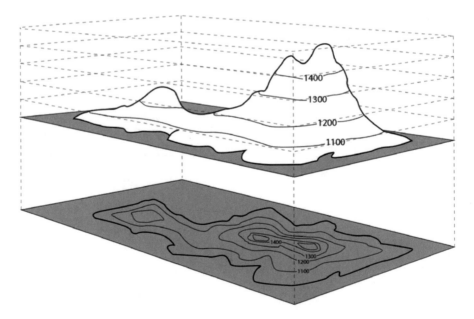

**Fig. 6** Typical contours for hillocks. *Source* www.greenbelly.co/pages/contour-lines

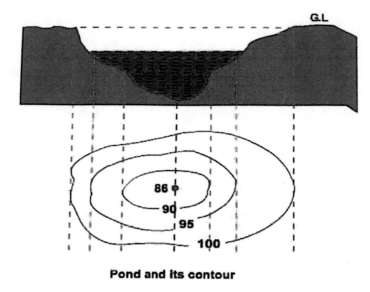

**Pond and its contour**

**Fig. 7** Typical contours for ponds. *Source* http://surveyingestimating.blogspot.com/2018/11/con touring.html

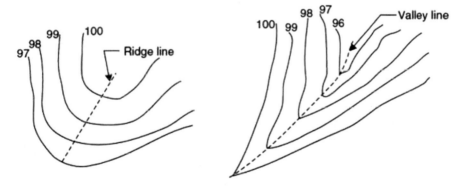

**Fig. 8** Ridge line and valley line. *Source* https://civilblog.org/2015/07/04/15-principal-character istics-of-contour-line

# Assessment

## Part A
1. In a sprinkler installed area, if much variation exists in the number of sprinklers operated every occasion and if the pump is driven by electric motor, _____ of the delivery valve must be resorted to during every irrigation

   a. **Throttling**
   b. Reversing

    c.   Arresting the leaks

    d.   Provision

2.    In a sprinkler installed area, if much variation exists in the number of sprinklers operated every occasion and if the pump is driven by diesel engine, the _____ of the engine must be adjusted during every irrigation

    a.   **Speed**

    b.   Oil

    c.   Vibration

    d.   Governor

    e.   Exhaust

3.    How much is the representative area for one sprinkler if spacing between laterals within a lateral row is 15 m and sprinkler spacing between successive laterals is also 15 m

    a.   **225 m²**

    b.   30 m²

    c.   1 m²

    d.   15 m²

4.    The recommended allowable pressure variation in a sprinkler lateral is _____.

    a.   **20%**

    b.   5%

    c.   100%

    d.   1%

5.    Which of the following is the most favourable layout of a sprinkler lateral?

    a.   **Along a mild downslope so that pressure loss due to friction and pressure gain due to elevation is equal**

    b.   On a mild upslope

    c.   On a level land

    d.   On a steep downslope

## Part B

1.    What do you understand from the term paired laterals?

2.    What do you understand from the term split line operation of laterals and mention the advantages of split line operation?

3.    Draw different typical lateral layouts of sprinklers for different topographical situations?

4.    Draw at least three typical lateral layouts of sprinklers for different topographical situations and at least two typical main line and pumping plant layouts?

5.    Discuss the situations under which booster pumps are recommended for use?

**Problems**

1. Yield of a well is 0.7 l/s. The selected sprinkler discharge rate has 0.5 l/s at the optimal operating pressure. In this case, find out the maximum number of sprinklers that can be operated simultaneously?
2. For an area to be installed with sprinkler irrigation system, the available discharge rate is 2 l/s. During the peak water use period, daily reference crop evapotranspiration is found to be 10 mm/day, the crop coefficient is 1.3, the application efficiency of sprinkler irrigation is 80% and the electric power availability in the area is 4 h/day. Find out the maximum area that can be brought under sprinkler irrigation for the area?

# References and Further Reading

1. Keller, J., & Bliesner, R. D. (1990). *Sprinkle and trickle irrigation.* Van Nostrand Reinhold.
2. Martin, D. L., Kincaid, C. D., & Lyle, W. M. (2007). *Design and operation of sprinkler irrigation systems.* https://doi.org/10.13031/2013.23699
3. USDA-NRCS-Part623-NEH. (2006). Sprinkler irrigation, Chapter-11. In *National engineering hand book.*

# Design of Periodic Move and Solid Set Sprinkler Systems

## Introduction

Sprinkler pipes are multi-outlet pipes for which estimation of loss of energy due to friction is a primary issue. It is known that friction loss occurs due to relative velocity difference between water particles and also between water and pipe inner wall. Flow in the sprinkler and drip irrigation pipes is predominantly turbulent. For such conditions, following equation is suggested by Keller and Bliesner [1] for finding friction loss when pipe diameter is less than 125 mm:

$$\Delta h_f = \frac{K Q^{1.75} L}{D^{4.75}} \tag{1}$$

where $\Delta h_f$ is friction loss (m), $Q$ is discharge rate (l/s), $L$ is the length of pipe (m) and $D$ is the diameter of pipe (mm). The value of $K$ for this set of units is $7.89 \times 10^5$.

This equation is independent of the type of pipe material. Note the mix of units for each variable. Convention in sprinkler irrigation is to express the discharge rate in l/s, length of the pipe in m and diameter in mm.

**Problem** A pipe of inside diameter 44.8 mm of length 50 m is installed to operate a sprinkler of discharge rate 2 l/s. Find out the friction loss in the pipe?

$$\Delta h_f = \frac{7.89 \times 10^5 \, 2^{1.75} \, 50}{44.8^{4.75}}$$

$$\Delta h_f = 1.91 \, \text{m}$$

**Problem** A pump connected to a sprinkler system is as shown in Fig. 1. 10 sprinklers are operated. The characteristic of each sprinkler is that it discharges at a rate of 0.5 l/s when the pressure head is 25 m. Sprinklers are spaced at 12 m interval. The lateral

V. Ravikumar, *Sprinkler and Drip Irrigation*,
https://doi.org/10.1007/978-981-19-2775-1_4

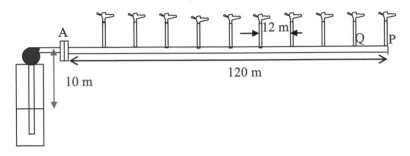

**Fig. 1** Layout of sprinkler lateral, well and pump

**Table 1** Characteristics of the pump system

| No. | Discharge rate (l/s) | Total head (m) | Head loss from foot valve to point $A$ (m) |
|-----|----------------------|----------------|--------------------------------------------|
| 1   | 4                    | 44.6           | 4.6                                        |
| 2   | 6                    | 41.2           | 3.2                                        |

is laid on a horizontal ground. Height of each riser pipe is 0.9 m and the friction and other minor losses occurring through the riser pipe are taken as 0.1 m. Diameter of the lateral is 44.8 mm.

The pump characteristics are as in Table 1.

1.  Find out the discharge rate of each sprinkler?
2.  Find out the average of pressure head at the points of connection of riser pipes with the lateral and average friction loss in the sprinkler pipe downstream of point '$A$'?
3.  Find out the average of pressure heads at the sprinkler heads?
4.  Which sprinkler discharges at the minimum rate and how much is the discharge rate of minimum discharging sprinkler?
5.  Which sprinkler discharges at the maximum rate and how much is the discharge rate of the maximum discharging sprinkler?

**Solution**
It is convenient to divide the system into two parts to solve. One part is the pump system from the foot valve to point '$A$' and another part is the sprinkler system (called as system) starting from point '$A$' to the last sprinkler head.

Assume at the point '$A$' (Fig. 1), the flange is disconnected from the sprinkler irrigation system and a control valve is fitted at point '$A$'. When we partially close the control valve, the total head developed by the pump increases and the discharge rate of the pump gets decreased. When the delivery valve is totally shut off, the total head developed by the pump will be the maximum. Assume at the point '$A$', pump system is removed and the sprinkler system is connected with a water tank. If the water level in the tank is increased, the pressure head at the sprinklers would be

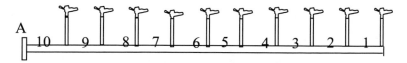

**Fig. 2** Segment count procedure for back-step calculation

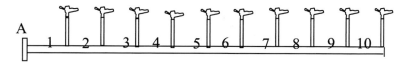

**Fig. 3** Segment count procedure for forward calculation

more and hence the sprinklers would discharge more. If the water level in the tank is reduced, the discharge rate of sprinklers would get reduced. Therefore, we must keep in mind that if the pump system without SIS is operated, as the pressure head developed gets increased, the discharge rate delivered by the pump gets decreased. On the other hand, if SIS is operated without a pump, the discharge rate gets increased if the pressure head at point 'A' is increased. Therefore, when both SIS and also the pump system are connected together, the operating pressure head at 'A' at the operating discharge rate can be found out by matching the SIS characteristic curve with the pump system curve. This matching procedure is demonstrated as follows:

## Pump Characteristics
Available pressure head from the pump at 'A' when discharging at the rate of 4 l/s
= Total head—Static lift—Head loss up to point $A$ = 44.6–10-4.6 = 30 m.

Similarly, available pressure head from pump at 'A' when discharging at the rate of 6 l/s = 41.2 – 10 – 3.2 = 28 m (Fig. 2 and Fig. 3).

## System Characteristics
Relationship between pressure head at point 'A' and discharge into the sprinkler system is found out by a method popularly called as back-step method. In this method, as the first step, a pressure head is assumed for the last discharging point $P$ (Fig. 1). Let us assume it as 25 m. For this assumed pressure head at '$P$', we have to find out the discharge rate from the sprinkler connected at point '$P$'.

Discharge rate at the sprinkler is a function of the pressure head available at the sprinkler head. The pressure head versus discharge rate of the sprinkler is given by the following equation:

$$q = C_d A \sqrt{2gh}$$

where '$q$' is discharge rate from nozzle of cross-sectional area '$A$', $C_d$ is the coefficient of discharge and '$h$' is pressure head. The preceding equation can be written as follows:

$$q = k\sqrt{h}$$

where $k$ is obviously equal to $C_d A(2g)^{0.5}$. In the data of this problem, for one pressure head value of 25 m, the discharge rate is given as 0.5 l/s. So the value of '$k$' for this sprinkler is found out using the preceding equation as follows:

$$k = \frac{0.5}{\sqrt{25}} = 0.1\frac{1}{s}/m^{0.5}$$

Pressure head at the last sprinkler head

= 25−elevation of riser pipe−friction and minor losses in the riser pipe

= 25 − 0.9 − 0.1 = 24 m

Discharge rate of the sprinkler head at point '$P$'

$$q = k\sqrt{h}$$

$$q = 0.1\sqrt{24} = 0.49\frac{1}{s}$$

The friction loss in the pipe segment $QP$ is as follows:

$$\Delta h = \frac{7.89 \times 10^5 \times 0.49^{1.75} \times 12}{44.8^{4.75}} = 0.04 \text{ m}$$

The pressure head at $Q$ = 25 m + Friction loss in segment $QP$

= 25 + 0.04   = 25.04 m

Based on the same procedure outlined above, the calculations proceed upstream until point '$A$'. The pressure heads at the points of connections of each riser pipe and also the discharge rate of each sprinkler are found out and presented in Table 2.

Once again, the pressure head at point '$A$' is assumed as 20 m and calculations are done as earlier and the results are presented in Table 3.

The system characteristics and pump characteristics are plotted as shown in Fig. 4 using the data in Tables 4 and 5. The point of intersection between both the characteristics shows that the system will operate so that the pressure head at '$A$' is 29.3 m and the discharge rate at '$A$' is 4.7 l/s.

Now again calculation should proceed from point '$A$' in forward direction to find out sprinkler discharge rate in each sprinkler. The results of forward calculation are presented in Table 6. Note that in backward step method, the segments are counted from the last sprinkler and in the forward step method; the segments are counted from the first sprinkler (Figs. 2 and 3). Table 7 shows the calculation procedure for the average pressure head loss in the pipes and average pressure head at the points of connection of riser pipes.

**Table 2** Back-step calculation for 25 m pressure head at the extreme end

| Segment No. (i) | Pressure head in the far end of segment $h_i = h_{i-1} + \Delta h_i$ (m) | Discharge rate of each sprinkler $q_i = 0.1\sqrt{h_i - 0.9} - 0.1$(l/s) | Discharge rate in each lateral segment $Q_i = \sum_{i=1}^{i} q_i$ (l/s) | Pressure head loss in the lateral segment $\Delta h_i = \frac{KQ_i^{1.75}}{D^{4.75}}s$ (m) |
|---|---|---|---|---|
| 1 | 25.00 | 0.49 | 0.49 | 0.04 |
| 2 | 25.04 | 0.49 | 0.98 | 0.13 |
| 3 | 25.17 | 0.49 | 1.47 | 0.27 |
| 4 | 25.44 | 0.49 | 1.97 | 0.44 |
| 5 | 25.88 | 0.50 | 2.46 | 0.66 |
| 6 | 26.54 | 0.51 | 2.97 | 0.91 |
| 7 | 27.45 | 0.51 | 3.48 | 1.21 |
| 8 | 28.66 | 0.53 | 4.01 | 1.54 |
| 9 | 30.20 | 0.54 | 4.55 | 1.92 |
| 10 | 32.12 | 0.56 | 5.11 | 2.36 |
| | 34.48 | Total = 5.11 | | |

**Table 3** Back-step calculation for 20 m pressure head at the extreme end

| Segment No. (i) (Counted forward) | Pressure head in the far end of segment $h_i = h_{i-1} + \Delta h_i$ (m) | Discharge rate of each sprinkler $q_i = 0.1\sqrt{h_i - 0.9} - 0.1$(l/s) | Discharge rate in each lateral pipe segment $Q_i = \sum_{i=1}^{i} q_i$ (l/s) | Pressure head loss in the lateral pipe segment $\Delta h_i = \frac{KQ_i^{1.75}}{D^{4.75}}s$ (m) |
|---|---|---|---|---|
| 1 | 20.00 | 0.44 | 0.44 | 0.03 |
| 2 | 20.03 | 0.44 | 0.87 | 0.11 |
| 3 | 20.14 | 0.44 | 1.31 | 0.22 |
| 4 | 20.36 | 0.44 | 1.75 | 0.36 |
| 5 | 20.72 | 0.44 | 2.19 | 0.54 |
| 6 | 21.25 | 0.45 | 2.64 | 0.74 |
| 7 | 22.00 | 0.46 | 3.10 | 0.98 |
| 8 | 22.98 | 0.47 | 3.57 | 1.26 |
| 9 | 24.24 | 0.48 | 4.05 | 1.57 |
| 10 | 25.81 | 0.50 | 4.55 | 1.92 |
| | 27.74 | Total = 4.55 | | |

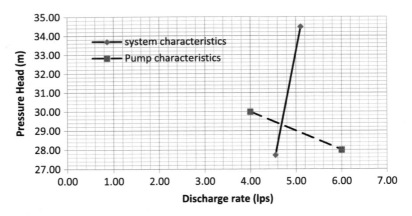

**Fig. 4** Matching pump and sprinkler system characteristics

**Table 4** Pump system characteristics with reference to point 'A'

| No. | Discharge rate (l/s) | Pressure head developed by the pump at point 'A' (m) |
|-----|----------------------|------------------------------------------------------|
| 1 | 4 | 30 |
| 2 | 6 | 28 |

**Table 5** Sprinkler system characteristics with reference to point 'A'

| No. | Discharge rate into sprinkler system (l/s) | Pressure head at point 'A' (m) |
|-----|--------------------------------------------|--------------------------------|
| 1 | 4.55 | 27.74 |
| 2 | 5.11 | 34.48 |

**Table 6** Computations for forward step method

| Segment ($i$) (Counted forward) | Pressure Head in pipe at near end $h_i = h_{i-1} - \Delta h_i$ (m) | Discharge rate $Q_i = Q_{i-1} - q_{i-1}$ (l/s) | Head loss $\Delta h_i = \dfrac{K Q_i^{1.75}}{D^{4.75}} s$ (m) | Discharge rate of sprinkler $q_i = 0.1\sqrt{h_i - \Delta h_i - 0.9} - 0.1$ (l/s) |
|----|-------|------|------|------|
| 1 | 29.30 | 4.70 | 2.04 | 0.51 |
| 2 | 27.26 | 4.19 | 1.66 | 0.50 |
| 3 | 25.60 | 3.69 | 1.33 | 0.48 |
| 4 | 24.27 | 3.21 | 1.04 | 0.47 |
| 5 | 23.22 | 2.74 | 0.79 | 0.46 |
| 6 | 22.43 | 2.27 | 0.57 | 0.46 |
| 7 | 21.86 | 1.82 | 0.39 | 0.45 |
| 8 | 21.47 | 1.37 | 0.23 | 0.45 |
| 9 | 21.24 | 0.92 | 0.12 | 0.45 |
| 10 | 21.12 | 0.47 | 0.04 | 0.45 |
| | 21.09 | | | Total = 4.68 |

**Table 7** Average pressure head and loss in the lateral

| Segment ($i$) | Pressure loss in segment '$i$' (m) | Pressure loss till the end of segment '$i$' from point $A$ (m) | Pressure head in pipe at the point of connection of riser pipe for each segment '$i$' (m) |
|---|---|---|---|
| 1 | 2.04 | 2.04 | 27.26 |
| 2 | 1.66 | 3.70 | 25.60 |
| 3 | 1.33 | 5.03 | 24.27 |
| 4 | 1.04 | 6.08 | 23.22 |
| 5 | 0.79 | 6.87 | 22.43 |
| 6 | 0.57 | 7.44 | 21.86 |
| 7 | 0.39 | 7.83 | 21.47 |
| 8 | 0.23 | 8.06 | 21.24 |
| 9 | 0.12 | 8.18 | 21.12 |
| 10 | 0.04 | 8.21 | 21.09 |
| | | $\overline{\Delta h_f} = 6.34$m | $\overline{h} = 22.96$m |

Note that, for calculating average pressure head in the sprinkler pipe, the pressure heads at the connection points of riser pipe alone are considered and the pressure head at the start of the lateral pipe (Point $A$) is not considered.

Graphically the average pressure head at the points of connection of each riser pipe and the average friction loss for carrying water to each point of connection of riser pipe are shown in Fig. 5. Based on the preceding discussions, we can write down the following relationship:

$$h_{\text{start}} - \overline{h} + \overline{\Delta h}$$

where $h_{\text{start}}$ is the pressure head at the start of the lateral, $\overline{h}$ is the average of pressure heads at the points of connection of riser pipe with lateral pipe and $\overline{\Delta h}$ is the average of friction losses for carrying water to each point of connection of riser pipe from the start of the lateral. Conventionally, $\overline{h}$ is just called the average pressure head of lateral pipe and $\overline{\Delta h}$ is called average friction loss in the pipe.

$$h_{\text{start}} = 22.96 + 6.34 = 29.30 \text{ m}$$

Let us imagine that there exists one sprinkler for which the pressure head is the average of all the pressure heads at the sprinkler heads ($\overline{h}_{\text{sprink}}$) and if the sprinkler is connected at that location with a riser pipe, then following relationships can be written:

$$\overline{h}_{\text{sprink}} = \overline{h} - h_r$$
$$\overline{h} = \overline{h}_{\text{sprink}} + h_r$$

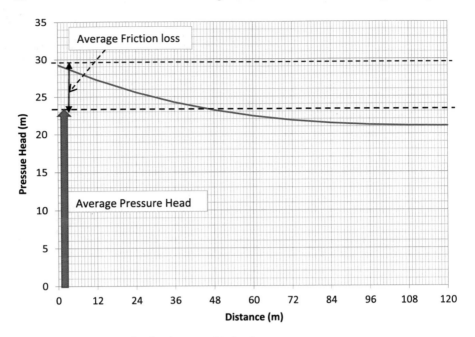

**Fig. 5** Average pressure head and average friction loss

where $h_r$ is the sum of height of riser pipe and friction and other losses in the riser pipe. From the preceding equations, following equation is written:

$$h_{\text{start}} = \overline{h}_{\text{sprink}} + h_r + \overline{\Delta h}$$

Average pressure head at the sprinkler head $(\overline{h}_{sprink})$ can be expressed as follows (Fig. 5).

$$\overline{h}_{\text{sprink}} = h_{\text{start}} - h_{\text{r}} - \overline{\Delta h}$$
$$\overline{h}_{\text{sprink}} = 29.3 - 0.9 - 0.1 - 6.34 = 21.96\,\text{m}$$

The minimum discharge rate sprinkler in this problem would occur for the last sprinkler for which the pressure head at the sprinkler head is the minimum.

Minimum pressure head on the lateral pipe $(h_{\min})$ can be found out using the following relationship:

$$h_{\min} = h_{\text{start}} - \Delta h_L$$
$$h_{\min} = 29.3 - 8.21 = 21.09\,\text{m}$$

where $\Delta h_L$ is the friction loss for the total length of lateral pipe.
The pressure head at the last sprinkler head

$$= 21.09 - (0.9 + 0.1) = 20.09\,\text{m}$$

Discharge rate for the minimum discharging sprinkler $= 0.45$ l/s.

The maximum discharge rate sprinkler in this problem would occur for the first sprinkler for which the pressure head at the sprinkler head is the maximum.

Maximum pressure head on the lateral pipe $(h_{max})$ can be found out using the following relationship:

$$h_{max} = h_{start} - \Delta h_1$$

where $\Delta h_1$ is the friction loss for the first segment of the lateral pipe from point $A$.

$$h_{max} = 29.3 - 2.04 = 27.26\,\text{m}$$

The pressure head at the first sprinkler head $= h_1 - h_r = 27.26 - (0.9 + 0.1) = 26.26\,\text{m}$

Discharge rate from the maximum discharging sprinkler $= 0.51$ l/s.

## Important Observations from This Problem

For any specific configuration of sprinklers and pump, the discharges of each sprinkler can be found out by doing head loss calculations by assuming a pressure head at the last sprinkler. Pressure loss calculations are to be done recursively for each pipe segment. The application of back-step method must be applied few times to get a pressure head versus discharge rate characteristics for a sprinkler system.

Matching the pressure head versus discharge characteristics of sprinkler system and pump system must be done to get the operating pressure head and discharge rate at the point of connection of sprinkler and pump systems. For doing such calculations, assistance of computer is needed and at least spreadsheet software package like MS-EXCEL is needed.

Forward step method must be applied after operating discharge rate into the sprinkler system and pressure head at the point of connection of sprinkler system is found out.

In real-life situations, before arriving at a feasible configuration of sprinkler system, many configurations are usually examined by the designer. Therefore only forward step calculations are done by assuming a discharge rate into the sprinkler system. Even in forward calculations, head loss calculations are not done for each pipe segment in a lateral. Head loss calculations are done for the total length of lateral only. In the succeeding sections, equations which are used conventionally in the design of sprinkler systems are derived.

## Friction Loss Equation in Sprinkler Lateral for a Uniform Diameter Pipe

Let us derive an expression for a horizontal sprinkler lateral pipe with each sprinkler delivering at a constant discharge rate of $q$. Let $N$ be the number of sprinklers and total length of pipe be $L$. Let $Q$ be the discharge rate entering into the lateral pipe. Let $D$ be the uniform diameter of the pipe. The lateral pipe is closed at the far end.

The total discharge $Q$ would gradually decrease with a discharge rate of $q$ after flowing through each segment of pipe. In Fig. 6, it can be seen that the fall of pressure head is steeper at the start and the steepness gradually gets reduced because the flow velocity gets decreased along the length of the flow due to reduction in discharge rate.

$$q = \frac{Q}{N} \quad \text{and} \quad s = \frac{L}{N}$$

We know that friction loss occurring when a uniform discharge rate of $Q$ (l/s) occurs for a length of $L$ (m) through a uniform pipe diameter of $D$ (mm) is found out using following equation:

$$\Delta h_f = \frac{K Q^{1.75} L}{D^{4.75}}$$

Let the total friction loss (m) for the total length of the pipe is $\Delta h_L$.

**Fig. 6** Pressure distribution in horizontal laterals

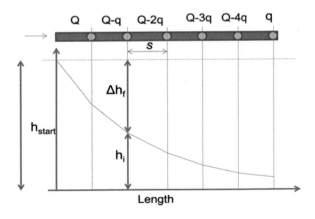

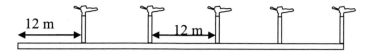

**Fig. 7** Lateral layout for friction loss computation numerical problem

$$\Delta h_L = \sum_{i=1}^{n} \Delta h_f^i$$

$$\Delta h_L = \frac{K \times s}{D^{4.75}} \sum_{i=1}^{N} (Q - (i-1)q)^{1.75}$$

$$\Delta h_L = \frac{K \times s}{D^{4.75}} \sum_{i=1}^{N} \left( Q - (i-1)\frac{Q}{N} \right)^{1.75} \tag{2}$$

$$\Delta h_L = \frac{Q^{1.75} K L}{N N^{1.75} D^{4.75}} \sum_{i=1}^{N} (N \quad i+1)^{1.75}$$

Equation 2 can also be written equivalently as follows:

$$\Delta h_L = \frac{Q^{1.75} K L}{N^{2.75} D^{4.75}} \sum_{i=1}^{N} i^{1.75}$$

$$\Delta h_L = \frac{Q^{1.75} K L}{D^{4.75}} \left( \frac{1}{N^{2.75}} \sum_{i=1}^{N} i^{1.75} \right)$$

where the term in the bracket on the right side is called as Christiansen's [2] friction factor ($F$) and is approximately expressed as follows for easiness in calculation:

$$F = \left( \frac{1}{N^{2.75}} \sum_{i=1}^{N} i^{1.75} \right) = \frac{1}{2.75} + \frac{1}{2N} + \frac{0.75}{6N^2} \tag{3}$$

$$\Delta h_L = \frac{K Q^{1.75} L}{D^{4.75}} F \tag{4}$$

**Problem** Find out the friction loss that would occur in a sprinkler lateral with following data (Fig. 7).

| | |
|---|---|
| Length of lateral | = 60 m. |
| Diameter of the lateral | = 57.4 mm. |
| Discharge rate of each sprinkler | = 0.5 l/s. |
| Spacing between each sprinkler | = 12 m. |

(continued)

(continued)

Length of the first sprinkler from the submain    $= 12$ m.

## Solution

$$\Delta h_L = \frac{K Q^{1.75} L}{D^{4.75}} F$$

$$F = \frac{1}{2.75} + \frac{1}{2N} + \frac{0.75}{6N^2}$$

$$F = \frac{1}{2.75} + \frac{1}{2 \times 5} + \frac{0.75}{6 \times 5^2}$$

$$F = 0.469$$

$$\Delta h_L = \frac{(7.89 \times 10^5)(5 \times 0.5)^{1.75} 60}{57.4^{4.75}} \times 0.469$$

$$\Delta h_L = 0.488 \text{ m}$$

The equation developed for uniform diameter with closed end can be used for calculating friction loss in multi-diameter laterals (also called as Telescopic laterals). Following problem illustrates how it is done:

**Problem** One lateral consists of two segments. The first segment is 36 m length with inside diameter of 82.4 mm. The second segment is 96 m length with inside diameter of 68.6 mm. The sprinklers are placed at a spacing of 12 m, and the first sprinkler is also placed at the spacing of 12 m. Discharge rate of each sprinkler is 0.5 l/s. Find out the friction loss that would occur in the sprinkler lateral.

## Solution
It should be recalled that the equation we derived is on the assumption that the sprinklers are placed at a uniform spacing and the end of the pipe is closed.

If 82.4 mm lateral is laid for the entire length of 132 m, the friction loss that would occur for the entire length is shown as '$x$' in Fig. 8.

If 82.4 mm lateral is laid for the 96 m segment also, the friction loss that would occur in the 96 m segment alone  is shown as '$z$'.

If 68.6 mm lateral is laid for the second segment, the friction loss that occurs for the 96 m length alone is shown as '$y$'.

Hence by observing the geometry in Fig. 8, the friction loss that would occur if 82.4 mm lateral is laid for the first segment of length 36 m and if 68.6 mm lateral is laid out for the second segment of length 96 m, the total friction loss would be equal to $x + y - z$ (Fig. 8).

## Finding x
$D = 82.4$ mm

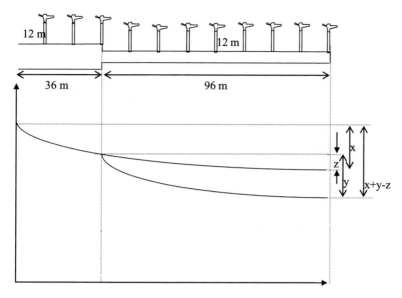

**Fig. 8** Friction loss curves for the telescopic pipe numerical problem

$N = 11$

$L = 132$ m

Christiansen's $F$ for 11 sprinklers is 0.41

$$\Delta h = \frac{(7.89 \times 10^5)(11 \times 0.5)^{1.75} 132}{82.4^{4.75}} \times 0.41$$

$$\Delta h = 0.669 \text{ m}$$

**Finding z**

$D = 82.4$ mm

$N = 8$

$L = 96$ m

Christiansen's $F$ for 8 sprinklers is 0.428

$$\Delta h = \frac{(7.89 \times 10^5)(8 \times 0.5)^{1.75} 96}{82.4^{4.75}} \times 0.428$$

$$\Delta h = 0.291 \text{ m}$$

**Finding y**

$D = 68.6$ mm

$N = 8$

$L = 96\,\text{m}$

Christiansen's F for 8 sprinklers is 0.428

$$\Delta h = \frac{(7.89 \times 10^5)(8 \times 0.5)^{1.75}96}{68.6^{4.75}} \times 0.428$$

$$\Delta h = 0.695\,\text{m}$$

$$x + y - z = 0.669 + 0.695 - 0.291 = 1.074\,\text{m}$$

Hence, the total friction loss for the entire length of 136 m when two pipe segments with diameter of 82.4 mm for a length of 36 m and with a diameter of 68.6 mm for a length of 96 m is 1.074 m.

## Scallopi's Correction Factor for Christiansen's F

In many field situations, the first sprinkler alone would not be installed with a common spacing 's' and the first sprinkler may be at a fractional spacing. Let that spacing be 'αs' and α is a constant varying between 0 and 1 (Fig. 9).

For such situations, Scaloppi [3] introduced a correction to Christiansen's factor (F) which is as follows:

$$F(\alpha) = \frac{NF - (1 - \alpha)}{N - (1 - \alpha)} \tag{5}$$

where $F(\alpha)$ is Scallopi factor and α is the ratio of first sprinkler spacing to normal spacing.

**Fig. 9** Illustration for Scallopi's correction factor

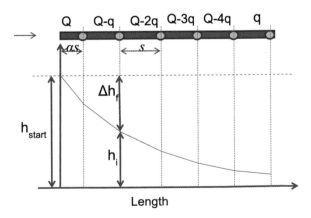

## *Average Friction Loss*

It can be noticed that the friction loss curve is approximately a curve. An expression for finding the average friction loss in a sprinkler pipe was found out by Anwar [4, 5] which is as follows:

$$\overline{\Delta h} = \overline{F} \times \Delta h_{\text{L}} \tag{6}$$

where $\overline{F}$ is called as Anwar's average factor for finding average friction loss in a lateral. The value of $\overline{F}$ can be found out using the following formula:

$$\overline{F} = 1 - \left( \frac{1}{N} \frac{\sum_{i=1}^{N} (N - i) i^{1.75}}{\sum_{i=1}^{N} i^{1.75}} \right) \tag{7}$$

If the ratio of first sprinkler spacing to the normal sprinkler spacing in a lateral is not equal to 1, $\overline{F}$ must be corrected as follows:

$$\overline{F}(\alpha) = 1 - \left( \frac{NF(1 - \overline{F})}{NF + \alpha - 1} \right) \tag{8}$$

Table 8 lists the values of Christiansen's factor and Anwar's average factor for full spacing as well as half spacing of the first sprinkler in order to help in doing calculations faster.

## Pressure Head Distribution in Sloping Laterals

In real-life situations, land is mostly sloping and either lateral is laid on upslope or downslope. The pressure head at any location of a sloping lateral depends on the pressure head at the start and elevation loss or gain until that location with reference to the start of the lateral and the friction loss occurring from the start of the lateral to that location. The pressure head at any location '$i$' for the sloping lateral can be found out by the following equation (Fig. 10):

$$h_i = h_{\text{start}} - \Delta h_i + p \Delta Z_i \tag{9}$$

where '$p$' is an indicator and takes a value of minus one for upslope and plus one for downslope and $\Delta Z_i$ is the change in elevation from the start until the point '$i$'.

In case of uniformly sloping laterals, the pressure head relationship in terms of averages is as follows (Fig. 11):

$$h_{\text{start}} = \overline{h} + \overline{\Delta h} - p\overline{\Delta Z} \tag{10}$$

**Table 8** Christiansen F and Anwar $\overline{F}$ values

| Number of outlets | Christiansen $F$ values | | Anwar $\overline{F}$ values | |
|---|---|---|---|---|
| | Full spacing | Half spacing | Full spacing | Half spacing |
| 1 | 1.00 | 1.00 | 1.000 | 1.000 |
| 2 | 0.650 | 0.533 | 0.885 | 0.813 |
| 3 | 0.546 | 0.456 | 0.840 | 0.770 |
| 4 | 0.498 | 0.426 | 0.816 | 0.754 |
| 5 | 0.469 | 0.410 | 0.801 | 0.747 |
| 6 | 0.451 | 0.401 | 0.790 | 0.742 |
| 7 | 0.438 | 0.395 | 0.783 | 0.741 |
| 8 | 0.428 | 0.390 | 0.777 | 0.739 |
| 9 | 0.421 | 0.387 | 0.772 | 0.737 |
| 10 | 0.415 | 0.384 | 0.768 | 0.736 |
| 11 | 0.410 | 0.382 | 0.765 | 0.736 |
| 12 | 0.406 | 0.380 | 0.763 | 0.736 |
| 13 | 0.403 | 0.379 | 0.761 | 0.736 |
| 14 | 0.400 | 0.378 | 0.759 | 0.736 |
| 15 | 0.398 | 0.377 | 0.757 | 0.736 |
| 16 | 0.395 | 0.376 | 0.756 | 0.736 |
| 17 | 0.394 | 0.375 | 0.754 | 0.736 |
| 18 | 0.392 | 0.374 | 0.753 | 0.736 |
| 19 | 0.390 | 0.374 | 0.752 | 0.736 |
| 20 | 0.389 | 0.373 | 0.751 | 0.736 |
| 25 | 0.384 | 0.371 | 0.748 | 0.736 |
| 30 | 0.380 | 0.370 | 0.745 | 0.736 |
| 35 | 0.378 | 0.369 | 0.747 | 0.736 |
| 40 | 0.376 | 0.368 | 0.742 | 0.736 |

where $\overline{\Delta Z}$ is average loss or gain due to elevation difference between the start and the end. It is half of the total elevation difference between the start and end of the lateral.

In case of horizontal and upslope laterals, the location of the maximum pressure head is the point of connection of the first sprinkler riser pipe and the minimum pressure head is the point of connection of the last sprinkler pipe. Hence finding the magnitude of the extent of maximum pressure head and the minimum pressure head is straightforward. But in the case of downsloping laterals, the minimum pressure head and maximum pressure head can anywhere and they can be found out only when the pressure heads at all the connection points of riser pipe are found out. Figure 12 shows a typical pressure head curve for a downslope lateral.

**Fig. 10 a** Pressure
head distribution in uniform
upslope lateral. **b** Pressure
head distribution in uniform
downslope lateral

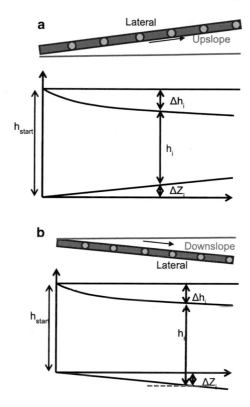

# Design of Laterals

The procedure presented here is for a single diameter lateral closed at the end. The
discharge rate from a sprinkler head can be found out using the following equation:

$$q = k\sqrt{h^{\mathrm{sp}}} \tag{11}$$

where $q$ is sprinkler discharge rate; $k$ is discharge coefficient and $h^{\mathrm{sp}}$ is the pressure
head at the sprinkler head.

The pressure head variation ($V_h$) at the sprinkler heads is to be defined as follows:

$$V_h = \frac{h^{\mathrm{sp}}_{\mathrm{max}} - h^{\mathrm{sp}}_{\mathrm{min}}}{h^{\mathrm{sp}}_{\mathrm{a}}} \times 100 \tag{12}$$

where $h^{\mathrm{sp}}_{\mathrm{max}}$ the maximum pressure head at the sprinkler is head; $h^{\mathrm{sp}}_{\mathrm{min}}$ is the minimum
pressure head at the sprinkler head; $h^{\mathrm{sp}}_{\mathrm{a}}$ is the average pressure head at the sprinkler
head.

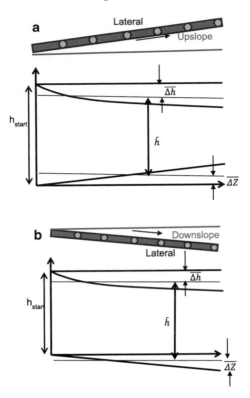

**Fig. 11** **a** Average pressure head in uniform upslope lateral, **b** Average pressure head in uniform downslope lateral

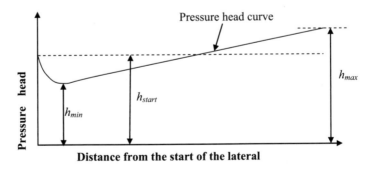

**Fig. 12** Pressure head distribution in a typical uniform downslope lateral

If $h_r$ represents the height of riser pipe, then the following equation can be written:

$$h_i^{\text{sp}} = h_i - h_{\text{r}} \tag{13}$$

where $h_i^{sp}$ is the pressure head at the sprinkler head '$i$' and $h_i$ is pressure head at the lateral pipe and $h_i$ is the height of riser pipe. Equation 12 can be written as follows after substituting Eq. 13:

$$V_h = \frac{(h_{max} - h_r) - (h_{min} - h_r)}{\overline{h} - h_r} \times 100$$
$$V_h = \frac{h_{max} - h_{min}}{h_a^{sp}} \times 100 \tag{14}$$

where $h_{max}$ is the maximum pressure head in the lateral; $h_{min}$ is the minimum pressure head in the lateral; $\overline{h}$ is the average pressure head in the lateral;

$$\overline{h} = \frac{1}{N} \sum_1^N h_i \tag{15}$$

where $h_i$ is the pressure head in the pipe at the point of connection of riser pipe '$i$'.

Convention is to have an allowable pressure variation value of 20%, and therefore Eq. 14 can be written as follows:

$$0.2h_a^{sp} \geq (h_{max} - h_{min}) \tag{16}$$

For horizontal and upslope lateral $h_{max}$ would occur at the first outlet and $h_{min}$ would occur at the last outlet. Therefore,

$$0.2h_a^{sp} \geq (h_1 \quad h_N) \tag{17}$$

where $h_1$ and $h_N$ are the pressure heads at the first outlet and the last outlets, respectively.

$$0.2h_a^{sp} \geq (\Delta h)_{1 \text{ to } N} - p(\Delta Z)_{1 \text{ to } N} \tag{18}$$

where $(\Delta h)_{1 to N}$ is the head loss due to friction in the lateral pipe from the first outlet to the last outlet; '$p$' is an indicator and takes a value of minus one for upslope and plus one for downslope; $(\Delta Z)_{1 \text{ to } N}$ is the change in elevation from the first outlet to the the last outlet.

Equation 18 can be written as follows:

$$0.2h_a^{sp} \geq \frac{K(Q - q_a)^{1.75}(L - S_1)F_{N-1}}{D^{4.75}} - p(\Delta Z)_{1 \text{ to } N} \tag{19}$$

where $q_a$ is the average discharge rate of sprinkler, $S_1$ is the first sprinkler spacing and $F_{N-1}$ is Christiansen's reduction coefficient for $(N - 1)$ outlets.

In practice, Eq. 19 is further simplified and conventionally used as follows:

$$0.2h_a^{\text{sp}} \geq \frac{KQ^{1.75}LF_N}{D^{4.75}} - p\Delta Z$$

$$0.2h_a^{\text{sp}} \geq \frac{KQ^{1.75}LF_N}{D^{4.75}} - p\frac{SL}{100} \tag{20}$$

In Eq. 20, $S$ is slope of the land in per cent.

It should be kept in mind that this design equation is based on the assumption that the maximum pressure head and the minimum pressure head are at the extremes of the lateral. But we have seen already that in case of downslope, the locations of the maximum and the minimum pressures can be anywhere and they can be found out only after the length of the lateral is decided. Hence, the preceding equations are not completely valid for downslope laterals. However these equations are used as a first level approximation. More advanced analysis for downslope laterals are discussed in the drip lateral design section, and the same principles can be used for sprinklers also.

**Problem** In a horizontal sprinkler lateral at the first sprinkler, the pressure head is 25 m, and at the last sprinkler, the pressure head is 27.5 m. The sprinkler discharges at a rate of 0.5 l/s at the pressure head of 25 m and the discharge coefficient of the sprinkler ($k$) are 0.1 l/s/m$^{0.5}$. Find out variation of pressure head as well as the variation of discharge rate.

Average of the maximum and minimum pressure heads = 26.25 m.
Variation of pressure head ($V_h$)

$$V_h = \frac{27.5 - 25}{26.25} \times 100 = 9.524\%$$

Discharge rate at 27.50 m pressure head = 0.52 l/s (By Eq.1 of Chapter 2)
Discharge rate at 25.00 m pressure head = 0.5 l/s (By Eq.1 of Chapter 2)
Average discharge rate at 26.25 m pressure head = 0.512 l/s. (By Eq.1 of Chapter 2)
Variation of discharge rate ($V_q$)

$$V_q = \frac{0.520 - 0.500}{0.512} \times 100 = 47.6\%$$

One inference that should be noted from this problem is that when the pressure variation is 10%, approximately the discharge rate variation is 5%. The following equation holds good relating pressure head variation and discharge rate variation.

$$V_q = xV_h \tag{21}$$

where '$x$' is called as flow exponent of sprinklers and is equal to 0.5.

**Problem** A sprinkler lateral has to be designed for a horizontal land. The average design discharge rate is 0.47 l/s. The sprinklers have a discharge coefficient of

0.1 l/s/m$^{0.5}$. The length of lateral is 114 m. 10 sprinklers have to be operated with 12 m spacing, and the first sprinkler spacing is 6 m. Examine two diameters of lateral pipe; 44.8 mm and 57.4 mm. Take the height of the riser pipe as 0.9 m and the friction and other minor losses occurring through the riser pipe are taken as 0.1 m.

Select the diameter for which the variation of pressure head is less than 20%. Also find the pressure head required to be maintained at the start of the lateral for 44.8 mm diameter and also for the 57.4 mm diameter lateral in order to get the average design discharge rate of 0.47 l/s from the sprinklers.

Total discharge rate into the lateral $(Q) = 0.47 \times 10 = 4.7$ l/s.

Average pressure head at the sprinkler head $\left(h_a^{sp}\right)$:

$$q = k\sqrt{h}$$
$$0.47 = 0.1\sqrt{h_a^{sp}}$$
$$h_a^{sp} = 21.9 \text{ m}$$

Average pressure head in the pipe $(\overline{h})$:

$$\overline{h} = h_a^{sp} + \text{friction and other minor losses in riser pipe}$$
$$= 21.9 + 0.9 + 0.1$$
$$= 22.90 \text{ m}$$

By using Table 8, Christiansen's and Anwar's correction factors are found out as follows:

$F_{10} - 0.415 \quad F_{10}(0.5) - 0.384 \quad \overline{F_{10}} = 0.768 \quad \overline{F_{10}}(0.5) = 0.736.$

The computations for each pipe diameter are done and presented in Table 9. From the preceding calculations, it is obvious that 57.4 mm diameter pipe has to be selected for use, since the pressure head variation is less than 20%. Note that Eq. 14 is used for finding pressure head variation in the lateral which gives more accurate results. But if we make use of the widely used approximate Eq. 20, the pressure head variation in the lateral for horizontal lateral would be the ratio of total head loss due to friction for the total length of the lateral to the average pressure head at the sprinkler head expressed in per cent. If we make use of that concept, we get pressure head variation of 33.56% for 44.8 mm pipe diameter and 10.34% for 57.4 mm diameter.

**Problem** Design optimal length of a lateral for an average sprinkler discharge rate of 0.5 l/s operating at an average pressure head of 25 m with an inner diameter of lateral pipe of 57.4 mm. All the sprinklers are placed at a uniform spacing of 12 m, and the height of riser pipe is 1 m. Design for the following 2 cases. Case 1: Slope of the land is 1% upward and Case 2: Slope of the land is 1% downward.

**Case-1. Slope of the Land is 1% Upward**

Equation 20: $0.2h_a^{sp} \geq \frac{KQ^{1.75}LF_N}{D^{4.75}} - p\frac{SL}{100}$.

**Table 9** Hydraulic analysis for two different pipe diameters

| Variable | Formula | For pipe diameter 44.8 mm | For pipe diameter 57.4 mm |
|---|---|---|---|
| Head loss due to friction for total length of pipe (m) | $\Delta h = \frac{KQ^{1.75}L}{D^{4.75}}F_{10}(0.5)$ | $= 7.37$ m | $= 2.27$ m |
| Average head loss due to friction (m) | $\overline{\Delta h} =$ $\frac{KQ^{1.75}LF_{10}(0.5)}{D^{4.75}}\overline{F_{10}}(0.5)$ | $= 5.43$ m | $= 1.67$ m |
| Pressure head needed at the start of the lateral (m) | $h_{start} = \overline{h} + \overline{\Delta h}$ | $= 22.90 + 5.43$ $= 28.33$ m | $= 22.90 + 1.67$ $= 24.57$ m |
| Head loss due to friction for the first segment of the lateral of 6 m length | $(\Delta h)_1 = \frac{KQ^{1.75}6}{D^{4.75}}$ | $= 1.01$ m | $= 0.31$ m |
| Maximum pressure head in the lateral ($h_1$) | $h_1 = h_{start} - (\Delta h)_1$ | $= 28.33 - 1.01$ $= 27.31$ m | $= 24.57 - 0.31$ $= 24.26$ m |
| Minimum pressure head in the lateral ($h_{10}$) | $h_{10} = h_{start} - \Delta h_L$ | $= 28.33 - 7.37$ $= 20.96$ m | $= 24.57 - 2.27$ $= 22.30$ m |
| Pressure head variation in the lateral (%) | $V_h = \frac{h_{max} - h_{min}}{h_a^{sp}}100$ | $= 28.92\%$ | $= 8.93\%$ |

For 12 sprinklers; $L = 144$ m and $F_{12} = 0.406$.

$$0.2 \times 25 \geq \frac{789000 \times 6^{1.75} \times 144 \times 0.406}{57.4^{4.75}} - (-1)\frac{1 \times 144}{100}$$

$5\,\text{m} > 6.13\,\text{m (Not satisfactory)}$

For 11 sprinklers; $L = 132$ m and $F_{11} = 0.410$.

$$0.2 \times 25 \geq \frac{789000 \times 5.5^{1.75} \times 132 \times 0.410}{57.4^{4.75}} - (-1)\frac{1 \times 132}{100}$$

$5 < 5.05$ m (Near to satisfactory)

For 10 sprinklers; $L = 120$ m and $F_{10} = 0.415$.

$$0.2 \times 25 \geq \frac{789000 \times 5^{1.75} \times 120 \times 0.415}{57.4^{4.75}} - (-1)\frac{1 \times 120}{100}$$

$5 > 4.10\,\text{m (Satisfactory)}$

From the preceding calculations, it can be seen that the optimal length can be 132 m because the quantity on the right-hand side of the equation is approximately equal to the left-hand side. But if one wants to be more cautious, one can select the optimal length to be 120 m also.

**Case-2. Slope of the Land is 1% Downward**

For 15 sprinklers; $L = 180$ m and $F_{15} = 0.398$.

$$0.2 \times 25 \geq \frac{789000 \times 7.5^{1.75} \times 180 \times 0.406}{57.4^{4.75}} - (1)\frac{1 \times 180}{100}$$

$5 < 6.68$ m(Not satisfactory)

For 14 sprinklers; $L = 168$ m and $F_{14} = 0.400$.

$$0.2 \times 25 \geq \frac{789000 \times 7^{1.75} \times 168 \times 0.400}{57.4^{4.75}} - (1)\frac{1 \times 168}{100}$$

$5 < 5.38$ (Near to Satisfactory)

For 13 sprinklers; $L = 156$ m and $F_{13} = 0.403$.

$$0.2 \times 25 \geq \frac{789000 \times 6.5^{1.75} \times 156 \times 0.403}{57.4^{4.75}} - (1)\frac{1 \times 156}{100}$$

$5 < 4.24$ m (Satisfactory)

Hence the optimal length of the lateral for downward slope of 1% is 156 m.

# Design of Submains

It is usual to have the total allowable pressure variation equally divided between submain and lateral. During design initially laterals are designed first. If a lateral is designed for say 5% pressure variation due to some reasons, then the submain can be designed for 15% pressure variation so that the total pressure variation is equal to or less than 20%.

**Problem** Design the diameter of a submain which must have 5 laterals with an average lateral discharge of 1.5 l/s each and the laterals are spaced at 15 m spacing with first lateral also at the full spacing of 15 m. The submain is to be laid on a land of 0.5% downslope. The average design pressure of sprinkler heads is 25 m. The laterals were designed for a pressure variation of 10%.

**Solution**
Allowable pressure variation of submain = 20%—pressure variation in lateral = 10%

$$0.1 h_a^{sp} \geq \frac{K Q^{1.75} L F_N}{D^{4.75}} - P \frac{SL}{100}$$

$$0.1 \times 25 \geq \frac{789000 \times 7.5^{1.75} \times 75 \times 0.469}{D^{4.75}} - (1)\frac{0.5 \times 75}{100}$$

$$D \geq 62.41 \text{ mm}$$

From the pipe sizes provided in Table 1 of the introduction Chap. 1, it can be seen that 75 mm diameter of 0.4 MPa strength pipe has the inside diameter of 68.6 mm. Hence, this pipe can be selected.


## Design of Main

A main pipe in a sprinkler irrigation system is that section of pipe connecting the water source to the inlet of either submain or lateral. In this section, main pipe design is presented for a simple case with the assumption that the pump has been already installed and the available pressure head and discharge rate from the pump is known.

The main pipe may have filters, valves and fittings like bends and tees. When water passes through these accessories, there will be loss of head. Hence when designing main pipe three following issues should be properly addressed:

1. Elevation difference between water source and the point of delivery (Submain/Lateral)
2. Friction loss in main pipe (Major loss)
3. Head loss in fittings and accessories (Minor loss).

The minor losses are conventionally calculated based on the following formula:

$$h_m = \frac{k v^2}{2g} \tag{23}$$

where $h_m$ is the minor loss of an accessory (m); $k$ is the constant for each specific accessory whose value can be referred from Table 10; $v$ is the velocity of flow and $g$ is the acceleration due to gravity.

Let us convert Eq. 23 into a form convenient for our use by substituting for velocity of flow as flow rate divided by cross-sectional area of pipe.

$$h_m = k \frac{8 Q^2}{g \pi^2 D^4}$$

If discharge rate is substituted in l/s and diameter in mm, the preceding equation becomes as follows:

$$h_m = 82627 \frac{k Q^2}{D^4} \tag{24}$$

**Table.10** Coefficients for minor losses (www.edl.pumps.org)

| Fitting Type | K | | Fitting Type | K |
|---|---|---|---|---|
| **Pipe Entry Losses** | | | **Gradual Enlargements** | |
| | | | Ratio d/D   q = 10° typical | |
| Square Inlet | 0.50 | | 0.9 | 0.02 |
| | | | 0.7 | 0.13 |
| | | | 0.5 | 0.29 |
| Re-entrant Inlet | 0.80 | | 0.3 | 0.42 |
| | | | **Gradual Contractions** | |
| | | | Ratio d/D   q = 10° typical | |
| Slightly Rounded Inlet | 0.25 | | 0.9 | 0.03 |
| | | | 0.7 | 0.08 |
| | | | 0.5 | 0.12 |
| Bellmouth Inlet | 0.05 | | 0.3 | 0.14 |
| | | | **Valves** | |
| **Pipe Intermediate Losses** | | | Gate Valve (fully open) | 0.20 |
| Elbows R/D < 0.6   45° | 0.35 | | | |
|                    90° | 1.10 | | | |
| | | | | |
| Long Radius Bends (R/D > 2)   11¼° | 0.05 | | Reflux Valve | 2.50 |
|                               22½° | 0.10 | | | |
|                               45° | 0.20 | | | |
|                               90° | 0.50 | | Globe Valve | 10.00 |
| **Tees** | | | | |
| (a) Flow in line | 0.35 | | | |
| (b) Line to branch flow | 1.00 | | Butterfly Valve (fully open) | 0.20 |
| | | | | |
| | | | Angle Valve | 5.00 |
| **Sudden Enlargements** | | | | |
| Ratio   d/D | | | | |
| 0.9 | 0.04 | | | |
| 0.8 | 0.13 | | | |
| 0.7 | 0.26 | | | |
| 0.6 | 0.41 | | Foot Valve with strainer | 15.00 |
| 0.5 | 0.56 | | | |
| 0.4 | 0.71 | | | |
| 0.3 | 0.83 | | | |
| 0.2 | 0.92 | | | |
| <0.2 | 1.00 | | Air Valves | zero |
| **Sudden Contractions** | | | | |
| Ratio   d/D | | | | |
| 0.9 | 0.10 | | | |
| 0.8 | 0.18 | | Ball Valve | 0.10 |
| 0.7 | 0.26 | | | |
| 0.6 | 0.32 | | **Pipe Exit Losses** | |
| 0.5 | 0.38 | | Square Outlet | 1.00 |
| 0.4 | 0.42 | | | |
| 0.3 | 0.46 | | | |
| 0.2 | 0.48 | | Rounded Outlet | 1.00 |
| <0.2 | 0.50 | | | |

**Problem** A sprinkler system is to be installed at an elevated field from a water source (Fig. 13). After the point of connection of sprinkler system at A, one primary filter, one reflux valve and one butterfly valve are connected. Both the valves have dimensions as follows: 63 mm outside diameter and 55 mm inside diameter. Allowable pressure head loss in the filter is 3 m.

The pump is capable of delivering 4 l/s when pressure head at A is 36 m. Four laterals are to be operated with a ball valve control for each lateral and all the laterals take off at the same elevation. Each lateral discharge at 4 l/s and only one lateral is operated at a time. The vertical level difference between the point A and the laterals

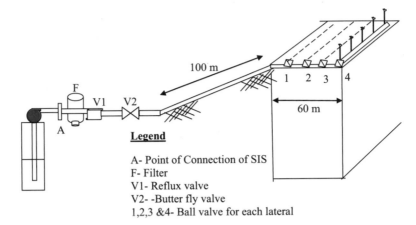

**Fig. 13** Layout for the design of main pipe

take off is 4.954 m. The pressure head needed at the start of each lateral is 25 m. The main pipe runs on a slope for 100 m on the land, and there is also a segment of 60 m pipe laid on level ground as shown in the figure for connection of laterals. Design the diameter and strength of main pipe.

**Solution**

$$\text{Velocity of flow in the valves} = \frac{Q}{A} = \frac{4/1000}{\left(\frac{\pi}{4}\right) \times 0.055^2} = 1.68 \, \text{m/s}$$

| Fitting | $K$ value (Table 10) |
|---|---|
| Reflux valve | 2.5 |
| Butterfly valve | 0.2 |
| Ball valve | 0.1 |

Minor head loss in reflux valve, butterfly valve and 4 ball valves

$$h_m = 82627 \frac{kQ^2}{D^4}$$

$$\sum k = (2.5 + 0.2 + (0.1 \times 4)) = 3.1$$

$$h_m = 82627 \frac{3.1 \times 4^2}{55^4} = 0.446 \, \text{m}$$

Total loss in fittings = Loss in filter + Minor loss = 3 + 0.446 = 3.446 m.

The most critical situation is obviously to provide water supply to the lateral number 4, and the main design should be done to provide water supply for the 4th lateral.

For the design of main following equation is used:

Elevation Gain(−)/Loss(+) + Total loss in Fittings + Allowable friction loss in main = Available head at A − Pressure head needed at lateral 4.

Note in the preceding equation, gain in pressure head would occur if water in main pipe flows downslope and its magnitude should be deducted. In this case, it is upslope and hence is added.

4.954 + 3.446 + Allowable Friction Loss in Main = 36 − 25

Allowable friction loss in main = 2.60 m. By using Eq. 1, we get as follows:

$$\frac{789000 \times 4^{1.75} 160}{D^{4.75}} = 2.60\,\text{m}$$

$$D = 69.23\,\text{mm}$$

The head at the water source is 35 m. So at least 0.35 MPa strength pipe is needed. The water from the water source travels upslope, and it has more chances of encountering water hammer because when power goes off, water in the upslope pipe will come back and hit on the control head. Only to safeguard the system from backflow, reflux valve is provided here. In order to withstand water hammer, the strength of pipe recommended in this case is 0.6 MPa. If we select pipe outside diameter of 90 mm with 0.6 MPa strength, the maximum pipe thickness is 6.1 mm. This pipe will have at least 77.8 mm inside diameter which is higher than our design requirement of 69.23 mm.

**Problem** Design a sprinkler irrigation system (SIS) and lateral diameter if a farmer prefers to install periodic move laterals. Field data is as below:

**Data**

| | |
|---|---|
| Crop | Potato |
| Peak water requirement | 7 mm/day |
| Power available per day | 12 h |
| Irrigation interval during peak water use period | 3 days |
| Maximum well discharge | 3 l/s |
| Shifting time for moving one sprinkler setting | 0.25 h |
| Height of riser pipe | 1 m |

Highly windy area, soil is clay and runoff occurs for a precipitation rate more than 20 mm/h.

See Fig. 14

**Solution**

Manufacturer's catalogue as shown in Table 12 is available from many companies. One sample page of a Catalogue (Courtesy: Netafim Irrigation) is shown. From the

**Fig. 14** Field map

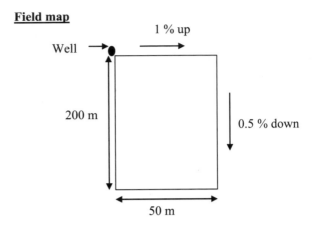

table, select arbitrarily some sprinkler head and check whether, it serves our purpose. Select 5.5 mm * 2.2 mm nozzle from the table.

Operating pressure = 2.5 atm = 25 m (less pressure selected since area is windy)

Spacing—12 * 12 m (Selected in multiples of 6 m—Pipe availability in market)

Precipitation rate—13.2 mm/h—(less than 20 mm/h)

$$\text{Flow rate/sprinkler} = 1905 l/h$$
$$= 0.529 l/s$$

$$\text{Total area of land} = 50 \times 200$$
$$= 10000 \, \text{m}^2$$

$$\text{Area to be irrigated in one day} = \text{Total area/Irrigation interval}$$
$$= 10000/3 \, \text{m}^2$$
$$= 3333 \, \text{m}^2$$

Total number of sprinklers that can be operated simultaneously
= Well yield/ discharge per sprinkler
= 5.67 sprinklers
= 5 sprinklers(Rounded)

$$\text{Area irrigated by 5 sprinklers} = 5 \times \text{Area irrigated by one sprinkler head } (12 \times 12)$$
$$= 720 \, \text{m}^2$$

Number of movements needed/day

= Area irrigated per day/Area irrigated per setting

= 3333/720

= 4.68 moves

= 5 moves (Rounded)

Irrigation depth/irrigation = Irrigation Interval

× Peak water requirement/Uniformity of Sprinkling

Uniformity of sprinkling from Table 12 = 84%

= 3 × 7/0.84

= 25 mm

Time of operation per irrigation = Irrigation depth per irrigation/Precipitation rate

= 25/13.2

= 1.89 h

Time of operation per day = Number of moves per day

× Time of operation per irrigation

= 5 × 1.89 = 9.47 h

Total shifting time = 5 × 0.25 = 1.25 h

Total time needed = Total Irrigation time + Total shifting time

= 9.47 + 1.25

= 10.72 h

Total hours of power available are 12 h, whereas the total time needed for operation of sprinklers per day is only 10.72 h. Hence with the above selection of sprinkler head, operations can be done to serve the requirement.

**Layout of Sprinkler Systems**

Figure 15 shows, how the lateral layout can be designed to fit the topography.

The sprinkler lateral lines can be laid out along the dotted lines shown in figure above. It can be noted that at the eastern and western ends of the field, the spacing provided is only 7 m. This is because, at those ends, the sprinkling will not have overlapping.

**Fig. 15** Lateral layout

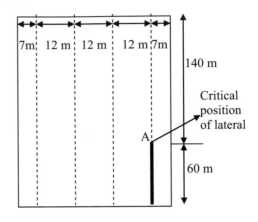

## Design of Lateral Diameter

$$\text{By Eq. 20: } 0.2 h_a^{sp} \geq \frac{K Q^{1.75} L F_N}{D^{4.75}} - p \frac{SL}{100}$$

$$0.2 \times 25 \geq \frac{789000 \times 2.645^{1.75} \times 60 \times 0.46}{D^{4.75}} - (1) \frac{0.5 \times 60}{100}$$

$$D = 35.31 \text{ mm}$$

In the market, 63 mm to 90 mm diameter pipes are easily available. Hence we will select 63 mm (Inside diameter −57.4 mm) pipes with 0.4 Mpa strength of internal diameter 57.4 mm (Table 11).

**Table 11** Inner and outer diameters of HDPE pipes

| S. No. | Outer diameter (mm) | Inside diameters (mm)' | | | |
|--------|---------------------|-----------|---------|---------|---------|
|        |                     | 2.5 ksc   | 4 ksc   | 6 ksc   | 8 ksc   |
| 1      | 32                  | −         | −       | −       | 28.6    |
| 2      | 40                  | −         | −       | 34.8    | 33.6    |
| 3      | 50                  | −         | 44.8    | 43.6    | 41.0    |
| 4      | 63                  | −         | 57.4    | 55.2    | 53.0    |
| 5      | 75                  | −         | 68.6    | 65.6    | 63.0    |
| 6      | 90                  | 84.8      | 82.4    | 79.0    | 75.6    |
| 7      | 110                 | 104       | 100.6   | 96.6    | 92.6    |
| 8      | 125                 | 118       | 114.6   | 110.00  | 105.2   |

**Table 12** Sample manufacturer's catalogue (*Courtesy*—Netafim Irrigation)

| Nozzles | | Pressure (atm) | Flow rate (l/h) | Wetted* diameter (m) | Precipitation (mm/h) Spacing (m) | | | | | | |
| Front (mm) | Rear (mm) | | | | 12 × 12 | 12 × 16 | 12 × 18 | 14 × 16 | 16 × 16 | 18 × 18 | 20 × 20 |
|---|---|---|---|---|---|---|---|---|---|---|---|
| 3.5 | 2.2 | 2.5 | 950 | 25 | 6.6 | 4.9 | 4.4 | 4.2 | 3.7 | 2.9 | 2.4 |
| | | 3 | 1035 | 25 | 7.2 | 5.4 | 4.8 | 4.6 | 4.0 | 3.2 | 2.6 |
| | | 3.5 | 1115 | 26 | 7.7 | 5.8 | 5.2 | 5.0 | 4.4 | 3.4 | 2.8 |
| | | 4.0 | 1190 | 26 | 8.3 | 6.2 | 5.5 | 5.3 | 4.6 | 3.7 | 3.0 |
| 4.0 | 2.2 | 2.5 | 1135 | 26 | 7.9 | 5.9 | 5.3 | 5.1 | 4.4 | 3.5 | 2.8 |
| | | 3.0 | 1240 | 26 | 8.6 | 6.5 | 5.7 | 5.5 | 4.8 | 3.8 | 3.1 |
| | | 3.5 | 1335 | 27 | 9.3 | 7.0 | 6.2 | 6.0 | 5.2 | 4.1 | 3.3 |
| | | 4.0 | 1420 | 27 | 9.9 | 7.4 | 6.6 | 6.3 | 5.5 | 4.4 | 3.6 |
| 4.5 | 2.12 | 2.5 | 1330 | 29 | 9.2 | 6.9 | 6.2 | 5.9 | 5.2 | 4.1 | 3.3 |
| | | 3.0 | 1450 | 30 | 10.1 | 7.6 | 6.7 | 6.5 | 5.7 | 4.5 | 3.6 |
| | | 3.5 | 1560 | 30 | 10.8 | 8.1 | 7.2 | 7.0 | 6.1 | 4.8 | 3.9 |
| | | 4.0 | 1660 | 31 | 11.5 | 8.6 | 7.7 | 7.4 | 6.5 | 5.1 | 4.2 |
| 4.75 | 2.2 | 2.5 | 1540 | 30 | 10.7 | 8.0 | 7.1 | 6.9 | 6.0 | 4.7 | 3.8 |
| | | 3.0 | 1680 | 30 | 11.7 | 8.8 | 7.8 | 7.5 | 6.6 | 5.2 | 4.2 |
| | | 3.5 | 1810 | 31 | 12.6 | 9.4 | 9.4 | 8.1 | 7.1 | 5.6 | 4.5 |
| | | 4.0 | 1930 | 31 | 13.4 | 10.1 | 8.9 | 8.6 | 7.5 | 6.0 | 4.8 |
| 5.0 | 2.2 | 2.5 | 1610 | 32 | 11.2 | 8.4 | 7.5 | 7.2 | 6.3 | 5.0 | 4.0 |
| | | 3.0 | 1760 | 32 | 12.2 | 8.2 | 8.1 | 7.9 | 6.9 | 5.4 | 4.4 |
| | | 3.5 | 1895 | 33 | 13.2 | 9.9 | 8.8 | 8.5 | 7.4 | 8.5 | 4.7 |
| | | 4.0 | 2020 | 33 | 14.0 | 10.5 | 9.4 | 9.0 | 7.9 | 6.2 | 5.1 |

(continued)

**Table 12** (continued)

| Nozzles | | Pressure (atm) | Flow rate (l/h) | Wetted* diameter (m) | Precipitation (mm/h) Spacing (m) | | | | | | |
|---|---|---|---|---|---|---|---|---|---|---|---|
| Front (mm) | Rear (mm) | | | | 12 × 12 | 12 × 16 | 12 × 18 | 14 × 16 | 16 × 16 | 18 × 18 | 20 × 20 |
| 5.5 | 2.2 | 2.5 | 1905 | 32 | 13.2 | 9.9 | 8.8 | 8.5 | 7.4 | 5.9 | 4.9 |
| | | 3.0 | 2080 | 33 | 14.4 | 10.8 | 9.6 | 9.3 | 8.1 | 6.4 | 5.2 |
| | | 3.5 | 2240 | 33 | 15.6 | 11.7 | 10.4 | 10.0 | 8.7 | 6.9 | 5.6 |
| | | 4.0 | 2385 | 33 | 16.6 | 12.4 | 11.0 | 14.6 | 9.3 | 7.4 | 6.0 |
| 6.0 | 2.2 | 2.5 | 2180 | 32 | 15.1 | 11.4 | 10.1 | 9.7 | 8.5 | 6.7 | 5.5 |
| | | 3.0 | 2380 | 33 | 16.5 | 12.4 | 11.0 | 10.6 | 9.3 | 7.3 | 6.0 |
| | | 3.5 | 2560 | 33 | 17.8 | 13.3 | 11.9 | 11.4 | 10.0 | 7.9 | 6.4 |
| | | 4.0 | 2730 | 33 | 19.0 | 14.2 | 12.6 | 12.2 | 10.7 | 8.4 | 6.8 |
| 6.5 | 2.2 | 2.5 | 2495 | 33 | 17.3 | 13.0 | 11.6 | 11.1 | 9.7 | 7.7 | 6.2 |
| | | 3.0 | 2720 | 34 | 18.9 | 14.2 | 12.6 | 12.1 | 10.6 | 8.4 | 6.8 |
| | | 3.5 | 2930 | 34 | 20.3 | 15.3 | 13.6 | 13.1 | 11.4 | 9.0 | 7.3 |
| | | 4.0 | 3120 | 35 | 21.7 | 16.3 | 14.4 | 13.9 | 12.2 | 9.6 | 7.8 |
| 7.0 | 2.2 | 2.5 | 2815 | 34 | 19.5 | 14.7 | 13.0 | 12.6 | 11.0 | 8.7 | 7.0 |
| | | 3.0 | 3070 | 35 | 21.3 | 16.0 | 14.2 | 13.7 | 12.0 | 9.5 | 7.7 |
| | | 3.5 | 3305 | 36 | 23.0 | 17.2 | 15.3 | 14.8 | 12.9 | 10.2 | 6.3 |
| | | 4.0 | 3520 | 36 | 24.4 | 18.3 | 16.3 | 15.7 | 13.8 | 10.9 | 8.8 |

*Height of sprinkler head above ground 1 m

**Inlet Head Required at the Start of the Lateral**

$$\overline{\Delta h} = \Delta h\,\overline{F}$$

$$\overline{\Delta h} = \frac{789000 \times 2.645^{1.75} \times 60 \times 0.46}{57.4^{4.75}} \times 0.801$$

$$= 0.42\,\text{m}$$

$$h_{\text{start}} = \overline{h} + \overline{\Delta h} - p\overline{\Delta Z}$$

$$\overline{h} = \overline{h}_{\text{sprink}} + h_r$$

$$h_{\text{start}} = (25 + 1) + 0.42 - (0.5 \times 60/100)/2$$

$$= 26.27\,\text{m}$$

**Pressure Needed at the Well**

Calculate friction loss from the well to the point $A$, which is assumed as the most critical point.

$$\text{Length of pipe from well to point } A = 7 + 12 + 12 + 12 + 140\,\text{m}$$

$$= 183\,\text{m}$$

$$\text{Friction loss from well to point } A = \frac{789000 \times 2.645^{1.75} \times 183}{57.4^{4.75}}$$

$$= 3.49\ \text{m}$$

$$\text{Net elevation gain from well to point } A = \text{Elevation gain for 140 m in 0.5\% slope}$$

$$- \text{Elevation loss for 43 m in 1\% slope}$$

$$= (0.5/100) \times 140 - (1/100) \times 43$$

$$= 0.27\,\text{m}$$

$$\text{Pressure head required at the well} = 26.27 + 3.49 - 0.27$$

$$= 29.49\,\text{m}$$

# Summary

For accurate prediction of pressure variation in a lateral, back-step method can be used. But this method is computationally intensive and also computer assistance is needed. In the design of sprinkler pipes, many alternative designs have to be examined. Hence forward step of finding friction loss is conventionally used. Design equations useful for lateral laid on uniform slope with telescopic pipe diameters are derived and demonstrated with numerical examples. Christiansen's friction factor for finding friction loss in sprinkler laterals and Scallopi's correction factor to account for the variation in location of first sprinkler in the lateral are derived. The use of Anwar's factor for finding average friction loss is demonstrated with numerical examples. Design of lateral, main and submain pipes in uniform slope is demonstrated with numerical examples.

## Appendix

The Darcy–Weisbach equation normally used in fluid mechanics is as follows:

$$h_f = \frac{fLv^2}{2gD} \qquad (25)$$

where

$h_f$ is head loss (m).

$f$ is called Darcy–Weisbach friction factor.

$L$ is length of pipe (m).

$V$ is velocity of flow (m).

$g$ is acceleration due to gravity (m/s$^2$).

$D$ is diameter of pipe (m).

The flow in microirrigation pipes is generally turbulent flow and the pipes are smooth. For these conditions, the expression for finding friction factor f was given by Blasius, which is as follows:

$$f = 0.32\text{Re}^{-0.25} \qquad (26)$$

where Re is Reynolds number of flow. This equation is valid for the Reynolds number between 2000 to 100,000. It should be noted that this equation is valid for any pipe material which is smooth.

The expression for Reynolds number (Re) is as follows:

$$\text{Re} = \frac{\rho v d}{\mu} \qquad (A.3)$$

where $\rho$ is mass density of water and $\mu$ is dynamic viscosity of water.

Substitute $\rho = 1000$ kg/m$^3$ and $\mu = 0.001002$ NS/m$^2$ for temperature of $20^0$ centigrade. If velocity of flow ($v$) is substituted with flow rate ($Q$) divided by cross-sectional area of flow as circle, we get the following equation:

$$\Delta h = \frac{K Q^{1.75}}{D^{4.75}} L \qquad (A.4)$$

In microirrigation, head loss is expressed in metres (m), flow rate is in litres per second (l/s) and length of pipe in metres (m) and diameter of pipe in millimetres (mm). For such situation, the value of $K = 7.89 \times 10^5$.

Let us see how the relationship would turn out to be if the flow is laminar. For laminar flow due to Poiseuille's law, the friction factor '$f$' can be expressed as follows:

$$f = \frac{64}{R_e} \qquad (A5)$$

If we substitute the preceding relationship in A.1, we get the following relationship:

$$\Delta h = \frac{K Q^2 L}{D^4}$$

where $K = (128\mu / \pi \rho g)$.

## Assessment

### Part A

1. For calculating average pressure head in the sprinkler pipe, the pressure heads at the connection points of riser pipe are used and the pressure head at the start of the lateral pipe is not used—Say True or False. True

2. The expected variation of discharge rate in the sprinklers for 20% variation of pressure is _____

   a. **10%**
   b. 20%
   c. 5%
   d. 2%

3. For finding total friction loss in a sprinkler lateral _____ factor is useful

   a. **Christiansen**
   b. Anwar
   c. Darcy
   d. Scallopi

4. For finding average friction loss in a sprinkler lateral _____ factor is useful

   a. **Anwar**

    b.   Scallopi

    c.   Darcy

    d.   Weisbach

5.    _____ method is useful to find out friction loss in every segment of lateral if the total discharge rate into the sprinkler lateral as well as pressure head at the start of lateral is known

    a.   **Forward step**

    b.   Back step

    c.   Rolling

    d.   Sleeping.

## Part B. Descriptive Answers

1.    Draw with a neat sketch showing locations of maximum pressure and minimum pressure in a downslope lateral?

2.    Draw with a neat sketch showing locations of maximum pressure and minimum pressure in an upslope lateral?

## Part C. Problems

1.    A pipe of inside diameter 44.8 mm of length 25 m is installed to operate a sprinkler of discharge rate 1.5 l/s. Find out the friction loss in the pipe?

2.    In a horizontal sprinkler lateral, at the first sprinkler, the pressure head is 25 m and at the last sprinkler, the pressure head is 27.5 m. The sprinkler discharges at a rate of 0.4 l/s at the pressure head of 25 m, and the discharge coefficient of the sprinkler (k) is 0.1 $l/s/m^{0.5}$. Find out variation of pressure head as well as the variation of discharge rate.

3.    Find out the friction loss that would occur in a sprinkler lateral with following data:

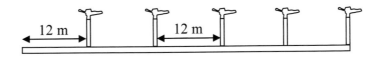

| Length of lateral | 60 m |
|---|---|
| Diameter of the lateral | 44.8 mm |
| Discharge rate of each sprinkler | 0.6 l/s |
| Spacing between each sprinkler | 12 m |
| Length of first sprinkler from submain | 12 m |

Christiansen's friction factor $(F)$ for $N$ number of outlets is given by the formula:

$$F = \frac{1}{2.75} + \frac{1}{2N} + \frac{0.75}{6N^2}$$

4. In a horizontal sprinkler lateral at the first sprinkler, the pressure head is 25 m, and at the last sprinkler, the pressure head is 27.5 m. The sprinkler discharges at a rate of 0.5 l/s at the pressure head of 25 m. Find out variation of pressure head as well as the variation of discharge rate.

5. Design optimal length of a lateral for an average sprinkler discharge rate of 0.4 l/s operating at an average pressure head of 25 m with an inner diameter of lateral pipe of 57.4 mm. All the sprinklers are placed at a uniform spacing of 12 m, and the height of riser pipe is 1 m. Slope of the land is 0.5% upward. Christiansen's friction factor $(F)$ for N number of outlets is given by the formula:

$$F = \frac{1}{2.75} + \frac{1}{2N} + \frac{0.75}{6N^2}$$

6. A sprinkler lateral has to be designed for a horizontal land. The average design discharge rate is 0.4 l/s. The sprinklers have a discharge coefficient of 0.1 l/s/m$^{0.5}$. The length of lateral is 114 m. 10 sprinklers have to be operated with 12 m spacing. Take the height of the riser pipe is 0.9 m, and the friction and other minor losses occurring through the riser pipe are taken as 0.1 m. Christiansen's F for 10 sprinklers is 0.415 and Anwar's $\overline{F} = 0.768$ for 10 sprinklers. Examine whether 44.8 mm diameter of lateral pipe can be selected.

# References and Further Reading

1. Keller, J., & Bliesner, R. D. (1990). *Sprinkle and trickle irrigation.* Chapman & Hall.
2. Christiansen, J. E. (1942). Irrigation by sprinkling: *California Agricultural Experiment Station Bulletin* No. 670, University of California, Davis, Calif.
3. Scaloppi, E. J. (1988). Adjusted "F" factor for multiple-outlet pipes. *Journal of Irrigation & Drainage Engineering ASCE, 114*(1), 169–174.
4. Anwar, A. A. (1999a). Factor G for pipelines with equally spaced multiple outlets and outflow: *Journal of Irrigation & Drainage Engineering,* ASCE *125*(1), 34–38.
5. Anwar, A. A. (1999). Adjusted factor Ga for pipelines with multiple outlets and outflow. *Journal of Irrigation and Drainage Engineering ASCE, 125*(6), 355–360.
6. Anwar, A. A. (2000a). Inlet pressure for tapered horizontal laterals: *Journal of Irrigation & Drainage Engineering ASCE, 126*(1), 57–63, 38.
7. Anwar, A. A. (2000). Adjusted average correction factors for sprinkler laterals. *Journal of Irrigation and Drainage Engineering ASCE, 126*(5), 296–303.
8. Sadeghi, S. H., & Peters, T. R. (2011). Modified G and $G_{AVG}$ correction factors for laterals with multiple outlets with outflow: *Journal Journal of Irrigation & Drainage Engineering ASCE, 137*(11), 769–704.
9. Sprinkler Irrigation. (2016). 210-VI-NEH: NRCS, USDA.

# Selection of Pumps

## Introduction

To study this section, basic understanding of fluid mechanics is essential. Following are the some of the points to be recalled:

- Head is energy possessed by unit weight of water.
- Gravitational head is due to its elevation from a datum and work can be derived from water when waterfalls freely from its position to the datum due to gravitation.
- Kinetic head is due to its velocity. Kinetic head $= v^2/2g$ where v is velocity of water and $g$ is acceleration due to its gravity.
- Pressure head is due to the addition of pressure to water. Pressure head $= P/\gamma = h$, where $P$ is pressure and $\gamma$ is weight density of water and '$h$' is height water standing in a vertical pipe inserted at the point.
- All three forms of energy when expressed in terms of head have unit in terms of length (e.g. metre).
- Total head is the sum of gravitational, kinetic and pressure heads.
- Power of flowing fluid is rate of transfer of energy through fluid per unit time.

## Reference States for Components for Energy

The reference state for each component of energy is very important in understanding the energy transfer in water. For gravitation energy, datum may be anywhere. The datum energy possessed by a body is the amount of work that can be extracted from the body when the body is brought to the datum. If the body is below the selected datum, if we want to bring the body to datum, we have to do work on the body against gravitational pull of the earth. So, when a body is below the datum, the gravitational energy possessed by the body is termed as negative.

When an object is at 3 m height from the datum, the gravitational head is 3 m and when an object is at 2 m below the datum, the gravitational head is $(-2)$ m (Fig. 1).

V. Ravikumar, *Sprinkler and Drip Irrigation*,
https://doi.org/10.1007/978-981-19-2775-1_5

3m

Datum

**Fig. 1** Gravitational energy

For kinetic energy, the reference is zero velocity. When a moving object of velocity v is brought to rest, the work done by the object is the kinetic energy possessed by the body.

For pressure energy, the reference is atmospheric pressure. A pressurized object when brought to atmospheric pressure, the work that can be derived from the object is the pressure energy possessed by the body.

### Problem

A pump delivers at a discharge rate of 2 m$^3$/s. The pump operates against a static lift of 50 m (Fig. 1). What is the power to be delivered by the pump for lifting water? Neglect the friction losses through the pipe. Assume the kinetic head at the delivery is negligible (Fig. 2).

### Solution

Static Water Lift $(Z) = 50$ m
$P/\gamma = 0$
$V^2/2\,g = 0$
$H = 50 + 0 + 0$
$Q = 2$ m$^3$/s
$\gamma = 9810$ N/m$^3$
Power $= 9810 \times 2 \times 50$
$\qquad = 981{,}000$ N m/s (or)
Watts $= 981$ kW

**Fig. 2** Illustration for static lift

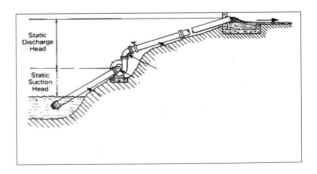

Static Discharge Head

Static Suction Head

**Fig. 3** Heads of water in a
flowing pipe for the problem

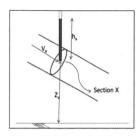

**Problem**

Water flows in a pipe with a velocity of 10 m/s, the height of water $Z_x$ at a section $x$
from the datum is 2 m. The discharge rate of water is 3 m$^3$/s. The pressure head at
the section '$x$' is 1 m. What is the total energy possessed by the water at section '$x$'
and what is the power of flowing fluid at section $x$? (Fig. 3).

**Solution**

Kinetic Energy head $(v^2/2g) = 10^2/(2 \times 9.81) = 5.1$ m.
Gravitational Energy Head $Z_x = 2$ m.
Pressure head $h_x = 1$ m.
Total head $= 5.1 + 2 + 1 = 8.1$ m.
Weight Density of Water $= 9810$ N/m$^3$.
Power $= \gamma QH = 9810 \times 3 \times 8.1 = 238{,}383$ N m/s or Watts.

# Pump

Pump is needed for lifting the water from the wells and to add sufficient pressure to
the water so as to flow through the pipes. Normally centrifugal, submersible and jet
pumps are used in sprinkler irrigation system (SIS) and drip irrigation system (DIS).
Airlift pumps cannot be used. The discussion about pumps here is common for both
SIS and DIS. When an engineer intends to install DIS/SIS, he faces two kinds of
situations. One is, a pump may already exist in the field and he should design DIS/SIS
to suit to the existing pump. Sometimes the designer should install both the pump
and the SIS/DIS. In that case, the selection of pump has to be done by considering
the design of SIS/DIS and also the water yield characteristics of the water source
(Fig. 3).

The discharge rate from the pump may be easily obtained by fitting an orifice
metre. Approximate measurements can be done by just measuring the distance of
throw at the delivery and the height of delivery pipe (coordinate method).

Yield of a well is the discharge rate that can be sustained from the well without
any appreciable change in the water level for sufficiently longer duration of pumping.
Usually the well drillers use v-notch during well drilling to find out the yield of a
well. Most of the time, this itself becomes a fairly useful data in the selection process

of pump. Using a water metre for measurement of discharge rate is also done and while doing so the delivery pipe is usually bent up to get a full a pipe flow (Fig. 4).

In centrifugal pumps and submersible pumps, radial flow, mixed flow and axial flow impellers are available (Fig. 5). Axial flow pumps are not suited for SIS/DIS. Mixed flow pumps are energy efficient than radial flow pumps. However, if the groundwater level fluctuation is more, mixed flow pumps are not suited because they are designed to operate in a narrow range of operating pressure head.

## Total Dynamic Head (TDH)

Total dynamic head is the head to be imparted to the water so that the designed discharge rate of water is delivered at all the outlets. This TDH is the sum of frictional loss of head in all the pipe sections, head losses in fittings and accessories, elevation heads and pressure heads required at the outlets. The process of finding TDH conventionally has following steps:

- **Identifying the critical submain/lateral**.

    Critical submain/lateral is one, which may fail first to deliver sufficient quantity of discharge rate and pressure head when the performance level of the pump goes down. The critical submain/lateral may be at the largest distance from the pump or at the highest elevation from the pump or in some arbitrary combination of distance and elevation. Identifying critical submain/lateral is usually done by trial-and-error process only and the designer gains insight in identifying the critical lateral as he develops experience. Doing a back-step calculation for the entire pipe network once is a foolproof way of identifying the critical lateral. But mostly designers do not do this because the process of doing back-step calculation is laborious and time-consuming.
- Finding elevation difference between the centroid of critical submain/lateral and water level in well
- **Deciding pressure head required at the outlet**
- Computing friction losses in all pipe sections
- Computing losses in fittings and accessories
- Adding miscellaneous losses in order to have a factor of safety for unaccounted losses. It is normally taken as 10–20% of pipe fitting and frictional losses.
- Summing up all the head losses, elevation losses and head required at the outlets to get TDH.

**Fig. 4** Ways of
measurement of discharge
rate from a bore well

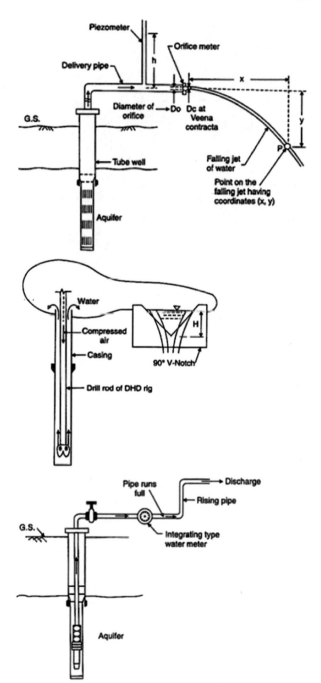

**Radial flow**                 **Mixed flow**                 **Axial flow**

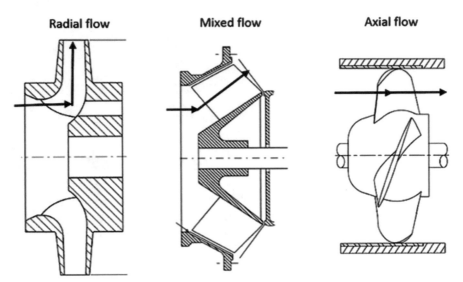

**Fig. 5**  Pump impeller types (pumpsandsystems.com)

## *Power of Pump*

During the design process, yield of well may be a constraining factor or allowable electric load of the farmer may be a constraining factor. The design should be made for the constraining factor.

Let us assume that 7.5 HP power connections are available to the farmer and yield of the well is not a constraining factor and the depth of the water level during the lean period of irrigation during pumping is 26 m. Usually 10% of static lift is assumed as friction loss in suction and delivery pipe. Assume a DIS is to be connected, and 15 m pressure head is needed at the delivery of the pump. The TDH needed from the pump ($H$) if DIS is connected can be calculated as follows:

$H = 26$ m $+ 2.6$ m$(10\%$ of Static lift$) + 15$ m for DIS connection

$H = 43.6$ m

From the pump manufacturer's catalogue, models which are 7.5 hp and can develop pressure head more than 43.6 m should be selected. Among those models, the pump which gives more discharge rate should be selected.

Let us see, how the pump selection must be done when the yield of the well is the constraining factor through an example.

### Problem

A well has a yield of 1 l/s and a depth of water level during pumping in the lean irrigation period is 26 m. How much is the power of the pump to be selected?

$H = 26$ m $+ 2.6$ m $(10\%$ of static lift$) + 15$ m for DISconnection

$= 43.6$ m

$$\text{Power} = \frac{\gamma \, Q \, H}{\eta} \qquad (1)$$

where

$\gamma$ is the weight density of water (9810 N/m$^3$);
$Q$ is discharge rate (m$^3$/s);
$H$ is total dynamic head (m);
$\eta$ is the overall efficiency of the pump (Conventionally taken as 0.5).

$$\text{Power} = 9810 \times 1 \times 10^{-3} \times 43.6/0.5$$
$$= 855.43 \text{ W}.$$

From the pump manufacturer's catalogue, those pumps which are just above 855 watts with a total head more than 43.6 m and the discharge rate of 1 l/s or less can be selected.

## Operating Characteristics of Pump

Figure 6 shows typical operating characteristics of a centrifugal pump. From the figure, it must be observed that every pump is designed for delivering a specific discharge rate. At the design discharge rate and at the design head, efficiency of any pump is the maximum. Therefore, while selecting a pump, care should be taken in such a way that operation of the pump is around the best efficiency point.

**Fig. 6** Typical operating characteristics of a centrifugal pump

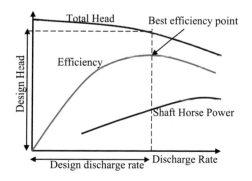

## *Net Positive Suction Head (NPSH)*

We know that a water mass can evaporate totally if the pressure surrounding water is reduced below the vapour pressure of water. It should also be kept in mind that the pressure on the suction side of the pump is below the atmospheric pressure. As pressure at the entry of impeller becomes lower, rate of evaporation increases, and at a certain critical level, cavitation is caused. Cavitation is a process of evaporated bubbles of water moving over the vanes. Bubbles break down due to increase of pressure along the way over the vanes. Cavitation reduces the pump efficiency and causes corrosion of impeller vanes.

The pressure head in excess of the vapour pressure head of water at the eye of the pump is theoretically called as net positive suction head (NPSH).

$$\text{NPSH} = \frac{P_{\text{eye}}}{\gamma} - \frac{P_v}{\gamma} \tag{2}$$

where $P_{\text{eye}}/\gamma$ is pressure head at the eye of impeller and $P_v/\gamma$ is the vapour pressure head in absolute pressure measurement system (0.24 m of water column). More NPSH during pumping is favourable environment for pumping.

The pressure measurement at the eye of impeller is very difficult, and it is possible to measure the pressure at the suction pipe near the impeller. Let this pressure be $Ps/\gamma$. As water goes near the impeller eye, the water velocity gets reduced significantly (assumed as zero) and the velocity head of water at the measurement location of pressure in the suction pipe would get converted into pressure head. Therefore, following equation can be written:

$$\frac{P_{\text{eye}}}{\gamma} = \frac{P_s}{\gamma} + \frac{v_s^2}{2g} \tag{3}$$

Therefore,

$$\text{NPSH} = \frac{P_s}{\gamma} + \frac{v_s^2}{2g} - \frac{P_v}{\gamma} \tag{4}$$

Applying Bernoulli's equation between water level in the well and the location of pressure measurement in the suction pipe (Fig. 7), following equation is written:

$$\frac{P_s}{\gamma} = \frac{P_a}{\gamma} - \left( \frac{v_s^2}{2g} + h_s + h_{fs} \right) \tag{5}$$

where $v_s$ is the velocity of water at the suction pressure measurement location, $h_s$ is the static suction lift and $h_{fs}$ is the head loss due to friction in suction pipe.

Combining the preceding two equations, we get the following equation:

**Fig. 7** Pressure
measurement in suction pipe

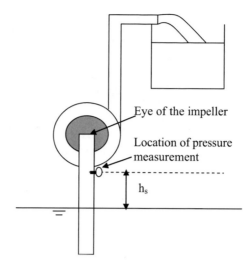

Eye of the impeller

Location of pressure
measurement

$h_s$

$$\text{NPSH} = \frac{p_a}{\gamma} - \left(\frac{p_v}{\gamma} + h_s + h_{fs}\right) \tag{6}$$

Pump manufacturers would test by increasing the $h_s$ gradually until the inception of cavitation. The cavitation can be identified when sudden drop in head is caused. The NPSH calculated with that $h_s$ is called as the required NPSH. So while operating the pump, the actual NPSH should be more than required NPSH. If less, cavitation would occur. For submersible pumps, NPSH need not be considered because they are operated with the highest NPSH level as the suction head is zero.

**Problem**

A pump installation site has an atmospheric pressure head of 10.1 m. The water level is at 5 m below the eye of the pump. Find out the NPSH available. Normally if no data about friction loss is available 10% pipe length can be assumed as friction loss.

$$\text{NPSH available} = 10.1 - (5 + 0.1 \times 5 + 0.24) = 4.36 \text{ m.}$$

Therefore, the pump purchased for this site should have the required NPSH less than 4.36 m.

## Selection of Pump Using Manufacturer's Catalogue

Nowadays pump manufacturers provide very easy to use charts to help in the selection of pumps. Figure 8 shows a chart to narrow down the search in which one can select a group of models after the user has decided about the TDH needed and the

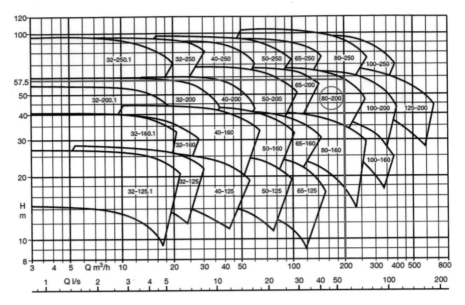

**Fig. 8** A sample Screening chart for pump selection for 2900 rpm. *Courtesy* KSB Pumps

corresponding discharge rate required. As a sample data the TDH is taken as 57.5 m and the discharge rate required is 200 m³/h. In Fig. 8, the group of model corresponds to 80 mm discharge nozzle and 200 mm impeller diameter. Then at the second level, corresponding to the selected group of models from Fig. 8, another chart (Fig. 9) relating the efficiency of operation, head developed, NPSH required and power of electric motor needed is selected. In Fig. 9, three sets of curve are presented. The top most set relates discharge rate versus total head developed for different impeller diameters. It should also be noted that, efficiency for each operating point can also be read from the top most set of Fig. 9.

For our data, discharge rate of 200 m³/h and 57.5 m TDH, on the 219 mm impeller diameter curve, a mark has been made. The efficiency of operation of this point can be seen as 83.5%. Incidentally that efficiency value is the maximum for this group of pumps. If the discharge rate is increased or decreased, the efficiency value gets reduced. Selecting a pump which exactly matches maximum efficiency like this is difficult to get. But the user's objective is to select a pump so that the selected pump has the operating point closer to the maximum efficiency value.

From Fig. 9, the corresponding NPSH required can also be read as 5.5 m which indicates that the actual site conditions of installing the pump should have NPSH more than 5.5 m. From Fig. 9, power of electric motor needed for operation of the pump can also be read as nearly 38 kW for 219 mm of impeller radius.

**Problem**
Compute the total dynamic head for a sprinkler irrigation system whose layout is as shown in Fig. 10 and select a suitable pump for operation of this system.

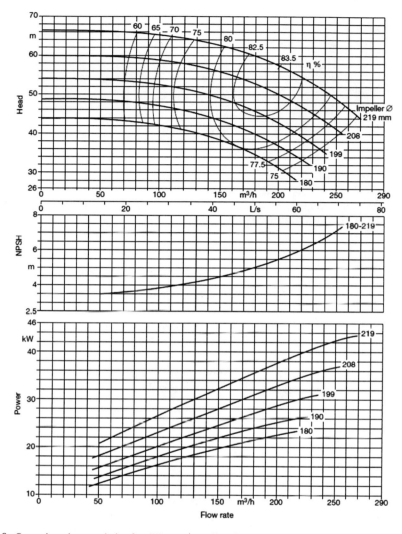

**Fig. 9** Operating characteristics for different impeller diameters. *Courtesy* KSB Pumps

## Data:

### Pump
Suction and delivery pipe length is 5 m and diameter is 65.6 mm and one foot valve is provided. Design discharge rate is 8 l/s.

### Control Head
Inside diameter of pipe and all the accessories are 79 mm. Pipe length is 10 m. One reflux valve is provided. One filter exists and is cleaned when the pressure differential between the inlet and outlet reaches 2 m, Two butterfly valves are also provided.

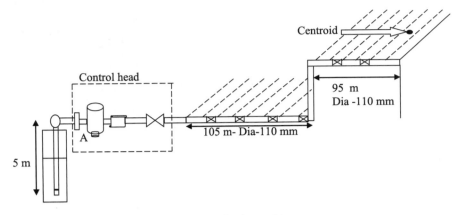

**Fig. 10** Sprinkler system layout for a pump selection problem

### Main, Submain and Laterals

Main is buried and not removable. The main pipe has provision for 14 laterals with 14 Tees. At a time two laterals can be operated. One lateral discharge rate is 4 l/s with 10 sprinklers operating per lateral. Lateral length is 120 m. Lateral diameter is 68.6 mm. Main has 6 butterfly valves, pressure head required at the sprinkler heads is 25 m. Length of main pipe is 210 m and diameter is 82.4 mm. Spacing between laterals is 15 m.

Centroid of the area irrigated at the last two laterals (critical laterals) has an elevation of 25 m above water level in well. Assume pressure head loss in filter is 2 m. Riser pipe is 1 m height.

## *Head Losses in Pipe Sections*

When the last two laterals are operating, the main pipe section of 30 m length serving the last two laterals can be treated as submain pipe. Hence main pipe length is taken as 180 m. The pipe head loss calculations are shown in Table 1. Table 2 shows fittings loss calculations.

Total Dynamic Head (TDH)

= Friction & fitting losses

+ Elevation difference between water level

in well and the centroid of critical submain

+ Loss in filter + Riser height + Head needed at the outlet

$$TDH = 7.12 + 5.32 + 25 + 2 + 1 + 25$$
$$= 65.44 \text{ m}$$

**Table 1** Pipe head loss calculations

| Item | Formula | Anwar $F$ | Discharge rate (l/s) | Length (m) | Diameter (mm) | Frictional loss (m) |
|---|---|---|---|---|---|---|
| Pump pipe | $\Delta h =$ | – | 8 | 5 | 65.6 | 0.35 |
| Control head pipe | $\dfrac{K Q^{1.75} L}{D^{4.75}}$ | | 8 | 10 | 79 | 0.29 |
| Main pipe | | | 8 | 180 | 82.4 | 4.29 |
| Submain pipe | $\overline{\Delta h} =$ $\dfrac{K Q^{1.75} L}{D^{4.75}} \overline{F}$ | 0.885 (2 outlets) | 8 | 30 | 82.4 | 0.63 |
| Lateral pipe | | 0.768 (10 outlets) | 4 | 120 | 68.6 | 1.56 |
| Total | | | | | | 7.12 |

**Table 2** Fittings loss calculations

| Item | Formula | Numbers | $k$ value | $Q$ (l/s) | Diameter (mm) | Head loss (m) |
|---|---|---|---|---|---|---|
| Foot valve | $h_m =$ | 1 | 15 | 8 | 65.6 | 4.28 |
| Reflux Valve | $82627 \dfrac{k Q^2}{D^4}$ | 1 | 2.5 | 8 | 82.4 | 0.29 |
| Butterfly valves | | 8 | 0.2 | 8 | 82.4 | 0.18 |
| Tees | | 14 | 0.35 | 8 | 82.4 | 0.56 |
| Total | | | | | | 5.32 |

Add head for a factor of safety $= 20\%$

Total head needed $= 78.5$ m with a discharge rate of 8 l/s

Let us assume that the site conditions and preference of the farmer are to install a submersible mixed flow pump. Figure 11 shows a typical sample catalogue of a submersible mixed flow pump.

From the catalogue, pump model TMS-5512 which can deliver a discharge rate of 8 l/s against a total head of 79.8 m. The pump can optimally work in the range of 98.4 m of head with 5 l/s and 72 m head with 9 l/s. Hence it is selected.

# Examining the Sufficiency of Existing Pump in the Field

Before and after the connection of SIS/DIS, the discharge rate delivered and the pressure head developed by a pump will be very much different. After installing the SIS/DIS, the discharge rate will definitely get reduced because the flow through the SIS/DIS has to overcome the resistance offered by the SIS/DIS.

## PERFORMANCE CHART

**TARO "TMS 55 / 60 SERIES" - THREE PHASE MIXED FLOW SUBMERSIBLE PUMPSETS FOR 150 mm (6") BOREWELLS**
Approximate performance values of TMS 55 / 60 series at 415 V (-15% to +6%), 2880 rpm, 50 Hz AC power supply

| Model Name | | | Motor Rating | | | | CAPACITY | | | | | | | | | | | | |
|---|---|---|---|---|---|---|---|---|---|---|---|---|---|---|---|---|---|---|---|
| | | Connection | | | Stages | Pipe Size (mm) | Gpm | 0.0 | 46.2 | 52.8 | 66.0 | 85.8 | 106 | 112 | 125 | 139 | 152 | 165 | FL Current (A) |
| | | | | | | | l/m | 0.0 | 210 | 240 | 300 | 390 | 480 | 510 | 570 | 630 | 690 | 750 | |
| | | | | | | | M³/hr | 0.0 | 12.6 | 14.4 | 18.0 | 23.4 | 28.8 | 30.6 | 34.2 | 37.8 | 41.4 | 45.0 | |
| Pump | Motor | | kW | HP | | | l/s | 0.0 | 3.5 | 4.0 | 5.0 | 6.5 | 8.0 | 8.5 | 9.0 | 10.5 | 11.5 | 12.5 | |
| TMS 5503 | TS 022 | DOL | 2.2 | 3 | 3 | | | 28.5 | 26.4 | 25.8 | 24.6 | 22.5 | 20.0 | 19.1 | 18.0 | 14.0 | | | 6.5 |
| TMS 5505 ⅄ | TS 037 | DOL | 3.7 | 5 | 5 | | | 47.5 | 44.0 | 43.0 | 41.0 | 37.5 | 33.3 | 31.8 | 30.0 | 23.3 | | | 10.0 |
| TMS 5508 | TS 056 | DOL/SD | 5.5 | 7.5 | 8 | | | 76.0 | 70.4 | 68.8 | 65.6 | 60.0 | 53.2 | 50.8 | 48.0 | 37.2 | | | 14.5 |
| TMS 5510 ⅄ | TS 075 | SD | 7.5 | 10 | 10 | 65 | | 95.0 | 88.0 | 86.0 | 82.0 | 75.0 | 66.5 | 63.5 | 60.0 | 46.5 | | | 19.5 |
| TMS 5512 ⅄ | TS 093 | SD | 9.3 | 12.5 | 12 | | | 114.0 | 105.6 | 103.2 | 98.4 | 90.0 | 79.8 | 76.2 | 72.0 | 55.8 | | | 25.0 |
| TMS 5515 ⅄ | TS 112 | SD | 11 | 15 | 15 | | | 142.5 | 132.0 | 129.0 | 123.0 | 112.5 | 99.8 | 95.3 | 90.0 | 69.8 | | | 29.0 |
| TMS 5520 ⅄ | TS 150 | SD | 15 | 20 | 20 | | | 190.0 | 176.0 | 172.0 | 164.0 | 150.0 | 133.0 | 127.0 | 120.0 | 93.0 | | | 39.0 |
| TMS 6004 ⅄ | TS 037 | DOL | 3.7 | 5 | 4 | | HEAD VALUES IN METRES | 41.0 | | 36.6 | 34.1 | 31.1 | 30.0 | 28.5 | 25.0 | 22.0 | 17.9 | | 10 |
| TMS 6005 ⅄ | TS 045 | DOL | 4.5 | 6 | 5 | | | 51.3 | | 45.8 | 42.6 | 38.9 | 37.5 | 35.6 | 31.3 | 27.5 | 22.4 | | 12 |
| TMS 6006 ⅄ | TS 056 | DOL/SD | 5.5 | 7.5 | 6 | | | 61.5 | | 54.9 | 51.2 | 46.7 | 45.0 | 42.8 | 37.5 | 33.0 | 26.9 | | 14.5 |
| TMS 6008 ⅄ | TS 075 | SD | 7.5 | 10 | 8 | | | 82.0 | | 73.2 | 68.2 | 62.2 | 60.0 | 57.0 | 50.0 | 44.0 | 35.8 | | 19.5 |
| TMS 6010 ⅄ | TS 093 | SD | 9.3 | 12.5 | 10 | 65 | | 102.5 | | 91.5 | 85.3 | 77.8 | 75.0 | 71.3 | 62.5 | 55.0 | 44.8 | | 25 |
| TMS 6012 ⅄ | TS 112 | SD | 11 | 15 | 12 | | | 123.0 | | 109.8 | 102.3 | 93.3 | 90.0 | 85.5 | 75.0 | 66.0 | 53.7 | | 29 |
| TMS 6014 | TS 130 | SD | 13 | 17.5 | 14 | | | 143.5 | | 128.1 | 119.4 | 108.9 | 105.0 | 99.8 | 87.5 | 77.0 | 62.7 | | 34 |
| TMS 6016 | TS 150 | SD | 15 | 20 | 16 | | | 164.0 | | 146.4 | 136.4 | 124.4 | 120.0 | 114.0 | 100.0 | 88.0 | 71.6 | | 39 |
| TMS 6020 | TS 187 | SD | 18.7 | 25 | 20 | | | 205.0 | | 183.0 | 170.5 | 155.5 | 150.0 | 142.5 | 125.0 | 110.0 | 89.5 | | 45 |

Performance confirming to IS : 8034 and 9283            D O L - Direct On Line            S D - Star Delta            Maximum outer diameter : 142 mm

**Fig. 11**   A sample pump catalogue. *Courtesy* Texmo Pump Industries, Coimbatore, India

**Fig. 12**   A typical pump label

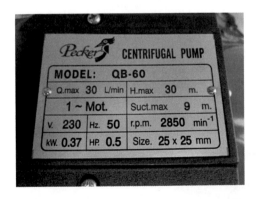

Figure 12 shows a typical pump label. In this pump label, an item H.Max is given as 30 m. The meaning of this statement is that this pump can generate a maximum pressure head of 30 m. That means, when the water level in the well is below 30 m, the pump can push water to the surface only and we will get only zero discharge rate. This pump can provide water only if the depth of water in the well is less than 30 m. If the water level in the well rises, the discharge rate would keep on increasing.

Another item in the pump label is Q.max. The meaning of Q.Max is the maximum discharge rate that the pump can deliver. This maximum discharge rate of 30 L/min would be realized when the water level in the well is at the surface of the land.

SIS functions well if the excess pressure head available is sufficient to operate sprinkler heads. Generally, at least a minimum excess pressure head of 25–30 m is expected for SIS. DIS functions well if the excess pressure head available for connection of DIS is between 10 and 15 m. Higher pressure head is needed when more number of filters are installed in the system and when the fertigation device adopted is venturi.

Let us assume that a pump has the maximum pressure head as 30 m. Let the water level in the well "during pumping" be 26 m (Fig. 13). We intend to connect DIS with the pump in the well. So for this pump, what is the pressure head we can expect to be available for connection of DIS?

Conventionally a 10% of total static lift (i.e. 26 m) is taken as a loss of pressure head occurring in the pipe sections within the well due to friction and other minor losses. This loss would amount here to 2.6 m. Therefore, we will get only 2.4 m residual pressure head for connection of DIS. (i.e. 30 m − 26 m − 2.6 m = 1.4 m). In the pump's ability to lift water for a height of 30 m, it's 26 m ability lifting is spent in lifting from the well to the land surface and while doing so 2.6 m head is lost as a friction and other losses in the pipe sections in the well.

Whether this 1.4 m is sufficient for connection for DIS? It is not sufficient for a satisfactory operation. But the system will perhaps function to a certain extent.

It is not practically possible to have a look at the pump label in case of submersible pumps installed in the wells. In such situations, let us see how can we find out the pressure that would be developed after connection of SIS/DIS right in the field itself. When the delivery valve is partially closed, the pressure on upstream side of the valve would increase and the jet issued will fall at a larger distance (x in Fig. 13). By observing this distance and also the height of the delivery pipe above the land (y in Fig. 13), we can measure the pressure developed by the pumping system at the

**Fig.13** Coordinate method

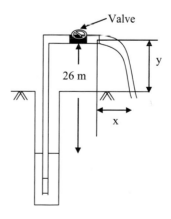

valve. This method is called as coordinate method. The derivation of the equation the pressure head developed at the upstream of the delivery valve is as follows:

The vertical distance travelled by a water particle issued out of the pipe is the product of average vertical velocity and time taken. Hence following equation is written:

$$y_t = \left(\frac{0 + v_y^t}{2}\right)t \tag{7}$$

where $y_t$ is the distance travelled by water particle from pipe outlet in y-direction after time $t$ and $v_y^t$ is the velocity in y-direction after time $t$. Vertical velocity change occurs at a uniform rate equal to the acceleration due to gravity ($g$) and the initial vertical velocity is zero. Hence,

$$v_y^t = 0 + gt \tag{8}$$

Substituting Eq. 8 in Eq. 7, we get,

$$y_t = \frac{gt^2}{2} \tag{9}$$

The horizontal distance travelled by the water particle from the pipe outlet ($x_t$) is related to horizontal velocity $v_x$ is as follows:

$$t = \frac{x_t}{v_x} \tag{10}$$

Note that in the preceding equation $v_x$ is constant with respect to time when air resistance against the horizontal velocity is taken as zero.

Substituting Eq. 10 in Eq. 9, we get

$$\frac{v_x^2}{2g} = \frac{x_t^2}{4y_t} \tag{11}$$

Applying Bernoulli's equation between just upstream of control valve and outlet of control valve, we get as follows:

$$H_{\text{valve}} = \frac{v_x^2}{2g} + k\frac{v_x^2}{2g}$$

where $H_{\text{valve}}$ is the total head on the upstream of the valve. '$k$' depends on the ratio of inside pipe diameter ($D$) and equivalent diameter of valve opening ($d$). For instance, for a ($d/D$) value of 0.2, $k = 0.5$

$$H_{\text{valve}} = 1.5\frac{v_x^2}{2g} \tag{12}$$

Measurements in the field are taken only when the water strikes the ground, and so the subscript $t$ in $x_t$ and $y_t$ can be removed and the preceding equation is conventionally written as follows:

$$H_{valve} = 1.5 \frac{x^2}{4y} \tag{13}$$

For instance, if $y = 1$ and $x = 6$ m, the pressure head developed is 9 m. We have to adjust the valve in such a way that the total head developed just upstream of the valve ($H_{valve}$) is sufficient for our purpose. Conventionally at least $H_{valve} = 15$ m is expected for DIS and 30 m is expected for SIS.

If gate valve is not available at the delivery end of the pipe, it is not possible for us to adjust the discharge and we can get only one discharge rate for which we can find out the pressure head developed at the exit of the delivery by observing the x- and y-coordinates. Even in such situations, it is possible for us to estimate the design discharge rate when connecting a DIS/SIS.

If we assume power added by the pump on the water is constant in both the situations such as before connecting DIS/SIS and after connecting DIS/SIS, following equation can be written:

$$Q_1(h_1 + h_s + h_L) = Q_2(h_2 + h_s + h_L) \tag{14}$$

where $h_L$ is the head loss in the well which is assumed as the same value in both the situations; $h_s$ is the static lift; $h_1$ is pressure head at the valve for $Q_1$ discharge rate before connecting DIS/SIS; and $h_2$ is the pressure head at the valve for $Q_2$ discharge rate after connecting DIS/SIS.

For instance, for the preceding data discussed, let us assume that when the delivery valve is fully opened, the discharge rate measured is 2 litres per second (l/s). We can find out the discharge rate when the pressure head developed at the valve is 10 m using the preceding equation.

2 l/s (0 m + 26 m + 2.6 m) = $Q_2$(10 m + 26 m + 2.6 m)

$Q_2 = 1.48$ l/s

Interestingly, the preceding equation tells us that if we are able to develop a pressure head at the valve to a level of 10 m, we can get a reduced discharge rate of 1.48 l/s. But this formula does not give guarantee that this pump would be capable of delivering a pressure head of 10 m at the valve.

## Pumps in Parallel

When additional water is to be pumped if the total dynamic head added to the water is the same, one can adopt connection of two or more pumps in parallel as shown in Fig. 14. In parallel pump connection, there can be two suctions or the suction for

**Fig.14** Pumps in parallel

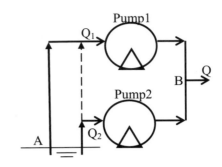

**Fig.15** Head versus
discharge rate for pumps in
parallel [1]

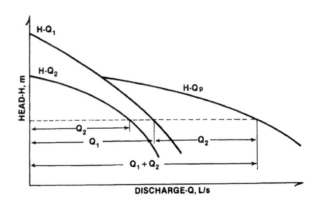

each pump can also be connected to common suction Fig. 14. But both the deliveries
have to be connected into one delivery as shown in the figure.

Let A be location of water level in well and B be point of connection of two
deliveries from the two pumps. If we apply Bernoulli's equation between points A
and B along Pump 1 and Pump 2 separately, we get as follows:

$$\frac{p_A}{\gamma} = \frac{p_B}{\gamma} + \frac{v_B^2}{2g} + Z_B - H_{P1} \tag{16}$$

$$\frac{p_A}{\gamma} = \frac{p_B}{\gamma} + \frac{v_B^2}{2g} + Z_B - H_{P2} \tag{17}$$

where $H_{P1}$ and $H_{P2}$ are the heads added by pumps 1 and 2, respectively, and $Z_B$ is
the elevation of point B above A.

From Eqs. 16 and 17, it is clear that the head imparted by pump1 ($H_{P1}$) and head
imparted by pump2 ($H_{P2}$) are one and the same (Fig. 15).

**Fig. 16** Pumps in series

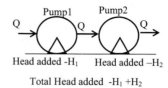

Head added -H₁     Head added –H₂

Total Head added -H₁ +H₂

## Pumps in Series

Multistage pumps are a typical example for pumps in series. In multistage pumps, water passes through successively one impeller after the other and along the way, the pressure imparted to the water keeps on increasing. Based on the same principle, if a farmer connects two separate pumps as shown in Fig. 14, then this system of connection is said to be pumps in series (Fig. 16).

When a single large size pump is used, the required NPSH may be higher. In such situations if we use pumps in series this problem gets reduced. Also in an undulating topography when sprinklers are operated at elevated locations, specifically only during that situation higher head may be needed. During such operation alone, one can use additional pump (booster pump) connected in series and in other situations the booster pump can be kept idle (Fig. 17).

Normally pump seals and casing are designed for a specific maximum pressure. If two pumps are connected in series, the pump in upstream side will be operating in a pressure range higher than the pressure for which the pump was designed. Hence care should be taken to provide seals which can operate at a higher pressure and examine the pump casing whether it can withstand the higher pressure.

### Problem

Two centrifugal pumps are connected in series. Data for both the pumps is as follows:

**Fig. 17** Head versus discharge rate for pumps in series [1]

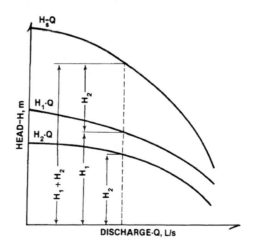

| Discharge rate (l/s) | 0 | 4 | 8 | 12 | 16 | 20 | 24 |
|---|---|---|---|---|---|---|---|
| Head from pump1 (m) | 50 | 51.8 | 50.8 | 48 | 42.5 | 32.5 | 18.3 |
| Head from pump 2 (m) | 46.7 | 45.9 | 44.2 | 40.3 | 34.3 | 26 | 17 |

Draw the resultant head versus discharge rate curve for the pumps in series in a graph sheet and find out when the pumps in series are discharging 15 l/s, how much head will be imparted by pump 1 and pump 2 individually?

| Discharge rate (l/s) | 0 | 4 | 8 | 12 | 16 | 20 | 24 |
|---|---|---|---|---|---|---|---|
| Total head from pump1 & 2 (m) | 46.7 | 45.9 | 44.2 | 40.3 | 34.3 | 26 | 17 |

From Fig. 18, it can be observed that when the pumps in series are discharging 15 l/s, pump 1 will impart head of 49 m and pump 2 will impart head of 42 m.

## Problem
Two centrifugal pumps are connected in parallel. Data for both the pumps is as follows:

| Head from the pump (m) | 20 | 25 | 30 | 40 | 45 |
|---|---|---|---|---|---|
| Discharge rate of pump 1 (l/s) | 23.4 | 22.3 | 20.8 | 19.2 | 17.2 |
| Discharge rate of pump 2 (l/s) | 22.7 | 20.5 | 18.2 | 15.6 | 12.2 |

Draw the resultant head versus discharge rate curve for the pumps in parallel in a graph sheet and find out if the pumps in parallel operate under a head of 30 m, how much will be the discharge rate delivered by each pump and how much will be the total discharge rate delivered by both the pumps in parallel ?

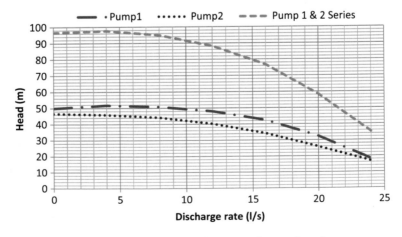

**Fig. 18** Discharge rate versus head for individual pumps and pumps in series

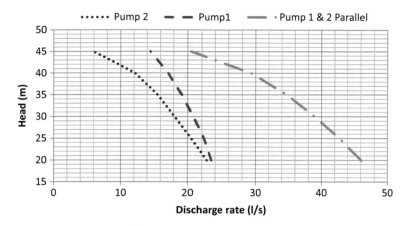

**Fig. 19** Discharge rate versus head for individual pumps and pumps in parallel

| Head from the pump (m) | 20 | 25 | 30 | 40 | 45 |
|---|---|---|---|---|---|
| Discharge rate of pump (l/s) | 46.1 | 42.8 | 39 | 34.8 | 29.4 |

From Fig. 19, it can be observed that when the pumps in parallel operate under the head of 30 m, pump 1 discharges 21 l/s and pump 2 discharges 18 l/s. The total discharge rate is 30 l/s when pump1 and pump 2 are operating in parallel.

## Selecting a Suitable Diesel Engine for Pump

Sometimes diesel engines are used as prime movers for the operation of pump. In such a situation, we need to know how selection of diesel engine must be done. Figure 20 shows a typical diesel engine characteristics containing torque, power delivered and fuel consumption with respect to rotational speed of the shaft. These curves are for the standard operating conditions. In actual operating conditions variation of two quantities, namely ambient temperature and altitude of the location, causes significant reduction in power delivered. Hence, correction must be made as per the following procedure:

### For Engines Without Turbochargers

- For every 300 m above sea level, reduce power output by 3%.
- For every 5% increase above 15 °C, reduce the power rate by 1%.

### For Engines with Turbochargers

- For every 300 m above sea level, reduce power output by 1%.
- For every 5% increase above 15 °C, reduce the power rate by 0.5%.

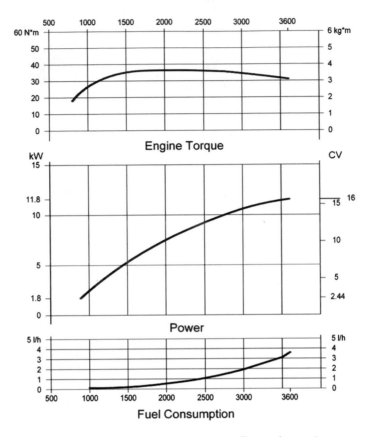

**Fig. 20** A typical diesel engine performance curves (www.allpa-marine.com)

Sometimes, the pump is not driven directly by the engine and belt or chain drives are used. In such cases, the efficiency of that drive must also be taken into account.

## Summary

Methods of measuring discharge rate from a well are discussed. Total dynamic head (TDH) and Net positive suction head (NPSH) are discussed. Importance of identifying critical lateral and submain is discussed. Procedure of selection of pumps using the necessary data is presented with numerical examples. Field method of examining the sufficiency of pump is presented. Determination method of pressure head and discharge rate for the pump connections in series and parallel are also presented. Method of determining the power of diesel engine coupled pumps is presented.

## Assessment

### Part A
### Choose the best option for the following questions:

1. _____is the sum of frictional loss of head in all the pipe sections, head losses in fittings & accessories, elevation heads and pressure heads required at the outlets.

   a. **Total Dynamic Head**
   b. Suction lift
   c. Suction head
   d. Total dynamic lift

2. If the range of groundwater level fluctuation is very high, _____ pumps are suited.

   a. **Radial flow**
   b. Mixed flow
   c. Submersible
   d. Jet

3. Mixed flow pumps are energy efficient than _____flow pumps.

   a. **Radial flow**
   b. Submersible
   c. Jet
   d. Airlift

4. The pressure on the suction side of the pump is _____ atmospheric pressure when a centrifugal pump pumps water from a well at a water depth of 2 m.

   a. **Lower than**
   b. Higher than
   c. Equal to
   d. None of these

5. Multistage pumps are a typical example for pumps in _____

   a. **Series**
   b. Parallel
   c. Pair
   c. Triple

### Part B
1. What do you understand from the term yield of a well?
2. What is net positive suction head of a pump?
3. Draw typical operating characteristics of a centrifugal pump.

4.  Show with illustration how pumps are connected for series and parallel connection. Discuss in terms of operating characteristics for pumps connected in parallel and pumps connected in series. What are the situations in which pumps in parallel and pumps in series are connected?
5.  Write down the step-by-step procedure of selecting a suitable diesel engine for pump.

## References and Further Reading

1.  Keller, J., & Bliesner, R. D. (1990). *Sprinkle and trickle irrigation.* Van Nostrand Reinhold.

# Maintenance and Operation of Sprinkler Irrigation Systems

## Introduction

It is known that SIS is designed and installed for managing the most difficult situations. But when the system is operated, the actual condition in the field may be different. For instance, on a level land when sprinklers are operated near the water source, the pressure head available will be higher and hence the discharge rate from the sprinklers will also be higher. During such situations sometimes throttling of the flow may be needed or the number of sprinklers that can be operated simultaneously may be increased. The other reasons for the change of operating conditions may be seasonal variation in depth to ground water level and undulating topography of the land. Under such varying operating conditions, developing the system characteristics and matching with pump characteristics is very useful in taking optimal decisions.

Needless to say that maintenance of the SIS needs to be done properly otherwise, the farmer may not be able to reap the full benefits from the SIS. Mostly maintenance needs to be done with the prime movers like electric motor and diesel engine. Since the maintenance aspects of electric motor and diesel engine are beyond the scope of this book, the reader is advised to consult other standard texts on this subject. However the maintenance aspects of SIS are presented here.

## Maintenance Tips for Sprinkler Irrigation System

During the operation of SIS, following checks are to be done at frequent intervals.

### At the Pumping Station and Control Head

- Noise
- Vibration
- Leakage
- Temperature of motor/engine

- TDH developed
- Clogging of filters/screens.

**Sprinkler Pipe Layout**

- Pressure developed during operation
- Pressure difference between the first sprinkler and the last sprinkler
- Rotational speed of all the sprinklers
    - Usually the last sprinkler in any lateral is the location at which the dirt gets accumulated. To solve this problem, End blocks of sprinkler laterals are opened up and flushing is done for some time and also the sprinkler heads are removed and blown up by using compressed air. The sprinkler heads need not be lubricated. Small and medium sprinklers are water lubricated. The bearings of large sprinklers come with pre-lubricated seals.
    - Mostly washers get worn out and need to be replaced.
    - Sometimes the top spring also gets worn out and needs to be replaced for proper rotation.
- Jet throw distances and breakage of jet into droplets.
    - Lower pressure causes lower throw distances and larger droplets. Higher pressure causes higher throw distances and smaller droplets causing more drift loss.
    - The washers in the pipes are designed in such a way that they do not leak when sufficient pressure is reached. If the pipes leak when the pressure is sufficient, it may be due to worn out washers and need to be replaced.
    - Pipe leaks.
- Non-verticality of riser pipe.
    - The saddle must be placed on a firm flat ground and riser pipe should be placed vertically.

# Operation of SIS

## System Characteristics

The system characteristic for a SIS is a relationship between the discharge rate and corresponding head needed to pass through the pump suction and delivery pipes, filters, all the pipe fittings and the sprinklers. It should be noted that system characteristic does not take into account the existence of pump. A typical system characteristics are shown in Fig. 1.

**Example** A pump operates with a static lift of 10 m from the water level in the well to the surface of the land. The length of pipe section operating inside the well is 15 m.

**Fig. 1** Typical system
characteristics

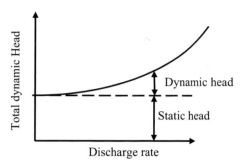

The riser pipe is 2 m height with a diameter of 44.8 mm. The system has fittings which are as follows:

- One foot valve with $k$ value of 15
- Two short bends with $k$ value of 1.1
- One gate valve with $k$ value of 0.2.

A main pipe of diameter 44.8 mm runs for a length of 150 m. A sprinkler of 4 l/s discharge rate with recommended operating pressure head of 30 m is used. In Fig. 2, location A is the nearest to the well which is at 25 m from the well and location B is the farthest to the well which is at 150 m from the well.

1. When the sprinkler is operating at the location A, what will be the discharge rate?
2. When the sprinkler is operating at the location B, what will be the discharge rate?
3. At the position A, if the farmer wants to operate the sprinkler at 30 m pressure head, how much pressure should he reduce by using the control valve?

The pump head versus discharge rate characteristics is as follows:

| $Q$ (l/s) | 0 | 3.5 | 4 | 5 | 6.5 | 8 | 8.5 | 9 | 10.5 |
|-----------|-----|------|------|------|-----|------|------|----|------|
| $H$ (m) | 76 | 70.4 | 68.8 | 65.6 | 60 | 53.2 | 50.8 | 48 | 37.2 |

**Fig. 2** One sprinkler
operated from a pump

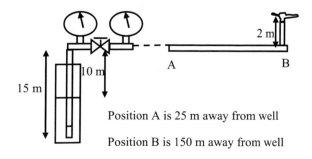

Position A is 25 m away from well

Position B is 150 m away from well

**Solution**
**System Characteristics for operating sprinkler at position A**

Total dynamic head (H) = Static lift in the well (10 m) + Riser pipe height (2 m)
$$+ \text{Friction loss in pipe}$$
$$+ \text{Friction loss in fittings} + \text{Head at the exit of the sprinkler}$$

$$\text{Friction loss in pipe} = 789000 \frac{Q^{1.75} L}{D^{4.75}}$$

$$= 789000 \frac{Q^{1.75}}{44.8^{4.75}} (15 + 25 + 2) = 0.4751 Q^{1.75}$$

The length comprises pipe in well, horizontal length and riser pipe.

$$\text{Head loss in fittings} = 82627 \frac{Q^2}{D^4} \sum k_i$$

$$= 82627 \frac{Q^2}{44.8^4} (15 + 2 \times 1.1 + 0.2) = 0.3569 Q^2$$

The discharge equation for sprinkler is as follows:

$$q = k_{\text{sprinkler}} \sqrt{h}$$
$$4 = k_{\text{sprinkler}} \sqrt{30}$$
$$k_{\text{sprinkler}} = 0.7303 \, \text{l/s/m}^{0.5}$$
$$\text{Head at the exit of the sprinkler} = \frac{Q^2}{k_{\text{sprinkler}}^2}$$

Note in the preceding equation, one sprinkler discharge rate and total discharge rate is one and the same quantity and hence $q = Q$.

$$\text{TDH} = 12 + 0.4751 Q^{1.75} + 0.3569 Q^2 + 1.8750 Q^2$$
$$\text{TDH} = 12 + 0.4751 Q^{1.75} + 2.2319 Q^2$$

**System Characteristics for Operating Sprinkler in Position B**
For position B, all the calculations are the same except friction loss in pipe

$$\text{Friction loss in pipe} = 789000 \frac{Q^{1.75}}{D^{4.75}} L$$

$$= 789000 \frac{Q^{1.75}}{44.8^{4.75}} (15 + 150 + 2) = 1.889 Q^{1.75}$$

$$\text{TDH} = 12 + 1.889 Q^{1.75} + 2.232 Q^2$$

By using the preceding head-discharge equations, following data can be developed:

| Q (l/s) | 0.0 | 0.5 | 1.0 | 1.5 | 2.0 | 2.5 | 3.0 | 3.5 | 4.0 | 4.5 | 5.0 |
|---|---|---|---|---|---|---|---|---|---|---|---|
| Head for position A | 12.00 | 12.70 | 14.71 | 17.99 | 22.53 | 28.31 | 35.34 | 43.60 | 53.09 | 63.80 | 75.74 |
| Head for position B | 12.00 | 13.12 | 16.12 | 20.86 | 27.28 | 35.34 | 45.01 | 56.26 | 69.08 | 83.46 | 99.38 |

The preceding data in the table and the head-discharge rate of the pump are plotted as shown in Fig. 3.

By observing the point of intersection of system curve for position A with pump curve, the discharge rate is approximately 4.62 l/s. By observing the point of intersection of system curve for position B with pump curve, the discharge rate is approximately 4.0 l/s.

At the position A, if the farmer wants to operate the sprinkler at 30 m pressure head, how much pressure should be reduced by using the gate valve?

As per the system curve A, when the pressure head is 53.09 m, the sprinkler will deliver 4.0 l/s. If we intentionally reduce the pressure head by throttling and bring the

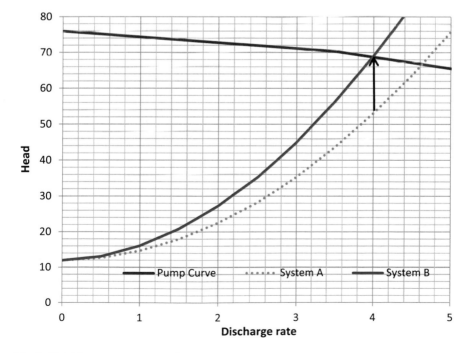

**Fig. 3** Matching system curves and pump curve

**Fig. 4** Field plan

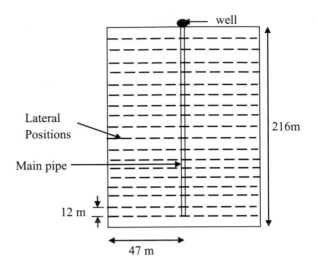

system curve to move along the arrow in Fig. 4, the system curve for A will cut the pump curve at 4 l/s discharge rate. If 15.99 m pressure reduction is done by throttling the valve and brought the system pressure head to 53.09 m, the farmer will be able to get 4 l/s discharge rate in position A also.

Following is another example which incorporates more real field situations and it is a little more complex example.

**Example** A sprinkler system operates with the following data (Fig. 4):

Sprinkler head-discharge rate—0.3 l/s at 20 m pressure head

Riser pipe height—2 m (Neglect friction losses in riser pipe)

Sprinkler spacing—12 m

Number of sprinklers/lateral—4 numbers

Lateral spacing—12 m

Length of lateral—42 m (The first sprinkler is at half spacing)

Lateral pipe diameter—44.8 mm

One foot valve with $k - 15$

Pipe in the well: Length—35 m, Diameter—68.6 mm.

One reflux valve with $k - 2.5$

A well yield test was conducted and the following equation was fitted relating the depth of water level in metres from the land surface ($s$) and the pumping discharge rate in l/s ($Q$):

During wet saeason: $s = 10 + Q + 0.01Q^2$;

during dry season: $s = 20 + 1.1Q + 0.02Q^2$

The pump installed in the well has the following characteristics:

| Q (l/s) | 0 | 3.5 | 4 | 5 | 6.5 | 8 | 8.5 | 9 | 10.5 |
|---------|-----|-------|-------|------|-----|------|------|----|------|
| Head (m) | 114 | 105.6 | 103.2 | 98.4 | 90 | 79.8 | 76.2 | 72 | 55.8 |

Land is perfectly levelled without any undulations. There is a horizontal main pipe of length 10 m and its diameter is 68.6 mm. Within 10 m length, there are three butter fly valves with $k$ value of 0.2. Find out how many laterals can be connected with the existing pump at a time during both the wet season and dry season in a year.

Assume that the minor losses at the point of connection of each lateral due to the tees are negligible.

Assume that the first sprinkler is operated at a spacing of 6 m from the main and the first lateral is at 12 m after the 10 m main pipe section. The field dimensions are as shown in Fig. 4.

**For Wet Season**

Total dynamic head (TDH) = Head needed at sprinkler head + Riser Pipe Height + Friction loss in lateral + level difference between water level in well and land + Friction loss in the pipe section in well + Friction loss in main + Friction loss in submain + Head loss in fittings (foot valve, reflux valve, butterfly valve) (Fig. 5).

Head needed at the sprinkler head = 20 m

Riser pipe height = 2 m

$$\text{Average friction loss in lateral} = 789000 \frac{Q^{1.75}}{D^{4.75}} L_{lat} \overline{F}_N$$

$$= 789000 \frac{1.2^{1.75}}{44.8^{4.75}} 42 \times 0.754$$

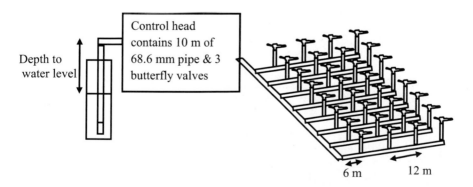

Fig. 5 Lay out of sprinkler system at the first set up during wet season

$$= 0.4928 \, \text{m}$$

Level difference between water level in well and land during wet season: $s = 10 + Q + 0.01Q^2$.

$$\text{Friction loss in the pipe section in well} = 789000 \frac{Q^{1.75}}{D^{4.75}} L_{\text{well}}$$

$$= 789000 \frac{Q^{1.75}}{68.6^{4.75}} 35 = 0.05231 Q^{1.75}$$

$$\text{Friction loss in main} = 78900 \frac{Q^{1.75}}{68.6^{4.75}} 10 = 0.01496 Q^{1.75}$$

$$\text{Average friction loss in submain} = 789000 \frac{Q^{1.75}}{D^{4.75}} L_{\text{sub}} \overline{F}_N$$

$$\text{Length of submain } (L_{\text{sub}}) = \frac{Q}{q_{\text{lat}}} \times s_{\text{lat}}$$

where $q_{\text{lat}}$ is discharge rate of one lateral and $s_{\text{lat}}$ is spacing of laterals. Therefore, $L_{\text{sub}} = 10Q$.

$$\text{Average friction loss in submain} = 789000 \frac{Q^{2.75}}{68.6^{4.75}} \overline{F}_N = 0.01496 Q^{2.75} \overline{F}_N$$

$$\text{Head loss in fittings} = 82627 \frac{Q^2}{D^4}$$

$$\sum k_i = 82627 \frac{Q^2}{D^4} \sum (1.5 + 2.5 + 3 \times 0.2) = 0.06753 Q^2$$

$$\text{TDH} = 32.4928 + Q + 0.077533 \, Q^2$$
$$+ 0.06727 \, Q^{1.75} + 0.01496 Q^{2.75} \overline{F}_N$$

**For Dry Season**

$$\text{TDH} = 42.4928 + Q + 0.08753 Q^2 + 0.06727 Q^{1.75} + 0.01496 Q^{2.75} \overline{F}_N$$

Following table can be developed using the preceding equations as follows:

| $\overline{F}_N$ | 1.000 | 0.885 | 0.840 | 0.816 | 0.801 | 0.790 | 0.783 | 0.777 | 0.772 |
|---|---|---|---|---|---|---|---|---|---|
| $Q$ (l/s) | 1.2 | 2.4 | 3.6 | 4.8 | 6.0 | 7.2 | 8.4 | 9.6 | 10.8 |
| TDH (m) (Wet) | 33.92 | 35.80 | 38.16 | 41.04 | 44.49 | 48.53 | 53.23 | 58.60 | 64.69 |
| TDH (m) (Dry) | 44.06 | 46.10 | 48.65 | 51.75 | 55.45 | 59.77 | 64.78 | 70.48 | 76.94 |

From Fig. 6, it can be seen that the matching point corresponding to wet season is approximately 10 l/s. That means eight laterals can be safely operated with 9.6 l/s. It

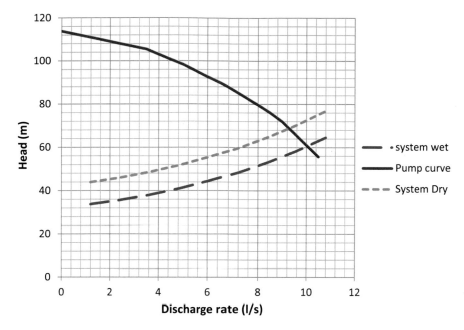

**Fig. 6** Matching pump curve with wet and dry system curves

can also be seen that the matching point corresponding to dry season is approximately 9.3 l/s. That means seven laterals can be safely operated safely with 8.4 l/s.

After taking a decision of connecting eight laterals during wet season, we will find out actually how much will be the total discharge rate and also average discharge rate of sprinklers. This can be done by following the preceding procedure as such with only one exception.

In the preceding procedure, the head needed at the sprinkler head is constantly taken as 20 m. That term has to be substituted with a different equation as explained below:

$$q = k_{\text{sprinkler}}\sqrt{h}$$
$$0.3 = k_{\text{sprinkler}}\sqrt{20}$$
$$k_{\text{sprinkler}} = 0.06708 \, \text{l/s/m}^{0.5}$$

Total number of sprinkler heads for eight laterals $= 8 \times 4 = 32$ sprinkler heads.

Therefore, average discharge rate of one sprinkler $q = \frac{Q}{32}$ l/s.

Head at the exit of the sprinkler $= \frac{q^2}{k_{\text{sprinkler}}^2} = \frac{(Q/32)^2}{0.06708^2} = 0.2170Q^2$

$$\text{TDH} = 12.4928 + Q + 0.2945Q^2 + 0.06727Q^{175} + 0.01496Q^{2.75}0.777$$

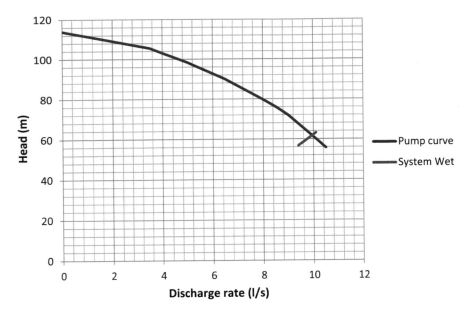

**Fig. 7** Matching the pump curve and system curve for 8 laterals

If the preceding equation is plotted as in Fig. 7, we get the matching point corresponding to 9.9 l/s and therefore average sprinkler discharge rate works out to 0.31 l/s.

## *Matching the Impeller Speed and Diameter for Variable requirements*

In a pump casing, we can make marginal changes in speed of rotation of impeller and/or in diameter of the impeller to get a desired change in pressure head and discharge rate. In a pumping system, the discharge rate ($Q$) can be expressed as follows:

$$Q = \pi D B V_f$$

where $D$ is outer diameter of the impeller; $B$ is the width of the impeller (Fig. 8) and $V_f$ is the radial velocity of the exiting water jet. Therefore if the width of the impeller is maintained constantly, we can write the preceding equation as follows:

$$Q \alpha D V_f$$

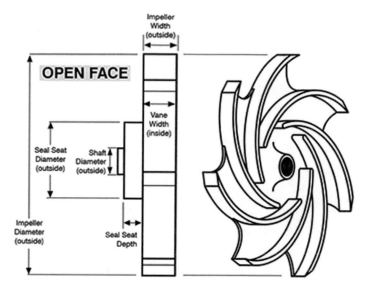

**Fig. 8** Illustration for impeller width (horizonparts.com)

The radial velocity is directly proportional to the speed of rotation per second ($N$) of impeller. Hence,

$$V_f \alpha N$$

Therefore, $Q \alpha DN$.

Let us study two kinds of situations which are as follows:

1. Changing the speed of rotation alone
2. Changing the diameter alone.

## Changing the Speed of Rotation Alone

For this case, the preceding equation gets transformed as follows:

$$Q \alpha N$$

If '$u$' is linear velocity of impeller periphery, then we can write the following equation:

$$u = \pi DN$$

$$u \alpha N$$

The linear velocity of impeller causes the velocity of water to gain velocity '$u$'. The relation between velocity of flow and pressure head ($H$) is as follows:

$$u \alpha \sqrt{H}$$
$$\sqrt{H} \alpha N$$
$$\sqrt{H} \alpha N^2$$

The minimum input power ($P$) needed for development of head $H$ is as follows:

$$P \alpha Q H$$

Substituting for $Q$ with $N$ and $H$ with $N^2$ in the preceding equation is as follows:

$$P \alpha N^3$$

From the preceding derivation, following equation can be written as follows:

$$\frac{Q_1}{Q_2} = \frac{N_1}{N_2}; \quad \frac{H_1}{H_2} = \frac{N_1^2}{N_2^2}; \quad \frac{P_1}{P_2} = \frac{N_1^3}{N_2^3} \tag{1}$$

## Changing the Diameter of Impeller Alone

For constant $N$, $Q \alpha D$.

For constant $N$, $u = \pi D N$ leads to $H \alpha D^2$

For constant $N$, $P \alpha Q H$ leads to $P \alpha D^3$.
From the preceding derivation, following equation can be written as follows:

$$\frac{Q_1}{Q_2} = \frac{D_1}{D_2}; \quad \frac{H_1}{H_2} = \frac{D_1^2}{D_2^2}; \quad \frac{P_1}{P_2} = \frac{D_1^3}{D_2^3} \tag{2}$$

Eq. 2 is valid only when the change of diameter is very less and not more than 20%. Equations 1 and 2 are called pump affinity or similarity laws.

**Example** A pump is operated with a discharge rate of 100 m³/h against a head of 100 m and the speed of rotation of impeller is 1750 rpm and power used is 5 kW. If the same pump set up has to deliver a discharge rate of 200 m³/h, how much must be speed of rotation of impeller and minimum power input need?

$$\frac{Q_1}{Q_2} = \frac{N_1}{N_2} \quad \frac{100}{200} = \frac{1750}{N_2}$$

$N_2 = 3500$ rpm

$$\frac{100}{H_2} = \frac{1750^2}{3500^2}$$

$H_2 = 400$ m

$$\frac{5}{P_2} = \frac{1750^3}{3500^3}$$

$P_2 = 40$ kW

Note that the doubling of discharge rate causes exponential increase in the head developed and the power input required.

## Time of Sprinkler Operation

The precipitation rate of sprinkler system is equal to the water applied with respect to time. This can be very easily obtained by placing a catch Can (collector) or a plastic tumbler in the field for a specified length of time. Table 1 provides maximum application rates for different soil textural classes.

Approximately, in moderately wet soils, depth of water wetted is equal to 20 times of depth of water applied and in dry soils; depth of water wetted is equal to 10 times of depth of water applied. In clay soils, the depth wetted is on lower side and in sandy soils; the depth wetted is on higher side. For any irrigation, the root depth of the crop can be examined and the irrigation to wet that much depth can be done. Normally in India, farmers apply a specified amount of water for any irrigation at a constant interval. Some farmers visually observe the soil moisture status and also the turgidity of crop leaves and do irrigation scheduling based on their experiences.

## Soil Moisture Characteristics

We know that when a soil is not saturated, the water is held by soil particles by the force of adhesion. If the force of adhesion is more, the plant has to exert sufficient suction pressure to extract water. For instance, let us say volumetric soil water content of loamy soil and also clay soil is at one level of volumetric water content say, 0.30. The suction pressure head needed to extract this water from loamy soil would be approximately 100 cm whereas for clayey soil; the suction pressure would be

**Table 1** Maximum sprinkler application rates [1]

| Soil texture | Maximum application rate (mm h$^{-1}$) | | | | Available water-holding capacity (mm m$^{-1}$) | Maximum depth of infiltration (mm) for a root zone 1 m deep with allowable depletions of (%) | | |
| --- | --- | --- | --- | --- | --- | --- | --- | --- |
| | Soil slope (%) | | | | | | | |
| | 0–5 | 5–8 | 8–12 | >12 | | 35 | 50 | 65 |
| Coarse sand | 50 | 40 | 30 | 20 | 50–70 | 21 | 30 | 39 |
| Fine sand | 40 | 32 | 24 | 16 | 75–85 | 28 | 40 | 52 |
| Loamy fine sand | 35 | 28 | 21 | 14 | 85–100 | 32 | 45 | 59 |
| Sandy loam | 25 | 20 | 15 | 10 | 110–125 | 42 | 60 | 78 |
| Fine sandy loam | 20 | 16 | 12 | 8 | 130–150 | 49 | 70 | 91 |
| Very fine sandy loan | 15 | 12 | 9 | 6 | 145–165 | 53 | 75 | 98 |
| Loam | 13 | 10 | 8 | 5 | 150–170 | 56 | 80 | 104 |
| Silt loam | 13 | 10 | 8 | 5 | 160–200 | 63 | 90 | 117 |
| Sandy clay loam | 10 | 8 | 6 | 4 | 140–170 | 53 | 75 | 98 |
| Clay loam | 8 | 6 | 5 | 3 | 150–180 | 56 | 80 | 104 |
| Silty clay loam | 8 | 6 | 5 | 3 | 140–180 | 56 | 80 | 104 |
| Clay | 5 | 4 | 3 | 2 | 130–180 | 53 | 75 | 98 |

approximately 1000 cm (Fig. 9). If say, paddy is grown in both the soil types at the same level of 0.30 of volumetric water content, the plant will be able to extract water from loamy soil whereas in the clayey soil at the same volumetric water content level, the plant will show wilt symptoms. This is because the paddy plant may not be able to exert 1000 cm of suction to extract water from clayey soil.

Owing to the reasons discussed in the preceding paragraph, the results of irrigation scheduling experiments done worldwide are normally provided in terms of optimal suction pressures as presented in Table 2. The optimal suction pressure values are dependent on only type of plant and they do not depend on the type of soil. For instance for beans, the optimal suction pressure head is 150 centibars. It means that the beans crop can exert a suction pressure of 150 centibars for extracting water. In a heavy soil like clay, even if water is available in soil above the suction pressure head of 150 centibars, the beans plant may not be able to extract water. Table 2 can be used for any type of soil for the crops mentioned in the Table.

Some useful facts about soil moisture are as follows:

- Water moves very little at soil moisture levels below field capacity. Figure 10 illustrates the relative hydraulic conductivities of different types of soil. The hydraulic

**Fig. 9** Water content versus suction head

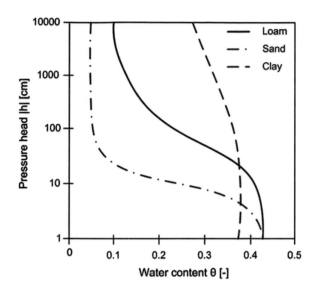

**Table 2** Optimal soil moisture suction heads for different crops [2]

| Crop | Suction pressure head (cm) | Crop | Suction pressure head (cm) | Crop | Suction pressure head (cm) |
|---|---|---|---|---|---|
| Beans | −1500 | Onion | −550 to −450 | Corn | −500 |
| Cabbage | −700 to −600 | Early growth | −650 to −550 | Vegetative | −12,000 to − |
| Tobacco | −800 to −300 | Bulbing time | | Ripening | 8000 |
| Sugarcane | − 500 to −150 | Carrot | −650 to −550 | Small grains | −500 to −400 |
| Cauliflower | − 700 to −600 | Sugar beet | −600 to −400 | Vegetative | −12,000 to − |
| Tomato | 800 to −1500 | Potato | −500 to −300 | Ripening | 8000 |
| Lemon | −400 | Orange | −400 | Grapes | −400 to −500 |
| Banana | −300 to −1500 | | | • Early | < −1000 |
| | | | | • Maturity | |

conductivity of clay soil does not fall down steeply whereas for the sand, conductivity falls down steeply and for silty (loamy) soil, hydraulic conductivity fall is in between.

- Plant can also extract water even when soil moisture level is above field capacity.
- Evaporation from the soil is limited to a few centimetres depth of soil.
- Plants do not pump water. They act more like living straws.
- The root zone of a crop grows progressively till the end of the vegetative growth.
- Water extraction pattern from the soil is as shown in Fig. 11. In the first 25% of root depth, 40% of total soil water uptake occurs and the soil water uptake keeps on reducing as the depth of roots increases as shown in the figure.

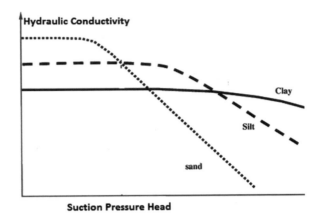

**Fig. 10**  Suction versus hydraulic conductivity

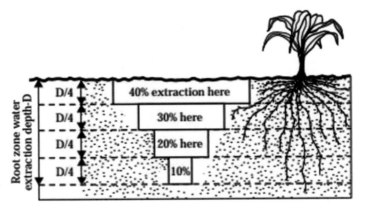

**Fig. 11**  Water extraction pattern of plants (University of Minnesota Extension)

In order to do efficient irrigation scheduling, soil moisture parameters such as field capacity, permanent wilting point, available water holding capacity and management allowable deficit have to be understood. Also the quantities such as crop water requirement, irrigation depth, gross irrigation depth and irrigation interval have to be estimated.

## Field Capacity

Field capacity is the moisture content of the soil after drainage of all the gravitational water from the soil. It is the sum of the capillary water and hygroscopic water.

**Table 3** Water holding capacities for different soils (http://eagri.org/eagri50/AGRO103/lec03.pdf)

| S. No. | Soil type | Per cent moisture—Dry basis | | Depth of available water per metre depth of soil (cm) |
|---|---|---|---|---|
| | | Field capacity (%) | Permanent wilting point (%) | |
| 1 | Find sand | 3–5 | 1–3 | 2–4 |
| 2 | Sandy loam | 5–15 | 3–8 | 4–11 |
| 3 | Silty loam | 12–18 | 6–10 | 6–13 |
| 4 | Clay loam | 15–30 | 7–16 | 10–18 |
| 5 | Clay | 25–40 | 12–20 | 16–30 |

Normally, the soil moisture suction pressure level at the field capacity is 0.1 to 0.3 atmospheres.

## Permanent Wilting Point

It is the moisture content at which the plant cannot draw sufficient moisture from the soil and the plant starts to wilt. Normally the soil moisture suction pressure at the permanent wilting point is around 15 atmospheres.

## Available Water Holding Capacity

It is the difference between the field capacity and the permanent wilting point. Table 3 provides the information on field capacity, permanent wilting point and available water holding capacities for different soils. From Table 3, it can be found that the clay has more available water than coarse fractions. But in coarse fractions, though the available water is less, plants can easily extract water than in clay soils.

## Management Allowable Depletion (MAD)

Normally any crop is to be irrigated, when the depletion from the available water in the soil gets reduced by 50%. But lower depletion levels to a level as low as 10% are also in use for high valued crops. At a lower depletion level, if irrigation is provided, it will be easy for plant to absorb water. The soil moisture depletion level at which irrigation is done is called as Management Allowable Deficit (MAD). Table 4 provides guideline values for medium to fine textured soil.

Lower MADs are used in following situations:

**Table 4** Management allowable deficits for medium to fine textured soil

| Crop | Crop growth stage | | | |
| --- | --- | --- | --- | --- |
| | Establishment | Vegetative | Flowering yield formation | Ripening maturity |
| Alfalfa hay | 50 | 50 | 50 | 50 |
| Alfalfa seed | 50 | 60 | 50 | 80 |
| Beans, green | 40 | 40 | 40 | 40 |
| Beans, dry | 40 | 40 | 40 | 40 |
| Citrus | 50 | 50 | 50 | 50 |
| Corn, grain | 50 | 50 | 50 | 50 |
| Corn, seed | 50 | 50 | 50 | 50 |
| Corn, sweet | 50 | 40 | 40 | 40 |
| Cotton | 50 | 50 | 50 | 50 |
| Cranberries | 40 | 50 | 40 | 40 |
| Garlic | 30 | 30 | 30 | 30 |
| Grains, small | 50 | 50 | 40 | 60 |
| Grapes | 40 | 40 | 40 | 50 |
| Grass pasture/hay | 40 | 50 | 50 | 50 |
| Grass seed | 50 | 50 | 50 | 50 |
| Lettuce | 40 | 50 | 40 | 20 |
| Milo | 50 | 50 | 50 | 50 |
| Mint | 40 | 40 | 40 | 50 |
| Nursery stock | 50 | 50 | 50 | 50 |
| Onions | 40 | 30 | 30 | 30 |
| Orchard, fruit | 50 | 50 | 50 | 50 |
| Peas | 50 | 50 | 50 | 50 |
| Peanuts | 40 | 50 | 50 | 50 |
| Potatoes | 35 | 35 | 35 | 50 |
| Safflower | 50 | 50 | 50 | 50 |
| Sorghum, grain | 50 | 50 | 50 | 50 |
| Spinach | 25 | 25 | 25 | 25 |
| Sugar beets | 50 | 50 | 50 | 50 |
| Sunflower | 50 | 50 | 50 | 50 |
| Tobacco | 40 | 40 | 40 | 50 |
| *Vegetables* | | | | |
| 1 to 2 ft root depth | 35 | 30 | 30 | 35 |
| 3 to 4 ft root depth | 35 | 40 | 40 | 40 |

- Shallow root zone crops such as onions
- Sandy soils
- When crop needs critically more water
- Salty water or salty soil.

## Irrigation Scheduling

Irrigation scheduling is the process of decision making about how much water to be applied for any irrigation and in what interval irrigation should be done.

1. Plant wilt symptoms method
2. Evapotranspiration estimation method.

## Plant Wilt Symptoms Method

In this method, the time of operation of drip or sprinkler system is decided based on wetting root zone to a depth of 75–100% of root depth. The depth of root zone development can be easily ascertained by cutting open the soil and observing the root zone. The time interval between successive irrigations can be decided based on the plant wilt symptoms, soil type and crop types. The maximum irrigation interval can be as big as the irrigation interval adopted during the conventional surface irrigation methods. The actual irrigation interval can be any interval equal to or less than the maximum irrigation interval. Farmers themselves can practically find this irrigation interval for the crops grown.

## Evapotranspiration Estimation Methods

Evapotranspiration (ET) can be estimated either from climatological data or from estimation of soil moisture. Methods for estimation of ET from meteorological data are discussed exclusively in a separate chapter number '15' namely 'Evapotranspiration for Microirrigated Crops'. The most simple method of soil moisture measurement is by gravimetric method. In this method, soil samples are taken from the field and oven dried to find out the soil moisture. Since this is a time consuming and laborious work, it is seldom practised in real field situations.

Other popular indirect methods of soil measurement which are popular are as follows:

1. Tensiometer
2. Electrical resistance method
3. Neutron probe
4. Time domain reflectometry.

Data obtained from these sensors are used to model the water movement in root zone to estimate evapotranspiration more accurately and these methods are exclusively dealt in a separate chapter number '16'namely 'Rhizosphere Modelling'.

## *Irrigation Requirements*

Irrigation requirements can be found out using following equations:

$$\text{Irrigation Requirement (IR)} = Et_c - \text{Rainfall} \tag{3}$$

If rain fall amount is greater than $ET_c$, then IR is taken as zero.

$$\text{Gross Irrigation Requirement (GIR)} = \text{IR/Application Efficiency} \tag{4}$$

Application efficiency is the ratio between the actual amount of water used for meeting ET and the actual amount of water delivered to the field. Application efficiency depends on type of irrigation. Normally application efficiency for drip irrigation is taken as 90%, for sprinkler irrigation, it is taken as 80%, for conventional furrow or flatbed irrigations, it is taken as 50%.

Table 4 provides the effective crop root depths that would contain approximately 80% of the feeder roots in a deep well-drained soil profile. The values in the table are useful for knowing the depth of soil to which irrigation should be done.

## *Irrigation Depth*

The irrigation depth to be applied per irrigation ($d_x$) is computed based on soil moisture using following equation:

$$d_x = \frac{\text{MAD}}{100} W_a Z \tag{5}$$

where MAD is the Management Allowable Deficit in %, $W_a$ is the available water holding capacity of the soil (cm/m depth of soil) and $Z$ is the effective root depth of soil (m).

## *Gross Irrigation Depth*

The gross irrigation depth (GID) is found out using the following equation:

$$GID = \frac{d_x}{\eta_a} \times 100 \tag{6}$$

where $\eta_a$ is application efficiency.

## Irrigation Interval

The irrigation interval (f) is found out using the following equation:

$$f = \frac{d_n}{U_d} \tag{7}$$

where $d_n$ is net depth of water application (mm) and $U_d$ is daily irrigation requirement (mm/day). The value of $d_n$ can be at the maximum equal to $d_x$. In places where deficit irrigation is practised, $d_n$ will be less than $d_x$. For normal irrigation practices, $d_n$ may be taken as equal to $d_x$.

Figure 13 is an illustration for understanding the different soil moisture constants.

**Example** Sweet corn crop is in vegetative stage grown in silty loam soil with water holding capacity of 10 cm per metre depth of soil. The recommended MAD for the crop as per Table 4 is 40%. Effective root depth is 0.4 m. Find out the gross irrigation depth of water to be applied by sprinkler irrigation. Take the application efficiency of sprinkler as 80% (Fig. 12; Table 5).

**Solution**

$$d_x = \frac{MAD}{100} W_a Z$$

$$d_x = \frac{40}{100} 0.4 \times 10 = 1.60 \, cm$$

$$GID = \frac{d_x}{\eta_a} \times 100$$

$$GID = \frac{1.60}{80} \times 100 = 2 \, cm$$

## Capacity of Pump

The capacity of pump to be selected depends on irrigation interval as well as how many days irrigation could be done within the irrigation interval. Let us assume, for an irrigation system, irrigation interval is four days and the total area is 8 ha. If all the days are used for irrigation, the total 8 ha could be divided into four divisions of

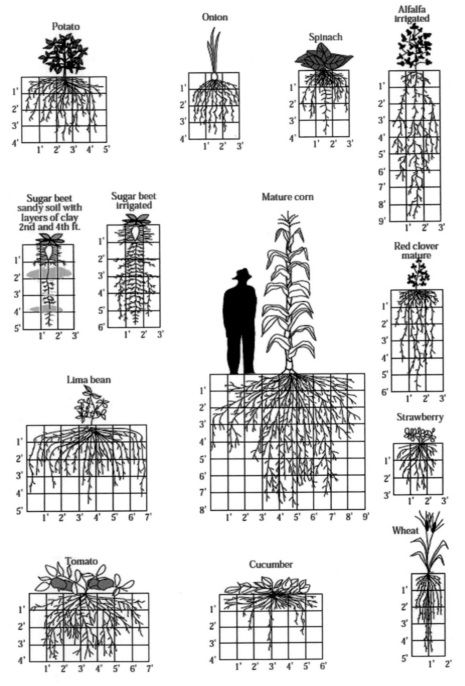

**Fig. 12** Illustration of root growth pattern of different crops (Irrigation Guide—USDA)

**Table 5** Effective root depths of different crops [1]

| Crop | Root depth (m) | Crop | Root depth (m) |
|---|---|---|---|
| Alfalfa | 1.2–1.8 | Lettuce | 0.2–0.5 |
| Almonds | 0.6–1.2 | Lucerne | 1.2–1.8 |
| Apple | 0.8–1.2 | Oats | 0.6–1.1 |
| Apricot | 0.6–1.4 | Olives | 0.9–1.5 |
| Artichoke | 0.6–0.9 | Onion | 0.3–0.6 |
| Asparagus | 1.2–1.8 | Parsnip | 0.6–0.9 |
| Avocado | 0.6–0.9 | Passion fruit | 0.3–0.5 |
| Banana | 0.3–0.6 | Pastures | 0.3–0.8 |
| Barley | 0.9–1.1 | Pea | 0.4–0.8 |
| Bean (dry) | 0.6–1.2 | Peach | 0.6–1.2 |
| Bean (green) | 0.5–0.9 | Peanuts | 0.4–0.8 |
| Bean (lima) | 0.6–1.2 | Pear | 0.6 to1.2 |
| Beet (sugar) | 0.6–1.2 | Pepper | 0.6–0.9 |
| Beet (table) | 0.4–0.6 | Plum | 0.8–1.2 |
| Berries | 0.6–1.2 | Potato (Irish) | 0.6–0.9 |
| Broccoli | 0.6 | Potato (sweet) | 0.6–0.9 |
| Brussels sprout | 0.6 | Pumpkin | 0.9–1.2 |
| Cabbage | 0.6 | Radish | 0.3 |
| Cantaloupe | 0.6–1.2 | Safflower | 0.9–1.5 |
| Carrot | 0.4–0.6 | Sorghum (grain and sweet) | 0.6–0.9 |
| Cauliflower | 0.6 | Sorghum (silage) | 0.9–1.2 |
| Celery | 0.6 | Soybean | 0.6–0.9 |
| Chard | 0.6–0.9 | Spinach | 0.4–0.6 |
| Cherry | 0.8–1.2 | Squash | 0.6–0.9 |
| Citrus | 0.9–1.5 | Strawberry | 0.3–0.3 |
| Coffee | 0.9–1.5 | Sugarcane | 0.5–1.1 |
| Corn (grain and silage) | 0.6–1.2 | Sudan grass | 0.9–1.2 |
| Corn (sweet) | 0.4–0.6 | Tobacco | 0.6–1.2 |
| Cotton | 0.6–1.8 | Tomato | 0.6–1.2 |
| Cucumber | 0.4–0.6 | Turnip (white) | 0.5–0.8 |
| Egg plant | 0.8 | Walnuts | 1.7–2.4 |
| Fig | 0.9 | Watermelon | 0.6–0.9 |
| Flax | 0.6–0.9 | Wheat | 0.8–1.1 |

(continued)

**Table 5** (continued)

| Crop | Root depth (m) | Crop | Root depth (m) |
|------|---------------|------|---------------|
| Grapes | 0.5–1.2 | | |

*Approximately 80% of the feeder roots are in the top 60% of the soil profile. Soil and plant environmental factors often offset normal root development; therefore, soil density, pore shapes and sizes, soil–water status, aeration, nutrition, texture and structure modification, soluble salts and plant root damage by organisms should all be taken into account

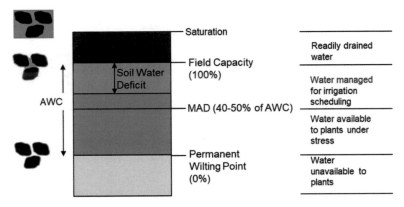

**Fig. 13** Illustration for soil moisture constants (MNWheat.org)

area equal to 2 ha and each division could be irrigated for one day. If two days are declared holidays in a week and in that case, though four days of irrigation interval is available, the total area should be divided into two divisions of 4 ha area and each 4 ha division should be irrigated within a single day.

The capacity of the pump (l/s) needed can be found out using the following equation:

$$Q_S = \frac{2.78\,\text{GID}}{T} \tag{8}$$

where $A$ is area to be irrigated per day (ha), GID is gross irrigation depth (mm) and $T$ is time of operation of pump in hours per day. Often the value of $T$ is the time period during which electricity is available for pumping.

**Example** Compute the gross irrigation requirement and irrigation system capacity requirement for a sprinkler system for a single crop in the design area.

Data

| | |
|---|---|
| Area | = 16 ha |
| Crop | = Cotton |
| | (continued) |

(continued)

| | |
|---|---|
| Root depth | = 1.5 m |
| Peak moisture use rate | = 5 mm/day |
| Available soil water per metre depth of soil | = 80 mm |
| Application efficiency | = 70% |
| Operating time/day | = 10 h/day |

Irrigation is given at 20% depletion of available soil water.
Weekly one day is holiday and irrigation cannot be done on that day.

**Solution**

Irrigation depth $= 0.2 \times 80 \times 1.5 = 24$ mm ( Eq. 5).

Gross irrigation depth $= 24 \times 100/70 = 34.28$ mm (Eq. 6).

| | |
|---|---|
| Irrigation interval | $= 24$ mm/(5 mm/day) (Eq. 7). |
| | $= 4.8$ days (Approximately 5 days) |

The entire area should be irrigated within five days. If one day is holiday, within four days all the area should be irrigated.

Divide the area into small units.

| | |
|---|---|
| No of units irrigated | = Irrigation interval—No. of holidays |
| | = 4 |
| Area to be irrigated / day | = 16 ha/4 days |
| | = 4 ha/day |

Capacity of pump needed

$$= 2.78 \times 4 \times 34.28/10 = 38.08 \, \text{l/s}$$

## *van-Genuchten's Equation*

Following is the popular equation relating soil suction pressure head and volumetric water content [6].

$$\theta(h) = \frac{\theta_s - \theta_r}{\left(1 + (\alpha h)^n\right)^{\frac{n-1}{n}}} + \theta_r$$

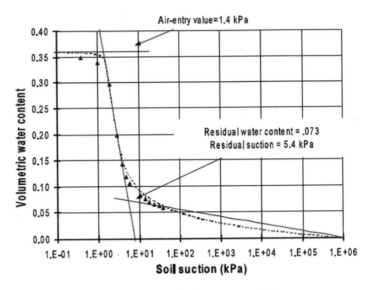

**Fig. 14** Illustration for air entry value and residual water content [3]

where $\theta(h)$ is volumetric water content (cm³/cm³) at the soil water pressure head of $h$ (cm); $\theta_s$ is saturated water content (cm³/cm³); $\alpha$ is called as inverse of air entry value which is the suction pressure head at which air is able to get into pores; $\theta_r$ is called as residual water content which is a fitting parameter which represents suction pressure head where soil water content curve flattens as shown in Fig. 14.

**Example** Find the water content for silty soil when the soil moisture suction is 800 cm. If the field capacity of the soil is 0.23, find out the depth of water to be applied for bringing 100 cm of soil depth to field capacity from the water content computed?

**Solution**

From Table 6, $\theta_r = 0.034$; $\alpha = 0.016\ \text{cm}^{-1}$; $\theta_s = 0.46$

$$\theta(h) = \frac{0.46 - 0.034}{\left(1 + (0.016 \times 800)^{1.37}\right)^{\frac{0.37}{1.37}}} + 0.034$$
$$= 0.199$$

Difference between the moisture content and field capacity $= 0.230 - 0.199 = 0.031$.

Depth of water to be applied $= 0.031 \times 100 = 3.1$ cm.

**Table 6** van-Genuchten parameter values for different soil textural classes.

| Texture | $\theta_s$ | $\theta_r$ | $\alpha$ (cm$^{-1}$) | $n$ | $K_s$ (cm day$^{-1)}$ |
|---|---|---|---|---|---|
| Sand | 0.43 | 0.045 | 0.145 | 2.68 | 712.8 |
| Loamy sand | 0.41 | 0.057 | 0.124 | 2.28 | 350.2 |
| Sandy loam | 0.41 | 0.065 | 0.075 | 1.89 | 106.1 |
| Loam | 0.43 | 0.078 | 0.036 | 1.56 | 24.96 |
| Silt | 0.46 | 0.034 | 0.016 | 1.37 | 6.00 |
| Silt loam | 0.45 | 0.067 | 0.020 | 1.41 | 10.80 |
| Sandy clay loam | 0.39 | 0.100 | 0.059 | 1.48 | 31.44 |
| Clay loam | 0.41 | 0.095 | 0.019 | 1.31 | 6.24 |
| Silty clay loam | 0.43 | 0.089 | 0.010 | 1.23 | 1.68 |
| Sandy clay | 0.38 | 0.100 | 0.027 | 1.23 | 2.88 |
| Silty clay | 0.36 | 0.070 | 0.005 | 1.09 | 0.48 |
| Clay | 0.38 | 0.068 | 0.008 | 1.09 | 4.80 |

## *Irrigation Scheduling Under Salinity*

When saline lands are irrigated and if leaching of salts by rainfall is not expected to occur during the crop season, leaching of salts must be done. Computation of irrigation requirement must take into account the water needed for leaching of salts. The irrigation requirement can be calculated using the following formula:

$$IR = ET_c \frac{2EC_e}{2EC_e - EC_i}$$

where $EC_e$ and $EC_i$ are electrical conductivity of soil saturation extract and electrical conductivity of irrigation water expressed in terms of dS/m, respectively. Derivation of the preceding equation is available in salinity chapter number 5 of Basic section.

Table 7 provides minimum and maximum values of $EC_e$ for various crops. Minimum $EC_e$ is the value at which yield reduction starts to occur and the maximum $EC_e$ is the value at which yield is zero.

**Exercise** EC of irrigation water of an agricultural field grown with onion is 2 dS/m. EC of saturation extract is 3 dS/m. The cumulative ET for two days before next irrigation is 10 mm. Find out the depth of irrigation and also the leaching requirement ratio. Also state whether yield reduction would occur for this field condition.

**Solution**

$$IR = ET_c \frac{2EC_e}{2EC_e - EC_i}$$

Depth of irrigation, IR $= 10 \frac{2 \times 3}{2 \times 3 - 2} = 15$ mm

**Table 7**  Minimum and maximum values of $EC_e$ for various crops [4]

| Crop | $EC_e$·ds/m | | Crop | $EC_e$· ds/m | |
|------|------|------|------|------|------|
|      | Min[1] | Max[2] |      | Min[1] | Max[2] |
| *Field crops* | | | | | |
| Cotton | 7.7 | 27 | Corn | 1.7 | 10 |
| Sugar beets | 7.0 | 24 | Flax | 1.7 | 10 |
| Sorghum | 6.8 | 13 | Broad beans | 1.5 | 12 |
| Soybean | 5.0 | 10 | Cowpeas | 1.3 | 8.5 |
| Sugarcane | 1.7 | 19 | Beans | 1.0 | 8.5 |
| *Fruit and nut crops* | | | | | |
| Date palm | 4.0 | 32 | Apricot | 1.6 | 6 |
| Fig olive | 2.7 | 14 | Grape | 1.5 | 12 |
| Pomegranate | 2.7 | 14 | Almond | 1.5 | 7 |
| Grapefruit | 1.8 | 8 | Plum | 1.5 | 7 |
| Orange | 17 | 8 | Blackberry | 1.5 | 6 |
| Lemon | 1.7 | 8 | Boysenberry | 1.5 | 6 |
| Apple, pear | 1.7 | 8 | Avocado | 1.3 | 6 |
| Walnut | 1.7 | 8 | Raspberry | 1.0 | 5.5 |
| Peach | 1.7 | 6.5 | Strawberry | 1.0 | 4 |
| *Vegetable crops* | | | | | |
| Zucchini squash | 4.7 | 15 | Sweet corn | 1.7 | 10 |
| Beets | 4.0 | 15 | Sweet potato | 1.5 | 10.5 |
| Broccoli | 2.8 | 13.5 | Pepper | 1.5 | 8.5 |
| Tomato | 2.5 | 12.5 | Lettuce | 1.3 | 8 |
| Cucumber | 2.5 | 10 | Radish | 1.2 | 9 |
| Cantaloupe | 2.2 | 16 | Onion | 1.2 | 7.5 |
| Spinach | 2.0 | 15 | Carrot | 1.0 | 8 |
| Cabbage | 1.8 | 12 | Beans | 1.0 | 6.5 |
| Potato | 1.7 | 10 | Turnip | 0.9 | 12 |

Leaching requirement ratio, $LR = \frac{D_d}{D_i} = \frac{15-10}{15} = 0.33$.

As per Table 7, minimum EC of saturation extract for onion is 1.2 dS/m, and maximum EC of saturation extract is 7.5 dS/m. So we can expect a yield reduction because the EC of irrigation water itself is above minimum EC.

## *Summary*

Maintenance issues of sprinkler irrigation systems is presented. When operating conditions vary in the field, how the pump operation needs to be adjusted is discussed with numerical examples. Analytical equations for impeller trimming and changing the impeller speed for optimizing the use of power are developed. Methods of estimation of irrigation requirement are presented with examples. Irrigation scheduling methods are also presented.

## Assessment

### Part A. Choose the best option for the following questions

1.  Usually the _____ sprinkler in any lateral is the location at which the dirt gets accumulated and causes clogging problem.

    a.  **last**
    b.  first
    c.  mid
    d.  none of these

2.  The saddle of a sprinkler head must be placed on a firm flat ground and riser pipe should be placed _____

    a.  **vertically**
    b.  horizontally
    c.  at a slight angle
    d.  at 90°

3.  If top spring in a sprinkler head gets worn out, _____of the sprinkler gets affected.

    a.  **rotation**
    b.  lubrication
    c.  life.
    d.  discharge rate.

4.  Pressure variation between the first sprinkler and the last sprinkler must be checked and it must be lower than _____

    a.  **20 %**
    b.  5 %
    b.  7 %
    b.  10 %

5.    Usually small and medium sprinklers are lubricated with ——

     a.    **water**
     b.    coconut oil
     c.    Castor oil
     d.    synthetic oil

**Part B. Descriptive Answers**

1.    In a periodic move lateral system, when the system is operating at design pressure, if the pipes are leaking what could be the reason?
2.    Why in a periodic move laterals, pipes do not leak at higher pressures and leaks at lower pressures?
3.    Explain briefly about how irrigation scheduling is done by farmers using sprinkler irrigation system in India?
4.    What checks are usually done at the pumping station and control head of a sprinkler irrigation system?
5.    Explain briefly with a sketch about system characteristics of a sprinkler irrigation system?
6.    What are the maintenance tips for a sprinkler irrigation system?

# References and Further Reading

1. Keller, J., & Bliesner, R. D. (1990). *Sprinkle and trickle irrigation.* Chapman & Hall.
2. Dasberg, S., & Or, D. (1999). *Drip irrigation.* Springer.
3. Vanapalli, S. K., Sillers, W. S., & Fredlund, M. D. (1998). The meaning and relevance of residual state to unsaturated soils. In *Proceedings of the Fifty-First Canadian Geotechnical Conference,* Edmonton, AB, October 4–7 (pp. 1–8)
4. Ayers, R. S., & Westcot, D. W. (1985) Water quality for agriculture. In *FAO Irrigation and Drainage Paper 29 Rev. 1,* FAO, Rome.
5. Irrigation Guide, Part-652. (1997). NRCS-USDA
6. Van Genuchten, M. T. (1980). A closed-form equation for predicting the hydraulic conductivity of unsaturated soils. *Soil Science Society America Journal, 44*(5), 892–898.

# Drip Irrigation Components

## Introduction

In drip irrigation water is discharged as drops at the root zone of plants. Irrigation is done at frequent time intervals. Plastic pipes are used for conveyance. Water advances on the soil around a dripper only after the amount of water applied exceeds the infiltration rate at a point. Typically, a dripper wets a diameter from 20 to 120 cm depending on the soil type. The depth of wetting for any specific soil primarily depends on the time of operation of the dripper.

Drip irrigation components include pump, pipes, tubes, filters, fertigation devices and water emitting devices, flow control equipment, many fittings and accessories (Fig. 1). For the first-time users, it can be a confusing array of components and gadgets. It would be a daunting task to select the right type of system and assemble the components suitable for irrigation needs. This section would primarily help in selection and installation of right components and their management.

Drip irrigation has many advantages and also has few disadvantages. They are listed as below:

### Advantages

- Only root zone is wetted and percolation losses are negligible
- Conveyance loss is less
- Weed growth is less
- Labourer requirement is less
- Salty soils can be irrigated
- Salty water can be irrigated
- Fertilizers can be applied through irrigation water
- Increase in yield of crops.

### Disadvantages

- Costly

V. Ravikumar, *Sprinkler and Drip Irrigation*,
https://doi.org/10.1007/978-981-19-2775-1_7

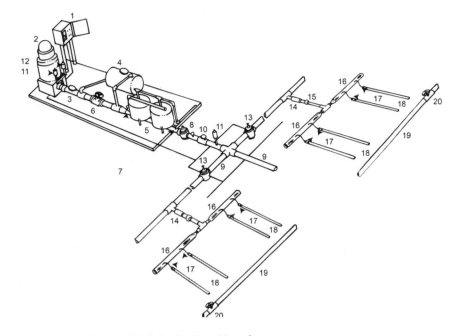

Courtesy: Roberts-Drip Irrigation User Manual

| | |
|---|---|
| 1. System controller | 11. Air vents at high points, after valves and at ends |
| 2. Pump | of lines |
| 3. Back flow prevention valve | 12. Pressure relief valve |
| 4. Fertilizer injector/tank | 13. Field control valve |
| 5. Filter tanks | 14. Submain secondary filters |
| 6. Butterfly valve or ball valve | 15. Pre-set pressure regulator |
| 7. Pressure gauges | 16. Submain |
| 8. Mainline control valve | 17. Lateral hookups |
| 9. Mainline | 18. Laterals |
| 10. Flow meter | 19. Flushing manifolds |
| | 20. Flush valves |

**Fig. 1** A typical layout of drip irrigation system

- The drippers get clogged over a period of time. So, periodic maintenance is necessary

## Salinity Under Drip Irrigation

Since frequent irrigation is possible in drip irrigation, the salt concentration of soil solution can be maintained nearly equal to the salt concentration of the irrigation water. Salt is transported to the periphery of wetted front in drip irrigation (Fig. 2). Water leaves the salt when it gets evaporated from the soil. Since salt build-up is away

**Fig. 2** Salt build-up in drip irrigation (www.fao.org)

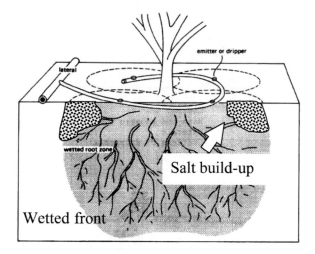

from the root zone, plants do not get affected very much. Sometimes in saline lands under drip irrigation when a light rainfall occurs, the salt build up tends to move into the root zone. Under such situations plants show wilt symptoms and farmers may get confused over this behaviour of crop in saline lands. Under such situations, after small rains, drip irrigating the crop sufficiently well would again transport the salts out of the root zone.

## Classification of Drip Irrigation Systems

### *Point and Line Source*

Typically, in drip irrigation, dripper is the fundamental component and each dripper when operated acts as a point source only (Fig. 3). When drippers are closer to each other, after operating the drippers for some time, if the point sources overlap and appear as a continuous wetted strip, then it is called as line source (Fig. 4). Generally, the dripper location and plant location need not be matched in line sources. Spacing between plants does not have any bearing with the spacing between drippers in line sources. The spacing between the drippers must be selected in such a way that the adjacent wetted fronts overlap before the wetted front moves deeper than root depth. Figure 5 illustrates the correct and wrong dripper spacing methods for line sources. It is always better to conduct wetting pattern studies in the field by cutting-open a trench after operating a line source for the duration of irrigation and examining the wetting fronts (Fig. 6).

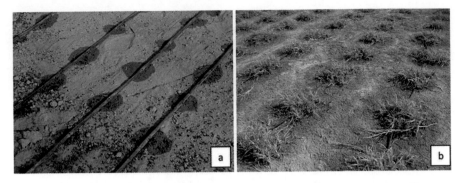

**Fig. 3**  Point source

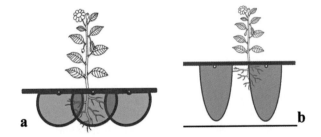

**Fig. 4**  Line source

**Fig. 5**  Dripper spacing in line source. **a** Correct dripper spacing. **b** Too long a dripper spacing

**Fig. 6**  Wetting pattern study in field (https://www.agric. wa.gov.au/water-manage ment/using-dye-show-water- movement-below-drip-irriga tion)

## *Online Drippers*

Online drippers are attached on the lateral by drilling a hole and inserting into the lateral by hand (Fig. 7). Mostly online drippers are used for widely spaced tree crops. Sometimes, a microtube start is inserted into the hole, microtube is attached to the microtube start, and dripper is attached at the end of the microtube (Fig. 8). This kind of microtube connections are needed when many drippers are provided to each tree. Many brands of online drippers have been designed in such a way that they can be opened up for cleaning.

For tree crops when large application rate is used, at the end of the microtubes, no dripper is fixed. The microtube discharges at around 50 litres per hour (l/h). This discharge rate is higher for the most soils and cause runoff. Hence in order to prevent runoff, trenches or bunds around the tree are formed (Fig. 9).

Sometimes, microsprinklers are also used at the end of micro-tubes which do not have the problem of runoff as the water is sprayed over an area around 2 m$^2$ and more (Fig. 10). The water distribution with in a spray radius of a micro-sprinkler may not be uniform and this non-uniformity of application within the spray area is usually not a cause of concern.

In some cases, bubblers are used for tree irrigation which normally have discharge rate ranging between 100 and 250 l/h. In Fig. 11, it can be seen that the discharged water is made to stagnate underground and water uptake occurs due to capillary rise

**Fig. 7**  Online dripper

**Fig. 8**  Microtube emitter (www.nmsu.edu)

**Fig. 9** Microtube emitter with a trench around tree

**Fig. 10** Microsprinkler

of water. The upper soil above the stagnating water serves as mulch. If water from the bubblers is applied on the land surface, a circular bund as in Fig. 9 is formed to arrest the runoff water.

## *Inline Drippers*

When drippers are either placed inside the lateral pipes or on the walls of lateral pipes, they are called as inline drippers. The spacing between them is fixed while manufacturing itself. They are available in the market with different spacings and discharge rates. In case of thicker walled lateral pipes, the inline drippers are made as hollow drippers and placed as shown in Fig. 12. In case of flexible thin walled

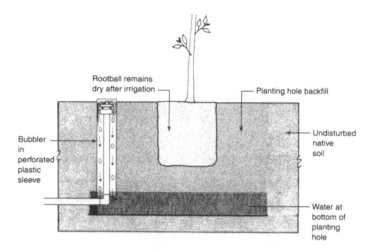

**Fig. 11** A typical bubbler installation

**Fig. 12** Inline dripper—hollow full round

lateral pipes, the inline drippers are placed on the wall of the lateral as shown in Fig. 13. They are very much useful in row field crops and small tree crops. They are also called as drip tapes.

**Fig. 13** Inline dripper—on the wall

## Subsurface Drip

The subsurface drip irrigation (SDI) system is much similar to the surface DIS with a primary difference of placing laterals and drippers underground (Figs. 14, 15, 16 and 17) The depth of placement often varies from 15 to 30 cm. For subsurface drip, mostly thinner tubes are used with thickness varying from 0.15 to 0.6 mm. However, thicker pipes can also be used for SDI.

Downstream end of all the laterals is usually connected to a manifold and provided with a control valve for easy flushing. Nowadays, surface drip laterals are also provided with flushing manifolds for easy flushing. Sometimes instead of providing

**Fig. 14**  Wetting pattern in subsurface drip

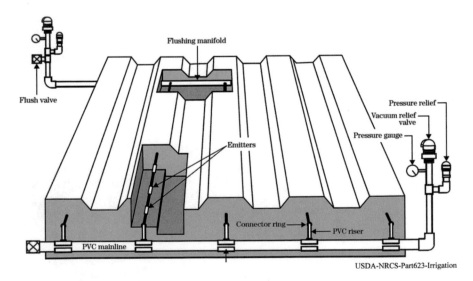

USDA-NRCS-Part623-Irrigation

**Fig. 15**  A typical submain layout of subsurface drip irrigation system

**Fig. 16** Subsurface lateral laid in a land

**Fig. 17** Wetting by
subsurface drip

flushing manifold, submains are provided at both the ends of the lateral. This kind of provision would have more uniformity.

A pressure regulator is needed because thinner tubes are used in SDI and excess pressure may harm the system. Properly designed vacuum breaker is an essential accessory; otherwise drippers will suck soil particles. Root intrusion into the drippers is possible, and this is prevented by chemical treatments. Regular irrigation would help reduce the problem of intrusion because, if the plant is made to starve for water, then the roots of plant would grow towards wetness.

Since the dripper is in contact with soil, there is a possibility of back pressure development on the dripper outlet passage from the water in the soil. This problem is predominant in clayey soils. Pressure-compensating drippers are used to overcome this problem.

**Advantages Over Surface Drip**

- Reduced evaporation from soil, surface runoff and deep percolation
- Soil surface is dry which increases infiltration during rain
- Reductions in weed germination and weed growth
- The laterals need not be removed between one crop season to the next season
- Mechanical harvesting and intercultural operations can be easily done.

**Disadvantages**

- If the depth of lateral is low, there is possibility of damage to laterals during mechanical harvesting and intercultural operations
- Rodent and insect damages
- Root intrusion
- Non-functioning is known very late, known only after the crop failure
- Back pressure of soil water around drippers, affect dripper outflow

# Emitters or Drippers

Emitters are the devices, through which water comes out of the pipe network. The emitters dissipate the pressure of outcoming water. The pressure dissipation is brought about by turbulence, vortex and friction along the path of the flow.

Normally the drippers are specified with dripper discharge rates 2 l/h, 4 l/h, etc. Many times, people misconstrue that the dripper would discharge the specified discharge rate, whatever is the pressure at the inlet of dripper. It is correct to say that 2 l/h at 10 m pressure head or any specified pressure head. It is conventional to express dripper discharge rate for 10 m pressure head.

Ideally a good emitter must have a larger diameter of passage and smaller discharge rate. Both the requirements are opposed to each other. This is a challenge, and hence, research on the improvement of dripper development technology continues to happen every day. Hence different dripper types have got evolved (Fig. 18).

Following is the relationship between emitter clogging and diameter of flow passage:

0.7 mm and less—Very sensitive to clogging
0.7 mm–1.5 mm—Sensitive
1.5 mm and more—Relatively insensitive.

## *Bernoulli's Law for Flow in Emitter*

If we apply Bernoulli's law between inlet (1) and outlet (2) of dripper (Fig. 19), we get

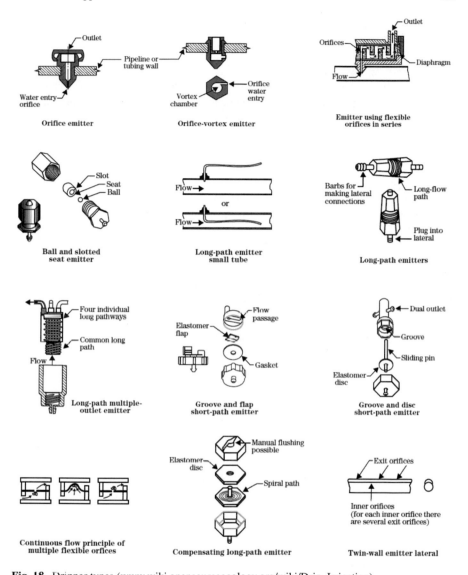

**Fig. 18** Dripper types (www.wiki.opensourceecology.org/wiki/Drip_Irrigation)

**Fig. 19** Bernoulli's law for flow in dripper

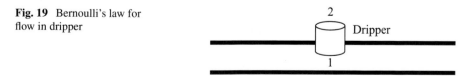

$$z_1 + \left(\frac{v^2}{2g}\right)_1 + h_1 = z_2 + \left(\frac{v^2}{2g}\right)_2 + h_2 + h_f + h_m \qquad (1)$$

where $z$ represents gravitational energy head, $h$ represents pressure head, $h_f$ is energy loss with in dripper and $h_m$ is loss of energy at the entry of inlet. Refer Chap. 1 of Basic section for a detailed discussion on application of Bernoulli's law and the explanation for the notations in Eq. 1.

In case of drippers, the difference between gravitational energy between point 1 and 2 can be neglected.

In case of long path emitter small tube (Fig. 18), the datum difference between inlet and outlet sometimes is significant. The inlet and outlet velocity will be one and the same. Pressure energy at the outlet is zero. Hence for the long path emitter small tube (microtube), Eq. 1 can be approximated as follows:

$$h_1 + Z_1 - Z_2 = h_f + h_m \qquad (1a)$$

**Example**

A microtube of inside diameter has 4 mm diameter of length 1 m which is connected to a drip lateral. The elevation difference between the inlet and outlet of the microtube is 0.5 m. Find out the flow in the microtube for a pressure head of 10 m at the inlet of the microtube? Assume turbulent flow in the microtube. Assume the entry loss $h_m = 0$. Also find out the Reynolds number of the flow? The dynamic viscosity of water is $1.002 \times 10^{-3}$ N s/m².

$$h_f = \frac{789000 \times q^{1.75}}{D^{4.75}} L$$

$$Z_1 - Z_2 = 0.5$$

By applying Eq. 1, we get

$$10 - 0.5 = \frac{789000 \times q^{1.75}}{4^{4.75}} \times 1$$

$$q = 0.067 \text{ l/s}$$

$$\text{Re} = \frac{\rho v d}{\mu}$$

$$\text{Re} = 21284$$

In case of drippers, the elevation difference between inlet and outlet is neglected due to the reason that its value is negligible. Hence application of Eq. 1 would be as follows:

$$h_f + h_m = h_1 \tag{2}$$

We know that

$$h_f = \frac{f L v^2}{2g D}$$

where

$h_f$ is head loss.
$f$ is called Darcy–Weisbach friction factor, and it is a function of Reynolds number.
$L$ is length of flow.
$q$ is flow rate of dripper.
$g$ is acceleration due to gravity.
$D$ is diameter of pipe.

$$h_m = \frac{k_m v^2}{2g}$$

$k_m$ is approximately equal to 0.5. After substituting for $h_m$ and $h_f$ in Eq. 2,

$$h_1 = \frac{v^2}{2g} \left( \frac{fl}{D} + k_m \right)$$

When microtubes are used as drippers, the value of $h_m$ is significant whereas when drippers with low discharge rate are used $h_m$ is neglected. Then the preceding equation can be written as follows:

$$h_1 = \frac{v^2}{2g} \left( \frac{fl}{D} \right)$$

The friction factor can be expressed as a function of Reynolds number. Substituting $v = q/(\pi D^2/4)$ in the preceding equation will introduce the dripper discharge rate in the preceding equation, and the preceding equation can be written as follows:

$$q = kh^x \tag{3}$$

where

$q$    Emitter discharge rate.
$k$    Discharge coefficient depends primarily on c/s area of flow in the dripper.
$h$    Pressure head at the inlet of emitter.
$x$    Emitter flow exponent.

**Fig. 20** Pressure versus
discharge relationship of
drippers

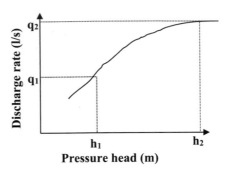

**Fig. 20** Pressure versus discharge relationship of drippers

The value of $k$ depends on the cross-sectional area of flow in the dripper. The value of $x$ depends on how the pressure is dissipated inside the dripper. The value of $x$ varies between 0 and 1. If the value is 0, value of $q$ is equal to $k$. It means that, whatever be the pressure head, the discharge remains constant. As the value of $x$ increases, the discharge variation with respect to head variation increases. Hence, it must be understood that lower value of flow exponent is better. The variation will be maximum when the flow exponent is 1.

For instance, if the discharge rate of a dripper is 4 l/h at 10 m pressure head and the flow exponent of dripper is say, 0.5, the value of discharge coefficient '$k$' can be easily found out using the dripper discharge equation as follows:

$$q = k\,h^x$$
$$4\,\text{l/h} = k \times 10^{0.5}$$
$$k = 1.27\ (\text{l/h})/\text{m}^{0.5}$$

Many manufacturers provide a relationship between pressure head and discharge rate of drippers (Fig. 20). From these relationships, two points can be selected as shown in the figure.

They can be substituted in the dripper discharge equation and two simultaneous equations can be obtained as follows:

$$q_1 = k\,h_1^x \ \& \ q_2 = k\,h_2^x$$

These two equations are solved for two unknowns $k$ and $x$ by converting them to logarithmic equations as below:

$$\log q_1 = \log k + x \log h_1 \ \& \ \log q_2 = \log k + x \log h_2.$$

**Example**

$q_1 = 4$ l/h at $h_1 = 10$ m and $q_2 = 4.2$ l/h at $h_2 = 12$ m. Find out $k$ and $x$?

$$x = \frac{\log\left(\frac{q_1}{q_2}\right)}{\log\left(\frac{h_1}{h_2}\right)} = 0.27$$

$$k = \frac{q_1}{h_1^x} = 2.14(\text{l/h})/\text{m}^x$$

## *Orifice Emitters*

This type of emitter has a very small opening. The flow exponent of this drippers is obviously 0.5 (similar to sprinkler nozzles) because Torricelli equation is the governing equation for flow through orifices.

$$\text{Torricelli equation} : q = C_d A\sqrt{2gh}.$$

where $q$ is discharge rate from orifice of cross-sectional area '$A$'. $C_d$ is coefficient of discharge approximately equal to 0.6 for small orifices and $h$ is pressure head. From the preceding equation it can be understood that the $k$ value for an orifice dripper is $C_d A(2g)^{0.5}$.

**Example**

Find out the diameter of orifice dripper for a 4 l/h dripper at 10 m pressure head? Find out the dripper discharge rate variation if the operating pressure head is increased to 12 m (20% variation in pressure head).

$$q = 4\text{l/h} = 1.11 \times 10^{-6}\,\text{m}^3/\text{s}; \ \ h = 10\,\text{m}$$

$$d = \left(\frac{4 \times 1.11 \times 10^{-6}}{0.6\pi\sqrt{2 \times 9.81 \times 10}}\right)^{0.5} = 0.00041\,\text{m}$$

$$= 0.41\,\text{mm}$$

It can be seen that the diameter obtained is very low and it gets classified in the class of very sensitive to clogging.

For the same emitter, if we change the pressure head to 12 m (20% increase from 10 m), the discharge rate from the dripper works out to be as follows:

$$q = 0.6\left(\frac{\pi}{4}\right)0.00041^2\sqrt{2 \times 9.81 \times 12} = 1.217 \times 10^{-6} \text{ m}^3/\text{s} = 4.38\,1/\text{h}$$

$$\text{The discharge variation} = \frac{q_{var}}{q} \times 100 = \frac{(4.38 - 4)}{(4 + 4.38)/2} \times 100 = 9.1\%$$

It can be seen that for a 20% variation of pressure head, the discharge rate variation is only approximately 10%.

## Vortex Emitter

Vortex emitter was developed in order to retain the goodness of the orifice dripper in terms of lower flow exponent value along with a larger diameter of flow passage than orifice emitter. In the vortex type, pressure dissipation is brought about by causing rotation of water in the dripper flow path. The '$x$' value of these drippers was found to be equal to 0.4. The total mass of the vortex dripper is the least of all the dripper types which makes this dripper more economical.

## Long Path Emitters

Long path emitters have an improvement over the orifice emitter in terms of clogging because the dissipation of pressure energy is brought about by passing the water through a narrow path for a longer distance and lose of pressure is by friction. Hence larger diameter flow passages are possible but the flow exponent value gets increased. If the flow is turbulent, $x$ value is near 0.57 and if the flow is laminar, $x$ value is near 1. The proof for these facts is as below:

For the turbulent flow, it is proved in Appendix that following equation can be used for estimating the friction loss:

$$\Delta h = \frac{K Q^{1.75}}{D^{4.75}} L$$

The term $\Delta h$ in the preceding equation can be substituted with pressure head at the inlet of dripper because the pressure head at the outlet of the dripper is zero (atmospheric). The preceding equation can be rearranged in the form of dripper discharge equation as below:

$$q = \left(\frac{D^{4.75}}{Kl}\right)^{0.57} h^{0.57}$$

For the laminar flow, it is proved in Appendix that following equation can be used for estimating the friction loss:

$$\Delta h = \frac{KQ}{D^4} L$$

The preceding equation can be rearranged in the form of dripper discharge equation as below:

$$q = \left(\frac{D^4}{Kl}\right) h$$

The discharge coefficient $k$ for laminar flow is $\left(\frac{D^4}{Kl}\right)$. As the viscosity of water depends on the temperature, due to temperature variations also there is a possibility of discharge variation in case of laminar flow.

## Tortuous Path Emitters

In tortuous path emitters, the energy loss is brought about by abrupt turns in the flow path along with friction loss. The flow path is shorter, and the diameter of flow passage is relatively larger in these emitters. The flow exponent is around 0.5–0.65.

## Pressure Compensating Dripper

Whatever is the type of dripper, if a flexible diaphragm exists at the exit of the dripper passage, and then it is called as pressure compensating dripper. The flexible diaphragm will adjust the flow passage according to the pressure so that always a constant discharge is issued out. The flow exponent of the pressure compensating dripper varies from 0 to 0.2. These drippers are costlier than the other drippers.

These drippers can be used for hilly areas with very high variations in elevation. The disadvantage of this dripper is that the flexibility of the diaphragm gets lost over a period of time and the discharge rate gets affected. So, unless necessary, better not to use this kind of dripper.

## Self-flushing Drippers

Self-flushing drippers have flexible pathways and when dirt gets accumulated in the flow path, the increase in pressure upstream of the dirt causes expansion of flow

**Fig. 21** Present-day drippers

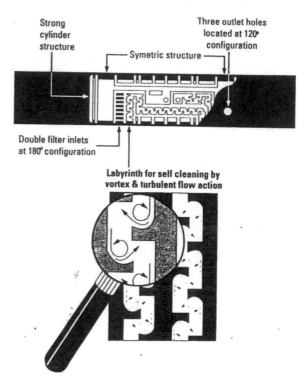

passage and hence the dirt gets removed gradually. Mostly the self-flushing emitters are pressure compensating drippers.

## Present-Day Drippers

The present-day drippers make use of all the processes such as turbulence, vortex, friction and tortuous flexible path ways to bring about pressure losses (Fig. 21.) Table 1 provides a synoptic view of the characteristics of different types of drippers.

## Variation in the Emitters Discharge Due to Manufacturing

For instance, if 2 l/h drippers of some specific type are produced, the cross-sectional area of flow cannot be produced with exactly the same diameter for all the drippers. This is due to the inherent difficulty in any production process. Therefore, value of '$k$' of any specific dripper in any bunch would have a minor variation. Usually manufacturer's provide the data regarding the values of '$k$' and '$x$' and also the

**Table 1** Characteristics of emission devices [3]

| Emission device[2] | TDR | | | | | Flushing ability |
|---|---|---|---|---|---|---|
| | $x^3$ | $CV^4$ | 113° | 149° | $MFPD^6$ | |
| Orifice | 0.42 | 0.07 | 0.92 | 0.88 | 0.024 | None |
| Vortex/orifice | 0.70 | 0.05 | 1.04 | 1.07 | – | Continuous |
| Multiple flexible orifices | 0.70 | 0.07 | 1.04 | 1.07 | – | Continuous |
| | 0.50 | 0.27 | 1.15 | 1.21 | (0.012) | Automatic |
| Ball and slotted seat | 0.49 | (0.25) | 0.83 | 0.79 | (0.012) | Automatic |
| | 0.15 | 0.35 | 0.85 | 0.81 | 0.012 | Automatic |
| Compensating ball and slotted seat | 0.25 | 0.09 | 0.90 | 0.89 | (0.012) | Automatic |
| Capped orifice sprayers | 0.56 | (0.05) | (1.03) | (1.05) | 0.04 | None |
| | 0.53 | (0.05) | (1.03) | (1.05) | 0.06 | None |
| *Long path* | | | | | | |
| Small tube | 0.70 | 0.05 | 1.08 | 1.13 | 0.039 | None |
| | 0.80 | 0.05 | 1.16 | 1.22 | 0.039 | None |
| Spiral path | 0.75 | 0.06 | 1.19 | 1.18 | 0.031 | Automatic |
| | 0.65 | 0.02 | (1.10) | (1.15) | 0.028 | None |
| Compensating | 0.40 | 0.05 | 1.19 | 1.33 | (0.030) | None |
| | 0.20 | 0.06 | 1.11 | 1.24 | (0.030) | Automatic |
| Tortuous | 0.50 | (0.08) | 1.40 | 1.70 | 0.031 | None |
| | 0.65 | 002 | 1.08 | 1.14 | (0.039) | None |
| *Short path* | | | | | | |
| Groove and flap | 0.33 | 0.02 | 1.00 | 1.00 | 0.012 | Automatic |
| Slot and disc | 0.11 | 0.10 | 1.06 | 1.08 | 0.012 | Automatic |
| *Line source* | | | | | | |
| Porous pipe | 1.0 | 0.40 | 2.70 | 3.80 | – | None |
| Twin chamber | 0.61 | 0.17 | (1.05) | (1.10) | (0.016) | None |
| | 0.47 | (0.10) | (1.04) | (1.08) | (0.016) | None |

1. Test data at a standard operating temperature of 68 °F Numbers in parentheses are estimates
2. Double entries indicate different devices of the same general type
3. Emitter discharge exponent
4. Emitter coefficient of manufacturing variation
5. Temperature-discharge ratio, the ratio of the emitter discharge at a temperature higher than 68 °F to that at 68 °F
6. Minimum flow-path dimension—not meaningful with continuous flushing

manufacturing coefficient of variation for each type of their drippers. Table 2 provides classification procedure of different classes of drippers based on the manufacturing coefficient of variation. It can be seen from the table that lower standards are set for inline drippers because in closely spaced water application, the effect of variation

**Table 2** Classification of coefficient of manufacturing variation (Vm)

| Drip and spray emitters CVs | Classification |
|---|---|
| CV < 0.05 | Excellent |
| 0.05 < CV < 0.07 | Average |
| 0.07 < CV < 0.11 | Marginal |
| 0.011 < CV < 0.15 | Poor |
| 15 < CV | Unacceptable |
| **Line source tubing CVs** | **Classification** |
| CV < 0.10 | Good |
| 0.10 < CV < 0.20 | Average |
| 0.20 < CV | Poor to unacceptable |

*Courtesy* (210-VI-NEH, October 2013)

in water distribution from drippers gets weakened as continuous wetted strip gets formed.

## Filtration

Clogging is the most undesirable aspect of drip irrigation. If sufficient care is not given in the selection of components and operation of drip irrigation, drip irrigation would become an undesirable option. But nowadays, the filtering technology combined with electronics has grown well to deal with this problem satisfactorily.

Filters are used for removing the physical impurities in the water to prevent clogging of emitters. The nature and the magnitude of impurities is essential information before filter selection is done. Table 3 shows different physical, chemical and biological impurities in water those are normally tested. Based on the magnitudes of the impurities, the water is classified into three classes. If the plugging potential of water falls in moderate and severe ranges, the filtration process must be carefully done.

Mostly algae and bacterial slime growth is significant in river and lake water. For surface water sources, water test for salinity need not be done normally. Rotten egg smell is an easy indication of hydrogen sulphide presence. Brown colour is an indication of iron presence.

Generally, all particles larger than 0.075 mm should be filtered. It is recommended that all the particles greater than 1/10th diameter of flow passage in dripper be removed if the water contains significant amount of biological impurities because individual particles tend to collect together and plug the emitters. When water contains significant amounts of particulate matter, at least the particles larger than the flow passage in emitters must be filtered.

Debris screens are simple screens at the inlet of suction pipe of pumps which prevents debris from going into the microirrigation system. When water is used from

**Table 3** Criteria for plugging potential of microirrigation water sources

| Factor | Plugging hazard based on concentration | | |
|---|---|---|---|
| | Concentrations | | |
| | Slight | Moderate | Severe |
| *Physical* | | | |
| Suspended solids (filterable) (ppm) | <50 | 50–100 | >100 |
| *Chemical* | | | |
| pH | <7.0 | 7.0–7.5 | >7.5 |
| Dissolved solids (ppm) | <500 | 500–2000 | >2000 |
| Manganese (ppm) | <0.1 | 0.1–1.5 | >1.5 |
| Iron (ppm) | <0.1 | 0.1–1.5 | >1.5 |
| Hydrogen sulphide (ppm) | <0.5 | 0.5–2.0 | >2.0 |
| Hardness[a] (ppm) | <150 | 150–300 | >300 |
| *Biological* | | | |
| Bacteria (population) | <10,000 | 10,000–50,000 | >50,000 |

[a] Hardness as ppm $CaCO_3$ [1]

surface sources such as canals, rivers and lakes water would contain silt, debris and trashes. Desilting basins are used for settling silt by allowing water in a surface tank where in velocity of flow is sluggish. Images of the desilting basin and debris screens are available in Chap. 1 on introduction of sprinklers.

Filters are ubiquitous in all the drip irrigation system installations. Filters are normally divided into four types.

1. Vortex sand separators or hydrocyclones
2. Media filter
3. Disc filter
4. Screen filter.

## Vortex Sand Separators (Hydro Cyclones)

If the water contains particles heavier than water like sand, these filters are used. The size of sand particles which are between 70 and 300 μm can be separated from water by these filters. A typical sand separator is shown in Fig. 22. In these filters, water is made to rotate, and because of the centrifugal force, sand particles are separated out. The sand particles hit on the wall of the filter and helically fall down in a collection chamber at the bottom of the filter.

When large discharge rates need to be handled, a battery of two or more sand separators are used (Fig. 23). Usually, air valves are not provided as an integral part of the filter. But it is highly recommended that two air valves are provided at the

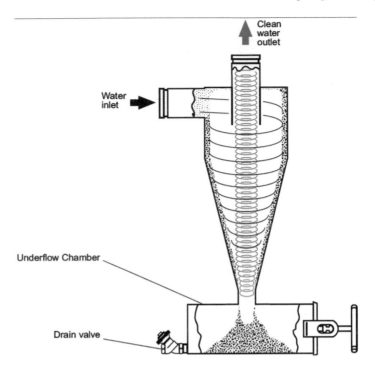

**Fig. 22** Sectional view of a vortex sand separator (www.hvi.com.au/)

**Fig. 23** Battery of vortex sand separators (www.hvi. com.au/)

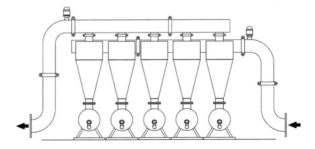

Note the air valves at the inlet and outlets of filters

inlet and outlets of the filter as shown in Fig. 23. Detailed discussion of locating air valves is available in the Chap. 1 Sprinkler irrigation types and components.

When these filters are used for discharge rates less than the rated discharge, the centrifugal force that would act on the sand particles may not be sufficient to throw the particles towards periphery and so sand separation may not be satisfactory. When vortex sand separators are used, there is a possibility of significant extent of residual impurities not getting filtered as the rate of discharge into the filters is rather a variable. Hence essentially a screen filter is used as a secondary filter to filter out the impurities that escape from it. This filter would cause a pressure head loss of 3–4 m.


## *Media Filter*

Sand filters are used when water contains primarily organic impurities like algae, slimes and fine suspended particles. In case of water with ferrous impurities, settling basins are used to convert ferrous form of iron into ferric form by the action of microbes in the atmosphere. They form colloidal slime like substances and they are usually filtered by media filters.

The process of filtration in the media filter is very simple. During normal filtration process, the water is made to pass from top to bottom through the media placed in a mild steel tank (Fig. 24). The cleaning of filters by removal of media from the filter cannot be done and so a provision is a necessity for backwashing of the filter by directing the water flow from the bottom to the top.

At least two parallel media filters are recommended because backwashing is to be done frequently to have trouble-free operation of drippers. Minimum of three filters in parallel has been a conventional optimum at present. If frequent backwashing is needed to be done, automatic backwashing provision is to be adopted (Fig. 25). Automation is discussed later in this chapter.

In media filters crushed granite or silica sand is used as filter media. The crushing process yields sharp edged particles capable of trapping linear filament like algae easily. Uniform sized particles are preferred rather than graded mixture. When graded mixture is used, there is a possibility of filling of larger pores by smaller sized particles. A uniformity coefficient of 1.5 and more is recommended. Uniformity coefficient is defined as the ratio of the size of screen that passes 60% of the mass to the size of the screen that passes only 10% of the mass.

Sometimes naturally available pea gravels of uniform size are also used but there is a slight disadvantage in such use that due to the smoothness of the particles the filtering efficiency is relatively less. Usually in media filters, air valve is an integral part due to the possibilities of heavy pressure surges during cleaning operation.

Table 4 shows the characteristics of different types of media normally used. When larger size particle is used, filtering efficiency is less and the pressure head loss between inlet and outlet of the filter is also less. The frequency of backwashing would also be less when larger size of particles is used. Therefore, a prudent grower would select the size of the particles of media in relation to the diameter of opening of the drippers. For instance, let us assume that the manufacturer's data of a dripper states that 0.75 mm is the diameter of opening of flow passage in the emitter. Then, one-tenth of diameter of opening is 0.075 mm, and #16 silica sand would be appropriate as the approximate size of aperture is 0.06 mm. Instead, if the grower opts for the most stringent filtration process by using #30 silica sand, he would incur more operating cost.

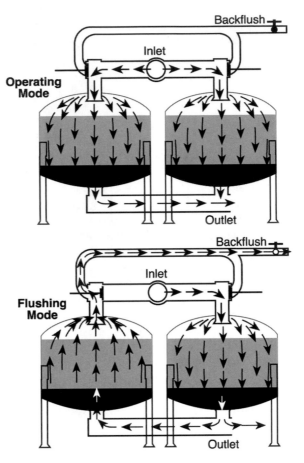

**Fig. 24** Sand filter in normal & backwash modes (https://www.ksre.k-state.edu/sdi/photos/dmc.html)

The diameter of the filter is usually larger of the order of 0.6–0.9 m. Larger diameter is needed because the recommended flow velocity during filtration is lower to improve the filtration efficiency. The recommended flow rate is between 14 and 20 l/s per m² area of the sand bed. For any specific operating pressure head, if larger diameter tanks need to be thicker than the smaller diameter tanks. This is because of the larger Hoop stress values in case of larger tanks.

Let us derive the Hoop stress equation as follows:

$$\text{Bursting Force} = \text{Tensile strength of the tank wall}$$

$$\pi r^2 P = 2\pi r t \sigma$$

where '$r$' is radius of tank (m); '$P$' is pressure (N/m²); '$t$' is thickness of the tank wall (m); $\sigma$ is tensile strength of steel N/m².

**Table 4** Characteristics of different media used in media filters

| No. | Sand media designation | Mean effective size (μm) | Screen equivalent (No of opening/inch²) | App. size of aperture (μm) | Pressure drop when clean (m) | Recommended discharge rate during back flow (l/s/m² of sand bed) |
|---|---|---|---|---|---|---|
| 1 | #8 crushed granite | 1900 | 100–140 | 160 | 1.5–2 | Material-escape less likely |
| 2 | #10 crushed granite | 1000 | 140–180 | 80 | | 17 |
| 3 | #16 silica sand | 825 | 150–200 | 60 | | 14 |
| 4 | #20 silica sand | 650 | 200–250 | 40 | 2–3.5 | 7–10 |
| 5 | #30 silica sand | 340 | 250 and more | 20 | | |

Normally the media filters are designed for a maximum pressure of $10^6$ N/m$^2$. Let us take the tensile strength of mild steel as $250 \times 10^6$ N/m$^2$. Then by applying the preceding equation the thickness of the tank needed works out to 6 mm for a 0.3 m tank radius and only 0.2 mm for a 0.1 m tank radius. Hence, media filters are usually very heavy and therefore the cost of the tank is higher.

This filter with crushed granite would cause a pressure head loss of 1.5–2 m and with silica sand, the pressure loss would be between 2 and 3.5 m.

## Disc Filters

Media filters have big and sturdy containers and they need more quantity of water for backwashing and also more duration for backwashing. In order to overcome these difficulties, the disc filter was invented.

They contain numerous discs with diagonal grooves, and the discs are placed one over the other; mechanical pressure is given by a spring-loaded nut, and narrow gaps get formed between the discs. The cross section of the gap changes along the flow direction of water between the discs (Fig. 26). Water moves radially inwards towards the centre. Due to the tortuous flow occurring in between the discs, the dirt keeps moving in between the discs and gets trapped between the discs somewhere (Fig. 27). Various types of disc filters are available to filter the dirt in the range of 25–800 μm.

Disc filters also come with a provision of backwash facility and during the process, the mechanical pressure on the discs is removed and the discs are made to rotate, and the dirt gets washed out of the discs. The disc filters are useful to filter if moderate level of organic matter exists in water. This filter may cause a pressure head loss of 2–3 m.

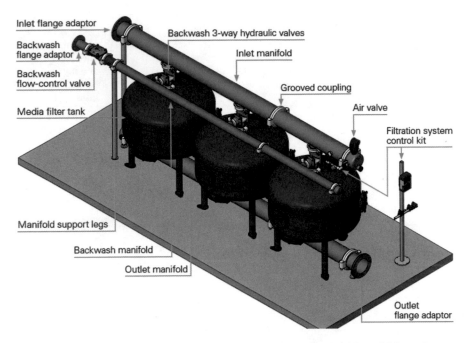

Inlet flange adaptor

Backwash 3-way hydraulic valves

Backwash flange adaptor

Inlet manifold

Backwash flow-control valve

Grooved coupling

Media filter tank

Air valve

Filtration system control kit

Manifold support legs

Backwash manifold

Outlet manifold

Outlet flange adaptor

**Fig. 25** A typical battery of sand filters with automatic backwash (www.ManualsLib.com)

## Screen Filter

Screen filters are the most commonly used type of filter and is the earliest invention. Even if the water does not contain any visible impurities, this filter is recommended. In this filter, the essential element is filter screen. Water enters into the filter body through filter element and passes through the outlet.

Nylon mesh or stainless-steel mesh is used. Nylon mesh can be made to flutter easily during backwash. The disadvantages of nylon mesh are that it does not last longer than stain less steel mesh and also the surface area needed for any specific discharge rate is approximately double that of stain less steel filter (Fig. 28).

Screen filters are prone to get clogged even when there is a slight increase in organic load. As an alternative to media filter and disc filter, screen filters also come with advanced sophistication to automatically remove filter cakes on the screen surface automatically. A typical auto rinse screen filter is shown in Fig. 29. The working of this filter is explained below:

Water entry in the filter is at the bottom and then through cylindrical screen element from inside out. Water from the inlet passes through nozzles and leaves the dirt over the screen and passes out of the screen. When the differential pressure between inlet and outlet goes above a threshold, a rinse cycle is started and is ended with in few seconds. During the rinse cycle, the rinse valve opens causing the pressure in the rinse chamber to fall down. Through the nozzles, water movement would occur from

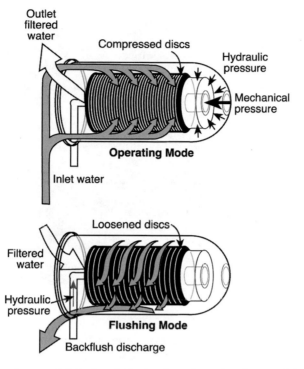

**Fig. 26** Disc filter in normal & backwash modes (https://www.ksre.k-state.edu/sdi/photos/dmc.html)

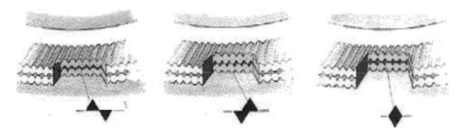

**Fig. 27** Flow passage shapes variation in between adjacent two discs (www.netafimindia.com)

**Fig. 28** A simple screen filter (https://www.ksre.k-state.edu/sdi/photos/dmc.html)

**Fig. 29** Automatic
self-cleaning filter (www.ori
val.com/case-studies/ama
zing-clarity)

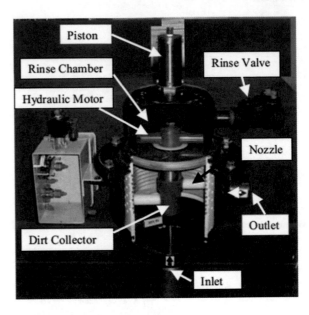

the high pressure-filtering chamber to the rinse chamber. Hydraulic motor in the rinse
chamber gets rotated due to the water flow through the rinse chamber. This rotation
causes the dirt collector also to rotate. When water flows in reverse direction with a
velocity through the screen, the deposited impurities on the inner side of the screen
goes through the nozzle and enter the rinse chamber. Pressurized water is used in
the piston on top of the filter to hold the dirt collector to its lowest position. Pressure
in the piston is released during rinse cycle. The dirt collector goes slowly up, and
during this upward movement, nozzle also is able to collect impurities throughout
the inner area of screen.

The maximum recommended flow rate for screen filters is 135 l/s per m$^2$ of screen
open area. Screen filters with various screen openings are available to filter the dirt
in the range of 70–300 µm. Screen filter may cause a pressure head loss of 1–2 m.

## Filter Selection

Table 5 can be used for the selection of filters based on the quality of the water. In
the table, recommendation is costlier than alternative but the efficiency is better for
the recommendation than the alternative.

**Table 5** Filter selection guide

| S. No. | Factors of water contaminants | Recommendation | Alternative |
|---|---|---|---|
| 1 | Sand Particles | Hydrocyclone & screen | Screen |
| 2 | Suspended solids (<50 mg/l) | Disc | Screen |
| 3 | Suspended solids (>50 mg/l) | Media & disc | Media & screen |
| 4 | Algae and organic matter (Low) | Disc | Screen |
| 5 | Algae and organic matter (High) | Media & disc | Media & screen |
| 6 | Iron & manganese (<0.5 mg/l) | Disc | Screen |
| 7 | Iron & manganese (>0.5 mg/l) | Media & screen | Disc |
| 8 | Hydrogen sulphide (<0.5 mg/l) | Disc | Screen |
| 9 | Hydrogen sulphide (>0.5 mg/l) | Media & screen | Disc |

## Hydraulics of Flow Through Filters

Following equation is used for relating discharge rate through filters ($Q$), pressure difference between inlet and outlet ($h$).

$$h = ke^{mQ}$$

where $k$ and $m$ are constants specific to the filter type and size. This data is available from manufacturers' catalogue (Source: Jain filters and Fertigation).

**Problem**

Find out the head loss through a sand filter with $k = 0.6917$, $m = 0.074$ for the discharge rate expressed in $m^3/h$ and pressure loss expressed in m. $Q = 20\ m^3/h$.

$$h = 0.6917e^{0.074 \times 20} = 3\ m$$

## Chemical Injection Devices

Fertilizers, acids, chlorine compounds, plant growth regulators, insecticides, herbicides and nematicides can be effectively applied through irrigation water through drip irrigation. There are many fertilizer injection mechanisms. They are broadly classified as follows:

1. Venturi fertilizer tank
2. Pressure differential fertilizer tank
3. Fertilizer injection pump
4. Automated fertigation systems.

## *Venturi Assembly*

A venturi has a converging section, neck (throat) and a diverging section (Fig. 30). Usually, the length of converging section is relatively smaller than the diverging section. In between the converging section and the diverging section, a throat section exists at which a pipe connection from the chemical solution to be injected joins. When water is passed through the narrow cross section, the velocity of flow increases and the pressure drops down. When the pressure at the throat section drops lower than atmosphere, chemical solution gets sucked in to the motive flow (Fig. 31). It must be noted that, if the direction of flow is reversed in the venturi, the venturi will not be able to create the suction in the throat section because when the flow is diverged in a shorter length, flow separation takes place and drop in pressure does not occur at the throat.

Venturi is fitted in the main line itself when the flows handled are very less (Fig. 32). Because of convergence and divergence of flow, pressure loss occurs between inlet and out of the venturi. Maximum venturi suction rate has been found to occur when there is a pressure loss of 50% of inlet pressure. Reduction of pressure loss more than 50% of inlet pressure does not increase the suction rate significantly (Table 6). The start of suction would occur when the pressure loss is 20–30%.

It can also be observed from the table that for different pressure heads at the outlet of venturi injector for any specific inlet pressure head, the motive flow through the venturi is approximately constant. Actually when pressure head loss between inlet

**Fig. 30** Components of venturi

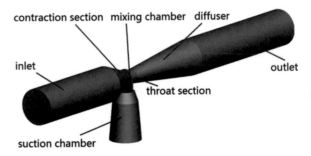

**Fig. 31** Working of venturi

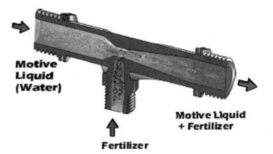

**Fig. 32** Venturi in main
line. *Courtesy* vander Gulik
and Evans (2007)

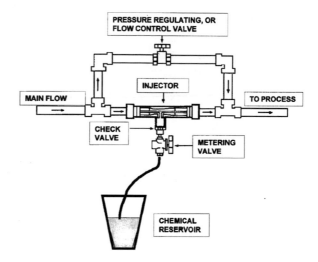

and outlet increases, the motive flow increases only to a smaller extent. Hence in the
table, a constant motive flow value is provided.

One of the difficulties faced by the farmers when using venturi injector is that
when the area irrigated is significantly less, the pressure head difference between the
inlet and outlet would not be sufficient to cause suction. In that case, inlet pressure
itself must be reduced sufficiently (mostly less than the design pressure head) to
cause the suction in venturi to happen. In such a case, the flow variation occurring
in drippers along a lateral may also be more than the allowable limits.

A simple way to increase the venturi suction rate is to place the fertilizer solution
tank at the maximum elevation possible (Fig. 33). Installation of venturi at a too higher
elevation would reduce the suction rate (Fig. 34). When the elevation of fertilizer
solution is higher than the venturi elevation, siphon effect would aid in increasing
the suction rate.

It is usual to fit the venturi in a bypass line and to create pressure difference
between inlet and outlet by a manual control valve as in Fig. 35. Fitting the venturi as
in Fig. 35b would help in marginally reducing the pressure loss compared to Fig. 35a.

Instead of providing all the filters upstream of venturi, a filter may be brought
inside the venturi loop as in Fig. 36a. This set up helps to use the inevitable pressure
loss occurring in the filter to be used for venturi suction. If the pressure is not sufficient
to cause suction additional pressure may be added by providing a very small booster
pump (Fig. 36b).

Pressure loss between inlet and outlet can also be adjusted by providing a pressure
regulating valve in main line (Fig. 36c). The use of pressure regulator in place of
manual control valve is useful if minimum operation requirements are to be met
at the control head. Venturi can also be connected between suction and delivery of
pump (Fig. 36d). This setup saves energy because the existing pressure differential
between the suction and delivery of the pump is used. Orientation of venturi must be

**Table 6** Venturi suction rates for different motive flows

| Operating pressure | | Model ¾″ × 0.9 | | Model 2″ × 12 | |
|---|---|---|---|---|---|
| Injector inlet (m) | Injector outlet (m) | Motive flow (l/h) | Suction flow (l/h) | Motive flow (m³/h) | Suction flow (l/h) |
| 14 | 3 | 522 | 215 | 6.7 | 1170 |
| | 7 | | 121 | | 905 |
| | 8 | | 78 | | 735 |
| | 10 | | – | | 282 |
| 21 | 3 | 636 | 190 | 79 | 1190 |
| | 7 | | 190 | | 1066 |
| | 10 | | 138 | | 1080 |
| | 14 | | 54 | | 590 |
| 28 | 3 | 726 | 176 | 9 | 1180 |
| | 7 | | 176 | | 1073 |
| | 10 | | 176 | | 1081 |
| | 14 | | 162 | | 1075 |
| | 17 | | 66 | | 864 |
| | 21 | | – | | 105 |
| 35 | 7 | 817 | 167 | 9.7 | 1106 |
| | 10 | | 167 | | 1091 |
| | 14 | | 167 | | 1069 |
| | 17 | | 167 | | 999 |
| | 21 | | 95 | | 643 |
| | 24 | | 19 | | 1071 |
| 42 | 7 | 885 | 162 | 10.8 | 1102 |
| | 14 | | 162 | | 1099 |
| | 21 | | 158 | | 1101 |
| | 24 | | 99 | | 982 |
| | 28 | | 44 | | 728 |
| 49 | 7 | 953 | 158 | 11.5 | 1121 |
| | 14 | | 158 | | 1123 |
| | 21 | | 157 | | 1124 |
| | 24 | | 157 | | 1115 |
| | 28 | | 127 | | 1124 |
| | 31 | | 61 | | 1075 |
| | 35 | | 9 | | 706 |
| | 38 | | – | | 213 |

**Fig. 33** Fertilizer solution tank higher than venturi elevation

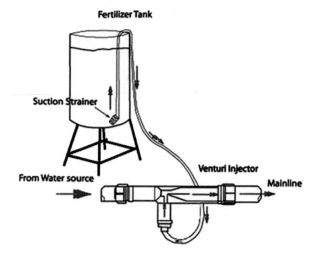

**Fig. 34** Too high venturi installation

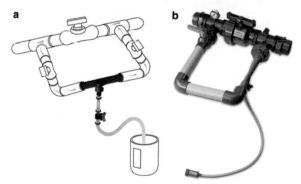

**Fig. 35** Venturi in bypass

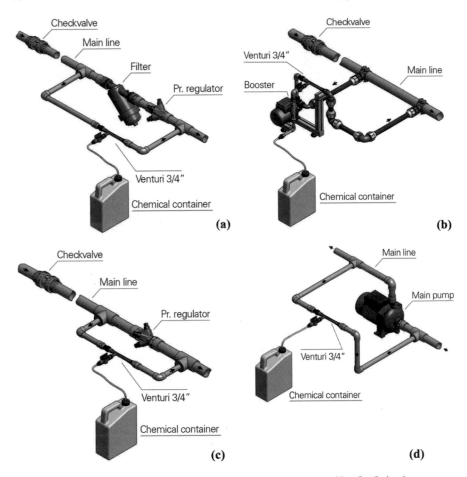

**Fig. 36** Different configurations of venturi in a bypass line. *Courtesy* Netafim Irrigations

horizontal rather than vertical and vertical installations are reported to have problems in uniform suction rate.

## *Pressure Differential Fertilizer Tank*

A closed tank containing fertilizer is placed in a bypass line as in Fig. 37a. The bypass is brought about by a control valve in the main line, which creates pressure gradient between the entry and exit of the fertilizer tank. Note that inlet pipe into the

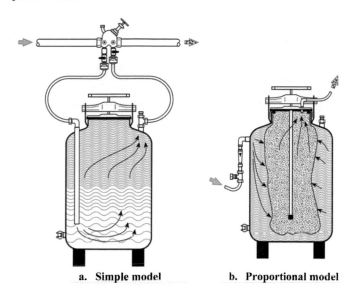

a.  **Simple model**                    b.  **Proportional model**

**Fig. 37**  Pressure differential fertilizer tank. *Courtesy* ODIS Irrigation Equipments

fertilizer tank extends until the bottom of the tank. This provision ensures complete removal of fertilizer solution from the tank.

### Example
Find out the rate of discharge through fertilizer tank (as in Fig. 37a) if the pressure head at the inlet is 14 m and at the outlet is 12 m. The inner diameter of pipe leading to fertilizer tank is 14 mm. Total length of inlet pipe is 1.2 m and outlet pipe is 0.8 m. There are two valves: one for inlet and another for outlet. There are two bends: one in inlet line and another in out let line. If we apply Bernoulli's equation between inlet (1) and outlet (2), we get the following equation:

$$z_1 + \left(\frac{v^2}{2g}\right)_1 + h_1 = z_2 + \left(\frac{v^2}{2g}\right)_2 + h_2 + h_f + \sum h_m^i$$

where $h_f$ is energy loss due to friction in 14 mm pipe. Friction loss would be negligible for the flow within the tank and $\sum h_m^i$ is total minor loss of energy due to a contraction at the inlet within the main pipe ($k = 0.5$), an inlet valve ($k = 0.2$), a bend in inlet ($k = 1.0$), an expansion into the tank ($k = 1.0$), a contraction at the outlet inside the tank ($k = 0.5$), a bend in the outlet ($k = 1.0$), a valve in the outlet ($k = 0.2$) and an expansion at the outlet in the main pipe ($k = 1$). $k$ value in the bracket are minor loss coefficients.

The inlet and outlet are at the same elevation and hence gravitational head difference is zero. Velocity of flow at inlet and outlet are the same and hence kinetic head difference is zero. Hence the preceding equation becomes as follows:

$$h_1 = h_2 + h_f + \sum h_m^i$$

$$\sum h_m^i = 82627 \frac{Q^2}{D^4} \sum k$$

$$\sum k = 0.5 + 0.2 + 1.0 + 1.0 + 0.5 + 1.0 + 0.2 + 1.0 = 5.4$$

$$\sum h_m^i = \frac{5.4 \times 82627 Q^2}{D^4}$$

It is customary to use an approximate form of friction loss equation in pipes where in the exponent for $Q$ is approximated to 2 and exponent for $D$ is approximated to 5. This assumption helps in doing the calculation easily without incorporation of much error.

$$h_f = 789000 \frac{Q^2}{D^5} L$$

$$14 = 12 + \frac{789000}{14^5} \times 2 + \frac{5.4 \times 82627 Q^2}{14^4}$$

$$Q = 0.370 \, l/s$$

Manufacturers provide the data of flow rate through the fertilizer tank for their pipe and valve connection configurations. A sample data is in Table 7.

Fertilizer tank can also be treated similar to the drippers and we can fit equations similar to dripper discharge rate equations. For the data in Table 7, following equations can be fitted using any spread sheet software. The equations obtained are $q = 676\Delta h^{0.5}$ for diameter of connecting pipe of 0.5 inches and $q = 327\Delta h^{0.6}$ for diameter of connecting pipe of 3/8 inches. It is interesting to observe that 0.5 inch pipe connection behaves as orifice type dripper and 3/8 inch pipe connection behaviour tends towards long path dripper.

Following mass balance equation can be written for the fertigation tank (Fig. 38):

**Table 7** Typical flow rates through fertilizer tanks [6]

| No. | Pressure head difference (m) | For diameter of connecting pipe = 0.5 inches (l/h) | For diameter of connecting pipe = 3/8 inches (l/h) |
|-----|------------------------------|----------------------------------------------------|----------------------------------------------------|
| 1   | 1                            | 660                                                | 320                                                |
| 2   | 2                            | 990                                                | 500                                                |
| 3   | 3                            | 1200                                               | 650                                                |
| 4   | 4                            | 1350                                               | 760                                                |
| 5   | 5                            | 1500                                               | 850                                                |
| 6   | 6                            | 1650                                               | 940                                                |
| 7   | 7                            | 1800                                               | 1030                                               |

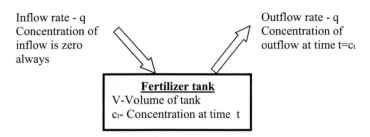

Inflow rate - q
Concentration of
inflow is zero
always

Outflow rate - q
Concentration of
outflow at time t=$c_t$

**Fertilizer tank**
V-Volume of tank
$c_t$- Concentration at time  t

**Fig. 38**  Schematic of flows through a fertilizer tank

Rate of change of mass of fertilizer $=$ Inflow mass rate $-$ Outflow mass rate

It can be seen that the inflow mass rate is zero. Hence the differential equation for mass transport with the assumption of complete mixing in the tank is as follows:

$$\frac{V \, dc_t}{dt} = -qc_t$$

where $V$—volume of tank; $c_t$—concentration at time $t$ and $q$—flow rate through the fertigation tank. Integrating the preceding equation from time zero to any time t and after substituting the limits, we get the following equation:

$$t = \frac{V}{q} \ln\left(\frac{C_0}{C_t}\right)$$

where $C_0$ is the concentration inside the tank at time zero.

**Example**
A 30 L fertilizer tank is filled with an initial concentration of 100 mg/l. A pressure difference of 2 m between inlet and outlet causes a discharge rate of 990 l/h through fertilizer tank. Find out the time of operation needed to bring the concentration of fertilizer solution in the tank to 1 mg/l.

$$t = \frac{30}{990} \ln\left(\frac{100}{1}\right)$$

$$t = 0.140 \text{ h}$$

$$t = 8.40 \text{ min}$$

Volume of water that would pass through the tank in 8.4 min is 138 L.

When fertilizer tank is used, the concentration of fertilizer solution will be the maximum at the start and tend to zero as time increases. Normally, for all the fertilizer solution to get out of the tank, it takes 4 to 10 times volume of fertilizer tank displacement through the tank. If liquid fertilizers are used, the volume displacements needed is around 4. If solid fertilizers are used, the volume displacements

**Table 8**  Minimum operation time for fertilizer tank (minutes)

| No. | Pressure head difference between inlet and outlet (m) | Tank volume (litres) | | | |
|-----|---|---|---|---|---|
| | | 30 | 60 | 90 | 120 |
| 1 | 2 | 12 | 15 | 25 | 30 |
| 2 | 3 | 9 | 10 | 15 | 15 |
| 3 | 5 | 7 | 8 | 10 | 10 |

needed is around 10. The data regarding minimum operation time needed are also given by manufacturers (Table 8).

If the total quantity of fertilizer applied only is the criterion in fertigation, the setup as in Fig. 37a is sufficient. If concentration of fertilizer solution is to be maintained at an approximately constant level, a minor modification in the fertilizer tank would help achieve it. Figure 37b shows a flexible bag in the tank in which fertilizer would be placed. The water that enters into the tank dissolves the fertilizer and joins the main pipe. The concentration of outflow from fertilizer tank is governed by the solubility characteristics of the fertilizer. This provision would greatly improve the uniformity of concentration of fertilizer through out the fertigation.

## *Fertilizer Injection Pump*

Fertilizer injection pumps are either reciprocating or diaphragm pumps. These pumps operate with the energy of flowing water in main pipe itself. It does not need any external source of power. The concentration of fertilizer in irrigation water can be maintained relatively more accurate and hence they are very much used in green houses. Figure 39 shows a reciprocating injector unit along with the typical accessories. The minimum pressure requirements in the main line are to be met for satisfactory operation the pump. These are costly and needs relatively more maintenance compared to the other methods of fertigation.

## *Pipes*

Pipe section which carries water from control unit to field is called as Main pipe. The pipeline connecting the Main for each field is called as Submain or Manifold. Usually, just before the connection of any Submain, a control valve and an Air valve need to be provided. Additionally, a pressure control valve and pressure gauge along with a secondary screen filter would also help improve the performance of the system (Fig. 40).

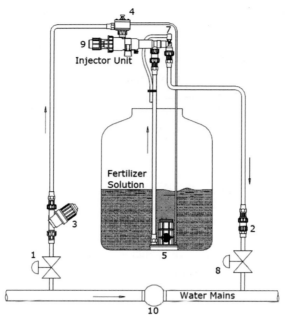

1.  Drive water hand valve
2.  End connector
3.  Filter
4.  Automatic cut-out
5.  Suction head
7.  Air-release valve
8.  Injection line hand valve
9.  Water exhaust
10. Check valve (optional)

**Fig. 39**  Fertigation pump (www.amiad.com)

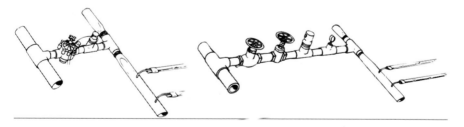

**Fig. 40**  Main to submain connection options. *Courtesy* Roberts Irrigation Products Inc.

Laterals are the pipes connected with the Submain. A Submain along with a group of Laterals and Drippers is called as one Subunit. A Subunit can be operated separately independent of the other subunits.

Main and Submains are usually PVC pipes. PVC pipes are specified by external diameter sizes of 20 mm, 25 mm, 30 mm, etc. Usually 0.25 MPa to 4 MPa strength of pipes are used for drip irrigation. The Bureau of Indian Standards (BIS) for different pipe strengths is provided in Table 9.

The PVC pipes have to be buried under the land. It should not be laid over the ground. If it is laid over the ground, it gets affected by ultra violet rays of the Sun and it will become brittle and will get broken in a few years.

Sometimes High-Density Poly Ethylene (HDPE) pipes are also used. They can be laid over the land. But they are costly compared to the PVC pipes.

Laterals are always made of Linear Low-Density Polyethylene (LLDPE) pipes. The laterals are laid over the ground in surface drip irrigation and laid below the ground in subsurface drip irrigation.

Holes are drilled in the submains with drill bits. Then a start washer is inserted in to the hole and a start connector is driven into the washer using a wooden hammer. The lateral pipe is inserted over the start connector. The laterals are closed at the end by folding the lateral pipes at the ends and fixing an accessory called as lateral end block.

Lateral pipes are specified with outer diameters. They are available in sizes of 8 mm, 12 mm, 16 mm, 20 mm and 25 mm. Lateral pipes are available with pressure rating of 0.1 Mpa, 0.125 Mpa, 0.25 Mpa and 0.4 Mpa. The Bureau of Indian Standards (BIS) for different emitting pipe strengths is provided in Table 10.

## PVC Fittings & Accessories

Different pipe fittings are shown in Fig. 41. Normally the delivery ends of pumps are Galvanized Iron (GI) pipes. GI connections are threaded. All the DIS fittings are made of PVC and these connections are made using an adhesive known as solvent cement. So, in order to connect DIS with the delivery end of pump either an FTA (Female Thread Adaptor) or an MTA (Male Thread Adaptor) is needed. If the GI delivery end has an external thread, FTA is connected. FTA, on one side has internal thread. On the other side, PVC pipe can be fitted by solvent cement. If the GI delivery end has an internal thread, MTA is used. The use of bends, tees are obvious. Reducer is used to connect pipes with different diameters. The saddle clamps are used for taking off small outlets from a large diameter pipe.

Union is useful at places where it is expected that a pipe section needs to be attended for a repair work and also in looped pipes as is available in backwash provisions of filters. For fitting pressure gauges and for taking off -takes for fertigation devices, saddle clamp is used.

In Fig. 42, lateral accessories are shown. The uses of them are obvious from the names. The goof plug is used to close the unwanted holes in the lateral.

**Table 9** Dimensions of unplasticized PVC pipes

(Clauses 7.1.1. and 7.1.2) All dimensions in millimetres

| Nominal outside diameter | Mean outside diameter | | Outside diameter at any point | | Working pressure, MPa | | | | | | | | | | | | | | | | | | |
|---|---|---|---|---|---|---|---|---|---|---|---|---|---|---|---|---|---|---|---|---|---|---|---|
| | | | | | Class 1 0.25 | | | Class 2 0.40 | | | Class 3 0.60 | | | Class 4 0.80 | | | Class 5 1.00 | | | Class 6 1.25 | | |
| | Min | Max | Min | Max | Avg max | Min | Max | Avg max | Min | Max | Avg max | Min | Max | Avg Max | Min | Max | Avg Max | Min | Max | Avg Max | Min | Max |
| (1) | (2) | (3) | (4) | (5) | (6) | (7) | (8) | (9) | (10) | (11) | (12) | (13) | (14) | (15) | (16) | (17) | (18) | (19) | (20) | (21) | (22) | (23) |
| 20 | 20.0 | 20.3 | 19.5 | 20.5 | | | | | | | | | | | | | 1.5 | 1.1 | 1.5 | 1.8 | 1.4 | 1.8 |
| 25 | 25.0 | 25.0 | 24.5 | 25.5 | | | | | | | | | | 1.6 | 1.2 | 1.6 | 1.8 | 1.4 | 1.8 | 2.1 | 1.7 | 2.1 |
| 32 | 32.0 | 32.3 | 31.5 | 32.5 | | | | | | | | | | 1.9 | 1.5 | 1.9 | 2.2 | 1.8 | 2.2 | 2.7 | 2.2 | 2.7 |
| 40 | 40.0 | 40.3 | 39.5 | 40.5 | | | | | | | 1.8 | 1.4 | 1.8 | 2.2 | 1.8 | 2.2 | 2.7 | 2.2 | 2.7 | 3.3 | 2.8 | 3.3 |
| 50 | 50.0 | 50.3 | 49.4 | 50.6 | | | | | | | 2.1 | 1.7 | 2.1 | 2.8 | 2.3 | 2.8 | 3.3 | 2.8 | 3.3 | 4.0 | 3.4 | 4.0 |
| 63 | 63.0 | 63.3 | 62.2 | 63.8 | | | | 1.9 | 1.5 | 1.9 | 2.7 | 2.2 | 2.7 | 3.3 | 2.8 | 3.3 | 4.1 | 3.5 | 4.1 | 5.0 | 4.3 | 5.0 |
| 75 | 75.0 | 75.3 | 74.1 | 75.9 | | | | 2.2 | 1.8 | 2.2 | 3.1 | 2.6 | 3.1 | 4.0 | 3.4 | 4.0 | 4.9 | 4.2 | 4.9 | 5.9 | 5.1 | 5.9 |
| 90 | 90.0 | 90.3 | 88.9 | 91.1 | 1.7 | 1.3 | 1.7 | 2.6 | 2.1 | 2.6 | 3.7 | 3.1 | 3.7 | 4.6 | 4.0 | 4.6 | 5.7 | 5.0 | 5.7 | 7.0 | 6.1 | 7.1 |
| 110 | 110.0 | 110.4 | 108.6 | 111.4 | 2.0 | 1.6 | 2.0 | 3.0 | 2.5 | 3.0 | 4.3 | 3.7 | 4.3 | 5.6 | 4.9 | 5.6 | 7.0 | 6.1 | 7.1 | 8.5 | 7.5 | 8.7 |
| 125 | 125.0 | 125.4 | 123.5 | 126.5 | 2.2 | 1.8 | 2.2 | 3.4 | 2.9 | 3.4 | 5.0 | 4.3 | 5.0 | 6.4 | 5.6 | 6.4 | 7.8 | 6.9 | 8.0 | 9.6 | 8.5 | 9.8 |
| 140 | 140.0 | 140.5 | 138.3 | 141.7 | 2.4 | 2.0 | 2.4 | 3.8 | 3.2 | 3.8 | 5.5 | 4.8 | 5.5 | 7.2 | 6.3 | 7.3 | 8.7 | 7.7 | 8.9 | 10.7 | 9.5 | 11.0 |
| 160 | 160.0 | 160.5 | 158. | 162.0 | 2.8 | 2.3 | 2.8 | 4.3 | 3.7 | 4.3 | 6.3 | 5.4 | 6.2 | 8.2 | 7.2 | 8.3 | 9.9 | 8.8 | 10.2 | 12.2 | 10.9 | 12.6 |
| 180 | 180.0 | 180.6 | 177.8 | 182.2 | 3.1 | 2.6 | 3.1 | 4.9 | 4.2 | 4.9 | 7.0 | 6.1 | 7.1 | 9.0 | 8.0 | 9.2 | 11.1 | 9.9 | 11.4 | 13.7 | 12.2 | 14.1 |
| 200 | 200.0 | 200.6 | 197.6 | 202.4 | 3.4 | 2.9 | 3.4 | 5.3 | 4.6 | 5.3 | 7.7 | 6.8 | 7.9 | 10.0 | 8.9 | 10.3 | 12.3 | 11.0 | 12.7 | 15.2 | 13.6 | 15.7 |
| 225 | 225.0 | 225.7 | 222.3 | 227.7 | 3.9 | 3.3 | 3.9 | 6.0 | 5.2 | 6.0 | 8.6 | 7.6 | 8.8 | 11.2 | 10.0 | 11.5 | 13.9 | 12.4 | 14.3 | 17.1 | 15.3 | 17.6 |
| 250 | 250.0 | 250.8 | 247.0 | 253.0 | 4.2 | 3.6 | 4.2 | 6.5 | 5.7 | 6.5 | 9.6 | 8.5 | 9.8 | 12.4 | 11.2 | 12.9 | 15.4 | 13.8 | 15.9 | 18.9 | 17.0 | 19.6 |

(continued)

**Table 9** (continued)

(Clauses 7.1.1. and 7.1.2) All dimensions in millimetres

| Nominal outside diameter | Mean outside diameter | | Outside diameter at any point | | Working pressure, MPa | | | | | | | | | | | | | | | | |
|---|---|---|---|---|---|---|---|---|---|---|---|---|---|---|---|---|---|---|---|---|---|---|
| | | | | | Class 1 0.25 | | | Class 2 0.40 | | | Class 3 0.60 | | | Class 4 0.80 | | | Class 5 1.00 | | | Class 6 1.25 | | |
| | Min | Max | Min | Max | Avg max | Min | Max | Avg max | Min | Max | Avg min | Min | Max | Avg Max | Min | Max | Avg Max | Min | Max | Avg Max | Min | Max |
| 280 | 280.0 | 280.9 | 276.6 | 283.4 | 4.8 | 4.1 | 4.8 | 7.3 | 6.4 | 7.4 | 10.7 | 9.5 | 11.0 | 14.0 | 12.5 | 14.4 | 17.2 | 15.4 | 17.8 | 21.1 | 19.0 | 21.9 |
| 315 | 315.0 | 316.0 | 311.2 | 318.8 | 5.3 | 4.6 | 5.3 | 8.2 | 7.2 | 8.3 | 12.0 | 10.7 | 12.4 | 15.6 | 14.0 | 16.1 | 19.3 | 17.3 | 19.9 | 23.8 | 21.4 | 24.7 |
| 355 | 355.0 | 356.1 | 350.7 | 359.3 | 5.9 | 5.1 | 5.9 | 9.2 | 8.1 | 9.4 | 13.4 | 12.0 | 13.8 | 17.6 | 15.8 | 18.2 | 21.8 | 19.6 | 22.6 | 26.8 | 24.1 | 27.8 |
| 400 | 400.0 | 401.2 | 395.2 | 404.8 | 6.6 | 5.8 | 6.7 | 10.3 | 9.1 | 10.5 | 15.1 | 13.5 | 15.6 | 19.8 | 17.8 | 20.5 | 24.4 | 22.0 | 25.3 | 30.2 | 27.2 | 31.3 |
| 450 | 450.0 | 451.4 | 444.6 | 455.4 | 7.4 | 6.5 | 7.5 | 11.6 | 10.3 | 11.9 | 17.0 | 15.2 | 17.5 | 22.2 | 20.0 | 23.0 | 27.5 | 24.8 | 28.6 | 33.8 | 30.5 | 35.1 |
| 500 | 500.0 | 501.5 | 494.0 | 506.0 | 8.2 | 7.2 | 8.3 | 12.8 | 11.4 | 13.2 | 18.8 | 16.9 | 19.5 | 24.8 | 22.3 | 25.7 | 30.5 | 27.5 | 31.7 | 37.5 | 33.9 | 39.0 |
| 560 | 560.0 | 561.7 | 553.2 | 566.8 | 9.2 | 8.1 | 9.4 | 14.3 | 12.8 | 14.8 | 21.0 | 18.9 | 21.8 | 27.6 | 24.9 | 28.7 | 34.1 | 30.8 | 35.5 | 42.0 | 38.0 | 43.7 |
| 630 | 630.0 | 631.9 | 622.4 | 637.6 | 10.3 | 9.1 | 10.5 | 16.1 | 14.4 | 16.6 | 23.7 | 21.3 | 24.5 | 31.0 | 28.0 | 32.2 | 38.4 | 34.7 | 40.0 | 47.2 | 42.7 | 49.2 |

Notes

1. The table is based on metric series of pipe dimensions given in ISO 161/I in respect of pipe dimensions and ISO DIS 4422

2. The wall thickness of pipes is based on a safe working stress of 8.6 MPa at 27 °C and the working pressure gets reduced at sustained higher temperatures. Occasional rise in temperature as in summer season with concurrent corresponding reduction in temperature during nights has no deleterious effect on the life working pressure of the pipes considering the total life of pipes

For classes 1, 2 and 3 of all sizes, this requirement need not to be satisfied as the ratio of minimum wall thickness to nominal outside diameter does not exceed 0.035 in these cases

**Table 10** Dimensions of polyethylene emitting pipes

Clauses 6.1.1., 8.3.1.1. and 8.3.2.1
All dimensions in millimetres

| S. No. | Nominal diameter | Inside diameter | Tolerance on I.D | Wall thickness | | | |
|--------|-------------------|------------------|-------------------|----------------------|----------------------|----------------------|----------------------|
| | | | | Class 1 0.100 MPa | Class 2 0.125 MPa | Class 3 0.250 MPa | Class 4 0.400 MPa |
| (1) | (2) | (3) | (4) | (5) | (6) | (7) | (8) |
| (i) | 12 | 10.50 | +0.20 | 0.4–0.5 | 0.6–0.7 | 0.8–1.0 | 1.1–1.3 |
| | | | −0.00 | | | | |
| (ii) | 16 | 14.20 | +0.20 | 0.5–0.6 | 0.7–0.9 | 1.0–1.2 | 1.3–1.5 |
| | | | −0.00 | | | | |
| (iii) | 20 | 18.00 | +0.20 | 0.7–0.8 | 0.9–1.1 | 1.2–1.4 | 1.5–1.7 |
| | | | −0.00 | | | | |
| (iv) | 25 | 22.60 | +0.20 | 0.9–1.1 | 1.2–1.4 | 1.5–1.7 | 1.8–2.2 |
| | | | −0.00 | | | | |

*Note* The wall thicknesses of pipes are based on safe working stress of 2.5 MPa at 20 °C. Occasional rise in temperature has no deleterious effects on the life and working pressure of the pipe. Clamping such as bands and screws, shall be of non-corrosive materials or of materials protected against corrosion

| Female Thread Adaptor | Male Thread Adaptor | L Bend | Long Bend |
|---|---|---|---|

| Tee | Reducer Tee | Reducer | Union |
|---|---|---|---|

| Saddle | End Cap | Flush Valve | Threaded Nipple |
|---|---|---|---|

**Fig. 41** PVC fittings & accessories (www.jains.com)

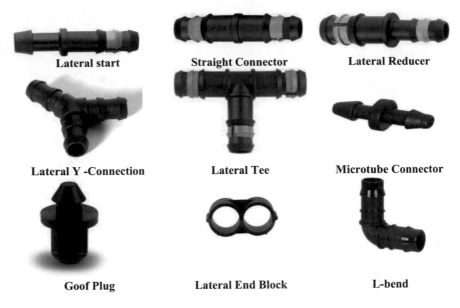

**Fig. 42** Lateral accessories

## *Pump*

A detailed account of knowledge on pump for the design and operation of drip irrigation and sprinkler irrigation was discussed in an earlier chapter. General recommended pressure head needed at the control head upstream of filters is 15–20 m. In places where groundwater fluctuation is very significant during the days of monsoon, due to rise of ground water table sometimes pressure developed would rise above the design pressure. In such cases, provision of bypass valve as shown in Fig. 43 would help in diverting a part of delivery water back to the well itself. If GI pipe fittings are

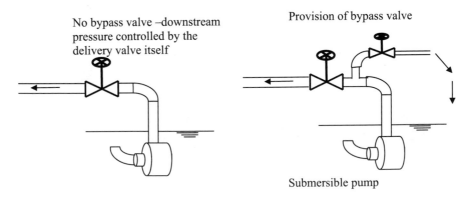

**Fig. 43** Provision of bypass valve

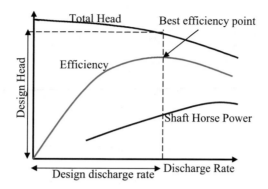

**Fig. 44** Typical operating characteristics of centrifugal pumps

used in upstream of delivery valve, it is advised not to opt for bypass mechanism. This is because, electric power consumed would be more if we pump more water. From Fig. 44, it can be seen that as the discharge rate increases the shaft power also increases. Instead, if we throttle the delivery discharge rate using the delivery valve, the problem can be managed well. This process would increase the pressure in the casing of pump. A small problem that may arise when the pressure inside casing is increased is that the packing ropes may show water leaks.

## Automation Components

Automation is a niche area in the field of irrigation systems management due to its immense advantages. Different components and systems used in automation are sequentially presented here.

## *Relay Circuit*

Normally, the electric motorized pumps and control valves are operated by manual operations. When automated, pumps as well as valves are switched ON and OFF by relay switches (Fig. 45). Relay switches are contained in a separate circuit working with direct current (DC) or alternating current (AC). The potential of the electric current in the relay circuit is very low. Mostly 12 V DC supply or 25 V AC supply is used. Due to the low voltage, these circuits do not cause any electric shock to the user. The relay circuits are electrically isolated from the primary circuits which are to be put ON and OFF. The relay circuit essentially consists of a soft iron core wound over with a copper coil which is called as relay. When this relay is supplied with electric power, the relay gets magnetized and attracts the moveable arm of the

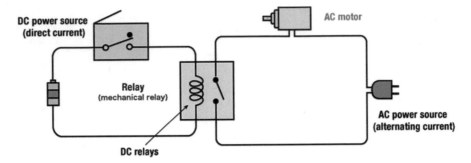

**Fig. 45** A simple relay switching circuit for automatic operation of motors

switch and contact of the switch in the primary circuit gets enabled. After the switch establishes the contact, the power supply already in the primary circuit gets closed and hence the electric motor in the primary circuit would run. Impulse for enabling the flow of current in the relay circuit comes from timers or sensors. Timers turn on/off the power in the relay circuit after a preset time is reached. Sensors like soil moisture sensors or water level sensors turn on the power when a preset level of soil moisture or water level is reached.

The relay in Fig. 45 is normally called as electro-mechanical relay and has few disadvantages. Because of the presence of moving parts, the life of the relay gets reduced due to wear and tear. Also an electric arc is produced during the operation, and it oxidizes the contacts due to temperature increase. Though this kind of electro-mechanical relays are used even now in many applications, solid-state electronic relays are also used for the same purpose. The solid-state electronic relays are made of semiconductor materials, and they do not have any moving part but they switch ON and OFF the primary circuit when the solid state electronic relay is energized with a low voltage electric power (Fig. 46).

**Fig. 46** Solid-state relay

## *Solenoid Valve*

The principle of operation of a solenoid valves is similar to a electro-mechanical relay. A solenoid is an electromagnet with a moveable plunger at the centre (Fig. 47). When electric power is supplied to the coil windings in the solenoid, the plunger will be attracted upwards and the gate stopping the flow will get opened up. If the solenoid valve opens the gate in the water flow line directly, then those valves are called as directly operated solenoid valves. Figure 48 shows a solenoid valve with a controller controlling only one valve. This controller can be programmed to open the valve on specific days for a specific duration. For instance, Monday and Thursday 8 A.M. to

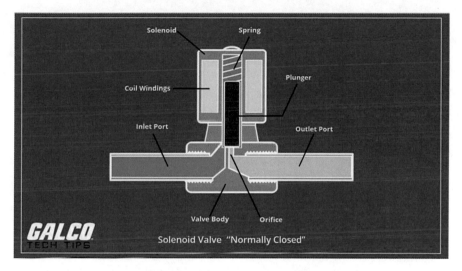

**Fig. 47**   Direct acting solenoid valve—sectional view

**Fig. 48**   Direct acting solenoid valve with controller (www.hunterind ustries.com)

**Fig. 49**  Pilot operated solenoid valve—external view

10 A.M. Whether water flows through the pipe or not, the controller will open up the valve as per our specified requirement.

In larger pipe diameters, pilot-operated solenoid valves are used (Fig. 49). Valve is in closed state when solenoid is at rest. B is an elastic diaphragm and a spring load pushes the diaphragm always down. Water enters through a central pinhole in the diaphragm and occupies the chamber C above the diaphragm. During the closed position, the solenoid blocks the water in the Chamber C from going to the outlet of valve (F). The diaphragm is so designed that the total force on upper side due to spring load and water on the upper side of diaphragm is slightly higher than the total force on the lower side of diaphragm during closed position (Fig. 50).

During opening of the valve, electric current is supplied to the solenoid. This raises the plunger of the solenoid due to a magnetic force developed in the solenoid. Then water from the Chamber C passes out through the passage D to F. Outflowing quantity of water from chamber C is more than the inflowing quantity of water through pinhole of the diaphragm. Partial emptiness in chamber C occurs which reduces the total downward force on the upper side of diaphragm. Therefore, the diaphragm totally gets lifted up and provides direct entry of water from inlet A to outlet F.

For closing the valve, electric current to the solenoid is cut-off. This process closes the flow path D to F. Water again gets filled in chamber C and downward force on the upper side of the diaphragm increases to bring down the diaphragm. This completes the closing operation of the valve.

From the preceding discussion it is obvious that at least a certain specified minimum pressure is necessary for lifting the diaphragm of pilot operated solenoid valve. If the pressure of water in the pipe is not sufficient, pilot operated valves cannot operate.

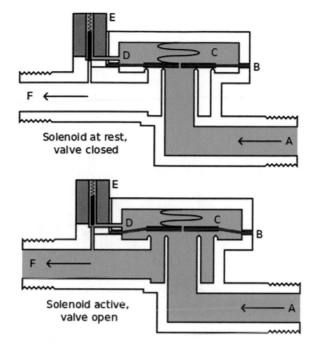

**Fig. 50** Pilot operated solenoid valve-sectional view (Wikipedia—solenoid valve)

## *Automatic Metering Valve*

Automatic metering valves do not need any electricity and they are purely mechanical devices. They are inbuilt with a water meter and a mechanically operated diaphragm valve. Instead of a solenoid, a simple mechanical valve is operated mechanically by getting impulse from the water meter. The user has to specify a flow in the metering valve. To start with, the valve has to be opened for operation. After the specified flow passes through the valve, the impulse from the water meter would trigger a bypass (black small diameter pipe in Fig. 51) so that water from inlet side would go to the upper side of the diaphragm. Since the upper side of the diaphragm has an additional downward pressure due to a spring, the valve would get closed.

Controlling the valves individually with the stand-alone valves without any master controller is easily possible when we get water supply from over-head tanks because closure of valves would not increase the pressure in the pipe line beyond limits. In case, pumps exist in water line when all the valves are closed and if the pump is ON, there is a possibility of bursting of pipes due to pressure increase beyond the limits. When the pressure near the pump gets increased beyond the limit, the power supply to the pump must be put OFF. For such situations, a separate pump start/stop circuit must be used (Fig. 52).

**Fig. 51** Automatic metering
valve (www.bermad.com)

Quick response and reliable pressure relief valves are also available in the market, and they can be used to relieve the pressure when the pressure goes beyond the limit (Fig. 53).

## Pressure Management Valves

As drip and sprinkler irrigation work under pressure, most of the time it is advisable to maintain desired pressure heads in the pipelines. When pressure at the control head rises above the design pressure, bypass valves are used to sustain desired pressure in the system. Automatic pressure sustaining valves are useful as bypass valves. Pressure-sustaining valves are essential in automatic fertigation systems. Pressure-reducing valves are used when we want to maintain a desired pressure in downstream side of a valve. Pressure-reducing valves are very much useful in highly undulating areas like hilly areas (Figs. 54, 55 and 56).

In the pressure management valves, the cross-sectional area of opening of valve is altered by supplying a variable quantity of water on the upper side of diaphragm (Fig. 55). A small diameter pipe from downstream side of the valve gets connected to a flow restriction passage. From the flow restriction passage water gets divided into two directions. One way connects to upper side of the diaphragm and another way passes through a pilot control and to the downstream side of the valve. At the flow restriction and pilot control, controlling knobs are available to fix-up the size of openings during the operation. At any point of time, if water moves from the flow restriction towards the upper side of diaphragm, the valve would tend to close and if water comes out from the upper side of the diaphragm, the valve would tend to open.

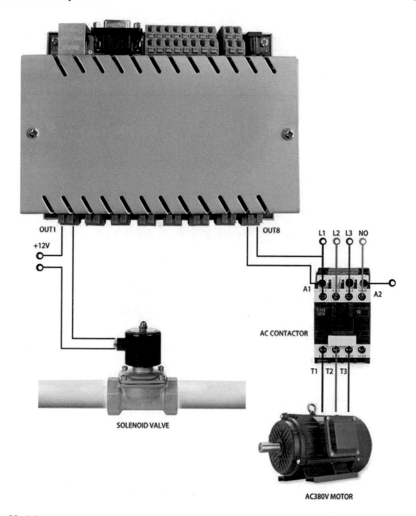

**Fig. 52** Master circuit for pump control

## Controllers for Backwashing of Sand and Disc Filters

When low-quality water or water with significant amount of algae is used, a battery of sand or disc filters are used. These filters are supposed to be cleaned quite often by back washing. The process of back washing is a tedious process because opening and closing of few valves have to be done simultaneously. These operations can be made automatic by installing solenoid valves along with a controller (Fig. 57). We can set flush time which is the time taken for each flushing, and we can also set wait time which is the time period between the successive flushes.

**Fig. 53**  Quick response pressure relief valves (www.bermad.com)

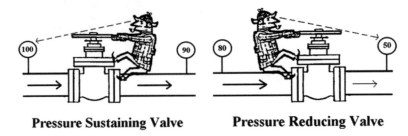

**Pressure Sustaining Valve**          **Pressure Reducing Valve**

**Fig. 54**  Principle of pressure management valves

**Fig. 55**  Cross-sectional
view of a pressure
management valve (Ragsdale
and Associates)

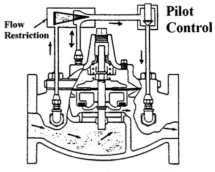

**Fig. 56** Pressure sustaining
valve (www.vag-group.com)

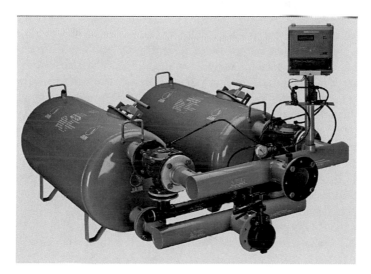

**Fig. 57** Automation of back washing in sand filters (www.jains.com)

## Sensor-Based Irrigation Systems

Sensors to measure soil moisture, evapotranspiration and rain are used in triggering the irrigation. These kind of systems are also called as closed loop systems, whereas the systems which work only on fixed time or volume as discussed in preceding sections are called as open loop systems.

Tensiometer, resistivity, capacitance and time domain reflectometry (TDR) are the types of sensors used in estimation of soil moisture (Fig. 58). Tensiometer and electrical resistivity sensors are widely used in automation due to their low cost and simplicity in operation. Capacitance and TDR sensors are fairly more accurate and versatile than the other sensors but they are costly. Hence these sensors are used only when a very high accuracy is needed and are mostly used in research only.

Tensiometers and electrical resistivity sensors provide the soil moisture in terms of tension with which the water is held by soil pores. If irrigation is done based on soil moisture tension, the farmers do not need to consider the soil type. For instance if the research data states that for sugarcane, the optimal soil moisture tension at which irrigation is to be done is 500 cm (of water column), then whether the crop is grown in sandy soil or loamy soil or clay soil, the irrigation is to be triggered when soil moisture tension reaches 500 cm.

Tensiometers work in a narrow range and the maximum tension until which tensiometers can work is only 1 atm (10 m of water column). It is very difficult to use the tensiometers in black soils because significant amount of moisture is available in black soil at the tensions above 1 atm. Maintenance of tensiometers is another big issue which tilts the scale towards resistivity sensors.

Simple resistivity sensors measure the electrical resistance between the two electrodes placed in soil. The resistance and the soil moisture tension are proportional to each other, and hence, the resistance values are easily converted into soil moisture tension values. If the contact area between the soil and the electrode is varied for the same soil at a specific soil moisture content, the measured resistance would vary. Existence of air pockets between the electrodes and soil would also affect the observations. When the soil is saline as well as during fertigation, the ionic atmosphere in soil affects the observations to a significant level. These problems are greatly reduced in case of gypsum block electrical resistivity sensors (Fig. 59). Gypsum block sensors are placed in gypsum block, and the observations are taken after the gypsum in the sensor equilibrates with the surrounding soil. As the gypsum has a good buffering capacity, the effect of salinity is very much reduced. As and when the gypsum block gets dissipated, the gypsum block has to be replaced.

In Fig. 59, a gypsum block resistivity sensor with a controller is shown. From the figure, it can be seen that the controller has 8 levels from wet to dry. The farmer

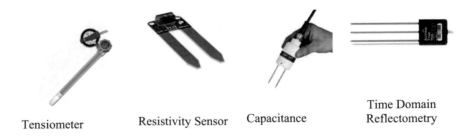

Tensiometer          Resistivity Sensor    Capacitance         Time Domain
                                                               Reflectometry

**Fig. 58**  Soil moisture sensors

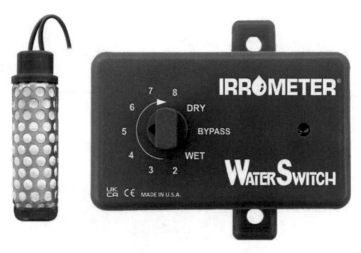

**Fig. 59** Gypsum block sensor with controller

has to specify the soil moisture tension at which, irrigation is to be started. After the irrigation is started, the pump will run for a specified duration and the pump will be stopped.

In Fig. 60, it can be seen that two tensiometers are connected to the controller. One tensiometer is placed at a higher level which corresponds to the active root zone. Another tensiometer is placed at a lower level which corresponds to the maximum rooting depth. When soil moisture tension in upper tensiometer drops below the

**Fig. 60** Automation with two tensiometers with a controller

**Fig. 61** Automation with tensiometer without controller

threshold level, the irrigation would be triggered. When irrigation is done, after sometime the wetting will reach the lower tensiometer. The stopping of irrigation will be done by the lower tensiometer after a prespecified level in the lower tensiometer is reached.

In Fig. 61, connection circuit without any controller is shown. When this system is used, we can only trigger the irrigation; but stopping of irrigation must be done separately by a timer. Wiring diagrams for connecting solenoid valves with controllers are supplied by many manufacturers so that even novice user can install the automation nowadays (Fig. 62).

The irrigation valves are normally placed at the field level, and the distance between the control head and the valves is usually higher. For such situations remote-controlled valves are used (Fig. 63). The remote controllers placed at the field level are powered using small solar panels, and communication between the control head and the remote control units is enabled by radio links. Since the computer placed at the control head can be easily connected with the Internet, all the operations can be done using any other computer or cell phone from any part of the world.

## *Automated Fertigation Systems*

Automated fertigation systems have typically five fertigation channels (Fig. 64). Major nutrients such as nitrogen, phosphorus, potash, micronutrients and acid would be supplied to the plants kept in separate containers. They have venturi coupled with

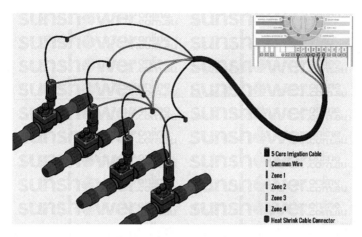

**Fig. 62** Wiring diagrams for connection of controllers with solenoid valves

**Fig. 63** Remote-controlled valve system

metering valves to control the proportion of fertilizer solution into the irrigation water. Electrical conductivity (EC) and pH of the irrigation water to be applied to the plants after mixing the raw irrigation water with fertilizer solution can be properly maintained by adding acid through a separate channel. All these are made possible by electronically monitoring flow rates as well as EC and pH with meters. Since these automated fertigation systems are computer controlled, they are integrated

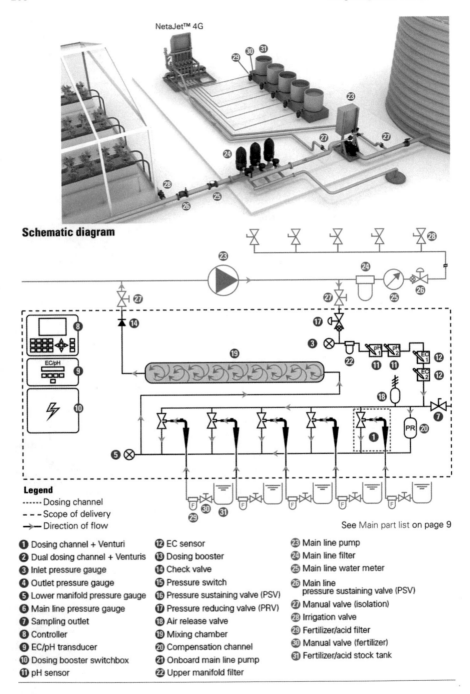

**Fig. 64** Automated fertigation system (www.netafim.com)

with automated filter backwashing also. These systems are very costly and mostly used for green house crops.

## Pollution Prevention Accessories in Chemigation

There are many possibilities of occurrence of pollution while chemigation is done. The chemical injected may get siphoned off due to power and mechanical failures during operation. When external pumps are used for fertigation, fertigation pump may continue to operate while the main pump has stopped. Considering all the possibilities, accessories such as double check valves, air/vacuum breakers, reduced pressure backflow prevention devices, pressure relief and sustaining valves have to be provided and properly controlled by controllers. A typical control loop with accessories is in shown in Fig. 65.

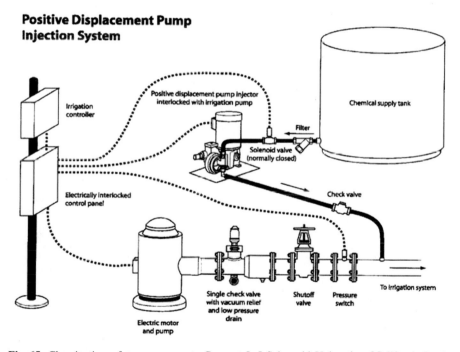

**Fig. 65** Chemigation safety arrangements. *Courtesy* L. J. Schwankl, University of California-Davis

## Variable Frequency Drive (VFD)

Variable frequency drive (VFD) is a device used to control the speed of the motor in pumps by adjusting the frequency of electric power. When different zones in an irrigation system need different levels of pressure heads and discharge requirements, VFD is very useful. Instead of using pressure control valves and bypass valves when VFD is used, the cost of electric power used would be less. VFD is akin to our domestic fan regulator which controls speed of the fan. It is also possible to integrate the VFD controllers in fully automated fertigation systems.

## Summary

In this chapter, drip irrigation is defined and the advantages of drip irrigation are presented. Classification of drip irrigation is done. Working principles and hydraulics of all the components such as dripper, filter and fertigation equipment are discussed. Drip irrigation pipes and their strengths are presented. Finally the components of automation of irrigation system and their working principles are also discussed.

## Appendix

The Darcy–Weisbach equation normally used in fluid mechanics is as follows:

$$h_f = \frac{f L v^2}{2g D} \tag{4}$$

where

$h_f$ is head loss (m)
$f$ is called Darcy–Weisbach friction factor
$L$ is length of pipe (m)
$V$ is velocity of flow (m)
$g$ is acceleration due to gravity (m/s$^2$)
$D$ is diameter of pipe (m).

The flow in microirrigation pipes is generally turbulent flow, and the pipes are smooth. For these conditions, the expression for finding friction factor $f$ was given by Blasius, which is as follows:

$$f = 0.32 \text{Re}^{-0.25} \tag{5}$$

where Re is Reynolds number of the flow. This equation is valid for the Reynolds number between 2000 and 100,000. It should be noted that this equation is valid for any pipe material which is smooth.

The expression for Reynolds number (Re) is as follows:

$$\text{Re} = \frac{\rho v d}{\mu} \tag{6}$$

where $\rho$ is mass density of water and $\mu$ is dynamic viscosity of water.

Substitute $\rho = 1000 \, \text{kg/m}^3$ and $\mu = 0.001002 \, \text{NS/m}^2$ for temperature of 20 °C. If velocity of flow ($v$) is substituted with flow rate ($Q$) divided by cross-sectional area of flow as circle, we get the following equation:

$$\Delta h = \frac{K Q^{1.75}}{D^{4.75}} L \tag{7}$$

In microirrigation, head loss is expressed in metres (m), flow rate is in litres per second (l/s) and length of pipe in metres (m) and diameter of pipe in millimetres (mm). For such situation, the value of $K = 7.89 \times 10^5$.

Let us see how the relationship would turn out to be if the flow is laminar. For laminar flow due to Poiseuille's law, the friction factor '$f$' can be expressed as follows:

$$f = \frac{64}{\text{Re}} \tag{8}$$

If we substitute the preceding relationship in (4), we get the following relationship:

$$\Delta h = \frac{K Q}{D^4} L$$

where $K = (128\mu/\pi\rho g)$. It is to be noted that head loss and discharge rate are linearly related in case of laminar flow.

## Assessment

### Part A

1. Which among the following is appropriate fertigation equipment when the operating pressure of drip irrigation system is very low?

   a. Fertigation pump
   b. **Fertigation tank**
   c. Venturi
   d. None of these

2.   If water contains significant amount of sand, the suggested filter is _____

   a.   Sand Filter
   b.   **Hydrocyclone**
   c.   Screen filter
   d.   Disc filter

3.   Bubblers are used in irrigation of trees. Usually the discharge rate in bubblers is ____than the infiltration capacity of the soil and therefore a small basin is formed around trees.

   a.   **larger**
   b.   Smaller
   c.   Equal to
   d.   None of these

4.   If very small rains such as less than 5 mm occur, there may be a possibility of salt build-up moving into the root zone in Drip irrigation systems. Because of this fact, in saline lands, plants appear _____after a small rain.

   a.   **wilted**
   b.   Healthy
   c.   Greenish
   d.   None of these

5.   In a venturi, at the throat section pressure _____atmospheric pressure.

   a.   **falls down below**
   b.   Goes above
   c.   Becomes equal to
   d.   None of these

## Part B

1.   What is the use of a Male Thread Adaptor?
2.   What is the use of a Female Thread Adaptor?
3.   What is the use of a Union?
4.   What are the uses of Saddle Clamp?
5.   Write down the dripper discharge rate equation and also mention the flow exponent value of orifice drippers?
6.   Write down the names of at least six lateral accessories?
7.   Distinguish between Inline dripper and Online Dripper based on their construction, working principle and use?
8.   Write a short note on pressure compensating drippers?
9.   Explain the features of subsurface drip irrigation with a neat sketch and also discuss about its advantages and disadvantages over surface drip?
10.  Draw a typical relay circuit used in Automation of DIS?
11.  Explain the working of a typical Solenoid valve with Pilot arrangement?

12. Why a Master circuit is needed to control the pump while automating a DIS?
13. Explain the working principle of Automatic Pressure Management Valves?
14. Write an account on sensor-based automation in DIS?
15. What is the use of a VFD?

# References and Further Reading

1. Bucks, A., Nakayama, F. S., & Gilbert, R. G. (1979). Trickle irrigation water quality and preventive maintenance. *Agricultural Water Management, 2*(2), 149–162. https://doi.org/10.1016/0378-3774(79)90028-3
2. Jain Irrigation systems Ltd, Filter and Fertigation Catalogue, India. www.jains.com
3. Keller, J., & Bliesner, R. D. (1990). *Sprinkle and trickle irrigation.* Chapman & Hall.
4. Micro irrigation, Part 623 Irrigation, National Engineering Hand book, NRCS-USDA, USA.
5. Sne, M. (2005). Drip irrigation. MASHAV-CINADCO, Israel
6. Nathan, R. (2005). Fertigation MASHAV-CINADCO, Israel.
7. Ragsdale and Associates. How to do stuff, distribution system control valves.
8. RO-Drip User Manual. www.robertsirrigation.com

# Drip Irrigation—Planning Factors

## Introduction

Before starting to design drip irrigation system (DIS), many factors such as the pump characteristics, soil characteristics and plant characteristics have to be studied. A discussion on the study of pump characteristics which is needed for sprinkler and drip irrigation is available in Chap. 5.

Peak crop water requirement data is useful in deciding the maximum irrigable area from any water source. In the process of selection of a suitable plant row-lateral-dripper configuration for adoption, various alternatives are usually evaluated. In order to evaluate the alternatives, the root spread of crops both in depth direction and sidewards direction needs to be studied. Then suitable plant row-lateral-dripper configuration can be decided by using standard procedures with the aid of analytical equations.

## Crop Water Requirement

Crop water requirement is usually calculated by estimating evapotranspiration based on the meteorological data available for any specific area, and it is usually expressed in terms of depth of water required per day. Sometimes water requirement data is also available in volumetric units as in Table 1. Peak water requirement data is needed at the design stage because if the DIS is designed for peak water requirement, then the DIS would safely function during the other periods.

**Table 1** A typical water requirement and spacing data for certain crops (*Courtesy* EPC Mahindra, India)

| No. | Crop | Spacing (ft × ft) | Peak water requirement (l/d/plant) |
|---|---|---|---|
| 1 | Grapes | 6 × 4 | 10–12 |
| | | 8 × 6 | 18–20 |
| | | 8 × 8 | 24 |
| | | 8 × 10 | 30 |
| 2 | Pomegranate | 10 × 10 | 30–40 |
| | | 12 × 12 | 40–50 |
| | | 15 × 15 | 70–75 |
| 3 | Guava | 15 × 15 | 70–80 |
| | | 18 × 18 | 100–120 |
| | | 25 × 25 | 120–130 |
| 4 | Mango | 25 × 25 | 120–140 |
| | | 30 × 30 | 150–170 |
| 5 | Sapota/Chiku | 25 × 25 | 120–140 |
| | | 30 × 30 | 150–170 |
| 6 | Orange/Lemon/Citrus | 16 × 16 | 75 |
| | | 18 × 18 | 85 |
| 7 | Custard apple | 10 × 10 | 40 |
| | | 12 × 12 | 50 |
| 8 | Ber | 10 × 10 | 30 |
| | | 12 × 12 | 55 |
| 9 | Banana | 6 × 4 | 22 |
| | | 6 × 4 | 25 |
| | | 5 × 5 | 22 |
| | | 3 × 6 × 5 | 25 |
| 10 | Papaya | 5 × 4 | 18 |
| | | 7 × 7 | 20 |
| 11 | Coconut | 25 × 25 | 90 |
| 12 | Cardamom | 10 × 10 | 15 |
| 13 | Rubber | 15 × 15 | 24 |
| 14 | Oil palm | 30 × 23 | 150 |
| 15 | Sugarcane | Lateral to lateral 8 ft | 20 l/d/m length |
| | | Lateral to lateral 7 ft | 18 l/d/m length |
| | | Lateral to lateral 6 ft | 16 l/d/m length |
| | | Lateral to lateral 5 ft | 14 l/d/m length |
| 16 | Cotton | Lateral to lateral 6 ft | 15 l/d/m length |

(continued)

**Table 1** (continued)

| No. | Crop | Spacing (ft × ft) | Peak water requirement (l/d/plant) |
|-----|------|-------------------|-------------------------------------|
| 17 | Vegetables/Flowers | Lateral to lateral 6 ft | 14 l/d/m length |
| | | Lateral to lateral 5 ft | 12 l/d/m length |
| | | Lateral to lateral 4 ft | 10 l/d/m length |
| 18 | Tea/Coffee | Lateral to lateral 8 ft | 15 l/d/m length |
| | | Lateral to lateral 7 ft | 13 l/d/m length |

# Maximum Area of the Land Irrigated

After evaluating the suitability of the pump available in the farmer's land, the next step is to ascertain the maximum extent of the land that can be irrigated. This depends on the discharge rate available from the pump during lean water availability period, peak crop water requirement and maximum duration of pumping possible. A relationship between the water availability and water demand on daily basis is useful to arrive at maximum irrigable area.

The maximum area that can be irrigated can be found out by balancing the water supply and demand of water which is as follows:

$$A = \frac{36 A_p Q T U_c}{v} \tag{1}$$

$Q$ is the pump discharge rate (l/s);

$T$ is the number of hours of power available or number of possible hours of irrigation per day;

$A_p$ is area allotted per plant (m²);

$U_c$ is uniformity coefficient (usually 90% for drip irrigation);

$V$ is the peak daily water requirement in litres.

Equation 1 is useful for a simple case of single crop grown by a farmer without giving attention to the rainfall in the area. A detailed water budgeting analytical method is presented in a separate Chap. 17 namely Water Budgeting and Economic Evaluation.

**Problem**

$$A_p = 1.8 \times 1.8\,\text{m}^2.\ Q = 1\,\text{l/s};\ T = 8\,\text{h};\ v = 25\,\text{l/d}$$

**Solution**

By applying the preceding formula, the maximum area that can be irrigated is calculated and is as follows:

$$A = 3359 \, \text{m}^2$$

If the water requirement data is available in terms of depth of water, the preceding formula gets converted into following formula:

$$A = \frac{36 \, Q T U_c}{E T_{cp} K_c} \tag{2}$$

where

Et$_{cp}$ is the peak reference crop water requirement in mm;

$K_c$ is crop coefficient.

**Problem**

$$Q = 1 \, \text{l/s}; \quad T = 8 \, \text{h}; \quad \text{Et}_{cp} = 8 \, \text{mm}; \quad K_c = 1; U_c = 90\%$$

$Q = 1 \, \text{l/s}; \quad T = 8 \, \text{h}; \quad \text{Et}_{cp} = 8 \, \text{mm}; \quad K_c = 1; \quad U_c = 90\%.$

**Solution**
By applying the above formula, the maximum area that can be irrigated is as follows:

$$A = 3240 \, \text{m}^2$$

## Maximum Area of One Subunit

One subunit means the area corresponding to one submain and its laterals with a control valve to control discharge into the submain. There is a general feeling existing in India that drip irrigation has to be done on daily basis. Though drip irrigation can be done even at less irrigation interval than a day, it is not economical and sometimes not feasible to irrigate on daily basis. The irrigation interval can be as large as for conventional irrigation methods. The irrigation interval must be decided based on the volume of water applied in each irrigation and crop water use between each irrigation.

If the irrigation interval is n days, the total area is divided into $(n - 1)$ parts, and for each part, n days water requirement can be given. The maximum area of one subunit can also be obtained by dividing the total area into n parts. The division of the total area into $(n - 1)$ parts is to have a factor of safety.

For the preceding data, if the irrigation interval is 4 days and if the area is divided into 3 parts, the maximum area of one subunit would work out to be 1080 m². While doing irrigation, the total depth of water applied for each part will be 3-day water requirement. So, when irrigation is done, the discharge rate of the well (1 l/s in this case) would be used for irrigating only 1080 m². Each 1080 m² part would be irrigated on successive days. Since, one day was taken as factor of safety, if the system is shut down for any maintenance and repair work, the irrigation interval of 4 days would not get affected.

## Wetting Pattern of Drippers

When water is applied by a dripper on land surface, the typical wetted sectional elevation profiles for each soil type is shown in Fig. 1. For clayey soils, the diameter of wetting is higher and depth of wetting is smaller. For loamy soils, depth and diameter of wetting are approximately equal. For sandy soils, the diameter of wetting is smaller than the depth of wetting. It must also be noted that the surface area of the soil wetted is little less than the area wetted few centimetres below the soil surface.

It is always better to pay attention to the relationship between wetting pattern and dripper discharge rates. Digging a small trench and operating a dripper and

**Fig. 1** Wetting profiles for different soil types

**Fig.2** Trench for wetting pattern study

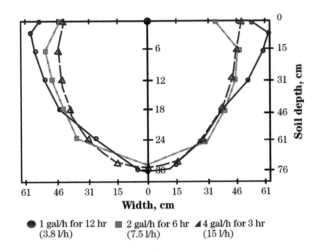

**Fig. 3** Wetted bulb for applying equal quantity of water with different rates in dry sandy soil [1]

● 1 gal/h for 12 hr    ■ 2 gal/h for 6 hr    ▲ 4 gal/h for 3 hr
(3.8 l/h)                (7.5 l/h)                (15 l/h)

observing the wetting pattern on the walls of the trench are very useful. When a dripper is operated near the wall of the trench, the direct entry of water on the walls must be arrested by some means. Figure 2 shows a provision of a wooden plank at the vertical wall so that the ponded water on the land surface does not flow over the surface of the wall. Since only one half of the soil is wetted, if we want to examine the wetting pattern for 4 l/h dripper, we must use 2 l/h dripper if we dig a trench before the conduct of the experiment.

Generally, it is believed that the dripper discharge rates affect the shape of wetted bulb. But from the results in Fig. 3, it can be seen that, when the total quantity of water applied is constant, different application rates produce more or less the same kind of wetting pattern. If clay content of soil increases, the wetted diameters would be more for larger discharge rates when total applied quantity of water is constant.

A very simple assumption for predicting the wetted dimension for a point source is to assume the wetted zone to be a hemisphere (i.e. diameter of wetting $(d) = 2 \times$ depth of wetting). For this assumption following equation can be written as follows [2]:

$$qt = \frac{1}{12}\pi d^3 (\Delta\theta) \qquad (3)$$

where $q$ is the application rate (m$^3$/s), t is the duration of application ($s$), and $\Delta\theta$ is the average change in volumetric water content in the wetted zone. In the absence of field data, it is recommended to substitute with a value of 50% of saturated water content if the initial soil moisture content is wetter which is usually the situation in drip irrigation. The preceding equation can be rearranged as follows:

$$d = \left(\frac{12qt}{\pi \Delta\theta}\right)^{0.33} \qquad (4)$$

**Fig. 4** Semi-ellipsoid
assumption of wetted bulb

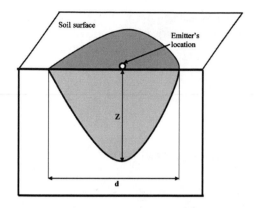

Usually, wetted bulb due to a point source is approximated as a semi-ellipsoid as shown in Fig. 4. Volume of semi-ellipsoid is given by the following formula:

$$V = \frac{\pi z d^2}{6} \tag{5}$$

where $z$ is maximum depth of wetting and d is the diameter of wetting on the surface.

The relationship between $d$ and $z$ depends on factors like discharge rate of dripper, volumetric soil water content before the irrigation, saturated hydraulic conductivity of soil and bulk density of soil.

## Empirical Equation for Wetting Pattern

One popular equation used to get the dimensions of $d$ and $z$ is Amin and Ekhmaj (2006) equation which is as follows:

$$d = 0.5 \Delta\theta^{-0.5626} V_w^{0.2686} q^{-0.0028} K_s^{-0.0344} \tag{6}$$

$$z = 2.03 \Delta\theta^{-0.383} V_w^{0.365} q^{-0.101} K_s^{0.195} \tag{7}$$

where $\Delta\theta$ is the average change in volumetric water content in the wetted zone. In the absence of field data, it is recommended to substitute with a value of 50% of saturated water content for $\Delta\theta$; $V_w$ is the volumetric water content in the wetted zone (m³); this can be found out from the value of initial soil moisture content and total amount of water applied; $q$ is the application rate (m³/s); and $K_s$ is saturated hydraulic conductivity (m/s).

This equation was derived from the published experimental data in the literature and using statistical regression analysis. Since this equation was not physically based, dimensional homogeneity need not be expected, and hence this equation is classified as purely empirical equation.

Equations 6 and 7 can be combined to get following Eq.:

$$d = 0.2971 \Delta \theta^{-0.2808} q^{0.0715} K_s^{-0.1779} Z^{0.7359} \tag{8}$$

## Semi-Empirical Equation for Wetting Pattern

Schwartzman and Zur [3] applied dimensional analysis for deriving wetting pattern relationship. Dimensional analysis is very useful in designing experiments and to arrive at a reasonable number of variables to be examined during experimentation.

$$d = 1.82 V_w^{0.22} \left( \frac{K_s}{q} \right)^{-0.17} \tag{9}$$

$$z = 2.54 V_w^{0.63} \left( \frac{K_s}{q} \right)^{0.45} \tag{10}$$

Equations 9 and 10 can be combined and approximated to get following Eq.:

$$d = 1.32 \left( \frac{zq}{K_s} \right)^{0.333} \tag{11}$$

where $V_w$ is the volumetric water content in the wetted zone ($m^3$), this can be found out from the value of initial soil moisture content and total amount of water applied, $q$ is the application rate ($m^3/s$), and $K_s$ is saturated hydraulic conductivity (m/s). Since this equation was derived based on dimensional analysis, it is expected that this equation would perform better than the Amin and Ekhmaj equation. But it is reported widely in the literature that the predictability of Amin and Ekhmaj equation is significantly better.

Either Eq. 8 or Eq. 11 can be used for finding wetted diameter for any desired minimum depth of wetting. With 20% factor of safety, 80% of maximum diameter of wetting obtained using these equations may be taken as the dripper spacing to get a continuous wetted strip. Table 2 was developed using Eq. 11 to get dripper spacing to get a continuous wetted strip for different type of soils.

Shallow rooted field crops may be wetted to a minimum depth of 0.3 m, medium rooted field crops may be wetted to a depth of 0.6 m, and tree crops may be wetted to a depth of 0.9 to 1.5 m.

**Problem** Compare the diameter of wetting by using Amin and Ekhmaj equation and Schwartzman and Zur equation for the following data:

$\Delta\theta = 0.25$; $K_s = 0.0000032$ m/s; $q = 1 \times 10^{-6}$ m$^3$/s; $z = 0.3$ m.

**Solution**

By Amin and Ekhmaj Equation: $d = 0.64$ m.

By Schwartzman and Zur Equation: $d = 0.60$ m
It can be seen that the values of the diameter do not differ much.

## Ponded Radius

When a dripper is operated, a small pool of water ponds over the land around dripper (Figs. 5 and 6). The ponded zone may not be clearly visible in sandy soils; however if clay proportion of a soil is more, the ponded zone diameter can be distinctly seen.

The diameter of ponding remains constant if a dripper is operated for sufficiently longer time. By measuring the diameter of ponded zone, it is possible to ascertain the saturated hydraulic conductivity of the soil by using the analytical solution developed by Wooding [4]. The Wooding equation is as follows:

$$\frac{q}{\pi r^2} = K_s \left( 1 + \frac{4}{\pi r \alpha} \right) \tag{12}$$

**Fig. 5** Ponded zone and wetted zone

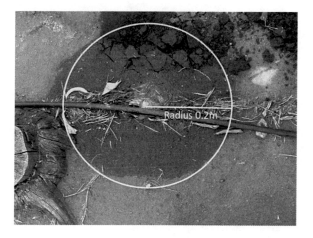

**Fig. 6** A typical sectional elevation view of wetting pattern for a single surface dripper

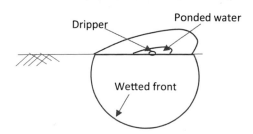

where

    $q$ is dripper discharge rate (cm$^3$/s);

    $K_s$ is saturated hydraulic conductivity of the soil (cm/s);

    $r$ is ponded radius (cm);

    $\alpha$ is Gardener's exponent which depends on soil characteristics (cm$^{-1}$);

    Equation 12 is a very easy way of finding saturated hydraulic conductivity of the soil, and $K_s$ can be used to estimate dripper spacing using Schwartzman and Zur equation or Amin and Ekhmaj equation.

**Problem**

1.    Find out the saturated hydraulic conductivity of the soil for the following data:

    Dripper discharge rate $= 3.6$ l/h; ponded radius $(r) = 30$ cm; Gardener's exponent $(\alpha) = 0.36$ cm$^{-1}$.

**Solution**

$q = 1$ cm$^3$/s;

    After substituting the data in the preceding formula, the $K_s$ value found out is as follows (Table 2):

$$K_s = 0.00032 \text{ cm/s}$$

# Lateral Configurations

While deciding on a suitable lateral configuration, factors such as crop spacing, crop water requirement, irrigation interval, rooting pattern, soil type and economics have to be considered. Different lateral configurations that are commonly used for closely grown field crops are shown in Fig. 7. One lateral per crop row is very common. Sometimes, the wetted volume obtained by providing one lateral per crop row is not sufficient. In such cases two laterals per crop row are provided. If installation cost reduction is important, one lateral for two or more crop rows is also used.

    In Fig. 7, for the case of one pipe line every other crop row has uniform crop row spacing. But in case of one lateral line per two crop rows, the spacing between crop rows adjacent to lateral will be smaller than the spacing between the crop rows which do not have lateral in between. They are called as paired or dual crop rows (Fig. 8).

    The spacing between emitters in Fig. 8 is denoted as $x$, whereas the spacing between crop rows is denoted as $y$. Usually, '$y$' is preferably less than '$x$'. This is to ensure that the wetted front reaches both rows of plants sufficiently.

    For widely spaced tree crops, two laterals per row of plants are sometimes used to provide more emission points per tree. Sometimes drippers are looped around a tree or one microsprinkler may also be provided. Layout types 1 and 2 in Fig. 9 are

**Table 2** Recommended dripper spacing to get a continuous wetted strip [5]

| Dripper discharge rate (l/h) | Minimum depth of wetting (m) | | | | | | |
|---|---|---|---|---|---|---|---|
| | 0.3 | 0.45 | 0.6 | 0.75 | 0.9 | 1.2 | 1.5 |
| *Coarse sand—$K_s = 0.0001$ m/s* | | | | | | | |
| 1 | 0.10 | 0.11 | 0.13 | 0.14 | 0.14 | 0.16 | 0.17 |
| 2 | 0.13 | 0.14 | 0.16 | 0.17 | 0.18 | 0.20 | 0.21 |
| 4 | 0.16 | 0.18 | 0.20 | 0.21 | 0.23 | 0.25 | 0.27 |
| 6 | 0.18 | 0.21 | 0.23 | 0.25 | 0.26 | 0.29 | 0.31 |
| 8 | 0.20 | 0.23 | 0.25 | 0.27 | 0.29 | 0.32 | 0.34 |
| 16 | 0.25 | 0.29 | 0.32 | 0.34 | 0.36 | 0.40 | 0.43 |
| *Sand—$K_s = 0.000058$ m/s* | | | | | | | |
| 1 | 0.12 | 0.14 | 0.15 | 0.16 | 0.17 | 0.19 | 0.20 |
| 2 | 0.15 | 0.17 | 0.19 | 0.20 | 0.22 | 0.24 | 0.26 |
| 4 | 0.19 | 0.22 | 0.24 | 0.26 | 0.27 | 0.30 | 0.32 |
| 6 | 0.22 | 0.25 | 0.27 | 0.29 | 0.31 | 0.34 | 0.37 |
| 8 | 0.24 | 0.27 | 0.30 | 0.32 | 0.34 | 0.38 | 0.41 |
| 16 | 0.30 | 0.34 | 0.38 | 0.41 | 0.43 | 0.48 | 0.51 |
| *Fine sand—$K_s = 0.000031$ m/s* | | | | | | | |
| 1 | 0.15 | 0.17 | 0.19 | 0.20 | 0.21 | 0.23 | 0.25 |
| 2 | 0.19 | 0.21 | 0.23 | 0.25 | 0.27 | 0.29 | 0.32 |
| 4 | 0.23 | 0.27 | 0.29 | 0.32 | 0.34 | 0.37 | 0.40 |
| 6 | 0.27 | 0.31 | 0.34 | 0.36 | 0.39 | 0.42 | 0.46 |
| 8 | 0.29 | 0.34 | 0.37 | 0.40 | 0.42 | 0.47 | 0.50 |
| 16 | 0.37 | 0.42 | 0.47 | 0.50 | 0.53 | 0.59 | 0.63 |
| *Loamy sand—$K_s = 0.000017$ m/s* | | | | | | | |
| 1 | 0.18 | 0.21 | 0.23 | 0.24 | 0.26 | 0.29 | 0.31 |
| 2 | 0.23 | 0.26 | 0.29 | 0.31 | 0.33 | 0.36 | 0.39 |
| 4 | 0.29 | 0.33 | 0.36 | 0.39 | 0.41 | 0.45 | 0.49 |
| 6 | 0.33 | 0.37 | 0.41 | 0.44 | 0.47 | 0.52 | 0.56 |
| 8 | 0.36 | 0.41 | 0.45 | 0.49 | 0.52 | 0.57 | 0.61 |
| 16 | 0.45 | 0.52 | 0.57 | 0.61 | 0.65 | 0.72 | 0.77 |
| *Loamy sand—$K_s = 0.00001$ m/s* | | | | | | | |
| 1 | 0.21 | 0.25 | 0.27 | 0.29 | 0.31 | 0.34 | 0.37 |
| 2 | 0.27 | 0.31 | 0.34 | 0.37 | 0.39 | 0.43 | 0.46 |
| 4 | 0.34 | 0.39 | 0.43 | 0.46 | 0.49 | 0.54 | 0.58 |
| 6 | 0.39 | 0.45 | 0.49 | 0.53 | 0.56 | 0.62 | 0.67 |
| 8 | 0.43 | 0.49 | 0.54 | 0.58 | 0.62 | 0.68 | 0.73 |
| 16 | 0.54 | 0.62 | 0.68 | 0.73 | 0.78 | 0.86 | 0.92 |

(continued)

**Table 2** (continued)

| Dripper discharge rate (l/h) | Minimum depth of wetting (m) | | | | | | |
|---|---|---|---|---|---|---|---|
| | 0.3 | 0.45 | 0.6 | 0.75 | 0.9 | 1.2 | 1.5 |
| *Sandy loam—$K_s = 0.0000072$ m/s* | | | | | | | |
| 1 | 0.24 | 0.27 | 0.30 | 0.32 | 0.34 | 0.38 | 0.41 |
| 2 | 0.30 | 0.34 | 0.38 | 0.41 | 0.43 | 0.48 | 0.51 |
| 4 | 0.38 | 0.43 | 0.48 | 0.51 | 0.55 | 0.60 | 0.65 |
| 6 | 0.43 | 0.50 | 0.55 | 0.59 | 0.63 | 0.69 | 0.74 |
| 8 | 0.48 | 0.55 | 0.60 | 0.65 | 0.69 | 0.76 | 0.82 |
| 16 | 0.60 | 0.69 | 0.76 | 0.82 | 0.87 | 0.96 | 1.03 |
| *Fine sandy loam—$K_s = 0.0000052$ m/s* | | | | | | | |
| 1 | 0.27 | 0.31 | 0.34 | 0.36 | 0.38 | 0.42 | 0.46 |
| 2 | 0.34 | 0.38 | 0.42 | 0.46 | 0.48 | 0.53 | 0.57 |
| 4 | 0.42 | 0.48 | 0.53 | 0.57 | 0.61 | 0.67 | 0.72 |
| 6 | 0.48 | 0.55 | 0.61 | 0.66 | 0.70 | 0.77 | 0.83 |
| 8 | 0.53 | 0.61 | 0.67 | 0.72 | 0.77 | 0.85 | 0.91 |
| 16 | 0.67 | 0.77 | 0.85 | 0.91 | 0.97 | 1.06 | 1.15 |
| *Loam—$K_s = 0.0000037$ m/s* | | | | | | | |
| 1 | 0.30 | 0.34 | 0.38 | 0.41 | 0.43 | 0.47 | 0.51 |
| 2 | 0.38 | 0.43 | 0.47 | 0.51 | 0.54 | 0.60 | 0.64 |
| 4 | 0.47 | 0.54 | 0.60 | 0.64 | 0.68 | 0.75 | 0.81 |
| 6 | 0.54 | 0.62 | 0.68 | 0.74 | 0.78 | 0.86 | 0.93 |
| 8 | 0.60 | 0.68 | 0.75 | 0.81 | 0.86 | 0.95 | 1.02 |
| 16 | 0.75 | 0.86 | 0.95 | 1.02 | 1.08 | 1.19 | 1.28 |
| *Silty loam—$K_s = 0.0000019$ m/s* | | | | | | | |
| 1 | 0.37 | 0.43 | 0.47 | 0.51 | 0.54 | 0.59 | 0.64 |
| 2 | 0.47 | 0.54 | 0.59 | 0.64 | 0.68 | 0.75 | 0.80 |
| 4 | 0.59 | 0.68 | 0.75 | 0.80 | 0.85 | 0.94 | 1.01 |
| 6 | 0.68 | 0.77 | 0.85 | 0.92 | 0.98 | 1.07 | 1.16 |
| 8 | 0.75 | 0.85 | 0.94 | 1.01 | 1.07 | 1.18 | 1.27 |
| 16 | 0.94 | 1.07 | 1.18 | 1.27 | 1.35 | 1.49 | 1.60 |
| *Sandy clay loam—$K_s = 0.0000012$ m/s* | | | | | | | |
| 1 | 0.43 | 0.50 | 0.55 | 0.59 | 0.63 | 0.69 | 0.74 |
| 2 | 0.55 | 0.63 | 0.69 | 0.74 | 0.79 | 0.87 | 0.94 |
| 4 | 0.69 | 0.79 | 0.87 | 0.94 | 0.99 | 1.09 | 1.18 |
| 6 | 0.79 | 0.90 | 0.99 | 1.07 | 1.14 | 1.25 | 1.35 |
| 8 | 0.87 | 0.99 | 1.09 | 1.18 | 1.25 | 1.38 | 1.48 |

(continued)

**Table 2** (continued)

| Dripper discharge rate (l/h) | Minimum depth of wetting (m) | | | | | | |
|---|---|---|---|---|---|---|---|
| | 0.3 | 0.45 | 0.6 | 0.75 | 0.9 | 1.2 | 1.5 |
| 16 | 1.09 | 1.25 | 1.38 | 1.48 | 1.58 | 1.74 | 1.87 |
| Clay loam—$K_s = 0.00000064$ m/s | | | | | | | |
| 1 | 0.54 | 0.61 | 0.67 | 0.73 | 0.77 | 0.85 | 0.92 |
| 2 | 0.67 | 0.77 | 0.85 | 0.92 | 0.97 | 1.07 | 1.15 |
| 4 | 0.85 | 0.97 | 1.07 | 1.15 | 1.23 | 1.35 | 1.45 |
| 6 | 0.97 | 1.11 | 1.23 | 1.32 | 1.40 | 1.54 | 1.66 |
| 8 | 1.07 | 1.23 | 1.35 | 1.45 | 1.54 | 1.70 | 1.83 |
| 16 | 1.35 | 1.54 | 1.70 | 1.83 | 1.94 | 2.14 | 2.30 |
| Silty clay loam—$K_s = 0.00000042$ m/s | | | | | | | |
| 1 | 0.62 | 0.71 | 0.78 | 0.84 | 0.89 | 0.98 | 1.05 |
| 2 | 0.78 | 0.89 | 0.98 | 1.05 | 1.12 | 1.23 | 1.33 |
| 4 | 0.98 | 1.12 | 1.23 | 1.33 | 1.41 | 1.55 | 1.67 |
| 6 | 1.12 | 1.28 | 1.41 | 1.52 | 1.61 | 1.78 | 1.91 |
| 8 | 1.23 | 1.41 | 1.55 | 1.67 | 1.78 | 1.95 | 2.10 |
| 16 | 1.55 | 1.78 | 1.95 | 2.10 | 2.24 | 2.46 | 2.65 |
| Sandy clay loam—$K_s = 0.00000033$ m/s | | | | | | | |
| 1 | 0.67 | 0.76 | 0.84 | 0.91 | 0.96 | 1.06 | 1.14 |
| 2 | 0.84 | 0.96 | 1.06 | 1.14 | 1.21 | 1.33 | 1.44 |
| 4 | 1.06 | 1.21 | 1.33 | 1.44 | 1.53 | 1.68 | 1.81 |
| 6 | 1.21 | 1.39 | 1.53 | 1.65 | 1.75 | 1.92 | 2.07 |
| 8 | 1.33 | 1.53 | 1.68 | 1.81 | 1.92 | 2.12 | 2.28 |
| 16 | 1.68 | 1.92 | 2.12 | 2.28 | 2.42 | 2.67 | 2.87 |
| Sandy clay—$K_s = 0.00000025$ m/s | | | | | | | |
| 1 | 0.73 | 0.84 | 0.92 | 0.99 | 1.06 | 1.16 | 1.25 |
| 2 | 0.92 | 1.06 | 1.16 | 1.25 | 1.33 | 1.46 | 1.58 |
| 4 | 1.16 | 1.33 | 1.46 | 1.58 | 1.68 | 1.84 | 1.99 |
| 6 | 1.33 | 1.52 | 1.68 | 1.80 | 1.92 | 2.11 | 2.27 |
| 8 | 1.46 | 1.68 | 1.84 | 1.99 | 2.11 | 2.32 | 2.50 |
| 16 | 1.84 | 2.11 | 2.32 | 2.50 | 2.66 | 2.93 | 3.15 |
| Clay—$K_s = 0.00000017$ m/s | | | | | | | |
| 0.83 | 0.95 | 1.05 | 1.13 | 1.20 | 1.32 | 1.42 | |
| 1.05 | 1.20 | 1.32 | 1.42 | 1.51 | 1.66 | 1.79 | |
| 1.32 | 1.51 | 1.66 | 1.79 | 1.91 | 2.10 | 2.26 | |
| 1.51 | 1.73 | 1.91 | 2.05 | 2.18 | 2.40 | 2.58 | |
| 1.66 | 1.91 | 2.10 | 2.26 | 2.40 | 2.64 | 2.84 | |

(continued)

**Table 2** (continued)

| Dripper discharge rate (l/h) | Minimum depth of wetting (m) | | | | | | |
|---|---|---|---|---|---|---|---|
| | 0.3 | 0.45 | 0.6 | 0.75 | 0.9 | 1.2 | 1.5 |
| 2.10 | 2.40 | 2.64 | 2.84 | 3.02 | 3.33 | 3.58 | |

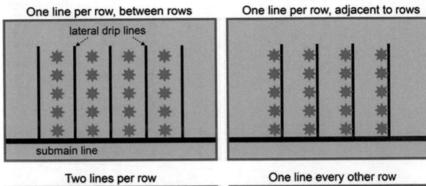

**Fig. 7** Lateral configurations for closely spaced crops

**Fig. 8** Lateral spacing and dripper spacing for paired or dual crop row

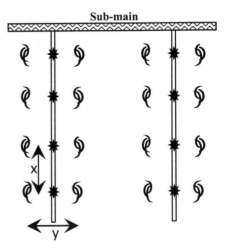

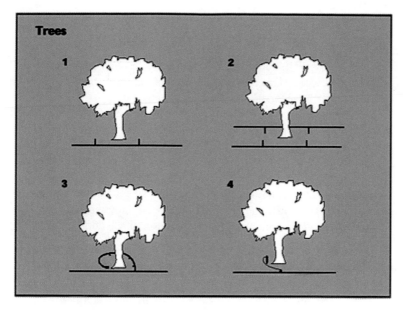

**Fig. 9** Lateral configurations for widely spaced tree crops

**Fig. 10** Wetted bulb and root zone for one lateral for two crop rows [6]

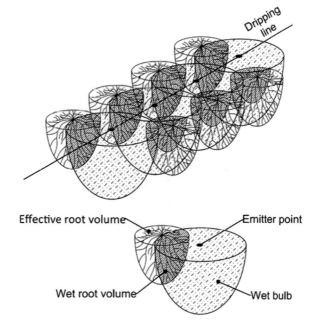

useful to roll the laterals easily, when intercultural operations are done in field.

The dripper spacing and discharge rate are the important design factors because the wetted zones and root zones must interact satisfactorily so that the applied water is optimally used by the plant with minimum wastage. Selection of a satisfactory configuration of drippers and number of laterals per crop row depends on crop characteristics, soil characteristics and climate condition.

In Fig. 10, one lateral is provided for two crop rows. The wetted bulb and the root zone are also shown. Notice the extent of wet bulb in relation to root spread. It is inevitable to irrigate some unwanted soil volume. So obviously a better design and operation are to minimize the wetting of unwanted soil volume where the root zone does not exist or grow.

Schwartzman and Zur equation or Amin and Ekhmaj equation is useful in selecting optimal dripper discharge rate and spacing of drippers. Following numerical example illustrates a very simple and commonly used method of selection of dripper spacing and discharge rate.

**Problem** Sugarcane crop is irrigated with drip irrigation. The crop rows are at a spacing of 1.5 m, and for each row, one lateral is laid. Discharge rate of each dripper is 2 l/h. The soil is clay loam. The clay loam soil has a saturated hydraulic conductivity of 0.000064 cm/s. The effective root depth on 30th day and 60th day is 20 cm and 30 cm, respectively. The effective root depth after 120th day is 40 cm constant throughout.

As per the irrigation programme, during the period till 90th day, irrigation is done when the soil moisture suction reaches −30 centibars or kPa. The soil moisture level at the -30 centibars has been found to be 20% (gravimetric). After 90 days, irrigation is done at the soil moisture suction of −50 centibars. The soil moisture level at the −50 centibars is 18% (gravimetric). Available soil moisture at the field capacity is 0.18 m per metre depth of soil. The permanent wilting point is 12.5% (gravimetric). The bulk density of the soil is 1.6 g/cm$^3$.

The minimum depth of wetting during irrigation is 20 cm. Three dripper spacings in the lateral have been proposed. They are 60, 75 and 90 cm. Recommend the dripper spacing which would suit well to this situation.

Find out the duration of irrigation during the periods near 30th and 45th day and also during the period near 120th day for the selected dripper spacing. Assume the cross section of wetted profile is rectangular (Fig. 11).

**Fig. 11** Rectangular wetted profile assumption

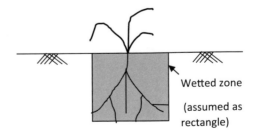

Wetted zone

(assumed as rectangle)

Wetted width–depth relationship

$$d = 1.32 \left( \frac{zq}{K_s} \right)^{0.33}$$

where $d$ is the maximum diameter of wetting (m)
   $z$ is wetted depth (m)
   $K_s$ is saturated hydraulic conductivity (m/s)
   $q$ is dripper discharge rate (m$^3$/s)

Diameter of wetting of the depth of 0.2 m $= d = 1.32 \left( \dfrac{0.2 \times 2 \times 10^{-3}/3600}{0.00000064} \right)^{0.33}$

$$= 0.74 \, \text{m}$$

With 20% factor of safety, spacing of dripper needed for wetting a depth of 0.2 m $= 0.59$ m.
   By similar calculations,

   Diameter of wetting for the depth of 0.3 m

   $= 1.32 \left( \dfrac{0.3 \times 2 \times 10^{-3}/3600}{0.00000064} \right)^{0.33}$

   $= 0.84 \, \text{m}$

   Spacing of dipper needed for wetting a depth of 0.3 m $= 0.67$ m
   Diameter of wetting for the depth of 0.40 m

   $= 1.32 \left( \dfrac{0.4 \times 2 \times 10^{-3}/3600}{0.00000064} \right)^{0.33}$

   $= 0.93 \, \text{m}$

   Spacing of dipper needed for wetting a depth of 0.40 m $= 0.74$ m

From the preceding calculations, following observations can be made:
   If depth of wetting is to be controlled at 0.2 m depth and more, the maximum dripper spacing cannot go beyond 0.59 m. Therefore, only 0.6 m spacing option can be used. If depth of wetting is to be controlled at 0.40 m depth and more, the dripper spacing can be 0.75 m. The 0.9 m dripper spacing is very difficult to use, because to get a continuous wetted strip, the drippers are to be operated till 0.9 m width/diameter is wetted. But when 0.9 m diameter is wetted, the depth of wetting would be beyond 0.40 m and there would be loss of water and nutrients below 0.40 m.

Soil moisture in volumetric basis (Sm) = specific gravity of bulk soil
                                         × Soil moisture(gravimetric)

Soil moisture in depth basis at permanent wilting point $= 2.5 \times 1.6/100$
$$= 0.2 \text{ m/m depth of soil}$$

*When the root depth is 20 cm*

Soil moisture in depth basis for 20%(gravimetric)
$$= 20 \times 1.6/100$$
$$= 0.32 \text{ m/m depth of soil}$$
Available soil water per metre depth of soil at this soil moisture level
$$= 0.32 - \text{permanent wilting point in depth basis}$$
$$= 0.32 - 0.2$$
$$= 0.12 \text{ m}$$

Depth of water needed to bring the soil
moisture to field capacity for 1 m depth of soil
$$= \text{Available soil water at field capacity} - \text{Available soil water at 18\%(gravimetric)}$$
$$= 0.18 - 0.12$$
$$= 0.012 \text{ m}$$

Depth of water needed to bring the soil moisture to field capacity for 1
m depth of soil
$$= \text{Available soil water at field capacity}$$
$$- \text{Available soil water at 18\%(gravimetric)}$$
$$= 0.18 - 0.12$$
$$= 0.012 \text{ m}$$

Depth of water needed to bring the soil moisture to field capacity for 1 m depth of soil
$$= \text{Available soil water at field capacity} - \text{Available soil water at 18 \% (gravimetric)}$$
$$= 0.18 - 0.12$$
$$= 0.06 \text{ m}$$
Depth of water needed to bring the soil moisture to field capacity for 0.2 m depth of soil
$$= 0.06 \times 0.2$$
$$= 0.012 \text{ m}$$

Volume of water needed per metre length of lateral $=$ width of wetting
$$\times \text{ Depth of water} \times 1$$

$$= 0.74 \times 0.012 \times 1$$
$$= 0.00888 \, \text{m}^3$$
$$= 8.88 \, \text{l}$$

Number of drippers per metre length of lateral $= 1/0.6$
$$= 1.67 \, \text{drippers}$$

Volume of water needed from each dripper $= 8.88/1.67$
$$= 5.32 \, \text{l}$$
Time of operation $= 5.32/2$
$$= 2.66 \, \text{h}$$

*When the root depth is 30 cm*

Depth of water needed to bring the soil moisture to field capacity for 0.3 metre depth of soil
$$= 0.06 \times 0.3$$
$$= 0.018 \, \text{m}$$

Volume of water needed per metre length of lateral
$$= \text{width of wetting} \times \text{Depth of water} \times 1$$
$$= 0.84 \times 0.018 \times 1$$
$$= 0.01512 \, \text{m}^3$$
$$= 15.12 \, \text{l}$$

Volume of water needed from each dripper $= 15.12/1.67$
$$= 9.05 \, \text{l}$$
Time of operation $= 9.05/2$
$$= 4.52 \, \text{h}$$

*When the root depth is 40 cm*

Soil moisture in depth basis for 18%(gravimetric) $= 18 \times 1.6/100$
$$= 0.288 \, \text{m/m depth of soil}$$

Depth of water needed to bring the soilmoisture to field capacity for 1 m depth of soil
= Available soil water at field capacity
− Available soil water at 20 % (gravimetric)
$$= 0.18 - 0.088$$

$= 0.092\,\mathrm{m}$

Depth of water needed to bring the soil moisture to field capacity for 0.4 m depth of soil

$= 0.092 \times 0.4$

$= 0.0368\,\mathrm{m}$

Volume of water needed per metre length of lateral $=$ width of wetting $\times$ Depth of water

$= 0.93 \times 0.0368 \times 1$

$= 0.0342\,\mathrm{m}^3$

$= 34.2\,\mathrm{l}$

$$\text{Volume of water needed from each dripper} = 34.2/1.67$$
$$= 20.47\,\mathrm{l}$$

Time of operation

$$= 20.47/2$$
$$= 10.23\,\mathrm{h}$$

In the preceding numerical example, it must be noted that the wetted volume is assumed as a simple rectangle rather than semi-ellipsoid. This assumption is to make the calculations simpler. In the preceding example, irrigation interval is not taken into account. Whenever soil moisture content falls below a critical level, irrigation is done.

Sometimes irrigation interval is fixed, and for every irrigation we must calculate the water to be applied to the soil to meet the evapotranspiration needs. In such situations, the wetted volume must have sufficient capacity to store the applied water. In the following example, we can see how the selection is made for such cases.

**Problem** Let us take a representative area as shown in Fig. 12. Spacing between

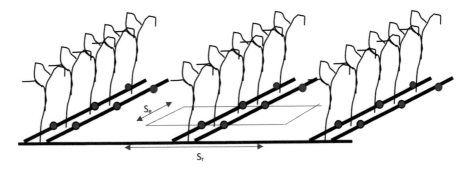

**Fig. 12** Representative area for selection of dripper-lateral configuration

crop rows is denoted as $S_r$, and spacing between drippers in a lateral is denoted as $S_e$. Two lateral pipes are provided per crop row. Let us denote evapotranspiration during peak water use as $Et_{cp}$ and minimum possible irrigation interval in which the farmer can do irrigation as $f_{min}$.

When a farmer irrigates during peak water use, he would apply an amount equal to $(Et_{cp} \times f_{min})$. The spacing between drippers, number of laterals and spacing between the laterals for each crop row have to be selected in such a way that when $(Et_{cp} \times f_{min})$ amount of water is applied, it should not go below the root zone. Also, the soil water holding characteristics and the wetted bulb must be able to hold this moisture without causing drainage and surface runoff.

Let us assume that the crop grown is sugarcane with following data:

Spacing between each crop row is 1.5 m; depth of wetting is 0.45 m; maximum expected evapotranspiration rate is 6 mm/d; minimum possible irrigation interval is 1 day; soil is sandy loam with saturated hydraulic conductivity of $7.2 * 10^{-6}$ m/s; water holding capacity of soil is 0.1; and management allowable deficit is 0.6.

We will select a suitable dripper spacing, dripper discharge rate and number of laterals per crop row for this data.

We will start with an arbitrary selection of 4 l/h dripper and use Schwartzman and Zur equation for our analysis.

$$q = 4\,l/h = 1.11 \times 10^{-6} m^3/s$$

Wetted diameter for wetting a depth of 0.45 m $= d = 1.32\left(\dfrac{zq}{K_s}\right)^{0.33}$

$$d = 0.543\,m$$

We will select a dripper spacing of 80% of diameter of wetting with a factor of safety of 20%

$$d = 0.543\,m$$

In the market, standard dripper spacing available is in multiples of 10 cm. Hence, we will select 40 cm spacing between drippers in the lateral.

Depth of irrigation water to be applied during peak crop water use $= Et_{cp} \times f_{min}$ $= 0.006 \times 1 = 0.006$ m.

Depth of Irrigation water to be applied during peak crop water use

$= Et_{cp} \times f_{min} = 0.006 \times 1 = 0.006\,m$

Depth of soil needed for storing 0.006 m depth of water

$\dfrac{ET_{cp}\,f_{min}}{WHC \times MAD}$

Representative area $= 0.40 \times 1.50 = 0.6\,m^2$

Volume of soil needed for storing 0.006 m depth

of water in the representative area $= 0.6 \times 0.1 = 0.06 \text{ m}^3$

Since we adopt drip irrigation, the wetted volume in the representative area is not as a rectangular prism with 0.20 m depth but it is rather a combination of semi-ellipsoids. Each ellipsoid will have a depth of 0.45 m and diameter of 0.4 m because the farmer has to operate only in this way to get a continuous wetted strip.

$$\text{Volume of soil wetted by one dripper} = V = \frac{\pi z d^2}{6}$$

$$= \frac{\pi \times 0.45 \times 0.4^2}{6} = 0.038$$

Number of drippers needed with in the representative area

$$= \frac{0.06}{0.038}$$

$$= 1.58$$

We need 1.58 numbers of drippers to irrigate the representative area. If we lay out only one lateral for one crop row, we will have 2 drippers irrigating the representative area. But one half of each dripper only would irrigate the representative area. Therefore, the total number of drippers that would irrigate the representative area is only one. So laying of one lateral per crop row is not sufficient. We must lay 2 laterals per crop row, and if we do so, we will have effectively 2 drippers (i.e. 4 × ½ drippers) in the representative area (Fig. 12).

We will start with another arbitrary selection of 2 l/h dripper and do a similar set of calculation as above:

$$q = 2 \, \text{l/h} = 5.55 \times 10^{-7} \text{m}^3/\text{s}$$

Wetted diameter for wetting a depth of 0.45 m

$$= d = 1.32 \left( \frac{zq}{K_s} \right)^{0.33}$$

$$d = 0.43 \, \text{m}$$

We will select a dripper spacing of 80% of diameter of wetting with a factor of safety of 20%

$$d = 0.34 \, \text{m}$$

In the market, standard dripper spacing available is in terms of multiples of 10 cm. Hence, we will select 30 cm spacing between drippers in the lateral.

Depth of irrigation water to be applied during peak crop water use $= \text{Et}_{cp} \times f_{min}$ $= 0.006 \times 1 = 0.006$ m.

Depth of soil needed for storing 0.006 m depth of water $= \frac{ET_{cp} f_{min}}{\text{WHC} \times \text{MAD}} = 0.10$ m.

Representative area $= 0.30 \times 1.50 = 0.45 \text{ m}^2$.

Volume of soil needed for storing 0.006 m depth of water in the representative area $= 0.45 \times 0.1 = 0.045$ m$^3$.

Volume of soil wetted by one dripper $= V = \frac{\pi z d^2}{6} = \frac{\pi 0.450.3^2}{6} = 0.02$ m$^3$.

Number of drippers needed with in the representative area $= \frac{0.045}{0.02} = 2.1$

Therefore for this case also we need to adopt 2 lateral pipes per crop row. Each lateral would have 2 l/h dripper with 0.3 m spacing. This selection would provide 2 drippers (or 4 times ½ dripper) for the representative area.

For option two, the dripper discharge per metre length of lateral is 6.66 which is lower than for the option one. Lower values of the dripper discharge per metre length of lateral lead to larger optimal lateral lengths, and the discussions and proof related to this are given in next chapter.

In option one, dripper discharge rate is higher than the option two. It must be recalled that higher dripper discharge rate would cause less clogging problem and less irrigation time. So, the choosing option lies with the farmer which depends on his preferences.

It is to be noted that in this method of selection of dripper and lateral configuration, the objective is to wet a specified volume of soil which stores water during peak water use period in the irrigation interval. The specified volume wetted is found using the data of peak crop water use.

## Emission Point Layouts for Widely Spaced Crops

Rather than deciding the lateral–dripper configuration based on volume of soil needed for accommodating peak crop water use, the lateral–dripper configuration selection is also done by arbitrarily fixing up the fraction area to be wetted by drippers. More the area is wetted, more the possibility of increase in yield due to improved drought resilience and also availability of more soil for nutrient extraction. But the cost of installation would increase if the fraction of area wetted is increased.

Keller and Bliesner recommend a method for choosing the lateral–dripper config- uration. This method is very much suited for widely spaced tree crops. Table 3 provides values of estimated area wetted by a 4 l/h emitter operated for different wetted soil depth in different types of soils.

They are based on the daily or alternate day irrigations that apply volumes of water sufficient to slightly exceed the crop water use rate. For wetting depths of 0.75 m and 1.5 m, the wetted diameter and optimal dripper spacing are given in Table 3. The second value is equal to the diameter of the wetted bulb. The first value is equal to the 80% of the first value. By multiplying the first value and the second value, the area wetted by the dripper is obtained. The reason why it is given like this is that the area obtained by multiplying the first value and the second value is approximately equal to the area of the circle obtained using the first value; for instance, $0.4 \times 0.5$ is approximately equal to $\frac{\pi}{4} \times 0.5^2$. (i.e. $0.2$ m$^2 \cong 0.196$ m$^2$). The table can be used for selecting the emitter spacing to get a continuous wetted strip. For instance,

**Table 3** Estimate of area wetted by a 4 l/h under different field conditions ( Adopted from Keller and Bliesner (1990)

| Wetted soil depth and soil texture | Degree of soil stratifications and equivalent wetted soil area (m × m) | | |
|---|---|---|---|
| | Homogeneous | Stratified | Layered |
| *Depth = 0.75 m* | | | |
| Coarse | 0.4 × 0.5 | 0.6 × 0.8 | 0.9 × 1.1 |
| Medium | 0.7 × 0.9 | 1.0 × 1.2 | 1.2 × 1.5 |
| Fine | 0.9 × 1.1 | 1.2 × 1.5 | 1.5 × 1.8 |
| *Depth = 1.5 m* | | | |
| Coarse | 0.6 × 0.8 | 1.1 × 1.4 | 1.4 × 1.8 |
| Medium | 1.0 × 1.2 | 1.7 × 2.1 | 2.2 × 2.7 |
| Fine | 1.2 × 1.5 | 1.6 × 2.0 | 2.0 × 2.4 |

if the wetted soil depth needed is 0.75 m and the soil type is coarse and stratified, then the emitter spacing may be chosen as 0.6 m. Most soils are stratified. Stratified refers to relatively uniform texture but having some particle orientation or some compaction layering that gives higher horizontal than vertical permeability. Layered refers to changes in texture with depth as well as particle orientation and moderate compaction. Though the table values correspond to 4 l/h discharge rate, the same value can be used approximately for any other discharge rate, because the time of operation can be adjusted to apply the total volume required.

Since the table values correspond to deeper depths of wetting, 0.75 m depth wetting can be used for smaller trees and 1.5 m depth wetting for larger trees. Figure 13 shows an emission point layouts with a single lateral for widely spaced crops. Definitions of the terms related to Fig. 13 are as follows:

| | |
|---|---|
| $P_d$ | Per cent shaded: Average horizontal area shaded by the crop canopy as a percentage of the total crop area |
| $P_w$ | Per cent wetted: Average horizontal area wetted in the top part of the crop root zone (15–30 cm depth) as a % of the total crop area. Generally a per cent wetted area between 30 and 70 % is used in design |
| $S_e$ | Emitter spacing |
| $S_l$ | Lateral spacing |
| $S_p$ | Plant spacing in row |
| $S_r$ | Row spacing (distance between plant rows) |
| $W$ | Width of the wetted strip |
| $S'e$ | Optimal emitter spacing: It is 80% of wetted diameter estimated from field tests or value taken from Table 3 |
| $N_p$ | Number of emission points per plant |

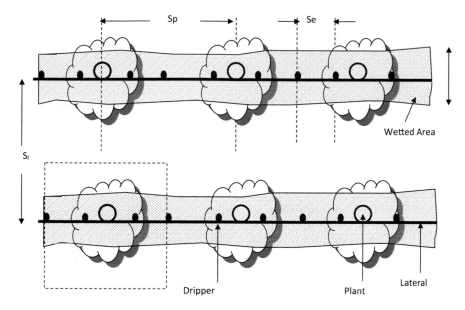

**Fig. 13** Emission layout with single lateral for widely spaced crops

## Computing Percentage Wetted Area

For straight single lateral systems with Se $\leq$ S$'$e, the percentage wetted can be computed as follows:

$$P_w = \frac{N_p S_e W}{S_p S_r} \times 100 \tag{13}$$

For double lateral systems and multi-exit layouts, if $S_e \leq S'_e$.

$$P_w = \frac{N_p S_e (S_e + W)/2}{S_p S_r} \times 100 \tag{14}$$

Microsprinkler wets a larger surface area of soil than drip emitters (Figure 14). They are often used on coarse textured soils where wetting a sufficiently large area would require a large number of drip emitters.

Figure 12 shows wetting pattern of a microsprinkler. Approximately the area wetted beyond the spray radius can be assumed as 50% of the S$'$e value from Table 3 for the homogeneous soils. Therefore the wetted surface area for spray emitter is as follows:

$$P_w = \frac{N_p \left(A_s + \frac{P_s S'_e}{2}\right)}{S_p S_r} \times 100 \tag{15}$$

where $A_s$ is the area of spray. This may be found out by allowing the microsprinkler to fall on a blacktop surface and finding the area, and also the perimeter of spray $(P_S)$ can be found out. The microsprinklers are assumed that they are placed without any overlap.

**Problem** An orchard with a tree spacing of 3.0 m $\times$ 5.0 m planted on a deep, medium textured homogeneous soil. Three emitter configurations are being considered.

1. A single row of 4 l/h emitters
2. A looped layout with 4 numbers of 4 l/h emitters/ tree
3. A microsprinkler wets a surface area with radius of 1.0 m.

Find the percentage area wetted by each emitter configuration and recommend a configuration which provides the area of wetting more than 33%.

From Table 3, for deep rooted crop, medium textured homogeneous soil, the value of S'e is 1.0 m and 'w' is 1.2 m. Therefore,

Let, spacing between emitters $(S_e) = 1.0$ m

**Case—1**

A single row of 4 l/h emitters

Spacing between trees in the row $= 3.0$ m

No. of emitters. Tree $= 3.0/1.0 = 3$

$$\text{Percentage area wetted (Pw)} = \frac{3 \times 1 \times 1.2}{3 \times 5} \times 100$$

$$= 24\%$$

**Case—2**

zig-zag configuration—4 emitters

$$P_w = \frac{4 \times 1 \times (1 + 1.2)/2}{3 \times 5} \times 100$$

$$= 29\%$$

**Fig. 14** Wetting pattern for microsprinklers

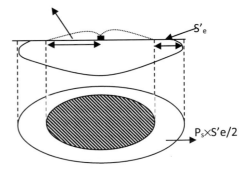

### Case—3
Microsprinkler

$$A_s = \pi r^2 = \pi \times 1^2 = 3.14 \, \text{m}^2$$
$$P_s = 2\pi r = 6.28 \, \text{m}$$

$$P_w = \frac{1\left(3.14 + \frac{(1.0 \times 6.28)}{2}\right)}{3.0 \times 5.0} \times 100$$
$$P_w = 42\%$$

Case-3 provides more than 33% of wetted area. Hence this is recommended for use.

**Table 4** Wetted area (%) for different soils with respect to lateral spacing and dripper discharge rates for applying 40 mm [4]

| Effective spacing between laterals (m)[a] | Effective emission point discharge rate[b] | | | | | | | | | | | | | | |
|---|---|---|---|---|---|---|---|---|---|---|---|---|---|---|
| | 1.5 lph (0.4 gph) | | | 2 lph (0.5 gph) | | | 4 lph (1 gph) | | | 8 lph (2 gph) | | | Over 12 lph (3 gph) | | |
| | Soil texture and recommended emission point spacing on the lateral m[c] | | | | | | | | | | | | | | |
| 1.0 m = 3.3 ft | C | M | F | C | M | F | C | M | F | C | M | F | C | M | F |
| | 0.2 | 0.5 | 0.9 | 0.3 | 0.7 | 1.0 | 0.6 | 1.0 | 1.3 | 1.0 | 1.3 | 1.7 | 1.3 | 1.6 | 2.0 |
| *Percentage of soil wetted*[d] | | | | | | | | | | | | | | | |
| 0.8 | 38 | 88 | 100 | 50 | *100* | 100 | *100* | 100 | 100 | 100 | 100 | 100 | 100 | 100 | 100 |
| 1.0 | *33* | 70 | *100* | 40 | 80 | 100 | 80 | 100 | 100 | 100 | 100 | 100 | 100 | 100 | 100 |
| 1.2 | 25 | 58 | 92 | 33 | 67 | *100* | 67 | *100* | 100 | *100* | 100 | 100 | 100 | 100 | 100 |
| 1.5 | 20 | 47 | 73 | 26 | 53 | 80 | 53 | 80 | *100* | 80 | *100* | 100 | *100* | 100 | 100 |
| 2.0 | 15 | *35* | 55 | 20 | *40* | 60 | 40 | 60 | 80 | 60 | 80 | *100* | 80 | *100* | 100 |
| 2.5 | 12 | 28 | 44 | 16 | 32 | 48 | 32 | 48 | 64 | 48 | 64 | 80 | 64 | 80 | *100* |
| 3.0 | 10 | 23 | 37 | 13 | 26 | 40 | 26 | 40 | 53 | 40 | 53 | 67 | 53 | 67 | 80 |
| 3.5 | 9 | 20 | 31 | 11 | 23 | *34* | 23 | *34* | 46 | *34* | 46 | 57 | 46 | 57 | 68 |
| 4.0 | 8 | 18 | 28 | 10 | 20 | 30 | 20 | 30 | 40 | 30 | 40 | 50 | 40 | 50 | 60 |
| 4.5 | 7 | 16 | 24 | 9 | 18 | 26 | 18 | 26 | *36* | 26 | *36* | 44 | *36* | 44 | 53 |
| 5.0 | 6 | 14 | 22 | 8 | 16 | 24 | 16 | 24 | 32 | 24 | 32 | 40 | 32 | 40 | 48 |
| 5.5 | 5 | 12 | 18 | 7 | 14 | 20 | 14 | 20 | 27 | 20 | 27 | *34* | 27 | *34* | 40 |

[a]Where double laterals (or laterals with multiple outlet emitters) are used in orchards, enter the table with both the spacing between outlets to either side of the row and across the space between the rows and proportion the percentages

[b]Where relatively short pulses of irrigation area applied, the effective emission point discharge rate should be reduced to approximately half of the instantaneous rate for safety

[c]The texture for the soils is designated by C—coarse, M—medium and F—fine. The emission point spacing is equal to approximately 80 per cent of the largest diameter of the wetted area of the soil underlying the point. Closer spacing on the lateral will not affect the percentage area wetted

[d]The percentage of soil wetted is based on the area of the horizontal section approximately 0.30 m beneath the soil surface. Caution should be exercised where less than 1/3 of the soil volume will be wetted

Table 4 developed by Keller and Karmeli [7] is also an useful research data in field by applying 40 mm of irrigation in different types of soils for different spacing between laterals and dripper discharge rates. This table can be used to know the approximate wetted area in % without doing any sort of calculation as done in preceding problem.

# Case Study

The author of this book selected spacing between laterals, dripper discharge rate and dripper spacing for a real-life situation, and his experiences are presented here [8]. Crop is sugarcane. Soil texture at the site is sandy clay loam. The quality of the water used has significant amount of organic matter as the source of water was a canal from a river. When the river water is not available, groundwater is used. During the period of canal flow the aquifers get recharged with river water and the electrical conductivity is around 1 dS/m, and during the periods of no canal flow, the electrical conductivity increases to around 3 dS/m.

## *Dripper Discharge Rate*

Lower dripper discharge rate has an advantage that the length of the lateral can be higher. Higher dripper discharge rate emitters have either higher velocity of flow or larger diameter of flow. Lower dripper discharge rate has a disadvantage of quicker clogging of emitters due to sluggish flow through the emitter passage. Generally, dripper discharge rate used is 4 l/h. In this case, 4 l/h dripper discharge rate is selected.

## *Spacing Between Drippers*

Dripper spacing is normally decided based on the minimum depth to be wetted when irrigation is done. The minimum depth to be wetted is decided based on the rooting pattern. Smith et al. [9] studied the root growth characteristics of sugarcane crop and found that majority of roots was within 0.35 m (Fig. 15). The rooting pattern of sugarcane crop was also investigated by the author by cutting-open the soil. The maximum distribution of roots was found to be within 0.40 m depth (Fig. 16). Hence, the minimum depth to be irrigated was fixed at 0.40 m.

Then a wetting pattern experiment was conducted with a 2 l/h dripper (Equivalent for 4 l/h dripper when wetting pattern is studied by cutting open a trench). For the depth of wetting of 0.40 m, the diameter of wetting was found to vary between 0.38 and 0.41 m. Hence a dripper spacing of 0.4 m was selected.

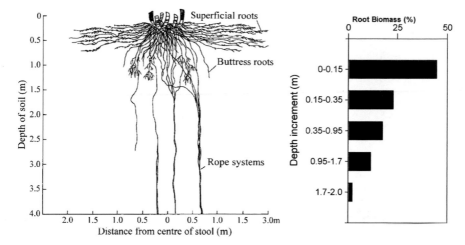

**Fig. 15** Rooting characteristics of sugarcane [8]

**Fig. 16** Rooting pattern of
sugarcane crop in
the investigated field [8]

## *Spacing of Lateral*

As the spacing between the laterals is increased, the total length of laterals needed to
be installed for an area decreases. Higher spacing between the plant rows also helps

**Fig. 17** Geometry of
laterals, drippers and plants
[5]

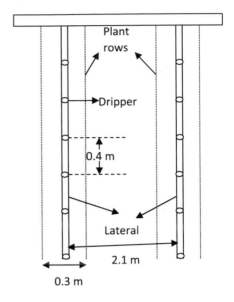

in improving the movement of inter-culture machines between crop rows. Garside et al. (2009) conducted field experiments with plant spacing of 1.5 m (single row spacing), 2.1 and 2.8 m (dual row spacing) for sugarcane crop. They reported that the yield for 2.1 m spacing with dual crop rows was on par with 1.5 m single row spacing. They also reported that sugarcane crop has a very good adaptability for the changes in crop spacing and with increased row spacing, appreciable yield reduction did not occur. Hence, the lateral spacing was selected as 2.1 m with dual crop rows.

The spacing between the crop rows of a dual row is also a design issue. If the distance is more than 0.40 m, the wetting front is not guaranteed to reach the plant rows on either side of lateral. Hence 0.30 m plant row spacing was adopted between both the crop rows of a dual row (Fig. 17).

## Summary

Formula for maximum area that can be irrigated by the farmer under drip irrigation based on the crop grown and water available is presented with examples. Method for finding maximum subunit area is also presented. Empirical formulae for finding the wetting pattern of drippers when operated on soil profile are presented. Dripper spacing selection procedure is discussed with examples. Dripper layouts for widely spaced crops are presented. Planning dripper spacing and spacing of laterals are demonstrated with a real-time case study.

## Assessment

### Part A

1. Normally under drip irrigation system the _____ varies between 30 to 70%.
   a. **surface area wetted**
   b. root zone
   c. foliage
   d. none of these

2. Choose the correct statement
   a. **The advantages of higher dripper discharge are that less clogging problem and less irrigation time.**
   b. The advantages of higher dripper discharge are that more clogging problem and less irrigation time.
   c. The advantages of higher dripper discharge are that more clogging problem and more irrigation time.
   d. The disadvantages of higher dripper discharge are that more clogging problem and more irrigation time.

3. Amin & Ekhmaj equation was obtained by doing statistical regression analysis using data available in literature. Say True or False—**True**
4. Higher dripper discharge rate would cause less clogging problem and less irrigation time. Say True or False—**True**

### Part B. Descriptive Answers

1. What is the procedure used for determining the spacing of drippers in a lateral to get continuous strip wetting?
2. How much must be the dripper discharge rate if a trench is excavated before wetting pattern experiment is done, for examining the wetting pattern of 4 l/h dripper?
3. Write down the Schwartzman and Zur equation for finding optimal dripper spacing of drippers in a lateral to get continuous strip wetting?
4. Draw a line sketch showing one lateral per two crop rows in a paired crop rows?
5. Discuss with a sketch for different lateral and dripper layouts normally used for tree crops?

## References and Further Reading

1. Roth, R. L. (1974). Soil moisture distribution and wetting pattern from a point-source. *Proc Second International Drip Irrigation Congress*, California, USA, 246–251.
2. Dasberg, S., & Dani Or. (1999). *Drip irrigation*. Springer.

3. Schwartzman, & Zur. (1986). Emitter spacing and geometry of wetted soil volume. *Journal of Irrigation & Drainage Division, 112*(3), 242–253
4. Wooding, R. A. (1968). Steady infiltration from large shallow circular pond. *Water Resource Research, 4*, 1259–1273.
5. Ravikumar, V., Ranghaswami, M. V., Appavu, K., & Chellamuthu, S. (2011). *Microirrigation and irrigation pumps*. Kalyani Publishers.
6. Amin Mohamad, S. M., & Ekhmaj Ahmed, I. M. (2015). *DIPAC—Drip irrigation water distribution pattern calculator*. https://www.researchgate.net/publication/266862730
7. Keller, J., & Karmeli, D. (1974). Trickle irrigation design parameters. *Transactions of American Society of Agricultural Engineers, 17*, 678–684.
8. Ravikumar, V., Angaleeswari, M., & Vallalkannan, S. (2021). *Design and Evaluation of drip irrigation system for sugarcane in India*. Springer. https://doi.org/10.1007/s12355-021-00983-7
9. Smith, D. M., Inman-Bamber, N. G., & Thorburn, P. J. (2005). Growth and function of the sugarcane root system, *Field Crops Research*. 92(2), 169–183. https://doi.org/10.1016/j.fcr.2005.01.017

# Hyraulics of Microirrigation Pipes

## Introduction

In both drip and sprinkler pipes multi-outlet pipes are encountered. But there are some significant differences between a drip and a sprinkler pipe. In drip irrigation, number of drippers is more than the number of sprinklers in a lateral pipe. The sprinklers are kept on a riser pipe whereas drippers are not kept on a riser pipe but on the land surface. Owing to these factors, the equations developed for drip and sprinkler design are slightly different.

In this chapter, formulas are derived for finding friction loss and average friction loss in multi-outlet pipe. Methods for finding locations and the magnitudes of average pressure heads, minimum and maximum pressure heads for the multi-outlet pipes laid on uniform slopes are presented. Analytical equations for finding coefficient of variation of pressure for a multi-outlet pipe laid in uniform slopes, and design procedure for optimal diameter and length of lateral/submain based on coefficient of variation of pressure is also introduced. The procedures for accounting pressure losses due to connection of emitters in lateral is also presented.

## Friction Loss Through Drip Laterals and Submains

The analytical equation developed for sprinkler irrigation to find out friction loss is often used as such for drip irrigation also. In case of sprinkler irrigation, it is to be noted that as the number of outlets increases, the Christiansen factor $F$ approaches to 0.3636. Mostly the number of drippers in a lateral or the number of laterals in submain is more than 30. In such cases, the Christiansen factor $F$ can be assumed as 0.3636.

Now we will also prove that the Christiansen factor $F$ is equal to 0.3636 if the number of drippers is infinite in a lateral. In order to derive the equation, we make following assumptions:

© The Author(s), under exclusive license to Springer Nature Singapore Pte Ltd. 2023     247
V. Ravikumar, *Sprinkler and Drip Irrigation*,
https://doi.org/10.1007/978-981-19-2775-1_9

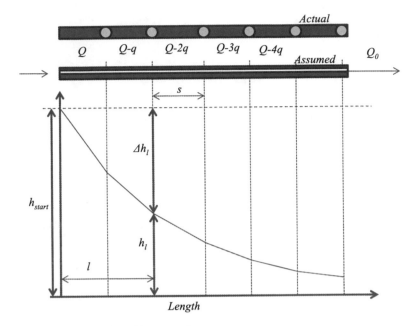

**Fig. 1** Multi-outlet drip pipe with open end

- Diameter of the pipe is uniform throughout the length.
- Out flow through drippers does not occur at discrete points.
- Outflow occurs throughout the pipe section uniformly. In Fig. 1, at the top part of the image, the circles with in the pipe indicate drippers if the pipe is assumed as lateral. But we assume instead, a continuous slot termed as 'Assumed'. Flow is assumed to occur uniformly through the slot.

Discharge rate ($q$) per unit length of lateral is called as specific discharge rate (SDR) which is equal to $q/s$; $q$ is dripper discharge rate and s is dripper spacing.

In Chap. 4, following equation was already introduced for finding friction loss when pipe diameter is less than 125 mm (Keller and Bliesner 1990):

$$\Delta h = \frac{K Q^{1.75} L}{D^{4.75}}$$

where $\Delta h$ is friction loss (m), $Q$ is discharge rate (litres per second), $L$ is the length of pipe (metres) and $D$ is the diameter of pipe (millimetres). The value of $K$ for this set of units is $7.89 \times 10^5$. For convenience of handling the terms, we will write the preceding equation as follows:

$$\Delta h = a Q^{1.75} L$$

where $a = K/D^{4.75}$. The differential form of the preceding equation is as follows:

$$\frac{dh}{dl} = -aQ_l^{1.75}$$

where $Q_l$ is the discharge rate at $l$ length from the start of the lateral. Minus is to be introduced in the preceding equation because differential calculus relies on reference axes, and it is known that the slope of the friction loss curve at all the points is negative. Note also that $Q$ is a function of $l$. Integrating the preceding equation from the start of the pipe to any length $l$ yields following equation:

$$\int_{h_{start}}^{h_l} dh = a \int_0^l Q_l^{1.75} dl \tag{1}$$

The discharge flowing through the pipe at any length $l$ from the start of the pipe can be expressed as follows:

$$Q_l = Q - \frac{q}{s}l$$

Substituting for $Q_l$ in Eq. 1 yields the following equation:

$$\int_{h_{start}}^{h_l} dh = -a \int_0^l \left( Q - \frac{q}{s}l \right)^{1.75} dl$$

$$[h]_{h_{start}}^{h_l} = -a \left[ \frac{\left( Q - \frac{ql}{s} \right)^{2.75}}{-2.75\left(\frac{q}{s}\right)} \right]_0^l$$

$$\Delta h_l = \frac{a}{2.75\left(\frac{q}{s}\right)} \left( Q^{2.75} - \left( Q - \frac{ql}{s} \right)^{2.75} \right) \tag{2}$$

Equation 2 can be used for finding friction loss in any lateral or submain pipe segment which need not to have been closed.

If end of the pipe is closed, we can write as follows:

$$\frac{q}{s} = \frac{Q}{L}$$

$$Q = \frac{qL}{s}$$

Then Eq. 2 becomes as below:

$$\Delta h_l = \frac{a}{2.75} \left( \frac{q}{s} \right)^{1.75} \left( L^{2.75} - (L-l)^{2.75} \right) \tag{3}$$

Friction loss for the full length of the pipe '$L$' is as follows:

$$\Delta h_L = \frac{a}{2.75}\left(\frac{q}{s}\right)^{1.75} L^{2.75} \tag{4}$$

Putting the following substitution in Eq. 4 yields,

$$\frac{q}{s} = \frac{Q}{L}$$

$$\Delta h_L = \frac{a}{2.75} Q^{1.75} L \tag{5}$$

Equations 4 and 5 are similar to the equation derived in the sprinkler design chapter with one exception. Here the Christiansen's factor is a constant and is equal to 1/2.75 or 0.3636. We can also write the preceding equation as follows:

$$\Delta h_L = \frac{\text{Friction loss for blind pipe}}{2.75} \cdot \tag{6}$$

Equation 3 can also be written as follows:

$$\Delta h_l = \Delta h_L\left(1 - \left(1 - \frac{l}{L}\right)^{2.75}\right)$$

## Average Friction Loss

We will derive here an equation for finding average friction loss for drip lateral/submain. From Fig. 2, it can be seen that the average friction loss is the average of ordinates of polygon ABC.

$$\overline{\Delta h_L} = \frac{\text{Area ABC}}{L}$$

$$\overline{\Delta h_L} = \frac{\int_0^L \Delta h_l dl}{L} \tag{7}$$

We know that Eq. 3 is as follows:

$$\Delta h_l = \frac{a}{2.75}\left(\frac{q}{s}\right)^{1.75}\left(L^{2.75} - (L - l)^{2.75}\right)$$

Therefore Eq. 7 can be written as follows:

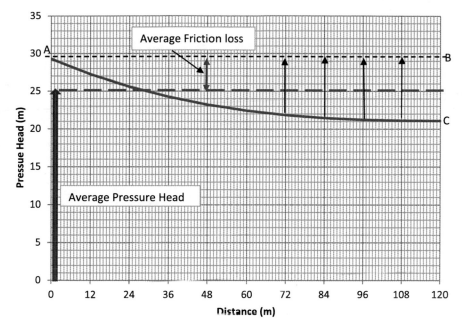

**Fig. 2** Illustration for finding average friction loss

$$\overline{\Delta h_L} = \frac{a}{2.75L}\left(\frac{q}{s}\right)^{1.75}\int\limits_{0}^{L}\left(L^{2.75} - (L-l)^{2.75}\right)\mathrm{d}l$$

$$\overline{\Delta h_L} = \frac{0.733a}{2.75}\left(\frac{q}{s}\right)^{1.75}L^{2.75}$$

$$\overline{\Delta h_L} = 0.733\Delta h_L \qquad\qquad (8)$$

Recall that in sprinkler irrigation design equations, Chrisiansen's '$F$' and Anwar's '$\overline{F}$' were used. But in the preceding design equation developed for drip irrigation, in place of Christiansen's factor and Anwar's factor, constants are used. These constants are independent of number of outlets in the pipe. Caution must be exercised when using these equations, because the extent of error introduced gets increased as the number of outlets in a lateral/submain pipe gets decreased.

Equation 2 is more generic form of equation and this can be used for many situations. For instance, in Fig. 3, a telescopic/tapered pipe is shown. A telescopic pipe has different diameters at different segments and the diameter is larger at the start and the diameters become progressively smaller along the direction of flow. For calculation of friction loss in telescopic pipes, Eq. 2 must be used for each pipe segment of uniform diameter separately.

**Problem** A submain is operating with an inlet discharge ($Q$) of 4.1667 l/s (Fig. 4). It runs for a total length ($L$) of 45 m. Diameter of the submain ($D$) is 58.6 mm. Laterals

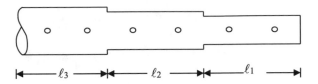

**Fig. 3** Telescopic (tapered) pipe

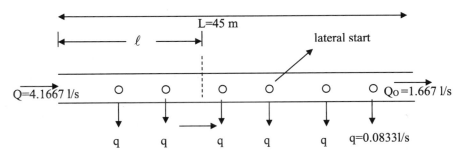

**Fig. 4** Multi-outlet drip pipe with open end for the problem

are fixed on the submain at a spacing ($s$) of 1.5 m. Each lateral discharges at rate ($q$) of 0.0833 l/s. There is an out flow ($Q_o$) of 1.667 l/s at the end of the submain. The pressure at the start of the submain ($h_{start}$) is 15 m. Find out the pressures at the locations of each lateral start.

**Solution** If Eq. 2 is applied for every length in multiples of 1.5 m, the pressure losses obtained are shown in Table 1.

**Problem** A submain is operating with an inlet discharge ($Q$) of 4.1667 l/s. It runs for a total length ($L$) of 75 m and closed at the end. Diameter of the submain ($D$) is 58.6 mm. Laterals are fixed on the submain at a spacing ($s$) of 1.5 m. Each lateral discharges at rate ($q$) of 0.0833 l/s. Find out the pressure losses at the location of 15, 30, 45, 60 and 75 m on the submain.

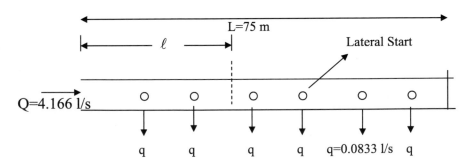

**Fig. 5** Multi-outlet drip pipe with closed end for the problem

**Table 1** Pressure losses for telescopic design problem

| Distance (m) | 1.5 | 3 | 4.5 | 6 | 7.5 | 9 | 10.5 | 12 | 13.5 | 15 |
|---|---|---|---|---|---|---|---|---|---|---|
| Pressure loss (m) | 0.06 | 0.11 | 0.16 | 0.21 | 0.26 | 0.31 | 0.36 | 0.40 | 0.44 | 0.48 |
| Pressure (m) | 14.94 | 14.89 | 14.84 | 14.79 | 14.74 | 14.69 | 14.64 | 14.60 | 14.56 | 14.52 |
| Distance (m) | 16.5 | 18 | 19.5 | 21 | 22.5 | 24 | 25.5 | 27 | 28.5 | 30 |
| Pressure loss (m) | 0.52 | 0.55 | 0.59 | 0.62 | 0.65 | 0.68 | 0.71 | 0.74 | 0.77 | 0.79 |
| Pressure (m) | 14.48 | 14.45 | 14.41 | 14.38 | 14.35 | 14.32 | 14.29 | 14.26 | 14.23 | 14.21 |
| Distance (m) | 31.5 | 33 | 34.5 | 36 | 37.5 | 39 | 40.5 | 42 | 43.5 | 45 |
| Pressure loss (m) | 0.81 | 0.83 | 0.85 | 0.87 | 0.89 | 0.91 | 0.92 | 0.94 | 0.95 | 0.96 |
| Pressure (m) | 14.19 | 14.17 | 14.15 | 14.13 | 14.11 | 14.09 | 14.08 | 14.06 | 14.05 | 14.04 |

**Solution** If Eq. 3 is substituted with the data, the pressure losses obtained are shown as below:

$$\Delta h_{l=15\,m} = 0.480\,m$$
$$\Delta h_{l=30\,m} = 0.790\,m$$
$$\Delta h_{l=45\,m} = 0.963\,m$$
$$\Delta h_{l=60\,m} = 1.035\,m$$
$$\Delta h_{l=15\,m} = 1.047\,m$$

# Ratio of Pressure Head Losses due to Friction Versus Ratio of Lengths

We already derived the following equation to find out friction loss in a multi-outlet pipe with the end closed:

$$\Delta h_L = \frac{a}{2.75}\left(\frac{q}{s}\right)^{1.75} L^{2.75}$$

Preceding equation can be written as follows:

$$\Delta h_L = bL^{2.75}$$

where $b = \frac{a}{2.75}\left(\frac{q}{s}\right)^{1.75}$

From the preceding equation following ratio can be written if SDR and diameter of pipe are treated as constants:

$$\frac{\Delta h_{L1}}{\Delta h_{L2}} = \frac{L1^{2.75}}{L2^{2.75}} \tag{9}$$

When a lateral is designed, usually length of lateral only is designed. During this process, friction loss calculation is done for different lengths of lateral without changing SDR and diameter of lateral. For such situations, the preceding equation is useful in doing calculations easily.

## Locations of Maximum and Minimum Pressures in Uniform Slopes

When a lateral is laid on upslope the pressure head would continuously decrease because both friction loss and elevation loss along the pipe length occur (Fig. 6). For this case, maximum pressure head will always be at the start and minimum pressure head will be at the end (Fig. 7).

When a lateral is laid on any uniform slope, the resultant pressure head curve is obtained by following equation:

**Fig. 6** Friction head loss and elevation loss in upslope

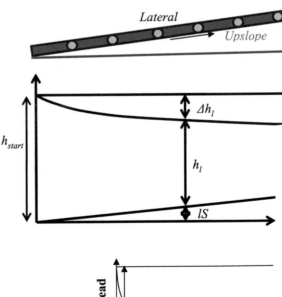

**Fig. 7** Pressure head curve in up slope (Type I)

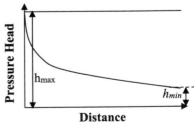

$$h_l = h_{start} - \Delta h_l + p\Delta Z_l \qquad (10)$$

where

$h_l$ is pressure head at any length $l$ from the start.

$h_{start}$ is pressure head at the start.

$\Delta h_l$ is total friction loss until length $l$ from the start.

$\Delta Z_l$ is total elevation difference between start and at $l$ ($\Delta Z_l = lS$).

$S$ is slope of land expressed in fraction.

$p = (-1)$ is for upslope and $p = 1$ for downslope.
When a lateral is laid on downslope, the nature of resultant pressure head curve depends the total friction loss in the lateral and the total pressure gain due to elevation along the pipe (Fig. 8). The location of minimum pressure head is where the pressure head loss gradient due to friction equals the pressure head gain gradient due to elevation.

In Fig. 9, the resultant pressure head curve alone is depicted separately. The shape of the curve in Fig. 9 corresponds to a downslope lateral.

We derived already an equation as below:

$$\Delta h_l = \Delta h_L \left(1 - \left(1 - \frac{l}{L}\right)^{2.75}\right)$$

We know that $\Delta Z_l = Sl$.

Substituting for $\Delta h_l$ and $\Delta Z_l$ in Eq. 10, we get as follows:

**Fig. 8** Friction loss and elevation gain curve for a downslope

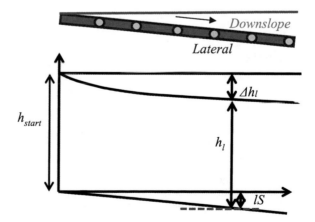

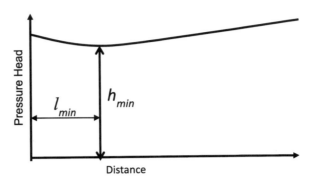

$$h_l = h_{\text{start}} - \Delta h_L \left( 1 - \left( 1 - \frac{l}{L} \right)^{2.75} \right) + Sl \tag{11}$$

Differentiating the preceding equation and equating it to zero are done for obtaining maximum or minimum pressure head.

$$S - 2.75 \frac{\Delta h_L}{L} \left( \left( 1 - \frac{l}{L} \right)^{1.75} \right) = 0$$

For lateral/submain, we will always get only minimum point when we use the preceding equation because maximum points are either at the start or at the end, and they are not also the points of inflection.

The term $\frac{\Delta h_L}{L}$ in the preceding equation is termed as average friction slope ($S_f$). Then the preceding equation can be rearranged as follows:

$$l_{\min} = L \left( 1 - \left( \frac{S}{2.75 S_f} \right)^{\frac{1}{1.75}} \right) \tag{12}$$

Equation 11 is written for find finding minimum pressure head as follows:

$$h_{\min} = h_{\text{start}} - \Delta h_L \left( 1 - \left( 1 - \frac{l_{\min}}{L} \right)^{2.75} \right) + Sl_{\min} \tag{13}$$

In the preceding equation $l_{\min}$ from Eq. 12 must be substituted.

Based on the nature of resultant pressure head curve, five types of slopes are recognized (Figs. 7 and 10). Type I profile would occur for the upslope, for the level field and for the very mild downslope situations.

Type IIa profile will get developed if the total friction loss in the lateral/submain is more than the total gain in pressure due to fall in elevation (i.e. $\Delta h_L > \Delta Z_L$). At the start of the pipe for Type IIa profile, the friction gradient is higher than slope and the friction gradient keeps on reducing as the length from the start increases. In this

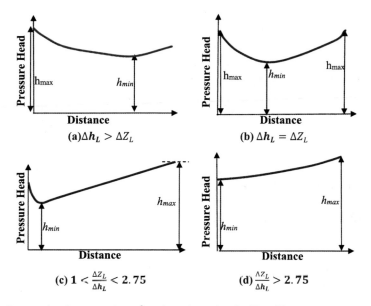

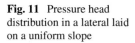

**Fig. 10** Pressure head curve in downslope lateral or submain (Type II)

**Fig. 11** Pressure head
distribution in a lateral laid
on a uniform slope

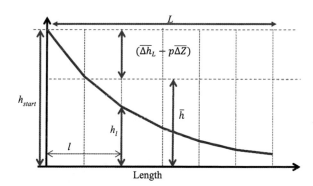

profile, minimum pressure occurs at a location where the friction gradient equals the slope.

Type IIb profile will get developed if the total friction loss in the lateral is equal to total gain in pressure due to fall in elevation (i.e. $\Delta h_L = \Delta Z_L$). For Type IIb, the minimum pressure head location will be towards the lateral start than Type IIa.

Type IIc profile gets developed when the ratio of total gain in pressure due to fall in elevation to the total friction loss in the lateral is greater than one. As this ratio increases the minimum pressure moves towards the lateral start and it reaches the lateral start when this ratio becomes 2.75. This can be proved by substituting $l_{min} = 0$ in Eq. 12.

$$0 = L\left(1 - \left(\frac{S}{2.75 S_f}\right)^{\frac{1}{1.75}}\right)$$

$$\frac{S}{S_f} = 2.75 = \frac{\Delta Z_L}{\Delta h_L}$$

Type IId profile would get developed when the ratio $\frac{\Delta Z_L}{\Delta h_L}$ is greater than 2.75 and $l_{min}$ value obtained will be negative which indicates that the solution is infeasible.

**Problem** Consider a lateral length of 150 m and diameter of 14 mm with a dripper spacing of 5 m. Pressure head at the start of the lateral is 10 m. Dripper discharge rate is 5 l/h. If the lateral is laid on downslopes of 0.2, 0.4, 0.8, 1 and 2%, find the locations and magnitudes of minimum and maximum pressure heads and also state the type of downslope for each case?

*Downslope 0.2%*

Total friction loss $= \Delta h_L = b L^{2.75}$

$$b = \frac{7.89 \times 10^5}{14^{4.75} 2.75}\left(\frac{5}{5 \times 3600}\right)^{1.75} = 6.17 \times 10^{-7}$$

$$\Delta h_L = 6.17 \times 10^{-7} \times 150^{2.75} = 0.595 \, m$$

Friction slope $= S_f = 0.595/150 = 0.004$

$$\frac{S}{S_f} = \frac{0.002}{0.004} = 0.50$$

The slope is Type IIa. Maximum pressure head location is at the start and minimum pressure head location is found out as follows:

$$l_{min} = 150\left(1 - \left(\frac{0.002}{2.75 \times 0.004}\right)^{\frac{1}{1.75}}\right) = 93.37 \, m$$

Friction loss at minimum pressure location $= \Delta h_{l min} = \Delta h_L\left(1 - \left(1 - \frac{l_{min}}{L}\right)^{2.75}\right)$

$$\Delta h_{l min} = 0.595\left(1 - \left(1 - \frac{93.37}{150}\right)^{2.75}\right) = 0.553 \, m$$

$$\Delta Z_{l min} = Sl_{min} = 0.002 \times 93.37 = 0.187 \, m$$

$$h_{min} = 10 - 0.553 + 0.187 = 9.634 \, m$$

The results of calculations done for all the other slope values are tabulated in Table 2.

**Table 2** Results of uniform downslope problem

| No. | Slope (%) | $S/S_f$ | Slope Type | $l_{min}$ (m) | $\Delta h_{lmin}$ (m) | $\Delta Z_{lmin}$ (m) | $h_{min}$ (m) | $h_{max}$ (m) | $h_{max}$ location |
|---|---|---|---|---|---|---|---|---|---|
| 1 | 0.2 | 0.500 | IIa | 93.37 | 0.553 | 0.186 | 9.63 | 10 | Start |
| 2 | 0.4 | 1.008 | IIa | 65.45 | 0.472 | 0.262 | 9.790 | 10 | Start |
| 3 | 0.8 | 2.017 | IIc | 24.36 | 0.229 | 0.195 | 9.965 | 10.61 | End |
| 4 | 1 | 2.521 | IIc | 7.27 | 0.076 | 0.073 | 9.997 | 10.91 | End |
| 5 | 2 | 5.042 | IId | ** | ** | 10.000 | 10.00 | 12.41 | End |

## Coefficient of Variation of Pressure and Discharge Rate in Uniform Slopes

Statistically the coefficient of variation of pressure heads ($CV_h$) is by the following formula:

$$CV_h = \frac{\sqrt{\sum_{i=1}^{n}(h_i - \overline{h})^2}}{\overline{h}\sqrt{n}} \tag{14}$$

where $h_i$ pressure is head at $i$th dripper; $\overline{h}$ is the average pressure head and $n$ is the total number of drippers.

Coefficient of variation is also a very useful criterion in design of laterals and submains laid on uniform slopes. The disadvantage of the above form of equation is that the pressure heads at all the dripper locations have to be evaluated in order to get the value of coefficient of variation of pressure heads.

We will derive an analytical equation to find out the coefficient of variation in a lateral/submain without using the data of pressure heads at all the dripper locations.

Let $S$ be the uniform slope (expressed in fraction) on which a lateral of length $L$ is laid. Let $h_l$ be the pressure head at any length $l$ and $h_{start}$ be the pressure head at the start of the lateral.

$$\Delta Z_L = SL$$
$$\Delta Z_l = Sl$$
$$\Delta h_L = \frac{a}{2.75}\left(\frac{q}{s}\right)^{1.75} L^{2.75}$$
$$\overline{\Delta h_L} = 0.733\Delta h_L$$
$$\Delta h_l = \Delta h_L\left(1 - \left(1 - \frac{l}{L}\right)^{2.75}\right) \tag{15}$$

Now following equations can be written:

$$h_l = h_{\text{start}} - \Delta h_l + plS$$

where $p = 1$ for downslope and $p = -1$ for upslope (Figs. 6 and 8).

$$\bar{h} = h_{\text{start}} - \overline{\Delta h_L} + p\overline{\Delta Z_L} \tag{16}$$

Let $D_l^2 = \left(h_l - \bar{h}\right)^2$.

Substitute for $h_l$ and $\bar{h}$ in the preceding equation:

$$D_l^2 = \left(-\Delta h_l + p\Delta Z_l + \overline{\Delta h_L} - p\overline{\Delta Z}\right)^2$$
$$D_l^2 = \left(\overline{\Delta h_L} - \Delta h_l\right)^2 + \left(p\Delta Z_l - p\overline{\Delta Z_L}\right)^2$$
$$+ 2\left(\overline{\Delta h_L} - \Delta h_l\right)\left(p\Delta Z_l - p\overline{\Delta Z_L}\right)$$

The coefficient of variation can also be expressed as follows using integral calculus:

$$\text{CV}_h = \frac{1}{\bar{h}}\left(\sqrt{\frac{1}{L}\int_0^L D_l^2 \, dl}\right) \tag{17}$$

$$\frac{1}{L}\int_0^L D_l^2 \, dl = \frac{1}{L}\int_0^L (\Delta h_l)^2 \, dl - \left(\overline{\Delta h_L}\right)^2$$

$$+ \left(\overline{\Delta Z_L}\right)^2 - \frac{p}{L}\int_0^L (\Delta Z_l)^2 \, dl$$

$$+ \frac{2}{L}\overline{\Delta h_L}\int_0^L \left(\Delta h_l - p\overline{\Delta Z_L}\right) dl$$

$$+ \frac{2p}{L}\int_0^L \Delta h_l\left(\overline{\Delta Z_L} - \Delta Z_l\right) dl \tag{18}$$

Put Eq. 18 in Eq. 17 and the result of the integration after simplification is as follows:

$$\text{CV}_h = \frac{1}{\bar{h}}\sqrt{\left(0.0827\,\Delta h_L^2 + 0.0833(SL)^2 - 0.1540\,p\,\Delta h_L\,SL\right)} \tag{19}$$

Put Eq. 15 in Eq. 19 and rearrange to get the following equation:

$$\frac{6.8076 \times 10^9}{D^{9.5}}\left(\frac{q}{s}\right)^{3.5}L^{5.5} - \frac{4.4184 \times 10^4}{D^{4.75}}PS\left(\frac{q}{s}\right)^{1.75}$$
$$L^{3.75} + 0.0833S^2L^2 - (\bar{h})^2CV_h^2 = 0 \tag{20}$$

Equation 19 is useful in designing the diameter of lateral/submain and Eq. 20 is useful in designing the length of lateral/submain.

**Problem** A 70 m length lateral with uniformly placed drippers with a dripper discharge rate of 4 l/h at a spacing of 0.3 m is laid on a uniform downslope of 0.06. The average pressure head of operation is 10 m. The maximum coefficient of variation of pressure allowed is 0.1 (i.e. 10%). Find out the optimal diameter of the lateral?

Equation 19 is as follows:

$$CV_h = \frac{1}{\bar{h}}\sqrt{\left(0.0827\ \Delta h_L^2 + 0.0833(SL)^2 - 0.154\ p\ \Delta h_L\ SL\right)}$$

The preceding equation can be rearranged as follows:

$$0.0827\ \Delta h_L^2 - 0.154\ p\ SL\ \Delta h_L$$
$$+ 0.0833(SL)^2 - (\bar{h}\ CV_h)^2 = 0$$
$$a\ \Delta h_L^2 - b\ \Delta h_L + c = 0$$

where $a = 0.0827$; $b = 0.154\ p\ SL$; $c = 0.0833(SL)^2 - (\bar{h}\ CV_h)^2$

$$\Delta h_L = \frac{-b \pm \sqrt{b^2 - 4ac}}{2a}$$
$$(\Delta h_L)_1 = 0.81\ \text{m}$$
$$(\Delta h_L)_2 = 7.01\ \text{m}$$

The diameter of lateral can be found out as follows:

$$\Delta h_L = \frac{7.89 \times 10^5}{D^{4.75}2.75}\left(\frac{q}{s}\right)^{1.75}L^{2.75}$$

$$0.81 = \frac{7.89 \times 10^5}{D^{4.75}2.75}\left(\frac{4}{0.3 \times 3600}\right)^{1.75}L^{2.75}$$
$$D_1 = 21.91\ \text{mm}$$

For the $(\Delta h_L)_2 = 7.01$ m., if we do similar kind of calculations, we get $D_2 = 13.91$ mm. We can select the lowest lateral diameter of 13.91 mm for adoption.

**Problem** A lateral of diameter 14 mm with uniformly placed drippers with a discharge rate of 4 l/h at a spacing of 0.3 m is laid on different uniform slopes. The slopes (vertical/horizontal) are as follows:

Up slopes: 0.03, 0.02 and 0.01

Horizontal: 0

Down slopes: 0.01, 0.02 and 0.03

The average pressure head of operation is 10 m. The maximum coefficient of variation of pressure allowed is 0.1 (i.e. 10%). Find out the maximum possible length of lateral for each slope?

For this kind of problems, as a first step, it is easy to calculate the optimal length for zero slope because the solution does not need trial-and-error calculation. For upslope, the search for optimal solution is done by reducing the length from the solution of zero slope and for downslope the search for optimal solution is done by increasing the length from the solution of zero slope. Use of solver in MS-Excel for this kind of calculation is advised. Results of the problem is presented in Table 3.

## Emitter Connection Head Losses

Emitter connections cause additional head losses due to the protrusions and subsequent local turbulence. It is customary to express the head loss due to emitter connection as an equivalent length of lateral pipe. An empirical relationship is widely used to find out this equivalent length ($f_e$ (in m)) which is as follows:

$$f_e = 3.5 \, b \, D^{-1.86} \tag{21}$$

where $b$ is the barb dimension in mm as shown in Fig. 12 and $D$ is the diameter of the lateral in mm. The value of equivalent length in metres ($f_e$) is for one emitter connection loss. In a lateral when $N$ is the number of emitters along the length of a lateral, the equivalent length of the lateral considering the emitter connection loss ($L_e$) is as follows:

$$L_e = L + (N \times f_e) \tag{22}$$

$f_e$ values range from 0.05 to 0.40 m for inline drippers and practically an average value of 0.20 m is adopted. Figure 12 shows the range of $f_e$ values for online emitter connections. Similar to emitter connection losses, there will be lateral connection losses in submains. Mostly this head loss is considered as negligible.

Equation 22 expressed in the following form is useful in calculations:

$$L_e = L\left(1 + \frac{f_e}{s}\right) \tag{23}$$

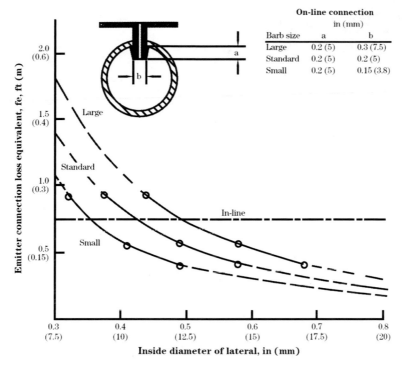

On-line connection
in (mm)

| Barb size | a | b |
|-----------|-------|-----------|
| Large | 0.2 (5) | 0.3 (7.5) |
| Standard | 0.2 (5) | 0.2 (5) |
| Small | 0.2 (5) | 0.15 (3.8) |

**Fig. 12** Equivalent length of lateral for emitter connection pressure loss (Keller and Bliesner 1990)

where $s$ is the spacing between emitters.

**Problem** A lateral of diameter 14 mm with uniformly placed drippers with a discharge rate of 4 l/h at a spacing of 0.3 m is laid on a level ground. The average pressure head of operation is 10 m. The equivalent length for one dripper is assumed as 0.2 m. The maximum coefficient of variation of pressure expected is 0.1 (i.e. 10%). Find out the maximum possible length of lateral?

The maximum length obtained was found to be 54.8 m from the preceding problem (Item No. 4 in Table 3). This length was obtained without considering the emitter connection losses. This length must be substituted in the place of $L_e$ in Eq. 23.

$$L_e = L\left(1 + \frac{f_e}{s}\right)$$

$$54.8 = L\left(1 + \frac{0.2}{0.3}\right)$$

$$L = 32.81 \text{ m}$$

The maximum possible length of actual lateral is 32.81 m. It is equivalent to 54.8 m length of lateral when the emitter connection losses are considered.

**Table 3** Equation and optimal length for each slope

| No. | Slope | Equation to be solved | $L$ (m) |
|-----|-------|------------------------|---------|
| 1 | 0.03 Up | $2.723 \times 10^{-10}L^{5.5} + 2.651 \times 10^{-7}L^{3.75} + 7.497 \times 10^{-5}L^2 - 1 = 0$ | 46.1 |
| 2 | 0.02 Up | $2.723 \times 10^{-10}L^{5.5} + 1.767 \times 10^{-7}L^{3.75} + 3.332 \times 10^{-5}L^2 - 1 = 0$ | 49.0 |
| 3 | 0.01 Up | $2.723 \times 10^{-10}L^{5.5} + 8.836 \times 10^{-8}L^{3.75} + 8.33 \times 10^{-6}L^2 - 1 = 0$ | 51.9 |
| 4 | 0.0 Level | $2.723 \times 10^{-10}L^{5.5} - 1 = 0$ | 54.8 |
| 5 | 0.01 Down | $2.723 \times 10^{-10}L^{5.5} - 8.836 \times 10^{-8}L^{3.75} + 8.33 \times 10^{-6}L^2 - 1 = 0$ | 57.7 |
| 6 | 0.02 Down | $2.723 \times 10^{-10}L^{5.5} - 1.767 \times 10^{-7}L^{3.75} + 3.332 \times 10^{-5}L^2 - 1 = 0$ | 60.6 |
| 7 | 0.03 Down | $2.723 \times 10^{-10}L^{5.5} - 2.651 \times 10^{-7}L^{3.75} + 7.497 \times 10^{-5}L^2 - 1 = 0$ | 63.4 |

It must be noted that the optimal length calculated without considering emitter connection loss has got reduced significantly. It must be kept in mind that not considering emitter connection loss would cause significant error. Care must be taken to get a correct data of equivalent length for emitter connection loss.

## Summary

A generic equation for estimating pressure losses in multi-outlet pipes is derived for which, the end of the pipe need not be in a closed condition. The pressure loss in multi-outlet drip pipe is found to be 0.3636 times the pressure loss in an equivalent length of blind pipe. Average friction loss in multi-outlet drip pipe is 0.73 times the total friction loss of the multi-outlet drip pipe. Pressure head curve in a downslope lateral or submain is divided into four types based on the total value of friction loss and total energy gain due to elevation. The magnitude and location of maximum and minimum pressure in a downslope lateral is found out based on the type of pressure head curve.

Equation for determining optimal length and optimal diameter of lateral/submain on uniform slopes based on coefficient of variation of pressure is derived. The equation corresponding to optimal diameter is quadratic in nature and hence direct analytical solution is available. The equation corresponding to optimal length is nonlinear and for which trial-and-error solution is easily obtained using an add-on tool called SOLVER in MS-Excel. Emitter connections cause additional head losses and these losses are accounted by adding an additional length to the actual length of the lateral.

## Assessment

### Part A. Choose the Best Option

1. The maximum pressure head in multi-outlet pipes is always at _____ of the pipe

   a. **the start**
   b. the end
   c. either at the start or at the end
   d. the middle

2. The average pressure head loss due to friction in a uniform multi-outlet pipe is _____ times the total friction loss in the pipe for infinite number of outlets.

   a. **0.733**
   b. 0.522
   c. 333
   d. 0.255

3. The total pressure head loss due to friction in a uniform multi-outlet pipe is _____ times the total friction loss in an equivalent blind pipe for infinite number of outlets.

   a. **0.3636**
   b. 0.522
   c. 333
   d. 0.255

4. When a lateral is laid on upslope the pressure head would continuously decrease because both friction loss and elevation loss along the pipe length occur. **Say true or false. True**

5. When a lateral is laid on downslope, the location of minimum pressure head is where the pressure head loss gradient due to friction equals the pressure head gain gradient due to elevation. **Say true or false. True**

### Part B. Descriptive Answers

1. What is the value of Christiansen's factor for drip irrigation pipes?
2. What is the relationship between average friction loss and total friction loss in drip irrigation pipes?
3. What are the assumptions involved in deriving friction loss in a multi-outlet pipe?

4.  How the emitter connection pressure losses are accounted in estimation of pressure loss in a lateral using equivalent length of pipe concept?
5.  Derive an analytical equation to find out the coefficient of variation in a lateral /submain without using the data of pressure heads at all the dripper locations?
6.  Derive expressions to find out the locations of maximum and minimum pressures in uniform slopes?

## Problems

1.  A 70 m length lateral with uniformly placed drippers with a dripper discharge rate of 2 l/h at a spacing of 0.6 m is laid on a uniform downslope of 0.04. The average pressure head of operation is 10 m. The maximum coefficient of variation of pressure allowed is 0.1 (i.e. 10%). Find out the optimal diameter of the lateral?
2.  A lateral of diameter 12 mm with uniformly placed drippers with a discharge rate of 4 l/h at a spacing of 0.3 m is laid on different uniform slopes. The slopes (vertical/unit of horizontal) are as follows:

    Upslopes: 0.03 0.02 and 0.01

    Horizontal: 0

    Downslopes: 0.01, 0.02 and 0.03

    The average pressure head of operation is 10 m. The maximum coefficient of variation of pressure allowed is 0.1 (i.e. 10 %). Find out the maximum possible length of lateral for each slope?
3.  Consider a lateral length of 50 m and diameter of 14 mm with a dripper spacing of 0.6 m. Pressure head at the start of the lateral is 10 m. Dripper discharge rate is 4 l/h. If the lateral is laid on downslopes of 0.2%, 0.4%, 0.8%, 1% and 2%, find the locations and magnitudes of minimum and maximum pressure heads and also state the type of downslope for each case?

## References and Further Reading

1.  Anyoji, H., & Wu, I. P. (1987). Statistical approach for drip lateral design. *ASAE, 30*(1), 187–192.
2.  Krishnan, M., & Ravikumar, V. (2002). A software for design of drip subunits with tapered pipes in non-uniform slopes. *Journal of Agricultural Engineering. Indian Council of Agricultural Research, 39*(4). http://epubs.icar.org.in/ejournal/index.php/JAE/article/view/14143
3.  Ravikumar, V., Santhana Bosu S., & Kumar, V. (2001). Drip irrigation system design by energy gradient line approach for non-uniform outflow. *Agricultural Engineering Journal, Bangkok, Thailand, 10*(3 & 4), 181–190

4. Ravikumar, V., Ranganathan, C. R., & Santhana, B. S. (2003). Analytical equation for uniformity of discharge in drip irrigation laterals. *Journal of Irrigation and Drainage Engineering American Society of Civil Engineers, 129*(4), 295–298.
5. Lamm, F. R., Ayars, J. E., & Nakayama, F. S. (2006). *Microirrigation for crop production-design, operation, and management*. eBook ISBN: 9780080465814

# Drip Subunit Design with Uniform Diameter Pipes in Uniform Slopes

## Introduction

An important objective in DIS is to design the system in such a way that all the drippers discharge uniformly in a subunit. But it is almost impossible to design and install a system with uniform discharge rate in all the drippers within any subunit. The reasons for variation of discharge are due to the pressure variations at the drippers due to friction and elevation differences and also difficulties in manufacturing drippers with perfectly uniform discharge rate in the production process in any industry.

There are many criteria being in use for designing of drip irrigation subunits. Designing based on each criterion is discussed in this chapter.

In drip irrigation system, a subunit usually has a control valve at the start of the submain and laterals are connected to the submain. While designing, both laterals and the corresponding submain are designed simultaneously. Figure 1 shows a typical subunit with pressure head distribution. The pressure variations are usually equally divided between submain and laterals.

## Relative Flow Difference Criterion

Relative flow difference criterion is the most widely used criterion due to its simplicity, and it is also easy to comprehend by the engineers and farmers.

$$\text{Relative Flow Difference } (V_q) = \frac{q_{max} - q_{min}}{\overline{q}} \tag{1}$$

where $q_{max}$ is the maximum discharge rate, $q_{min}$ is the minimum discharge rate, and $\overline{q}$ is the average discharge rate.

$$\text{Relative Pressure Head Difference } (V_h) = \frac{h_{max} - h_{min}}{\overline{h}} \tag{2}$$

V. Ravikumar, *Sprinkler and Drip Irrigation*,
https://doi.org/10.1007/978-981-19-2775-1_10

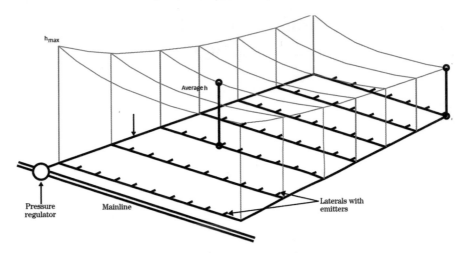

**Fig. 1** Typical pressure head distribution in a subunit (USDA-NRCS-Microirrigation)

where $h_{max}$ is the maximum pressure head, $h_{min}$ is the minimum pressure head, and $\overline{h}$ is the average pressure head. Generally, the average pressure head for which a subunit is designed is for 10 m. However, in low pressure systems, the average pressure may be as low as 1 m also.

A relationship between $V_q$ and $V_h$ is as follows:

$$V_q = x V_h \tag{3}$$

where $x$ is known as the flow exponent of dripper.

Allowable pressure head variation in a subunit $\Delta h_{su}^a$ is as follows:

$$\Delta h_{su}^a = h_{max}^a - h_{min}^a \tag{4}$$

where $h_{max}^a$ is the maximum allowable pressure head in the subunit and $h_{min}^a$ is the minimum allowable pressure head in the subunit.

From the value of $\Delta h_{su}^a$, allowable pressure head loss in submain/manifold ($\Delta h_m^a$) and allowable pressure head loss in lateral ($\Delta h_{lat}^a$) are equally divided into two. Then, the design of lateral and submain is proceeded.

Allowable relative pressure head variation in a subunit $\quad V_h^a(\text{subunit}) = \dfrac{\Delta h_{su}^a}{\overline{h}}$

$$\tag{5}$$

Allowable relative pressure head variation in a submain $\quad V_h^a(\text{submain}) = \dfrac{\Delta h_m^a}{\overline{h}}$

$$\tag{6}$$

Allowable relative pressure head variation in a lateral $V_h^a$(lateral) $= \dfrac{\Delta h_{lat}^a}{\overline{h}}$ (7)

Generally, design is done for $V_h^a$(subunit) $= 0.2$. In that case, $V_h^a$(submain) is designed for 0.1 and $V_h^a$(lateral) is designed for 0.1. This is because the dripper selected is assumed to be orifice dripper with a flow exponent of 0.5. If flow exponent is 0.5, the relative discharge rate variation of subunit would be 0.1 for the relative pressure variation of subunit of 0.2 as per Eq. 3. If the dripper selected has flow exponent greater or lower than 0.5, the allowable pressure head variation values have to be selected accordingly.

**Problem** For a design of drip irrigation system (DIS), the allowable relative discharge rate variation for a subunit $\left(V_q^a\right)$ is fixed as 0.1 and the dripper selected is a vortex dripper with flow exponent of 0.4. The DIS is designed f an average pressure head of 5 m. Find out the allowable pressure variation in the lateral and the submain?

$$V_q^a \text{(subunit)} = 0.1$$

If we divide the variation equally between lateral and submain we get,

$$V_q^a \text{(submain)} = 0.05$$

$$V_q^a \text{(lateral)} = 0.05$$

We know that $V_q = x V_h$. Hence,

$$V_h^a \text{(lateral)} = \frac{0.05}{0.4} = 0.125$$

Since,

$$V_h^a \text{(lateral)} = \frac{\Delta h_{lat}^a}{\overline{h}}$$

Allowablepressure head variation for lateral $\left(\Delta h_{lat}^a\right) = 0.125 \times 5 = 0.625\,\text{m}$.

When doing calculations for submain, laterals are treated as outlets and flow exponent '$x$' for laterals may be taken as 1.

$$V_h^a \text{(submain)} = 0.05/1 = 0.05$$

Since,

$$V_h^a \text{(submain)} = \frac{\Delta h_m^a}{\overline{h}}$$

Allowable pressure head variation for submain $\left(\Delta h_m^a\right) = 0.05 \times 5 = 0.25$ m.

## Maximum Length of a Horizontal Lateral

While designing a drip irrigation system in countries like India, availability of accurate topographical data about the land for the designer is seldom made available, and hence the design is made assuming that land is level to start with. If the lateral runs on an upslope, reduction of length of lateral/submain is done by thumb rules by the technicians working in the field. If the lateral runs on downslope, increasing the length of lateral/submain is seldom done due to avoidance of risk. Hence the maximum length of lateral/submain on a level land is a very important design data.

Equations 6 and 7 can be written by removing the subscripts for lateral/submain as follows:

$$V_h^a = \frac{\Delta h^a}{\overline{h}} \tag{8}$$

$$\overline{h} V_h^a = \Delta h^a \tag{9}$$

In case of horizontal laterals, variation of pressure head is caused only due to friction, and hence we can substitute the following total friction loss formula for a lateral/submain in the preceding equation.

$$\Delta h_L = \frac{7.89 \times 10^5}{2.75 \, D^{4.75}} \left(\frac{q}{s}\right)^{1.75} L^{2.75} \tag{10}$$

After substitution we get,

$$\overline{h} V_h^a = \frac{7.89 \times 10^5}{2.75 \, D^{4.75}} \left(\frac{q}{s}\right)^{1.75} L_{\max}^{2.75} \tag{11}$$

In the preceding equation $q$ is expressed in l/s. Often it is convenient to use litres per hour (l/h) for $q$ during design of lateral/submain. Hence, we convert the preceding equation for $q$ in l/h and rearrange as follows:

$$L_{\max} = \left(5.83 \frac{\overline{h} V_h^a \, D^{4.75}}{(q/s)^{1.75}}\right)^{0.3636} \tag{12}$$

where

$L$      Maximum length of horizontal lateral (m);

**Fig. 2** Drippers around a
tree

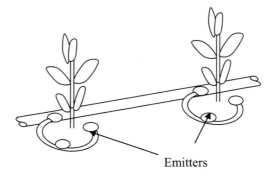

Emitters

$V_h^a$   Allowable pressure head variation of pressure for the lateral/submain.
    Normally 0.1 is taken;
$\overline{h}$   Average design pressure head. Normally it is taken as 10 m;
$D$   Inside diameter (mm);
q/s   Specific discharge rate (SDR) in (l/h/m), i.e. discharge rate per metre length
    of lateral.

For the tree crops where more than one dripper per tree is installed, following
formula is used for calculating SDR (Fig. 2):

$$(q/s) = \frac{1}{S_p} \sum_1^{N_p} q_i \tag{13}$$

where $N_p$ is the number of emitters per plant and $S_p$ is spacing between the plants
along the row.

**Problem** Find out the specific discharge rate for coconut plantation planted at a
spacing of 7 m. Two drippers of 4 l/h and a microsprinkler of 22 l/h are installed.

$$SDR = \frac{1}{7} \sum_1^{N_p} (4 + 4 + 22) = \frac{30}{7} = 4.29 \text{ lh}^{-1}\text{m}^{-1}$$

**Problem** A lateral is laid over level land. Allowable relative pressure head variation
in the lateral is 0.1. Uniform spacing of dripper is 0.5 m, dripper discharge rate is
2 l/h, and inside diameter of lateral is 14 mm. Find out the maximum length of the
lateral?

By applying Eq. 12, the maximum length of the lateral works out to be 75 m.
Table 1 provides data of maximum lengths of laterals for different specific
discharge rate values for four commonly available diameters of lateral for an average
pressure head of 10 m and relative pressure head variation of 0.1. Table 2 provides
a similar data for relative pressure head variation of 0.2.

**Table 1** Maximum lengths of level laterals with $\bar{h} = 10$ m and $V_h^a = 0.1$

| q/s (l/h/m) | 8 mm OD (6 mm ID) | 12 mm OD (10 mm ID) | 16 mm OD (13.7 mm ID) | 20 mm OD (17.2 mm ID) |
|---|---|---|---|---|
| 0.2 | 117 | 282 | 486 | 720 |
| 0.4 | 75 | 181 | 312 | 463 |
| 0.6 | 58 | 140 | 241 | 358 |
| 0.8 | 48 | 117 | 201 | 298 |
| 1.0 | 42 | 101 | 174 | 258 |
| 1.2 | 37 | 90 | 155 | 230 |
| 1.4 | 34 | 82 | 141 | 209 |
| 1.6 | 31 | 75 | 129 | 192 |
| 1.8 | 29 | 70 | 120 | 178 |
| 2.0 | 27 | 65 | 112 | 166 |
| 2.5 | 23 | 57 | 97 | 144 |
| 3.0 | 21 | 50 | 87 | 128 |
| 3.5 | 19 | 46 | 79 | 116 |
| 4.0 | 17 | 42 | 72 | 107 |
| 5.0 | 15 | 36 | 63 | 93 |
| 6.0 | 13 | 32 | 56 | 83 |
| 7.0 | 12 | 29 | 51 | 75 |
| 8.0 | 11 | 27 | 46 | 69 |
| 9.0 | 10 | 25 | 43 | 64 |
| 10.0 | 10 | 23 | 40 | 60 |
| 12.0 | 9 | 21 | 36 | 53 |
| 14.0 | 8 | 19 | 33 | 48 |
| 16.0 | 7 | 17 | 30 | 44 |
| 18.0 | 7 | 16 | 28 | 41 |
| 20.0 | 6 | 15 | 26 | 38 |
| 23.0 | 6 | 14 | 24 | 35 |
| 26.0 | 5 | 13 | 22 | 33 |

Table 3 provides data of maximum lengths of submains for different specific discharge rate values for four commonly adopted diameters of lateral for an average pressure head of 10 m and relative pressure head variation of 0.1.

**Table 2** Maximum lengths of level laterals with $\overline{h} = 10$ m and $V_h^a = 0.2$

| q/s (l/h/m) | 8 mm OD (6 mm ID) | 12 mm OD (10 mm ID) | 16 mm OD (13.7 mm ID) | 20 mm OD (17.2 mm ID) |
|---|---|---|---|---|
| 0.2 | 150 | 363 | 625 | 926 |
| 0.4 | 97 | 233 | 402 | 596 |
| 0.6 | 75 | 180 | 311 | 460 |
| 0.8 | 62 | 150 | 259 | 383 |
| 1.0 | 54 | 130 | 224 | 332 |
| 1.2 | 48 | 116 | 200 | 296 |
| 1.4 | 44 | 105 | 181 | 268 |
| 1.6 | 40 | 97 | 166 | 247 |
| 1.8 | 37 | 90 | 154 | 229 |
| 2.0 | 35 | 84 | 144 | 214 |
| 2.5 | 30 | 73 | 125 | 186 |
| 3.0 | 27 | 65 | 112 | 165 |
| 3.5 | 24 | 59 | 101 | 150 |
| 4.0 | 22 | 54 | 93 | 138 |
| 5.0 | 19 | 47 | 81 | 119 |
| 6.0 | 17 | 42 | 72 | 106 |
| 7.0 | 16 | 38 | 65 | 96 |
| 8.0 | 14 | 35 | 60 | 89 |
| 9.0 | 13 | 32 | 55 | 82 |
| 10.0 | 12 | 30 | 52 | 77 |
| 12.0 | 11 | 27 | 46 | 68 |
| 14.0 | 10 | 24 | 42 | 62 |
| 16.0 | 9 | 22 | 38 | 57 |
| 18.0 | 9 | 21 | 36 | 53 |
| 20.0 | 8 | 19 | 33 | 49 |
| 23.0 | 7 | 18 | 31 | 45 |
| 26.0 | 7 | 16 | 28 | 42 |

# Maximum Length of Lateral/Submain on Slopes

It is known that when lateral/submain is laid on slopes, the length of lateral must be shortened in case of upslope and may be increased in case of downslope. The equation for adjustment can be obtained by distributing the allowable pressure head variation between the friction and the elevation loss/gain. The following equation also has an inherent assumption that the maximum pressure head is at the start and the minimum pressure head is at the end.

The allowable friction loss ($\Delta_h^a$) can be expressed as follows:

**Table 3** Maximum lengths of level submains with $\bar{h} = 10$ m and $V_h^a = 0.1$

| q/s (l/h/m) | 40 mm OD (37.2 mm ID) | 50 mm OD (46.6 mm ID) | 63 mm OD (58.6 mm ID) | 75 mm OD (69.8 mm ID) |
|---|---|---|---|---|
| 25 | 126 | 186 | 277 | 374 |
| 50 | 81 | 120 | 178 | 241 |
| 75 | 63 | 93 | 138 | 186 |
| 100 | 52 | 77 | 115 | 155 |
| 200 | 34 | 50 | 74 | 100 |
| 300 | 26 | 38 | 57 | 77 |
| 400 | 22 | 32 | 47 | 64 |
| 500 | 19 | 28 | 41 | 56 |
| 600 | 17 | 25 | 37 | 50 |
| 800 | 14 | 21 | 31 | 41 |
| 1000 | 12 | 18 | 26 | 36 |
| 1200 | 11 | 16 | 24 | 32 |
| 1400 | 10 | 14 | 21 | 29 |
| 1600 | 9 | 13 | 20 | 27 |
| 1800 | 8 | 12 | 18 | 25 |
| 2000 | 8 | 11 | 17 | 23 |
| 2200 | 7 | 11 | 16 | 22 |
| 2400 | 7 | 10 | 15 | 21 |
| 2800 | 6 | 9 | 14 | 19 |
| 3000 | 6 | 9 | 13 | 18 |
| 3400 | 6 | 8 | 12 | 16 |
| 4000 | 5 | 7 | 11 | 15 |

$$(\Delta_h^a) = \text{Friction loss of the corrected equivalent length } (L_e)$$
$$+ \text{ Elevation loss in actual length}(L) \tag{14}$$

It is known that the ratio of friction loss for two different lengths of the lateral/submain pipes with the same diameter and SDR can be expressed as follows:

$$\frac{\Delta h_L}{\Delta h_{\text{Ltab}}} = \left(\frac{L}{L_{\text{tab}}}\right)^{2.75}$$

The length $L_{\text{tab}}$ is the maximum length for horizontal land for 1 m ($\Delta h_{Ltab} = 1$) friction loss, and these values are in Tables 1 and 3. Hence, the preceding equation can be written as follows:

$$\Delta h_L = \left(\frac{L}{L_{\text{tab}}}\right)^{2.75}$$

We know that the elevation gain is 'SL'. '$S$' is slope of the pipe, and '$L$' is the actual length of the pipe. The relationship between equivalent length of lateral (Le) and actual length of the pipe ($L$) is available in the preceding chapter which is as follows:

$$\Delta h_{Le} = \Delta h_L \left(1 + \frac{f_e}{s}\right)^{2.75}$$

$$\Delta h_{Le} = \left(\left(\frac{L}{L_{tab}}\right)\left(1 + \frac{f_e}{s}\right)\right)^{2.75}$$

where $f_e$ is equivalent length of lateral for one emitter connection head loss and '$s$' is the spacing between emitters. Using the preceding equation in Eq.14, we can write the following equation:

$$\Delta h^a = \left(\left(\frac{L}{I_{tab}}\right)\left(1 + \frac{f_e}{s}\right)\right)^{2.75} - pSL \tag{15}$$

where '$p$' $= 1$ for downslope and '$p$' $= -1$ for upslope.

The preceding equation is totally valid for upslope. For downslope also, as long as the maximum pressure head is at the start and minimum pressure is at the end, this equation is valid. Hence, this equation is used as a starting point for design. After adjusting the length with this equation, the relative pressure head variation must be assessed again by finding the maximum and minimum pressure head and the length must be adjusted by trial and error until the relative pressure head variation becomes less than the threshold level.

## Paired Lateral

A paired lateral is the one which has the lateral laid on both the sides. On level lands, the length of lateral on one side is equal to the other. In case of slopes, the length of downslope lateral will be longer than the upslope lateral (Fig. 3).

**Problem** A paired lateral has to be designed for a field as shown in Fig. 4. Design using inline lateral of diameter 13.7 mm. Dripper discharge rate of 4 l/h and uniform dripper spacing of 0.4 m were selected based on a field study of the soil. Average operating pressure is 10 m. Allowable relative pressure head variation is 0.1. Equivalent length for emitter connection loss for one dripper is 0.05 m. Pressure available at the start of the submain is 11 m.

Design suitable lengths of a paired lateral? Find also the average pressure heads on both the upslope and downslope laterals?

Specific discharge rate ($q/s$) $= 4/0.4 = 10$ l/h/m.

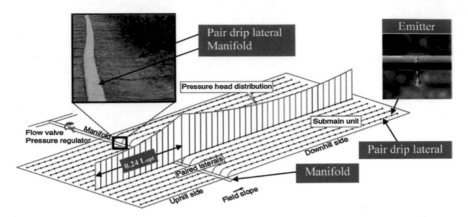

**Fig. 3** A typical paired lateral [4]

**Fig. 4** Lateral configuration
for paired lateral problem

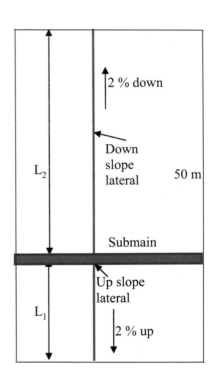

Maximum length of lateral for 13.7 mm diameter if laid on a level land = 40 m
(Table 1).

**Adjustment for 2% Downslope**

$$\Delta h^a = \left( \left( \frac{L}{L_{\text{tab}}} \right) \left( 1 + \frac{f_e}{s} \right) \right)^{2.75} - pSL$$

We can solve the preceding equation by Solver of Excel. While doing so, have the constraints on L to be as close as possible. The minimum constraint is fixed as 40 m for downslope, and the maximum constraint is fixed as 40 m for upslope while using Solver. The answer obtained using the Solver is as follows:

Maximum actual length of 13.7 mm diameter lateral on 2% downslope with emitter connection loss = 44.88 m.

**Adjustment for 2% Upslope**
If we do similar calculations as above, we get as follows:

Maximum actual length of 13.7 mm diameter lateral on 2% upslope with emitter connection loss = 26.86 m.

We will safely decide to select a downslope lateral of 13.7 mm diameter for a length of 30 m and upslope lateral of 20 m.

Hence $L_1 = 20$ m and $L_2 = 30$ m.

## *Average Pressure Head*

From the preceding chapter, we derived the following relevant equations:

$$\overline{h} = h_{\text{start}} - \overline{\Delta h_L} + p0.5\Delta z$$

$$\overline{\Delta h_L} = 0.733\Delta h_L$$

$$\Delta h_L = \frac{789000}{2.75 D^{4.75}} \left(\frac{q}{s}\right)^{1.75} L^{2.75}$$

$$\Delta z = LS$$

## *Average Pressure Head for Upslope*

For finding total friction loss, the length of lateral must be substituted considering the emitter connection loss.

$$L_e = L\left(1 + \frac{f_e}{s}\right)$$

$$L_e = 20\left(1 + \frac{0.05}{0.4}\right) = 22.5$$

$$\Delta h_L = \frac{789000}{2.75 D^{4.75}} \left(\frac{q}{s}\right)^{1.75} L^{2.75}$$

$$\Delta h_L = \frac{789000}{2.75 \times 13.7^{4.75}} \left(\frac{4}{3600 \times 0.4}\right)^{1.75} 22.5^{2.75} = 0.20 \text{ m}$$

For finding total elevation loss/gain, the actual length of lateral must be substituted without considering the emitter connection loss.

$$\Delta Z = LS = 20 \times 0.02 = 0.4 \ m$$

$$\overline{h} = 11 - 0.733 \times 0.2 + (-1)(0.5)0.4 = 10.65 \ m$$

## *Average Pressure Head for Downslope*

$$L_e = L\left(1 + \frac{f_e}{s}\right)$$

$$L_e = 30\left(1 + \frac{0.05}{0.4}\right) = 33.75 \text{ m}$$

$$\Delta h_L = \frac{789000}{2.75 D^{4.75}} \left(\frac{q}{s}\right)^{1.75} L^{2.75}$$

$$\Delta h_L = \frac{789000}{2.75 \times 13.7^{4.75}} \left(\frac{4}{3600 \times 0.4}\right)^{1.75} 33.75^{2.75} = 0.61 \text{ m}$$

$$\Delta Z = LS = 30 \times 0.02 = 0.6 \,\text{m}$$

$$\overline{h} = 11 - 0.733 \times 0.61 + (1)(0.5)0.6 = 10.85 \,\text{m}$$

If you look at the average pressure head values for upslope as well as downslope, they are slightly different. But if we had correctly selected the lengths of upslope and downslope lateral lengths, we could have got average pressure heads to be equal for both upslope and downslope. In the following sections we will derive equations to achieve this objective.

## Average Friction Loss in a Paired Lateral

Let $L_1$ be the length of lateral on right side and $L_2$ be the length of lateral on the left side (Fig. 5). OA and OB are friction loss curves for Lateral $L_1$ and $L_2$.

Let $Lp$ be the total length of the paired lateral. The average friction loss of a paired lateral $\left(\overline{\Delta h}_{LP}\right)$ can be found out by finding the weighted average of friction losses of both the laterals.

$$\overline{\Delta h}_{LPe} = \frac{\overline{\Delta h}_{L1e} L_{1e} + \overline{\Delta h}_{L2e} L_{2e}}{L_{pe}} \tag{16}$$

where $\overline{\Delta h}_{L1e}$ & $\overline{\Delta h}_{L2e}$ are the average friction losses in laterals $L_{1e}$ and $L_{2e}$, respectively.

Let $w = L_{1e}/L_{pe}$; then, $L_{2e}/L_{pe} = 1\text{-}w$.

As discussed in preceding sections, the ratio of total friction losses in two laterals if the specific discharge rate and diameter of pipe do not vary between the pipes, following equation can be written:

$$\frac{\Delta h_{L1e}}{\Delta h_{Lpe}} = \frac{L_{1e}^{2.75}}{L_{pe}^{2.75}} = w^{2.75}$$

$$\Delta h_{L1e} = \Delta h_{Lpe} w^{2.75}$$

$$\overline{\Delta h}_{L2e} = 0.733 \Delta h_{Lpe} w^{2.75} \tag{17}$$

Similarly,

$$\overline{\Delta h}_{L2e} = 0.733 \Delta h_{Lpe} (1 - w)^{2.75} \tag{18}$$

Substituting Eqs. 17 and 18 in Eq. 16 we get the following Equation

$$\overline{\Delta h}_{LPe} = 0.733 \Delta h_{LPe} (w^{3.75} + (1 - w)^{3.75}) \tag{19}$$

**Fig. 5** Average friction loss in a paired lateral

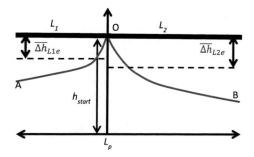

## Average Elevation Loss in a Paired Lateral

Average elevation loss in a paired lateral is the weighted average of elevation losses in both the laterals. Let $L_1$ and $L_2$ be the lengths of laterals 1 and 2, respectively (Fig. 6).

The average elevation loss for a paired lateral $\left(\overline{\Delta Z}_{LP}\right)$ can be found out by finding out the weighted average of elevation losses of both the laterals.

$$\overline{\Delta Z}_{LP} = \frac{\overline{\Delta Z}_{L1} L_1 - \overline{\Delta Z}_{L2} L_2}{L_p} \qquad (20)$$

where $\overline{\Delta Z}_{L1}$ & $\overline{\Delta Z}_{L2}$ be the average elevation losses in laterals $L_1$ and $L_2$, respectively. Note in the preceding equation that there is a pressure energy loss due to flow going up in $L_1$ and pressure energy gain due to flow going down in downslope in $L_2$.

If $w = L_{1e}/L_{pe}$ and $L_{2e}/L_{pe} = 1 - w$.
Then $w = L_1/L_p$ and then $L_2/L_p = 1 - w$.
Let $\Delta Z_{Lp} = \Delta Z_{L1} + \Delta Z_{L2}$.
Then it is also true that $\frac{\Delta Z_{L1}}{\Delta Z_{Lp}} = w$, Then

$$\Delta Z_{L1} = \Delta Z_{Lp} w$$

$$\overline{\Delta Z}_{L1} = 0.5 \Delta Z_{Lp} w \qquad (21)$$

$$\overline{\Delta Z}_{L2} = 0.5 \Delta Z_{Lp} (1 - w) \qquad (22)$$

Substituting Eqs. 21 and 22 in Eq. 20 we get the following equation:

$$\overline{\Delta Z}_{LP} = 0.5 \Delta Z_{Lp} (2w - 1) \qquad (23)$$

Sometimes the paired laterals are laid on ridges with varying degrees of slopes on either side (Fig. 7). Let $S_1$ and $S_2$ be the slopes of the laterals 1 and 2, respectively. $p_1$ and $p_2$ are indicators taking value of $(+1)$ for downslope and $(-1)$ for upslope. In

**Fig. 6** Average elevation loss in a paired lateral

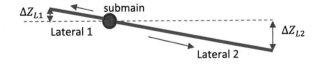

**Fig. 7** Average elevation loss in a paired lateral with varying slopes on both sides

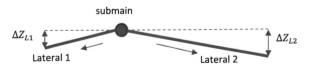

such cases Eq. 20 takes the following form:

$$\overline{\Delta Z}_{LP} = \frac{-0.5 p_1 S_1 L_1 L_1 - 0.5 p_2 S_2 L_2 L_2}{L_p} \tag{24}$$

**Problem** Find out the average friction loss and average elevation loss using the equations derived in preceding section for the data of the preceding problem. Find out the average pressure head in the subunit.

$$\overline{\Delta h}_{LP} = 0.733 \Delta h_{LPe}\left(w^{3.75} + (1-w)^{3.75}\right)$$

$$\overline{\Delta Z}_{LP} = 0.5 \Delta Z_{Lp}(2w - 1)$$

$L_{1c} = 22.5$ m and $L_{2e} = 33.75$ m and $w = 22.5/56.25 = 0.4$

$$\Delta h_L = \frac{789000}{2.75 D^{4.75}} \left(\frac{q}{s}\right)^{1.75} L^{2.75}$$

$$\Delta h_{Lpe} = \frac{789000}{2.75 \times 13.7^{4.75}} \left(\frac{4}{3600 \times 0.4}\right)^{1.75} 56.25^{2.75} = 2.50 \text{ m}$$

$$\overline{\Delta h}_{LPe} = 0.733 \times 2.50(0.4^{3.75} + 0.6^{3.75}) = 0.33 \text{ m}$$

$$\overline{\Delta Z}_{LP} = 0.5 \times .02 \times 50(2 \times 0.4 - 1) = -0.1 \text{ m}$$

Average pressure head of the subunit

$$\overline{h} = h_{start} - \overline{\Delta h}_{LPe} - \overline{\Delta Z}_{LP}$$

$$\overline{h} = 11 - 0.33 + 0.1 = 10.77 \text{ m}$$

**Problem** A submain with paired lateral system as in Fig. 7 has left-side lateral of 20 m with downslope of 1% and right-side lateral of 30 m with downslope of 2%. Find out the average elevation gain?

$$\overline{\Delta Z}_{LP} = \frac{-0.5 \times 0.01 \times 20^2 - 0.5 \times 0.02 \times 30^2}{50} = -0.22$$

## Equal Average Pressure Heads on Both Sides of a Paired Lateral

If we need average pressure head on either side of a paired lateral to be equal, the net pressure gain or loss with reference to submain on either lateral must also be equal. Hence following equation can be written:

$$\overline{\Delta h}_{L1e} - p_1 \overline{\Delta Z}_{L1} = \overline{\Delta h}_{L2e} - p_2 \overline{\Delta Z}_{L2} \tag{25}$$

We will assume a case of both the laterals laid on a single uniform slope (S). Then we can write as follows:

$$\overline{\Delta h}_{L1e} = 0.733 b L_{1e}^{2.75}$$

$$\overline{\Delta h}_{L2e} = 0.733 b (L_{pe} - L_{1e})^{2.75} \tag{26}$$

where $b = \frac{7.89 \times 10^5}{D^{4.75} 2.75} \left(\frac{q}{s}\right)^{1.75}$ and Eq. 25 becomes as follows:

$$0.733 b (L_{1e}^{2.75} - (L_{pe} - L_{1e})^{2.75}) + 0.5 S L_p = 0 \tag{27}$$

where $L_{pe}$ is the length considering emitter connection loss. Note that the $L_p$ term also exists in the elevation part of the preceding equation which is the length without considering emitter connection loss. When using this equation, $L_{1e}$ value obtained is also the length considering emitter connection losses.

If we consider two different diameters of pipes for upslope and downslope laterals, the preceding equation is modified as follows:

$$0.733 c \left( \frac{L_{1e}^{2.75}}{D_1^{4.75}} - \frac{(L_{pe} - L_{1e})^{2.75}}{D_2^{4.75}} \right) + 0.5 \, SL_p = 0 \tag{28}$$

where $c = \frac{7.89 \times 10^5}{2.75} \left(\frac{q}{s}\right)^{1.75}$.

**Problem** For the data of the preceding problem, find out the optimal location of the submain so that average operating pressure head on both upslope lateral and downslope lateral is the same? Also find out the average pressure head of both upslope and downslope laterals? Also assess the relative pressure head variation in downslope lateral?

$$0.733 b (L_{1e}^{2.75} - (L_{pe} - L_{1e})^{2.75}) + 0.5 S L_p = 0$$

where $b = \frac{7.89 \times 10^5}{D^{4.75} 2.75} \left(\frac{q}{s}\right)^{1.75}$.

$$b = \frac{7.89 \times 10^5}{13.7^{4.75} \times 2.75} \left(\frac{4}{3600 \times 0.4}\right)^{1.75} = 3.844 \times 10^{-5}$$

In this case, $L_{pe}$ value must be substituted with the equivalent length considering emitter connection losses.

$$L_{pe} = L\left(1 + \frac{f_e}{s}\right)$$

$$L_{pe} = 50\left(1 + \frac{0.05}{0.4}\right) = 56.25\,\text{m}$$

$$0.733 \times 3.844 \times 10^{-5}(L_{1e}^{2.75} - (56.25 - L_{1e})^{2.75}) + 0.5 \times 0.02 \times 50 = 0$$

$$L_{1e} = 18.95\,\text{m}$$

$$L_{2e} = 56.25 - 18.95 = 37.3\,\text{m}$$

## Average Pressure Head for Upslope

For finding the total friction loss, the length of lateral must be substituted considering the emitter connection loss.

$$\Delta h_{L1e} = \frac{789000}{2.75 D^{4.75}} \left(\frac{q}{s}\right)^{1.75} L_{1e}^{2.75}$$

$$\Delta h_{L1e} = \frac{789000}{2.75 \times 13.7^{4.75}} \left(\frac{4}{3600 \times 0.4}\right)^{1.75} 18.95^{2.75} = 0.125\,\text{m}$$

For finding total elevation loss, the length of lateral substituted must be the actual length of lateral.

$$L_{1e} = L_1\left(1 + \frac{f_e}{s}\right)$$

$$18.95 = L_1\left(1 + \frac{0.05}{0.4}\right)$$

$$L_1 = 16.84\,\text{m}$$

$$\Delta Z_1 = L_1 S = 16.84 \times 0.02 = 0.337\,\text{m}$$

$$\bar{h} = 11 - 0.733 \times 0.125 + 0.5(-1)0.337 = 10.73 \text{ m}$$

## Average Pressure Head for Downslope

$$\Delta h_{L2e} = \frac{789000}{2.75 D^{4.75}} \left(\frac{q}{s}\right)^{1.75} L_{2e}^{2.75}$$

$$\Delta h_{L2e} = \frac{789000}{2.75 \times 13.7^{4.75}} \left(\frac{4}{3600 \times 0.4}\right)^{1.75} 37.3^{2.75} = 0.807 \text{ m}$$

$$L_{2e} = L_2\left(1 + \frac{f_e}{s}\right)$$

$$37.3 = L_2\left(1 + \frac{0.05}{0.4}\right)$$

$$L_2 = 33.156 \text{ m}$$

$$\Delta Z_2 = L_2 S = 33.156 \times 0.02 = 0.663 \text{ m}$$

$$\bar{h} = 11 - 0.733 \times 0.807 + 0.5(1)0.663 = 10.73 \text{ m}$$

Hence, we find that both the average pressure heads on both the sides of the submain are one and the same.

## Relative Pressure Head Variation in Downslope

Location of minimum pressure head:

$$l_{min-e} = L_{2e}\left(1 - \left(\frac{SL_{2e}}{2.75\Delta h_{L2e}}\right)^{\frac{1}{1.75}}\right) = 37.3\left(1 - \left(\frac{0.02 * 37.3}{2.75 * 0.807}\right)^{\frac{1}{1.75}}\right) = 17.29 \text{ m}$$

Pressure head loss due to friction at minimum pressure head location:
Already it has been derived that

$$\Delta h_l = \Delta h_L\left(1 - \left(1 - \frac{l}{L}\right)^{2.75}\right)$$

$$= 0.807\left(1 - \left(1 - \frac{17.29}{37.3}\right)^{2.75}\right) = 0.66 \text{ m}$$

Find out the pressure head gain due to elevation at minimum pressure head location:

$$L_e = L\left(1 + \frac{f_e}{s}\right)$$

$$17.29 = L\left(1 + \frac{0.05}{0.4}\right)$$

$$l_{min} = 15.37 \text{ m}$$

$$\Delta Z_{l\,min} = 15.37 \times 0.02 = 0.31 \text{ m}$$

Net pressure head loss at the minimum pressure head location:

$$- \Delta h_{l_{min-e}} - \Delta Z_{l\,min} = 0.66 - 0.31 = 0.35 \text{ m}$$

## Slope Profile Type

$$\Delta h_{l.e} > \Delta Z_L$$

0.807 m > 0.663 m Slope is Type IIa.

Hence maximum pressure head is at the start. Hence the net pressure head loss at the minimum pressure head location is the maximum variation of pressure head.

$$V_h^a = \frac{\Delta h_l^a}{\bar{h}}$$

$V_h^a = \frac{0.35}{10.73} = 0.033$ which is less than 0.05. Hence the design is safe.

**Problem** A paired lateral has to be designed for a field whose slope details are provided in the map below:

- Dripper discharge rate of 4 l/h placed at uniform dripper spacing of 0.4 m in the lateral
- Equivalent length of lateral for one emitter connection loss—0.05 m
- Lateral spacing—1 m
- Average operating pressure head—10 m
- Lateral diameters available in the market—10 mm and 13.7 mm

**Fig. 8** Field map for a
subunit problem

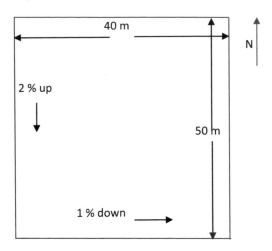

- Submain diameters available in the market—37.2 mm and 46.6 mm
- Allowable relative pressure head variation is 0.1 each for both lateral and submain.

**Solution**

Specific discharge rate $(q/s) = 4/0.4 = 10$ l/h/m.

Maximum length of lateral for 13.7 mm diameter if laid on a level land (emitter connection loss not accounted) = 40 m.

Maximum length of lateral for 10 mm diameter if laid on a level land (emitter connection loss not accounted) = 23 m.

We will follow the convention of laying the lateral along the higher slope direction and submain in the lower slope direction. Hence, we select north–south as the direction of the lateral. We will also select paired lateral system with downslope lateral from south to north and upslope lateral from north to south (Fig. 8).

**Maximum Lateral Lengths**

Following formula is used to find the maximum lateral lengths in 2% slope for 10 mm and 13.7 mm diameter laterals. We initially assume the slope profile to be Type I, and hence we use the following formula. At the end of the design, after doing all corrections, we would examine the slope profile and adjust the length if needed.

$$\Delta h^a = \left( \left( \frac{L}{L_{tab}} \right) \left( 1 + \frac{f_e}{s} \right) \right)^{2.75} - pSL$$

The results are tabulated as below:

| Diameter of lateral (mm) | Maximum horizontal length ($L_{tab}$) (m) | Maximum actual length (m) 2% downslope | Maximum actual length (m) 2% upslope |
|---|---|---|---|
| 10 | 23 | 23.55 | 17.45 |
| 13.7 | 40 | 44.85 | 26.85 |

We will choose a better option from different combinations possible.

Option1: 10 mm diameter for both upslope and downslope.

Total length $= 23.55 + 17.45 = 41.00$ m $< 50$ m. Infeasible.

Option 2: 10 mm diameter for downslope and 13.7 mm for upslope.

Total length $= 23.55 + 26.85 = 50.4$ m $> 50$ m. Very near to feasibility—less cost.

Option 3: 13.7 mm diameter for downslope and 10 mm for upslope.

Total length $= 44.85 + 17.45 = 62.30$ m $> 50$ m. Feasible—costlier to the preceding option.

Option 4: 13.7 mm diameter for both downslope and upslope.

Total length $= 44.85 + 26.85 = 71.70$ m $> 50$ m. Feasible—highest cost.

Option 2 is very near to feasibility. If cost reduction is the prime objective, option 3 can be adopted. We can select option 3 because it is feasible as well the cost is less compared to option 4. One more reason for rejecting option 2 is that in the following calculation we need to find out the optimal position of manifold by equalizing the pressure heads on both the sides of lateral. This process will additionally create a constraint which will infringe more on the permissible maximum lengths.

**Positioning the Manifold**

We will find out the position of the manifold so that the average pressure on both the sides of the manifold is the same. Following formula is used for this process.

$$0.733c\left(\frac{L_{1e}^{2.75}}{D_1^{4.75}} - \frac{\left(L_{pe} - L_{1e}\right)^{2.75}}{D_2^{4.75}}\right) + 0.5SL_p = 0$$

where

$$c = \frac{7.89 \times 10^5}{2.75}\left(\frac{q}{s}\right)^{1.75}$$

$$c = \frac{7.89 \times 10^5}{2.75}\left(\frac{10}{3600}\right)^{1.75} = 9.64$$

$$0.733 \times 9.64\left(\frac{L_{1e}^{2.75}}{10^{4.75}} - \frac{(56.25 - L_{1e})^{2.75}}{13.7^{4.75}}\right) + 0.5 \times 0.02 \times 50 = 0$$

$$L_{1e} = 15.70 \, \text{m}$$

$$L_1 = \frac{15.70}{1.125} = 13.96 \text{ m}$$

Therefore,

$$L_{2e} = 56.25 - 15.70 = 40.55 \text{ m}$$

$$L_2 = \frac{40.55}{1.125} = 36.04 \text{ m}$$

We can have $L_1 = 14$ m and $L_2 = 36$ m.

### Relative Pressure Head for Downslope Lateral

Relative pressure head ratio for upslope lateral need not be checked. Relative pressure head for the downslope lateral alone needs to be checked because our slope profile assumption of Type I may be wrong.

$L_{2e} = 36 \times 1.125 = 40.5$ m

$$\Delta h_{Le} = \frac{789000}{2.75 \times 13.7^{4.75}} \left( \frac{10}{3600} \right)^{1.75} 40.5^{2.75} = 1.01 \text{ m}$$

$$\Delta Z_L = L \times 0.02 = 36 \times 0.02 = 0.72 \text{ m}$$

### Slope Type

$\frac{\Delta Z_L}{\Delta h_{Le}} = \frac{0.72}{1.01} = 0.71 < 1$—Type I or IIa—maximum pressure at the start

$$l_{min-e} = L_{2e} \left( 1 - \left( \frac{SL_{2e}}{2.75 \Delta h_{L2e}} \right)^{\frac{1}{1.75}} \right)$$

$$l_{min-e} = 40.5 \left( 1 - \left( \frac{0.02 \times 40.5}{2.75 \times 1.01} \right)^{\frac{1}{1.75}} \right) = 20.47 \text{ m}$$

$$l_{min} = \frac{20.47}{1.125} = 18.19 \text{ m}$$

### Friction Loss and Elevation Gain Until Minimum Pressure Head Location

$$\Delta h_{l\,min-e} = \Delta h_{Le} \left( 1 - \left( 1 - \frac{l_{min-e}}{L_e} \right)^{2.75} \right)$$

$$\Delta h_{l\,\min-e} = 1.01\left(1 - \left(1 - \frac{20.47}{40.5}\right)^{2.75}\right) = 0.86 \text{ m}$$

$$\Delta Z_{l\,\min} = 18.19 \times 0.02 = 0.36 \text{ m}$$

Deviation between maximum and minimum pressure heads $= (0.86 - 0.36) = 0.50$ m.

Relative pressure head difference $= V_h^a$ (lateral) $= \frac{0.50}{10} = 0.05 < 0.1$

Design is accepted. If the $V_h^a$ computed is greater than 0.1, changing the pipe dimension or decreasing the length of the lateral must be done and again the calculation must be done until the relative pressure head difference is within the acceptable limit.

## Submain Design

Note that we have used up only 0.05 of relative pressure head variation, and therefore we can make use of residual 0.05 plus 0.1 relative pressure head variation allocated to submain for the submain design. That means the submain can be designed for a relative pressure head variation of 0.15. But as per this problem requirement, the submain is to be designed for a relative pressure head variation of 0.1. Hence, now the design for submain is done for a relative pressure head variation of 0.1 only.

$$\text{One lateral discharge rate} = \frac{50 \text{ m} \times 4 \text{ l/h}}{0.4 \text{ m}} = 500 \text{ l/h}$$

$$\text{Specific discharge rate} = \text{One lateral discharge rate/Lateral spacing}$$

$$= \frac{500}{1} = 500 \,\text{lh}^{-1}\text{m}^{-1}$$

Maximum Submain Lengths

| Diameter of submain (mm) | Maximum horizontal length ($L_{tab}$) (m) | Maximum length for downslope 1% (m) | Maximum length for upslope 1% (m) |
|---|---|---|---|
| 37.2 | 19 | 20.35 | 17.68 |
| 46.6 | 28 | 30.87 | 25.19 |

We can select the option of 37.2 mm for downslope and 46.6 mm for upslope.

The maximum total length available for this option is $20.35 + 25.19 = 45.54$ m $> 40$ m.

## Positioning the Location of Connection of Main pipe with Submain

For $D_1 = 46.6$ mm and $D_2 = 37.2$ mm, we get $L_1 = 22.43$ m and hence $L_2 = 40$–$22.43 = 17.57$ m.

## Relative Pressure Head for Downslope Submain

$L_2 = 17.57$ m

$$\Delta h_L = \frac{789000}{2.75 \times 37.2^{4.75}} \left( \frac{500}{3600} \right)^{1.75} 17.57^{2.75} = 0.83 \text{ m}$$

$$\Delta Z = 17.57 \times 0.01 = 0.18 \text{ m}$$

**Slope Type**

$\frac{\Delta Z}{\Delta h_L} = \frac{0.18}{0.83} < 1$. Hence, slope profile is Type IIa.

$$l_{min} = 17.57 \left( 1 - \left( \frac{0.18}{2.75 \times 0.83} \right)^{\frac{1}{1.75}} \right) = 14.24 \text{ m.}$$

**Friction Loss and Elevation Gain until Minimum Pressure Head Location**

$$\Delta h_{l \, min-e} = 0.83 \left( 1 - \left( 1 - \frac{14.24}{17.57} \right)^{2.75} \right) = 0.82$$

$$\Delta Z_{l \, min} = 14.24 \times 0.01 = 0.14 \text{ m}$$

Deviation between maximum and minimum pressure heads $= (0.82 - 0.14) = 0.68$ m.

Relative pressure head difference $= V_h^a$ (submain) $== \frac{0.68}{10} = 0.068 < 0.1$

Design is accepted. If the $V_h^a$ computed had been greater than 0.1, changing the pipe dimension or decreasing the length of the lateral must have been done and again the calculation must have been done until the relative pressure head difference is within the acceptable limit (Fig. 9).

**Problem:** The total length of a paired lateral is 110 m, and the diameter of the lateral is 14 mm with a dripper spacing of 0.6 m. Average dripper discharge rate is 4 l/h, and it is laid on a slope of 0.03. Find out the position of the submain so that the average pressure head of either side of lateral is equal. Assume that the emitter connection losses are negligible.

**Solution**

$$0.733 b (L_1^{2.75} - (L_p - L_1)^{2.75}) + 0.5 S L_p = 0.$$

where

$$b = \frac{7.89 \times 10^5}{D^{4.75} 2.75} \left( \frac{q}{s} \right)^{1.75}.$$

**Fig. 9** Layout and pipe size design for uniform pipe diameter and uniform slope problem

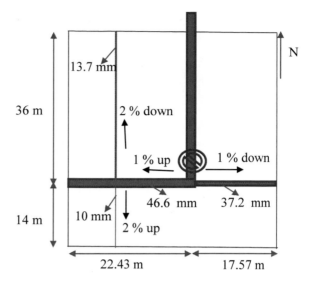

$$b = \frac{789 \times 10^5}{14^{4.75} 2.75}\left(\frac{4}{0.6 \times 3600}\right)^{1.75} = 1.706 \times 10^{-5}$$

$$0.733 \times 1.706 \times 10^{-5}(L_1^{2.75} - (110 - L_1)^{2.75}) + 0.5 \times 0.03 \times 110 = 0$$

Solving for $L_1$ by trial and error yields $L_1 = 34$ m.
Upslope lateral length is 34 m, and the downslope lateral length is 76 m.

## Statistical Variation Criterion

It is to be recalled that the emitter discharge rate variations are not only due to the pressure variations in a subunit. The variations occur due to manufacturing variations of drippers also. This variation is mostly significant and needs to be accounted. It is possible to account the manufacturing variation in the relative pressure difference criterion to a certain extent. Statistical variation method developed by Bralts et al. [5] is a robust and easily comprehensible method. In this method the variation of discharge due to pressure head differences is estimated using coefficient of variation of discharge. Usually, the manufacturing variation is also expressed by the coefficient of variation and is assumed as normally distributed. Therefore, determining variation of discharge due to the combination of both the factors namely pressure head and manufacturing could be determined by standard statistical methods assuming that both the factors are independently distributed.

In this method, the discharge variation due to pressure head is expressed in terms of coefficient of variation rather than relative flow difference.

The coefficient of variation of pressure head is defined as follows:

$$CV_h = \frac{\sqrt{\sum_{i=1}^{n}(h_i - \overline{h})^2}}{\overline{h}\sqrt{n}} \tag{29}$$

where $h_i$ pressure is head at $i$th dripper; $\overline{h}$ is the average pressure head; and n is the total number of drippers.

The coefficient of variation of dripper discharge rate $(CV_{qh})$ due to pressure head variation alone can be approximately related to $CV_h$ as follows:

$$CV_{qh} \cong xCV_h \tag{30}$$

where $x$ is flow exponent of the dripper. Theoretical proof for the preceding equation is available in [6].

When coefficient of variation of discharge due to pressure head in pipes $(CV_{qh})$ and the coefficient of variation of discharge due to manufacturing alone $(CV_{qm})$ are treated as independent random variables, the combined coefficient of variation due to pressure and manufacturing $(CV_q)$ is expressed as follows:

$$CV_q = \sqrt{(CV_{qh})^2 + (CV_{qm})^2} \tag{31}$$

In widely spaced crops, mostly more than one dripper is provided per tree. When more than one dripper is provided per tree, we need to consider the variation of discharge with respect to each tree rather than variation of discharge rate with respect to drippers. When we group greater number of emitters per tree, the net variation proportionally gets reduced. In such situations, following equation can be used to find out the coefficient of variation of discharge with respect to trees:

$$CV_q = \sqrt{\frac{(CV_{qh})^2 + (CV_{qm})^2}{N_p}} \tag{32}$$

where $N_p$ is number of emitters per plant. Table 4 provides the classification of emitters based on manufacturing variation. Tables 5 and 6 provide guidelines while designing with statistical variation criterion.

Usually both in lateral and submain, the same magnitude of coefficient of variation is maintained. However, if coefficient of variation is lower in lateral, the coefficient of variation in submain can be increased. The combined coefficient of variation for a subunit $(CV_q^{sub})$ is partitioned to the coefficient of variation for lateral $(CV_q^{Lat})$ and the submain $(CV_q^{man})$ using the following equation:

$$CV_q^{sub} = \sqrt{(CV_q^{Lat})^2 + (CV_q^{man})^2} \tag{33}$$

**Table 4** Standard for manufacturing coefficient of variation

| Dripper | Standard | $CV_{qm}$ Range (%) |
|---|---|---|
| Point Source | Excellent | 0 to 5 |
| | Very good | 5 to 7 |
| | Fair | 7 to 11 |
| | Poor | 11 to 15 |
| | Unacceptable | > 15 |
| Line source | Good | < 10 |
| | Average | 10 to 20 |
| | Poor | > 20 |

**Table 5** Guideline for combined coefficient of variation of discharge used for design and operation [1]

| Classification | $CV_q^{sub}(\%)$ |
|---|---|
| Excellent | < 10 |
| Good | 10 to 20 |
| Fair | 20 to 30 |
| Poor | 30 to 40 |
| Unacceptable | > 40 |

**Table 6** Design considerations with respect to nature of environment [9]

| No. | Guideline | $CV_q^{sub}$ Range (%) | EU Range (%) |
|---|---|---|---|
| 1 | Limited water resources with need for environmental protection | 5–15 | 80–95 |
| 2 | Limited eater resources but no environmental concerns | 15–25 | 70 -80 |
| 3 | Abundant water but environmental protection important | 10–30 | 75–85 |
| 4 | Abundant water with no environmental protection | 20–30 | 60–75 |

**Problem** The combined coefficient of variation for a subunit is fixed as 0.1. Find out the coefficient of variation for lateral and the submain separately if both the variations in lateral and submain are decided to be equal? If the dripper has a flow exponent of 0.6, find out the coefficient of variation of pressure head in lateral and submain?

$$(CV_q^{Lat}) = (CV_q^{man}).$$

Substitute this equality in following equation

$$CV_q^{sub} = \sqrt{(CV_q^{Lat})^2 + (CV_q^{man})^2}$$

$$0.1 = \sqrt{2(CV_q^{\text{Lat}})^2}$$

$$CV_q^{\text{Lat}} = \frac{0.1}{\sqrt{2}} = 0.071$$

$$CV_q^{\text{Lat}} = CV_q^{\text{man}} = 0.071$$

Coefficient of variation of pressure head in lateral is found out using following equation:

$$CV_{qh} \cong xCV_h$$

$$\text{CV}_h^{\text{Lat}} = 0.071/0.6 = 0.118$$

Coefficient of variation of pressure head in submain can be treated to be equal to the coefficient of variation of discharge in submain because the flow exponent for lateral discharge can be approximately assumed as equal to one.

$$CV_h^{\text{man}} = 0.071$$

From the preceding problem, it was demonstrated as to how the design coefficient of variation of pressure for lateral and submain must be calculated.

Using these quantities, design of laterals and submains can be done with the equations derived in the preceding chapter. The step-by-step procedure presented for the relative discharge rate difference criterion and statistical variation criterion is one and the same except with only one difference of using Eq. 20 of the preceding chapter in the place of Eq. 15.

## Relationship Between $CV_h$ and $V_h$

The equation of coefficient of variation for horizontal laterals is obtained by substituting zero for the slope in Eq. 19 of preceding chapter.

$$CV_h = \frac{\Delta h_L}{\bar{h}} \sqrt{0.0827} \tag{34}$$

By definition $V_h = \frac{\Delta h_L}{\bar{h}}$ for a level lateral/submain and hence,

$$V_h \cong 3.48 CV_h \tag{35}$$

**Problem** It was found from a computation earlier in a preceding problem that a submain designed had the following data:

Downslope—1%

Length—17.57 m.

Total friction loss—0.83 m

Total elevation gain—0.18 m

Deviation between maximum and minimum pressure heads = 0.68 m

Relative pressure head difference—0.068

Average pressure head—10 m

Find out the coefficient of variation of pressure head and also compare the value with Relative pressure head difference?

$$CV_h = \frac{1}{h}\sqrt{\left(0.0827\ \Delta h_L^2 + 0.0833(SL)^2 - 0.1543\ p\ \Delta h_L\ SL\right)}$$

$$CV_h = \frac{1}{h}\sqrt{\left(0.0827 \times 0.83^2 + 0.0833 \times 0.18^2 - 0.1543 \times 0.83 \times 0.18\right)}$$

$$CV_h = 0.019$$

From the data it can be seen that $V_h = 0.068$

So, it can be seen that the value of $V_h$ is 3.58 times more than the value of $CV_h$ for this data.

## Evaluation of Drip Irrigation Systems with Statistical Criterion

In DIS, for finding variation/uniformity of emitter flow, the discharge rates of all the emitters in a subunit need to be known. But in a subunit, measuring all the emitter flows are practically very difficult. Therefore, only a limited number of emitters is observed and their discharge rates are found out. The variation/uniformity of the discharges of an entire subunit is estimated from the sample data.

The variations in emitter discharge rates may be due to one or combinations of factors such as pressure variations at each emitter, the dripper manufacturing defects and plugging of emitters. In statistical variation method, all these variations are treated as normally distributed. Hence the properties of normal distribution are used in evaluation.

In this method, observations taken from 18 random locations are used. When the size of the sample is increased, the degree of confidence of evaluation would get improved. When the size of sample is increased, it is advised to increase in multiples of 6. This is helpful in doing computations. Following equation represents the coefficient of variation due to all the three factors:

$$CV_q = \sqrt{\frac{(CV_{qh})^2 + (CV_{qm})^2 + (CV_{qp})^2}{N_p}} \tag{36}$$

$$CV_q = \sqrt{\frac{(xCV_h)^2 + (CV_{qm})^2 + (CV_{qp})^2}{N_p}} \qquad (37)$$

where $CV_{qp}$ is the coefficient of variation of discharge due to emitter plugging.

It is easy to note down the time taken to fill a container rather than estimating dripper discharge rate. Hence time taken to fill a container of a constant volume (say, 50 ml) is found out for all the samples. Then the times are arranged in descending order. When evaluating plugged systems, sometimes it would take a very long time to get the container filled. In such occasions, one need not wait till all the containers are filled. For such containers, the time taken to fill must be estimated based on the discharge rate collected in a reasonable length of time.

Following formula is used to find out the value of coefficient of variation of discharge for a subunit $\left(CV_q^{sub}\right)$:

$$CV_q^{sub} = 0.667\frac{t_{us} - t_{ls}}{t_{us} + t_{ls}} \times 100 \qquad (38)$$

where $t_{us}$ is the sum of the top three values and $t_{ls}$ is the sum of the bottom three values.

**Problem** Following are the times (in seconds) taken to fill a container of 100 ml while operating a subunit with 2 l/h drippers for 18 drippers selected at random. Find out the coefficient of variation.

| 1800 | 1750 | 1600 | 1801 | 1755 | 1400 | 1568 | 1791 | 1890 |
|------|------|------|------|------|------|------|------|------|
| 1750 | 1866 | 1849 | 1768 | 1700 | 1740 | 1690 | 1709 | 1801 |

The times may be written in ascending order as follows:

| 1890 | 1866 | 1849 | 1801 | 1801 | 1800 | 1791 | 1768 | 1755 |
|------|------|------|------|------|------|------|------|------|
| 1750 | 1750 | 1740 | 1709 | 1700 | 1690 | 1600 | 1568 | 1400 |

$t_{us} = 1890 + 1866 + 1849 = 5605$
$t_{ls} = 1600 + 1568 + 1400 = 4568$
Applying Eq. 38, we get,

$$CV_q^{sub} = 0.068$$

If the value of coefficient of variation is less than the recommended threshold value, further analysis and field work are not needed to be done.

Let us assume that the evaluated coefficient of variation of a newly installed system is higher than the threshold level, and if the coefficient of variation due to manufacturing data is known, it is possible to find out the coefficient of variation due to pressure variation alone. This analysis would help in finding whether the newly installed system is properly designed in terms of pressure head.

**Problem** For the preceding problem, the coefficient of variation due to manufacturing is say 0.05, find out the coefficient of variation due to pressure head if the system installed is new. Assume that there is no grouping effect of emitters. The flow exponent of dripper is 0.5

Coefficient of variation due to plugging is assumed as zero for new systems.

$$0.068 = \sqrt{(0.5CV_h)^2 + (0.05)^2}$$

$$CV_h = 0.092$$

If the value of the $CV_h$ is above the threshold level, then we can understand that the design is faulty.

For plugged systems, the coefficient of variation due to plugging can be found out if the coefficient of variation due to pressure head is found out by measuring the pressure distribution in the subunit. Nowadays facilities are available to measure the pressure heads near drippers. At the start and the end of the lateral, measurement of pressure is easy by using lateral T-connector. While evaluating plugged systems the combined coefficient of plugging and manufacturing variation is found out. The combined coefficient of plugging and manufacturing variation would be taken as an indicator to know whether the systems is plugged or not. If the system is found to be plugged, corrective measures like acid treatment or chlorine treatment can be adopted to clear the plugging. After chemical treatment, the system must be evaluated again to know whether its performance has increased.

**Problem** Find out the combined coefficient of plugging and manufacturing variation for the following data:

Coefficient of variation due to pressure head—0.092
Combined coefficient of variation found out from field evaluation—0.068
Flow exponent of dripper—0.7

$$CV_q = \sqrt{(xCV_h)^2 + (CV_{qP\&m})^2}$$

$$CV_{qp\&m} = \sqrt{(0.068)^2 - (0.7 \times 0.092)^2}$$

$$CV_{qp\&m} = 0.0218$$

**Table 7** 90% confidence limits on coefficient of variation estimates from field observations

| Coefficient of Variation ($CV_q$) | Number of observations | | | |
|---|---|---|---|---|
| | 18 | 36 | 72 | 144 |
| 0.1 | 0.035 | 0.024 | 0.017 | 0.012 |
| 0.2 | 0.073 | 0.050 | 0.034 | 0.024 |
| 0.3 | 0.115 | 0.078 | 0.054 | 0.038 |
| 0.4 | 0.162 | 0.109 | 0.076 | 0.054 |

*Source* ASAE [1]

Table 7 provides confidence limits at 90% level for the field observation estimates of coefficient of variation of discharge. This table is very useful in statistically interpreting the field observations. For instance, if the field estimated $CV_q$ is 0.1 and number of observations is 18, the actual variation would be expected in the range of $0.1 \pm 0.035$. If the field estimated $CV_q$ is 0.4 and number of observations is 18, the actual variation would be expected in the range of $0.1 \pm 0.162$. Therefore, it must be kept in mind that when the field observed $CV_q$ is more, the number of observations must be increased to get a narrower confidence limit.

## *Keller and Karmeli Emission Uniformity Criterion*

Keller and Karmeli uniformity criterion [7] was one of the earliest propositions and is still popularly used in USA and in many other countries. This method is based on the premise that minimum discharge rate in a subunit must be given due attention as the effect of minimum discharge may show significant reduction in crop growth. In this method, emission uniformity ($EU_h$) due to pressure head variation alone without considering manufacturing variation is defined as follows:

$$EU_h = \frac{q\,min_h}{\bar{q}_h} \tag{39}$$

where $\bar{q}_h$ is the average discharge at the average pressure head location;
$q\,min_h$ is the minimum discharge at the minimum pressure head.

Emission uniformity due to manufacturing variation and grouping effect of emitters ($EU_{m\&e}$) is found out using following equation:

$$EU_{m\&e} = \left(1 - 1.27\frac{CV_{qm}}{\sqrt{N_p}}\right) \tag{40}$$

where $CV_{qm}$ is the coefficient of variation of discharge due to manufacturing variation alone and $N_p$ is the number of emitters per plant. The coefficient value 1.27 specifies the low quarter mean in the normal distribution due to manufacturing variation [3].

The combined emission uniformity (EU) is found out by multiplying Eqs. 39 and 40 which is as follows:

$$EU = \left(1 - 1.27\frac{CV_{qm}}{\sqrt{N_p}}\right)\frac{q\,\min_h}{\overline{q}_h} \tag{41}$$

From the minimum and the average pressure head data, the minimum and the average discharge rates are obtained using following equations:

$$q\,\min_h = k(h_{\min})^x \tag{42}$$

$$\overline{q}_h = k\overline{h}^x \tag{43}$$

where $k$ is emitter discharge coefficient and $x$ is emitter flow exponent. $h_{\min}$ and $\overline{h}$ are the minimum and average pressure heads.

Substituting preceding equations in Eq. 41 we get,

$$EU - \left(1 - 1.27\frac{CV_{qm}}{\sqrt{N_p}}\right)\frac{(h_{\min})^x}{\overline{h}^x} \tag{44}$$

When we start designing pipes, all the variables are known except $h_{\min}$ in the preceding equation. By substituting those known quantities, $h_{\min}$ would be found out from the preceding equation. Then an empirical equation is used to find out the allowable pressure head loss ($\Delta h_{su}^a$) in the subunit which is as follows:

$$\Delta h_{su}^a = 2.5(\overline{h} - h_{\min}) \tag{45}$$

If the calculated $\Delta h_{su}^a$ value is too small, the variables such as $CV_{qm}$, $N_p$, $h_{\min}$, $\overline{h}$ must be adjusted so that the $\Delta h_{su}^a$ obtained is reasonably bigger so that design could be done comfortably.

From the value of $\Delta h_{su}^a$, allowable pressure head loss in submain/manifold ($\Delta h_m^a$) and allowable pressure head loss in lateral ($\Delta h_l^a$) are equally divided into two. Then the design of lateral and submain is proceeded.

Generally, $EU$ value above 80% is recommended for design. Table 8 provides the recommended $EU$ ranges based on emitter type, spacing, topography and slope. Following equation provides an approximate relationship between $CV$ and $EU$ [3].

$$EU = 1 - 1.27\,CV_q \tag{46}$$

**Table 8** Recommended emission uniformity ranges in %[a]

| Emitter type | Spacing ft, (m) | Topography | Slope % | EU range % |
|---|---|---|---|---|
| Point source on perennial crops | > 13, (40) | Uniform steep or undulating | < 2<br>> 2 | 90 to 95<br>85 to 90 |
| Point source on perennial or semi-permanent crops | < 13, (4) | Uniform steep or undulating | < 2<br>> 2 | 85 to 90<br>80 to 90 |
| Line source on annual or perennial crops | All | Uniform steep or undulating | < 2<br>> 2 | 80 to 90<br>70 to 85 |
| Spray[b] | All | Uniform steep or undulating | < 2<br>> 2 | 90 to 95<br>80 to 90 |

[a]ASAE Engineering Practice Standard: ASAE EP 405.1 (1988) Design and Installation of MI Systems
[b]Keller and Bliesner, Sprinkle and Trickle Irrigation (2000)

## *Field Evaluation When Using Keller and Karmeli Criterion*

If Keller and Karmeli uniformity criterion is used following formula is used for estimating emission uniformity in the field:

$$EU = \frac{\overline{q}_{LQ}}{\overline{q}} \tag{47}$$

where EU is the field emission uniformity, $\overline{q}_{LQ}$ is the average rate of emitter flow rate of the lowest one-fourth of the emitter flow readings, and $\overline{q}$ is the average of emitter flow readings of the sample. The low quarter mean is assumed as an estimate of minimum emitter flow in the preceding equation.

A random sample size of 16 locations is used in Keller and Karmeli uniformity criterion. From the field estimated EU value, if we use Eq. 41, the uniformity of discharge due to pressure variation alone $\left(\frac{q_{min}^h}{q_h}\right)$ can be found out. The use of factorizing combined uniformity into uniformity due to manufacturing and uniformity due to pressure variation is same as that discussed in statistical variation criterion section.

**Problem** Coefficient of variation of pressure head is 0.1, coefficient of variation of discharge due to manufacturing is 0.05, number of emitters per plant is 4, and flow exponent of dripper is 0.5. Find the integrated variation of discharge due to pressure head, manufacturing variation and grouping of emitters? Estimate also the emission of uniformity from the coefficient of variation of discharge?

$$CV_q = \sqrt{\frac{(xCV_h)^2 + (CV_{qm})^2}{N_p}}$$

$$CV_q = \sqrt{\frac{(0.5 \times 0.1)^2 + 0.05^2}{4}}$$

$CV_q = 0.0353.$
$CV_q$ in % $= 3.53\%$
$EU = 1 - 1.27CV_q.$
EU $= 1 - 1.27 \times 0.0353 = 0.955.$
EU in % $= 95.5\%$

# Summary

Design methods based on relative flow difference criterion are easily understood by everyone, and hence this criterion is widely used around the world. In this chapter equations for finding average friction loss and optimal length of uniform diameter pipes in uniform slopes are derived. Procedures for designing lateral and submain in uniform slopes are presented with numerical examples. Paired lateral/submain hydraulic analysis is also presented. Relationship between relative flow difference criterion and other design criteria such as statistical variation criterion and Keller and Karmelli criterion are presented. Field evaluation methods of drip irrigation subunits are also presented.

## Assessment

### Part A Choose the Best Option

1. For a case of a paired lateral design one of the main objectives is to

   _____

   a. **Maintain the same average pressure head on each of the laterals**
   b. Maintain the same pressure head at the start of each of the laterals
   c. Maintain the same length for each of the laterals
   d. Maintain the same diameter for both the laterals.

2. _____lateral laying is economical.

   a. **Paired**
   b. Single
   c. Upslope
   d. None of these.

3. The maximum length of lateral for _____would be larger than the lateral in upslope if all the other factors are the same in both the sloped and horizontal laterals.

   a. **Downslope**
   b. Upslope

    c.   Level land

    d.   None of these.

4.    The maximum length of lateral for _____would be smaller than a downslope lateral if all the other factors are the same in both sloped and horizontal laterals.

    a.   **Upslope**

    b.   Downslope

    c.   Level land

    d.   None of these.

## Part B

1.    Define specific discharge rate of a drip lateral?

2.    What is the formula for finding average pressure head loss in a drip lateral?

3.    How do you find the specific discharge rate (SDR) for tree crops where more than one dripper per tree is installed?

4.    Explain how a paired lateral is designed?

5.    Why design method based on statistical variation concept is said to be more robust than the other design methods?

## Problems

1.    For a design of drip irrigation system (DIS), the allowable relative discharge rate variation for a subunit $(V_q^a)$ is fixed as 0.1 and the dripper selected is orifice type dripper with flow exponent of 0.5. The DIS is designed for an average pressure head of 12 m. Find out the allowable pressure variation in the lateral and the submain?

2.    Lateral is laid over level land. Allowable relative pressure head variation in the lateral is 0.2. Uniform spacing of dripper is 0.3 m, dripper discharge rate is 4 l/h, and inside diameter of lateral is 14 mm. Find out the maximum length of the lateral?

3.    A paired lateral has to be designed for a field as shown in Fig. 4. Design using inline laterals of diameter 13.7 mm. Dripper discharge rate of 2 l/h and uniform dripper spacing of 0.6 m were selected based on a field study of the soil. Average operating pressure is 10 m. Allowable relative pressure head variation is 0.1. Equivalent length for emitter connection loss for one dripper is 0.05 m. Pressure available at the start of the submain is 12 m. Design suitable lengths of a paired lateral along with the diameter of the lateral? Find also the average pressure heads on both the upslope and downslope laterals?

4.    The total length of a paired lateral is 90 m and the diameter of the lateral is 14 mm with a dripper spacing of 0.6 m. Average dripper discharge rate is 4 l/h, and it is laid on a slope of 0.03. Find out the position of the submain so that the average pressure head of either side of lateral is equal. Assume that the emitter connection losses are negligible.

# References and Further Reading

1. American Society of Agricultural Engineers (1996) EP458. *Field evaluation of microirrigation systems. ASAE Standards* (43rd ed., pp. 756–761). ASAE, St. Joseph.
2. Anonymous, USDA-NRCS. (2013). *Chapter 7. Microirrigation Part 623 Irrigation. National Engineering Handbook.*
3. Barragan, J., Bralts, V., & Wu, I. P. (2006). Assessment of emission uniformity for microirrigation design. *Biosystems Engineering, 93*(1), 89–97. https://doi.org/10.1016/j.biosystemseng.2005.09.010
4. Baiamonte, G. (2016) Simple relationships for the optimal design of paired drip laterals on uniform slopes. *Journal of Irrigation and Drainage Engineering,* 04015054, https://doi.org/10.1061/(ASCE)IR.1943-4774.0000971
5. Bralts, V. F., Wu, I.-P., & Gitlin, G. M. (1981). Manufacturing variation and drip irrigation uniformity. *Transactions of the ASAE. American Society of Agricultural Engineers, 24*(1), 0113–0119. https://doi.org/10.13031/2013.34209
6. Luis, J., Losada, A., M., Rodríguez-Sinobas, L., & Sánchez, R. (2004). Analytical relationships for designing rectangular drip irrigation units. *Journal of Irrigation and Drainage Engineering, 130*(1). https://doi.org/10.1061/~ASCE!0733-9437~2004!130:1~47
7. Keller, J., & Karmeli, D. (1974). Trickle irrigation design parameters. *Transactions of the ASAE, 17*(4), 678–684.
8. Keller, J., & Bliesner, R. D. (1990). *Sprinkle and trickle irrigation. An Avi Book.* Van Nostrand Reinheld.
9. Evans, R. G., Wu, I.-P., & Smajstrala, A. G. (2007). *Micro irrigation systems. Chapter 17 in Design and operation of Farm irrigation systems.* ASABE. (ISBN) 1-892769-64-6.
10. Ravikumar, V., Renuka, R., & Ranganathan, C. R. (2003). Design of tapered drip irrigation pipes in non-uniform slopes. *Journal of Agricultural Engineering. Indian Council of Agricultural Research, 40*(3), 63–73.
11. Wu, I.-P., Saruwatari, C. A., & Gitlin, H. M. (1983). Design of drip irrigation lateral length on uniform slopes. *Irrigation Science, 4,* 117–135. Wu, I.-P., & Gitlin, H. M. (1974). Drip irrigation design based on uniformity. *Transactions of ASAE, 17*(3), 429–432.
12. Wu, I. P., & Gitlin, H. M. (1975). Energy gradient line for drip irrigation laterals. *Journal of the Irrigation and Drainage Division, 101*(4), 323–326.
13. Zhang, L., Wu, P., & Zhu, D. (2013). Hydraulic design procedure for drip irrigation submain unit based on relative flow difference. *Irrigation Science, 31,* 1065–1073. https://doi.org/10.1007/s00271-012-0388-3

# Design with Telescopic Pipes in Uniform and Non-uniform Slopes

## Introduction

Telescopic pipe designs are done for mostly submains only rather than for laterals. In steep down slopes only, in order to handle high elevation gains, telescopic pipe designs are done for laterals. Derivation of direct design equations for maximum length under non-uniform slope situations is very difficult. Usually when design process is started, uniform pipe diameter and uniform slope are assumed and maximum length of lateral/submain is estimated. With this maximum length quantity as a base, telescopic pipe design is proceeded further. For this design methodology, equations are developed and demonstrated here.

## Trial-and-Error Method

Trial-and-error method is the simplest way of designing with telescopic pipes in slopes. For any specific non-uniform field slope, a set of diameters of lateral/submain must be known priory. Then for every location, calculation of friction loss ($\Delta h_i$) and elevation loss/gain ($\Delta Z_i$) is done. Then the following equations (discussed in preceding chapters) are used to find out the pressure heads at all the outlet locations.

$$h_i = h_{\text{start}} - \Delta h_i - \Delta z_i \tag{1}$$

where

$h_i$     is the pressure head at the $i$th section of the pipe (m);

$h_{\text{start}}$     is the pressure head at the start (m);

$\Delta h_i$     is the head loss due to friction until $i$th segment with reference to the start (m);

$\Delta z_i$     is the fall (minus sign would exist with the computed value) or rise until $i$th segment with reference to the start (m).

© The Author(s), under exclusive license to Springer Nature Singapore Pte Ltd. 2023    307
V. Ravikumar, *Sprinkler and Drip Irrigation*,
https://doi.org/10.1007/978-981-19-2775-1_11

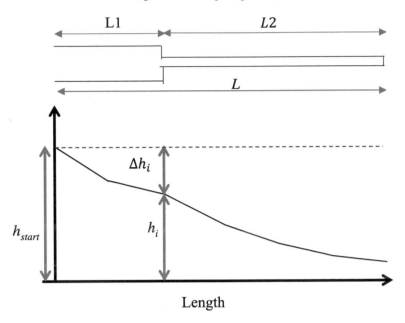

**Fig. 1** Friction loss in a telescopic lateral/submain

Following equation is used for finding the friction loss in lateral/submain for uniform pipe diameter whose end is closed (Fig. 1).

$$\Delta h_i = \frac{7.89 \times 10^5}{2.75 D^{4.75}} \times \left(\frac{q}{s}\right)^{1.75} \times \left(L^{2.75} - (L - l_i)^{2.75}\right) \tag{2}$$

where

$(q/s)$   is the specific discharge rate (l/s/m);
$L$      is the total length of the pipe (m);
$l_i$     is the length of pipe from the start until the end of $i$th segment (m);

For telescopic pipes, the preceding equation must be used for each diameter segment of lateral/submain separately. Following equation is used for finding elevation loss/gain:

$$\Delta z_i = \Delta z_{i-1} - p \times s \times S_i \tag{3}$$

where

'$p$'   is 1 for down slope and ($-1$) for upslope;
'$s$'   is spacing between emitter/laterals;
$S_i$   is slope of the $i$th segment of the pipe.

After finding out the pressure heads at every location in lateral/submain, the maximum pressure head, the minimum pressure head and the average pressure head are found out. Then the relative pressure head difference for the lateral/submain is found out. If the relative pressure head difference found out is below the allowable threshold level, the design is accepted or else the design is rejected.

The following problem demonstrates the simple trial-and-error method.

**Problem**

A submain of diameter 46.6 mm has a length of 50 m. The submain runs on a downslope of 1% for the first 20 m and runs on a 3% downslope for the remaining 30 m. Laterals are placed at a spacing of 2 m with each lateral discharging at a rate of 300 l/h. Pressure head at the start of the lateral is 10 m. Find out whether relative pressure head variation is below the threshold level of 0.1?

The friction loss at the start of all the laterals, elevation gain at the start of all the laterals in the submain and pressure head at the start of all the laterals were found out using the preceding equations and presented in Table 1 and Fig. 2. It can be seen that the minimum pressure head is 9.74 m and maximum pressure head is 10.48 m. The average pressure head is 9.99 m. When we find average pressure heads using analytical equations derived earlier, the average pressure head calculation takes into account the pressure head at the start of the submain also. But the correct calculation method is that the pressure head at the start of the submain must be omitted for finding average pressure head, maximum pressure head as well as minimum pressure head. The reasons for this way of calculating average pressure head are discussed in chap. 4.

The maximum pressure head = 10.48 m
The minimum pressure head = 9.74 m
The average pressure head = 9.99 m

$$\text{Relative Pressure Head Difference } (V_h) = \frac{h_{max} - h_{min}}{\bar{h}}$$

$$V_h = \frac{10.48 - 9.74}{9.99} = 0.074$$

The value of $V_h$ is less than the threshold value of 0.1 and hence the design can be accepted.

**Alternate Method of Solving The Problem**

In Fig. 3, it can be seen that the friction loss is treated as positive values and elevation gain is also treated as positive values. The reason for this way of depiction is to visualize how much of friction loss is balanced by elevation gain. The deviation between friction loss and elevation gain with reference to the submain start for every lateral start (offtake) location is found out (Table 1). In this method the reference pressure head at the start of the submain is assumed as zero (Do not think that how the submain will have discharge at zero pressure head). If the pressure head loss due to friction at every location is equal to the pressure head gain due to elevation, then

**Table 1** Pressure heads for a non-uniform slope problem

| Length (m) | Friction loss $(\Delta h_i)$ (m) | Fall in elevation $(\Delta Z_i)$ (m) | Gain in head due to fall in elevation (m) | Deviation between loss and gain ((2)–(4)) (m) | Pressure head $h_i$ (m) |
|---|---|---|---|---|---|
| (1) | (2) | (3) | (4) | (5) | (6) |
| 0 | 0.00 | 0 | 0 | 0.00 | 10.00 |
| 2 | 0.07 | −0.02 | 0.02 | 0.05 | 9.95 |
| 4 | 0.13 | −0.04 | 0.04 | 0.09 | 9.91 |
| 6 | 0.18 | −0.06 | 0.06 | 0.12 | 9.88 |
| 8 | 0.23 | −0.08 | 0.08 | 0.15 | 9.85 |
| 10 | 0.28 | −0.1 | 0.1 | 0.18 | 9.82 |
| 12 | 0.33 | −0.12 | 0.12 | 0.21 | 9.79 |
| 14 | 0.37 | −0.14 | 0.14 | 0.23 | 9.77 |
| 16 | 0.40 | −0.16 | 0.16 | 0.24 | 9.76 |
| 18 | 0.44 | −0.18 | 0.18 | 0.26 | 9.74 |
| 20 | 0.46 | −0.2 | 0.20 | **0.26** | **9.74** |
| 22 | 0.49 | −0.26 | 0.26 | 0.23 | 9.77 |
| 24 | 0.51 | −0.32 | 0.32 | 0.19 | 9.81 |
| 26 | 0.53 | −0.38 | 0.38 | 0.15 | 9.85 |
| 28 | 0.55 | −0.44 | 0.44 | 0.11 | 9.89 |
| 30 | 0.57 | −0.50 | 0.50 | 0.07 | 9.93 |
| 32 | 0.58 | −0.56 | 0.56 | 0.02 | 9.98 |
| 34 | 0.59 | −0.62 | 0.62 | −0.03 | 10.03 |
| 36 | 0.60 | −0.68 | 0.68 | −0.08 | 10.08 |
| 38 | 0.60 | −0.74 | 0.74 | −0.14 | 10.14 |
| 40 | 0.61 | −0.80 | 0.80 | −0.19 | 10.19 |
| 42 | 0.61 | −0.86 | 0.86 | −0.25 | 10.25 |
| 44 | 0.61 | −0.92 | 0.92 | −0.31 | 10.31 |
| 46 | 0.62 | −0.98 | 0.98 | −0.36 | 10.36 |
| 48 | 0.62 | −1.04 | 1.04 | −0.42 | 10.42 |
| 50 | 0.62 | −1.10 | 1.10 | **−0.48** | **10.48** |

there is no pressure head difference along the submain. Let us say, if the pressure head loss until a location is 2 m and pressure head gain due to elevation until the location is 0.5 m. Then the pressure head deviation from the pressure head at the start (assumed as zero) will be 1.5 m (i.e. $0 + 2 - 0.5 = 1.5$ m). If the pressure head loss for a location due to friction is 2 m and pressure head loss due to rise in elevation is 1 m, the pressure head deviation from the reference pressure head of zero metre will be 3 m (i.e. $2 - (-1) = 3$ m).

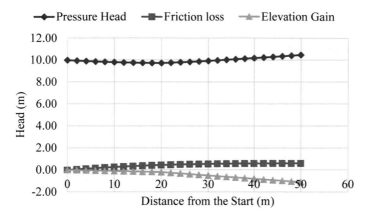

**Fig. 2** Pressure head, friction loss and elevation loss/gain for non-uniform slope problem

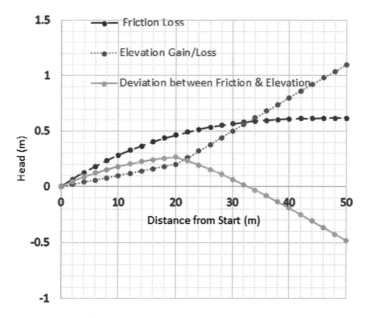

**Fig. 3** Deviation between friction and elevation losses for non-uniform slope problem

If friction loss curve and elevation gain/loss curves cut as in Fig. 3, there are two kinds of deviations; one is positive and another is negative. The maximum of positive deviations and minimum of negative deviation (i.e. biggest negative number) must be added up to get total deviation [2]. The cumulative deviation between the friction loss and elevation gain/loss calculated are shown in Table 1.

For this problem, the maximum positive deviation occurs at 20 m from the start of submain and the maximum negative deviation occurs at the end of the submain.

Hence, the total deviation obtained is 0.74 (i.e. $0.26 + 0.48 = 0.74$ m). If the curves do not cut each other, then the deviation may be either positive alone or negative alone. In that case, finding the maximum deviation is done easily.

Let us solve another problem with a telescopic pipe for the same non-uniform slope as that of the preceding problem.

**Problem**
A submain has a length of 50 m. Its diameter is 46.6 mm for the first 32 m and 37.2 mm for the next 18 m length. The submain runs on a downslope of 1% for the first 20 m and runs on a 3% downslope for the remaining 30 m. Laterals are placed at a spacing of 2 m with each lateral discharging at a rate of 300 l/h. Pressure head at the start of the lateral is 10 m. Find out whether relative pressure head variation is below the threshold level of 0.1?

Friction loss calculation for the first pipe section of diameter 46.6 mm for a length of 32 m

$$\Delta h_i = \frac{7.89 \times 10^5}{2.75 D^{4.75}} \times \left(\frac{q}{s}\right)^{1.75} \times \left(L^{2.75} - (L - l_i)^{2.75}\right)$$

Friction loss at the start of the first lateral:

$$\Delta h_1 = \frac{7.89 \times 10^5}{2.75 \times 46.6^{4.75}} \times \left(\frac{300/3600}{2}\right)^{1.75} \times \left(50^{2.75} - (50 - 2)^{2.75}\right) = 0.07 \, \text{m}$$

Do the same calculations until the starting point of 16th lateral (i.e. at 32 m). Friction loss at the start of 16th lateral:

$$\Delta h_{16} = \frac{7.89 \times 10^5}{2.75 \times 46.6^{4.75}} \times \left(\frac{300/3600}{2}\right)^{1.75} \times \left(50^{2.75} - (50 - 32)^{2.75}\right) = 0.579 \, \text{m}$$

Friction loss calculation for the second pipe section of diameter 37.2 mm for a length of 18 m
Friction loss at the start of 17th lateral:

$$\Delta h_{17} = 0.579 + \frac{7.89 \times 10^5}{2.75 \times 37.2^{4.75}} \times \left(\frac{300/3600}{2}\right)^{1.75}$$
$$\times \left(18^{2.75} - (18 - 2)^{2.75}\right) = 0.609 \, \text{m}$$

Friction loss at the start of the last lateral:

$$\Delta h_{25} = 0.579 + \frac{7.89 \times 10^5}{2.75 \times 37.2^{4.75}} \times \left(\frac{300/3600}{2}\right)^{1.75} \times \left(18^{2.75}\right) = 0.687 \, \text{m}$$

The results of the calculations are tabulated in Table 2 and drawn as in Fig. 4.

The maximum pressure head $= 10.41$ m

**Table 2** Pressure heads and deviations from the pressure head at the start for the telescopic non-uniform slope problem

| Length (m) | Friction loss ($\Delta h_l$) (m) | Elevation loss or gain ($\Delta Z_l$) (m) | Deviation between cumulative friction loss and elevation gain (m) ((2)-(-(3))) | Pressure head $h_l$ (m) |
|---|---|---|---|---|
| (1) | (2) | (3) | (4) | (5) |
| 0 | 0.000 | 0 | 0.00 | 10.00 |
| 2 | 0.065 | −0.02 | 0.05 | 9.95 |
| 4 | 0.126 | −0.04 | 0.09 | 9.91 |
| 6 | 0.183 | −0.06 | 0.12 | 9.88 |
| 8 | 0.235 | −0.08 | 0.15 | 9.85 |
| 10 | 0.283 | −0.1 | 0.18 | 9.82 |
| 12 | 0.326 | −0.12 | 0.21 | 9.79 |
| 14 | 0.367 | −0.14 | 0.23 | 9.77 |
| 16 | 0.403 | −0.16 | 0.24 | 9.76 |
| 18 | 0.436 | −0.18 | 0.26 | 9.74 |
| 20 | 0.465 | −0.2 | *0.26* | *9.74* |
| 22 | 0.491 | −0.26 | 0.23 | 9.77 |
| 24 | 0.514 | −0.32 | 0.19 | 9.81 |
| 26 | 0.534 | −0.38 | 0.15 | 9.85 |
| 28 | 0.552 | −0.44 | 0.11 | 9.89 |
| 30 | 0.567 | −0.5 | 0.07 | 9.93 |
| 32 | 0.579 | −0.56 | 0.02 | 9.98 |
| 34 | 0.609 | −0.62 | −0.01 | 10.01 |
| 36 | 0.633 | −0.68 | −0.05 | 10.05 |
| 38 | 0.652 | −0.74 | −0.09 | 10.09 |
| 40 | 0.666 | −0.8 | −0.13 | 10.13 |
| 42 | 0.676 | −0.86 | −0.18 | 10.18 |
| 44 | 0.682 | −0.92 | −0.24 | 10.24 |
| 46 | 0.686 | −0.98 | −0.29 | 10.29 |
| 48 | 0.687 | −1.04 | −0.35 | 10.35 |
| 50 | 0.687 | −1.1 | **−0.41** | **10.41** |

The minimum pressure head = 9.74 m
The average pressure head = 9.97 m

$$\text{Relative Pressure Head Difference } (V_h) = \frac{h_{max} - h_{min}}{\bar{h}}$$

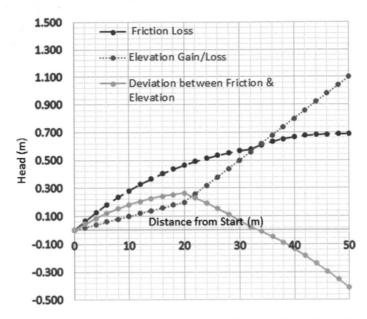

**Fig. 4** Deviation between friction and elevation losses for non-uniform slope telescopic pipe problem

$$V_h = \frac{10.41 - 9.74}{9.97} = 0.067$$

It must be observed that the relative pressure head difference has got reduced to 0.0607 whereas for the same data with uniform diameter of 46.6 mm throughout the length, we got the relative pressure head difference as 0.074 earlier. We have also reduced the cost incurred in pipe as well as relative pressure head variation. But it must also be noted that this is not the most optimal solution. By trial-and-error calculation one can improve the result still further depending on the need of the farmer.

## Telescopic Pipe Design

When design is done with telescopic pipes, trial-and-error method of selection of diameter of pipe segments is a laborious process and hence optimal methods were developed in order to help the designers to get a satisfactory design in a shorter time. We will learn a simple optimal method for telescopic design.

In Fig. 5, observe the nature of a pressure head curve for a uniform diameter submain pipe on a horizontal land. If allowable friction loss in the submain is $\Delta h_{mf}^a$, the diameter of the submain pipe must be selected so that the total friction loss in the uniform diameter pipe is equal to $\Delta h_{mf}^a$. Let us assume that we use a telescopic

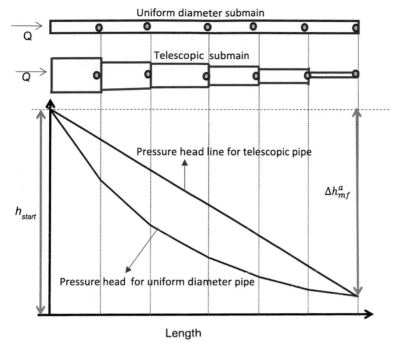

**Fig. 5** Pressure heads for uniform pipe and ideal telescopic pipe

submain for this situation. Assume that any pipe diameter of our interest is available in the market. We provide the first segment of the submain from the start with a larger diameter so that the pressure head of the telescopic pipe in the first segment is a straight line as shown in Fig. 5. After the delivery of water through the first lateral, the discharge rate in the submain gets reduced. For this discharge rate, we select a little smaller diameter of submain so that the pressure loss curve in the telescopic pipe is a continuation of the same straight line with the same slope. If we continue to use different diameters for each section for all the submain segments, we would get a straight pressure head curve for the telescopic submain. It has been proved by economic analysis that this way of designing telescopic pipe with a straight pressure head line as shown in Fig. 5 is economical and therefore the straight pressure head line here in this book is called as an ideal line.

Practically, only few pipe diameters are available in the market, and it is also not possible to change the diameter of submain after each outlet. Therefore, practically only few times the reduction of diameter is done in a multi-outlet pipe. Figure 6 shows three sections with three diameter pipes laid on an upslope. The diameters must be so selected that the pressure head curve for the designed telescopic submain must match closer to the ideal straight pressure head curve as in Fig. 6. It must also be noted that in each diameter pipe segment, the total allowable friction loss is distributed according to the length of the segment. For instance, for the length $L_1$,

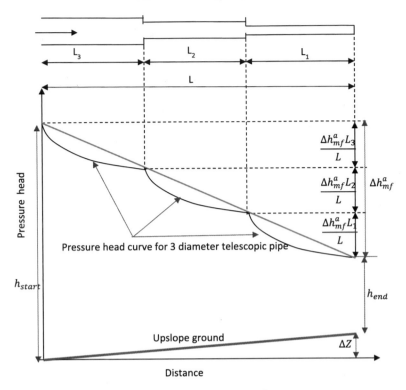

**Fig. 6** Pressure head for telescopic pipe with finite number of segments

the diameter of the pipe is selected so that friction loss in that segment is $\frac{\Delta h_{mf}^a L_1}{L}$. It is conventional to start the design from the closed end of the submain and hence the numbering of the segment starts from the closed end.

For the first segment from the last:

$$\frac{\Delta h_{mf}^a L_1}{L} = \frac{7.89 \times 10^5}{2.75 D_1^{4.75}} \times \left(\frac{q}{s}\right)^{1.75} \times L_1^{2.75}$$

$$L_1 = \left(\frac{\Delta h_{mf}^a \times D_1^{4.75}}{c \times L}\right)^{0.571} \tag{4}$$

where $c = \frac{7.89 \times 10^5}{2.75} \left(\frac{q}{s}\right)^{1.75}$.

For the remaining segments the equation to be used is as follows:

$$\frac{\Delta h_{mf}^a L_i}{L} = \frac{c}{D_i^{4.75}} \left((L_i + L_d)^{2.75} - L_d^{2.75}\right) \tag{5}$$

where $L_i$ is the length of the $i$th section for which design is made, $D_i$ is the diameter for the $i$th section and $L_d$ is the sum of designed lengths downstream of the $i$th section. Equation 5 is an implicit equation and solving the equation must be done by trial-and-error only.

Following equation must be used to find allowable friction loss in submain $(\Delta h_{mf}^a)$ :

$$\text{Allowable pressure variation in submain} = \Delta h_m^a = \Delta h_{mf}^a - p \Delta Z_L \qquad (6)$$

where $\Delta Z_L$ is total absolute elevation difference between the start and end of the submain. $p = 1$ if there is net downfall and $p = (-1)$ if there is net rise.

In the irrigation industry around the world, poly-plot method is very popular for designing in non-uniform slopes with telescopic/tapered pipes. This method is very much similar to the methods discussed earlier. But the method has been adapted here in such a way that manual designs are also done easily.

In this method, for each and every pipe diameter one graphical plot as shown in Fig. 7 is to be developed. Figure 8 shows poly-plots for 12 and 16 mm diameter lateral.

## Poly-Plot Method

The interpretation of the curves in Fig. 9 needs to be understood clearly and discussed here with reference to Fig. 9. The curve in Fig. 9 corresponds to one specific discharge rate. For the lengths of $L_1$ and $L_2$, the friction losses are $\Delta h_{L1}$, $\Delta h_{L2}$ respectively. Note that the closed end is always at the origin. When water is carried for a length of $(L_1 - L_2)$ from the start of the pipe of length $L_1$, the friction loss is $(\Delta h_{L1} - \Delta h_{L2})$. This way of representing the friction loss helps in simplifying calculations.

We will see now, how we can find out pressure heads at different locations of lateral/submain laid on slopes using a poly-plot [1]. In Fig. 10, an upslope pipe of length $L$ is shown. A tracing paper must be placed on the segment of friction loss curve corresponding to the specific diameter and specific SDR and the segment of the curve corresponding to the length of the pipe must be drawn and taken separately. That segment of curve OA is shown in the figure. From point A, draw on the tracing sheet for a depth of $h_{start}$ (i.e. AB) and the scale of drawing must be maintained as that of Poly-plot. Then draw the ground slope line BC as shown in the figure. Since, the case shown in the figure is for upslope the line BC is drawn as upslope. For a case of down slope or for any non-uniform slope, the ground slope line can be drawn accordingly. Between the curves OA and CB, the vertical lines shown in the figure give the pressure head in the lateral/submain at the respective locations.

Many a times, the designer needs to know only the deviation between cumulative head loss due to friction and cumulative elevation loss or gain rather than the pressure heads at each and every location.

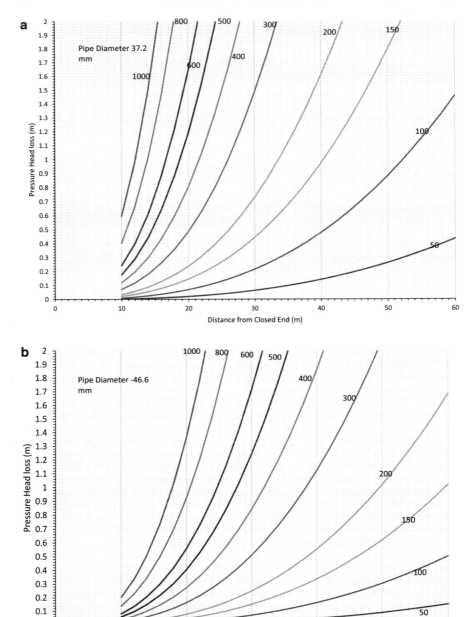

**Fig. 7** Poly-plots for different SDR for different submain pipe diameter

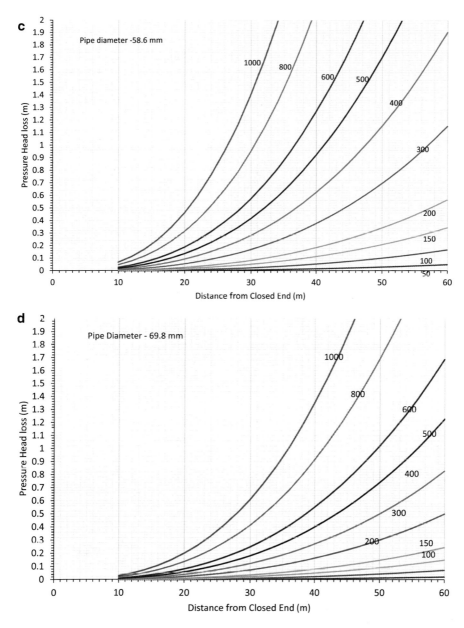

**Fig. 7** (continued)

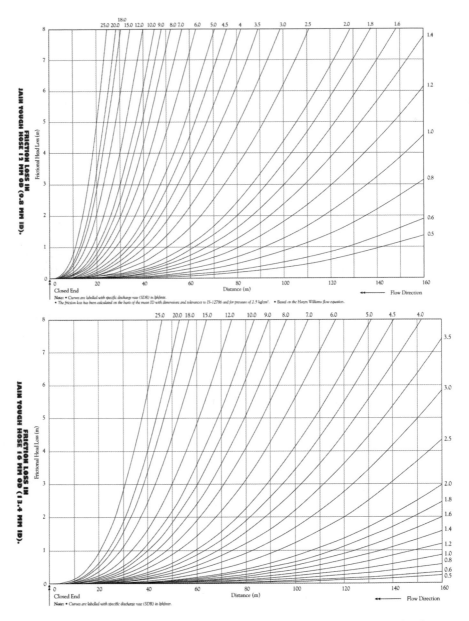

**Fig. 8** Poly-plot for different SDR for 12 and 16 mm laterals (*Courtesy* Jain Irrigation Systems Ltd.)

**Fig. 9** A sample friction
loss curve used in poly-plot

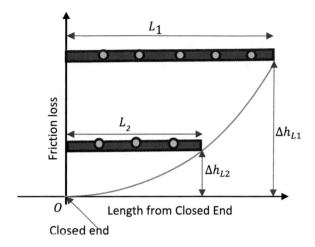

For doing analysis based on deviation of cumulative head loss due to friction and cumulative head loss/gain due to elevation, discussion is made now for a downslope case (Fig. 11). In this method, the ground slope curve AC is drawn from point A. The deviation between cumulative head loss due to friction and elevation loss/gain at any location is the vertical line drawn between the curve AC and curve AO. Among these deviations, maximum deviation is taken for calculating relative pressure head.

In case of downslopes, there is also a possibility that the ground slope line may cut the friction loss line as shown in Fig. 12. In such a case, the total deviation is the sum of maximum positive deviation and absolute negative deviation. In Fig. 12, the total deviation is (LM + OC).

Let us solve the same problem using poly-plot method which was solved earlier. The problem is repeated again here for convenience.

### Problem
A submain of diameter 46.6 mm has a length of 50 m. The submain runs on a downslope of 1% for the first 20 m and runs on a 3% downslope for the remaining 30 m. Laterals are placed at a spacing of 2 m with each lateral discharging at a rate of 300 l/h. Pressure head at the start of the lateral is 10 m. Find out total pressure head deviation??

When we adopt poly-plot graphical procedure, the image obtained is as shown in Fig. 13. If the ready-made plots are not available, the process of drawing Fig. 13 is explained as below:

Friction loss is calculated using the following formula:

$$\Delta h_L = \frac{7.89 \times 10^5}{2.75 D^{4.75}} \times \left(\frac{q}{s}\right)^{1.75} L^{2.75}$$

For different values of $L$, the frictional head losses are calculated and presented in Table 3. The friction loss value at point A in Fig. 13 is the same as the value

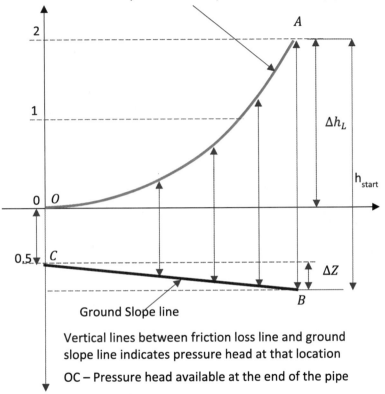

Fig. 10  Pressure head along lateral/submain for upslope in poly-plot

of elevation at point A. Therefore copy the value of 0.6162 m from column 2 of
Table 3 to the ordinate of elevation curve corresponding to 50 m. Then the ordinate
of elevation curve is calculated from 50 to 30 m for 1% down slope.

$$\Delta Z_{48} = \Delta Z_{50} - \left(\frac{1}{100}\right)2 = 0.6162 - 0.02 = 0.5961 \, \text{m}$$

$$\Delta Z_{46} = \Delta Z_{48} - \left(\frac{1}{100}\right)2 = 0.5961 - 0.02 = 0.5762 \, \text{m}$$

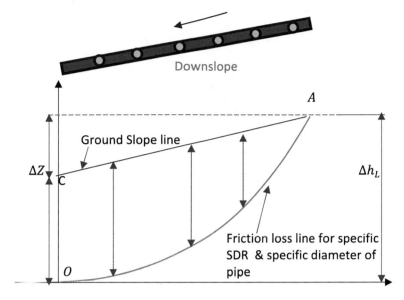

**Fig. 11** Deviation between friction loss and ground slope line in poly-plot (case a)

Do the same kind of calculations until point B corresponding to 30 m from closed end. Then slope changes to 3% down.

$$\Delta Z_{28} = \Delta Z_{30} - \left(\frac{3}{100}\right)2 = 0.4162 - 0.06 = 0.3562 \, \text{m}$$

Do the same kind of calculations until the closed end.

**Problem**

A horizontal submain must be designed for a length of 60 m. Laterals are placed at a spacing of 2 m with each lateral discharging at a rate of 300 l/h. Pressure head at the start of the lateral is 10 m. Design a telescopic submain with three diameters of pipe such as 58.6, 46.6 and 37.2 mm. The allowable relative pressure head variation is 0.1.

The methodology adopted for solving this problem uses the ideal straight-line friction loss which was discussed in the preceding sections. The ideal straight-line method is adapted here to suit to the poly-plot method.

$$\text{SDR} = \frac{\frac{300}{3600}}{2} = \frac{3}{72} \, \text{l/s/m}$$

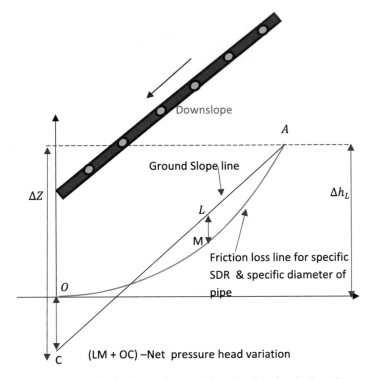

**Fig. 12** Deviation between friction loss and ground slope line in poly-plot (case b)

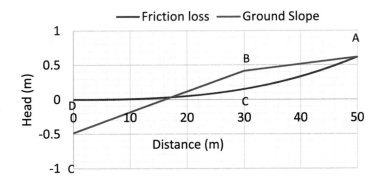

**Fig. 13** Poly-plot for a sample problem

**Allowable Friction Loss**

$$V_h^a(\text{submain}) = \frac{\Delta h_m^a}{\bar{h}}$$

$$0.1 = \frac{\Delta h_m^a}{10}$$

**Table 3**  Results of poly-plot of a non-uniform slope problem

| Distance (m) | Friction loss till the closed end (m) | Ordinate of elevation curve (m) | Deviation between friction loss and elevation (m) |
|---|---|---|---|
| 0 | **0.0000** | **−0.4838** | 0.4838 |
| 2 | 0.0001 | −0.4238 | 0.4239 |
| 4 | 0.0006 | −0.3638 | 0.3644 |
| 6 | 0.0018 | −0.3038 | 0.3056 |
| 8 | 0.0040 | −0.2438 | 0.2478 |
| 10 | 0.0074 | −0.1838 | 0.1912 |
| 12 | 0.0122 | −0.1238 | 0.1360 |
| 14 | 0.0186 | −0.0638 | 0.0824 |
| 16 | 0.0268 | −0.0038 | 0.0306 |
| 18 | 0.0371 | 0.0562 | −0.0191 |
| 20 | 0.0496 | 0.1162 | −0.0666 |
| 22 | 0.0644 | 0.1762 | −0.1118 |
| 24 | 0.0819 | 0.2362 | −0.1543 |
| 26 | 0.1020 | 0.2962 | −0.1942 |
| 28 | 0.1251 | 0.3562 | −0.2311 |
| 30 | **0.1512** | **0.4162** | −0.2650 |
| 32 | 0.1806 | 0.4362 | −0.2556 |
| 34 | 0.2134 | 0.4562 | −0.2428 |
| 36 | 0.2497 | 0.4762 | −0.2265 |
| 38 | 0.2897 | 0.4962 | −0.2065 |
| 40 | 0.3336 | 0.5162 | −0.1826 |
| 42 | 0.3815 | 0.5362 | −0.1547 |
| 44 | 0.4335 | 0.5562 | −0.1227 |
| 46 | 0.4899 | 0.5762 | −0.0863 |
| 48 | 0.5508 | 0.5962 | −0.0454 |
| 50 | **0.6162** | **0.6162** | 0.0000 |

$\Delta h_m^a = 1m = \Delta h_{mf}^a$ since horizontal.

Ordinates of ideal friction loss line is calculated and shown in Table 4. In Fig. 14, the ideal friction loss line is shown.

**First Segment from the Closed End**

$$L_1 = \left( \frac{\Delta h_{mf}^a \times D_1^{4.75}}{c \times L} \right)^{0.571}$$

**Table 4** Computation for friction loss curves for telescopic design problem using poly-plot

| Distance from closed end (m) | Ideal friction loss (m) $\frac{\Delta h_{mf}^a L_i}{L}$ | Case-1 | | Case-2 | |
|---|---|---|---|---|---|
| | | Diameter (mm) | Friction loss (m) | Diameter (mm) | Friction loss (m) |
| 0 | 0.000 | 37.2 | 0.000 | 37.2 | 0.000 |
| 2 | 0.033 | 37.2 | 0.000 | 37.2 | 0.000 |
| 4 | 0.067 | 37.2 | 0.002 | 37.2 | 0.002 |
| 6 | 0.100 | 37.2 | 0.005 | 37.2 | 0.005 |
| 8 | 0.133 | 37.2 | 0.012 | 37.2 | 0.012 |
| 10 | 0.167 | 37.2 | 0.021 | 37.2 | 0.021 |
| 12 | 0.200 | 37.2 | 0.035 | 37.2 | 0.035 |
| 14 | 0.233 | 37.2 | 0.054 | 37.2 | 0.054 |
| 16 | 0.267 | 37.2 | 0.078 | 37.2 | 0.078 |
| 18 | 0.300 | 37.2 | 0.108 | 37.2 | 0.108 |
| 20 | 0.333 | 37.2 | 0.145 | 37.2 | 0.145 |
| 22 | 0.367 | 37.2 | 0.188 | 37.2 | 0.188 |
| 24 | 0.400 | 37.2 | 0.239 | 37.2 | 0.239 |
| 26 | 0.433 | 37.2 | 0.297 | 37.2 | 0.297 |
| 28 | 0.467 | 37.2 | 0.365 | 37.2 | 0.365 |
| 30 | 0.500 | 37.2 | 0.441 | 37.2 | 0.441 |
| 32 | 0.533 | 37.2 | 0.527 | 37.2 | 0.527 |
| 34 | 0.567 | 46.6 | 0.559 | 46.6 | 0.559 |
| 36 | 0.600 | 46.6 | 0.596 | 46.6 | 0.596 |
| 38 | 0.633 | 58.6 | 0.609 | 46.6 | 0.636 |
| 40 | 0.667 | 58.6 | 0.624 | 46.6 | 0.680 |
| 42 | 0.700 | 58.6 | 0.640 | 46.6 | 0.727 |
| 44 | 0.733 | 58.6 | 0.658 | 58.6 | 0.745 |
| 46 | 0.767 | 58.6 | 0.677 | 58.6 | 0.764 |
| 48 | 0.800 | 58.6 | 0.697 | 58.6 | 0.784 |
| 50 | 0.833 | 58.6 | 0.719 | 58.6 | 0.806 |
| 52 | 0.867 | 58.6 | 0.743 | 58.6 | 0.830 |
| 54 | 0.900 | 58.6 | 0.768 | 58.6 | 0.855 |
| 56 | 0.933 | 58.6 | 0.795 | 58.6 | 0.882 |
| 58 | 0.967 | 58.6 | 0.824 | 58.6 | 0.911 |
| 60 | 1.000 | 58.6 | 0.854 | 58.6 | 0.942 |

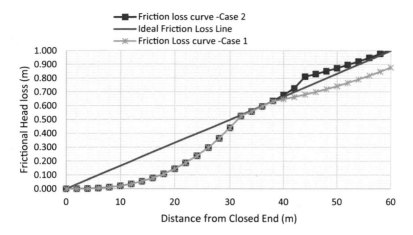

**Fig. 14** Friction loss curves for telescopic design problem using poly-plot

where $c = \frac{7.89 \times 10^5}{2.75} \left(\frac{q}{s}\right)^{1.75}$

$$c = \frac{7.89 \times 10^5}{2.75} \left(\frac{300/3600}{2}\right)^{1.75} = 1102.5$$

Let us assume the diameter of 37.2 mm for the first section from the closed end.

$$L_1 = \left(\frac{1 \times 37.2^{4.75}}{1102.5 \times 60}\right)^{0.571} = 32.15\,\text{m}$$

Let us take $L_1 = 32$ m.

If poly-plot of 37.2 mm is available in hand and used, the friction loss curve for SDR of (3/72) l/s/m would cut the ideal friction loss line as shown in Fig. 1.

Following formula is used to find friction loss for different values of $L$, to calculate the friction loss from 0 to 32 m and tabulated as in Table 4.

$$\Delta h_L = \frac{c}{D^{4.75}} L^{2.75}$$

**Second Segment from the Closed End**

By using the 46.6 mm diameter poly-plot, the friction loss curve segment corresponding to 32–36 m can be traced as in Fig. 13.

Instead, if we do the calculations, following formula can be used:

$$\Delta h_L = \frac{c}{D^{4.75}} \left(L^{2.75} - L_d^{2.75}\right) + \Delta h_{Ld}$$

where $L$ is the length from the closed end, $D$ is the diameter in that segment and $L_d$ is the sum of designed segment lengths downstream.

$$\Delta h_{34} = \frac{1102.5}{46.6^{4.75}} \left(34^{2.75} - 32^{2.75}\right) + 0.527 = 0.559$$

$$\Delta h_{36} = \frac{1102.5}{46.6^{4.75}} \left(36^{2.75} - 32^{2.75}\right) + 0.527 = 0.596$$

If we do calculation for $\Delta h_{38}$, it will be more than 0.633 m which corresponds to ideal friction loss line ordinate at 38 m. Hence we increase the diameter to 58.6 mm from 36 m.

**Third Segment from the Closed End**
From 38 m onwards, 58.6 mm diameter pipe is used and the calculation are as below:

$$\Delta h_{38} = \frac{1102.5}{58.6^{4.75}} \left(38^{2.75} - 36^{2.75}\right) + 0.596 = 0.632\,\text{m}$$

$$\Delta h_{40} = \frac{1102.5}{58.6^{4.75}} \left(40 - 36^{2.75}\right) + 0.596 = 0.647\,\text{m}$$

The calculation is done successively as above, and the results are tabulated.

It must be noticed that for the entire length of 60 m, the friction loss does not go above ideal friction loss. It must also be noted that the total friction loss is only 0.877 m whereas the allowable friction loss is 1 m. The design mentioned as case 1 in Table 4 can be accepted as such or the available residual friction loss of 0.133 m (i.e. $1 - 0.877 = 0.133$) may be used for increasing the length of 46.6 mm until 40 m and the remaining length in 58.6 mm so that the total friction loss is maintained as 1 m. The revised calculations are shown as case 2 in Table 4. Obviously, the case 2 is more economical than case 1.

The friction loss curves for case 1 and case 2 as well as ideal friction loss line are shown in Fig. 14. It must be noted that though the friction loss curve lies above the ideal friction line at some locations but provides economy. Hence case 2 is the most acceptable design.

We will do another telescopic design problem for a non-uniform slope situation.

**Problem**
A submain must be designed for a length of 50 m. Laterals are placed at a spacing of 2 m with each lateral discharging at a rate of 600 l/h. The submain is laid on a down slope of 0.5% for the first 20 m from the start and on a down slope of 1% for the next 30 m. The allowable relative pressure head variation is 0.1. The average design pressure head is 10 m. Design a telescopic submain with three diameters of pipe such as 58.6, 46.6 and 37.2 mm.

**Allowable Friction Loss**

$$V_h^a \text{(submain)} = \frac{\Delta h_m^a}{\bar{h}}$$

$$0.1 = \frac{\Delta h_m^a}{10}$$

Allowable pressure head difference in submain $(\Delta h_m^a) = 1$ m.

Allowable head loss due to friction: $\Delta h_{mf}^a = \Delta h_m^a + p\Delta Z$ where $\Delta Z$ is total absolute elevation difference between the start and end of the submain. $p = 1$ if there is net downfall and $p = (-1)$ if there is net rise.

$$\Delta Z = \frac{0.5}{100} \times 20 + \frac{1}{100} \times 30 = 0.4\,\text{m}$$

Allowable head loss due to friction $= 1 + 1 \times 0.4 = 1.4\,\text{m}$

Ordinates of ideal friction loss line is calculated and shown in Table 5. In Fig. 15, the ideal friction loss line is shown.

**First Segment from the Closed End**

$$L_1 = \left(\frac{\Delta h_{mf}^a \times D_1^{4.75}}{c \times L}\right)^{0.571}$$

where $c = \frac{7.89 \times 10^5}{2.75}\left(\frac{q}{s}\right)^{1.75}$

$$c = \frac{7.89 \times 10^5}{2.75}\left(\frac{600/3600}{2}\right)^{1.75} = 3708.32$$

Let us assume the diameter of 37.2 mm for the first section from the closed end.

$$L_1 = \left(\frac{1.4 \times 37.2^{4.75}}{3708.32 \times 50}\right)^{0.571} = 21.7\,\text{m}$$

Let us take $L_1 = 20$ m.

Following formula is used to find friction loss for different values of $L$, to calculate the friction loss from 0 to 20 m and tabulated as in Table 5.

$$\Delta h_L = \frac{c}{D^{4.75}}L^{2.75}$$

**Table 5** Computation for friction loss curves for telescopic design problem in non-uniform slopes using poly-plot

| Distance from closed end (m) | Ground slope (%) | Dia (mm) | Ideal friction loss (m) | Ground profile drawn from closed end (m) | Friction loss (m) | Ground profile drawn from A (m) | Deviation between friction loss curve and ground profile AB (m) |
|---|---|---|---|---|---|---|---|
| 0 | 1 | 37.2 | 0 | 0.000 | **0.000** | 1.000 | 1.000 |
| 2 | 1 | 37.2 | 0.056 | 0.020 | **0.001** | 1.020 | 1.019 |
| 4 | 1 | 37.2 | 0.112 | 0.040 | **0.006** | 1.040 | 1.034 |
| 6 | 1 | 37.2 | 0.168 | 0.060 | **0.018** | 1.060 | 1.042 |
| 8 | 1 | 37.2 | 0.224 | 0.080 | **0.039** | 1.080 | 1.041 |
| 10 | 1 | 37.2 | 0.28 | 0.100 | **0.072** | 1.100 | 1.028 |
| 12 | 1 | 37.2 | 0.336 | 0.120 | **0.119** | 1.120 | 1.001 |
| 14 | 1 | 37.2 | 0.392 | 0.140 | **0.182** | 1.140 | 0.958 |
| 16 | 1 | 37.2 | 0.448 | 0.160 | **0.263** | 1.160 | 0.897 |
| 18 | 1 | 37.2 | 0.504 | 0.180 | **0.364** | 1.180 | 0.816 |
| 20 | 1 | 37.2 | 0.56 | 0.200 | **0.486** | 1.200 | 0.714 |
| 22 | 1 | 46.6 | 0.616 | 0.220 | **0.536** | 1.220 | 0.684 |
| 24 | 1 | 46.6 | 0.672 | 0.240 | **0.595** | 1.240 | 0.645 |
| 26 | 1 | 46.6 | 0.728 | 0.260 | **0.663** | 1.260 | 0.597 |
| 28 | 1 | 46.6 | 0.784 | 0.280 | **0.740** | 1.280 | 0.540 |
| 30 | 1 | 46.6 | 0.84 | 0.300 | **0.828** | 1.300 | 0.472 |
| 32 | 0.5 | 58.6 | 0.896 | 0.310 | **0.861** | 1.310 | 0.449 |
| 34 | 0.5 | 58.6 | 0.952 | 0.320 | **0.899** | 1.320 | 0.421 |
| 36 | 0.5 | 58.6 | 1.008 | 0.330 | **0.940** | 1.330 | 0.390 |
| 38 | 0.5 | 58.6 | 1.064 | 0.340 | **0.985** | 1.340 | 0.355 |
| 40 | 0.5 | 58.6 | 1.12 | 0.350 | **1.035** | 1.350 | 0.315 |
| 42 | 0.5 | 58.6 | 1.176 | 0.360 | **1.089** | 1.360 | 0.271 |
| 44 | 0.5 | 58.6 | 1.232 | 0.370 | **1.148** | 1.370 | 0.222 |
| 46 | 0.5 | 58.6 | 1.288 | 0.380 | **1.212** | 1.380 | 0.168 |
| 48 | 0.5 | 58.6 | 1.344 | 0.390 | **1.281** | 1.390 | 0.109 |
| 50 | 0.5 | 58.6 | 1.4 | 0.400 | **1.355** | 1.400 | 0.045 |

## Second Segment from the Closed End

Friction loss from 22 to 30 m with diameter of 46.6 mm is calculated as below:

$$\Delta h_L = \frac{c}{D^{4.75}}\left(L^{2.75} - L_d^{2.75}\right) + \Delta h_{Ld}$$

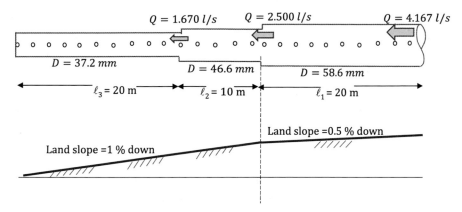

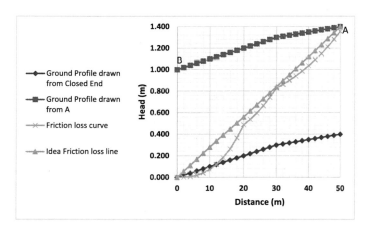

**Fig. 15** Friction loss curves for telescopic design problem in non-uniform slopes using poly-plot

$$\Delta h_{22} = \frac{c}{46.6^{4.75}} \left(22^{2.75} - 20^{2.75}\right) + 0.486 = 0.536 \, \text{m}$$

Similar calculations are done until 30 m. After 30 m, the friction loss is more than the ideal friction loss line.

**Third Segment from Closed End**

Friction loss is calculated from 32 to 50 m with 58.6 mm diameter as below:

$$\Delta h_{32} = \frac{c}{58.6^{4.75}} \left(32^{2.75} - 30^{2.75}\right) + 0.828 = 0.861 \, \text{m}$$

The friction loss curve is shown in Fig. 15. Ground profile curve drawn from point A in Fig. 15 in order to find the deviations between friction loss curve and ground profile BA. It can be seen that the deviation is at the maximum only one metre and at some places the deviation is marginally more than 1 m which can be neglected. Hence this design can be accepted.

## Telescopic Pipe with Only Two Sections

For the case of two diameters as shown in Fig. 16, the frictional head loss is obviously $(x - z + y)$. Hence if we equate this frictional loss to allowable frictional loss $(\Delta h^a_{mf})$ we get the following equation:

$$\Delta h^a_{mf} = c \left( \frac{L^{2.75} - L_1^{2.75}}{D_2^{4.75}} + \frac{L_1^{2.75}}{D_1^{4.75}} \right) \tag{7}$$

where $c = \frac{7.89 \times 10^5}{2.75} \left( \frac{q}{s} \right)^{1.75}$

$$L_1 = \left( \frac{c^{-1} \Delta h^a_{mf} - D_2^{-4.75} L^{2.75}}{D_1^{-4.75} - D_2^{-4.75}} \right)^{0.3636} \tag{8}$$

When using Eq. 8, it is to be known priorly about which two diameters are to be used. To handle this problem, we must find out the diameter of pipe needed assuming a uniform diameter pipe. Select the diameter of pipe available in the market as $D_2$

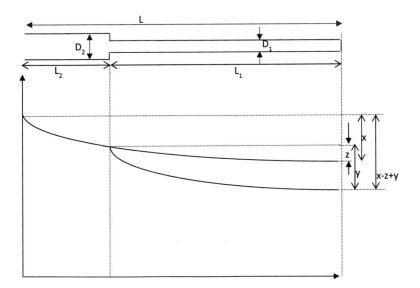

**Fig. 16** Telescopic pipe with only two diameters

which is just higher than the found-out diameter. Obviously, $D_1$ is the next smaller diameter available in the market.

**Problem**

A submain must be designed for a length of 50 m. Laterals are placed at a spacing of 2 m with each lateral discharging at a rate of 600 l/h. The submain is laid on a down slope of 0.5% for the first 20 m from the start and on a down slope of 1% for the next 30 m. The allowable relative pressure head variation is 0.1. The average design pressure head is 10 m. Design a telescopic submain with two diameters of pipe. The diameters of pipe available in the market are 58.6, 46.6 and 37.2 mm.

Note: The data of this problem is same as the preceding problem, with a constraint of using only two pipe diameters.

We have already calculated the allowable head loss due to friction $= 1.4$ m

**Diameter of Pipe Needed for Frictional Loss of 1.4 m**

$$\Delta h^a_{mf} = c \, L^{2.75} / D^{4.75}$$

$$D = \left( \frac{3708.32 \times 50^{2.75}}{1.4} \right)^{0.2105} = 50.59 \, \text{mm}$$

Hence, we would select 58.6 mm as $D_2$ and 46.6 mm as $D_1$.

$$L_1 = \left( \frac{3708.32^{-1} \times 1.4 - 58.6^{-4.75} \times 50^{2.75}}{46.6^{-4.75} - 58.6^{-4.75}} \right)^{0.3636} = 39.15 \, \text{m}$$

We will take $L_1 = 38$ m and $L_2 = 12$ m for easiness in calculation. Table 6 shows the detailed calculations, and Fig. 17 depicts the results. It can be seen from the last column of the table that the deviation between the friction loss and the loss/gain due to elevation is marginally more than 1 m at many places. Sometimes designers do not want to operate at the boundary like this, and the design is made sufficiently with some more factor of safety.

Let us assume that we design with $L_1 = 30$ m and $L_2 = 20$ m, the deviations get reduced to below 0.75 m as seen in Fig. 18

# Computer-Aided Design of Drip Subunits

Trial-and-error method of designing drip irrigation subunits is a cumbersome process when done manually. Most of the analytical calculation discussed in this book can be solved using freely downloadable software from the Web namely HYDROCALC of Netafim and NAANCAT of NaanDanJain.

**Table 6** Computation for friction loss curves for two diameter telescopic design problem in non-uniform slopes

| Distance from closed end (m) | Ground slope (%) | Dia (mm) | Ideal friction loss (m) | Ground profile drawn from closed end (m) | Friction loss (m) | Ground profile drawn from A (m) | Deviation between friction loss curve & ground profile AB (m) |
|---|---|---|---|---|---|---|---|
| 0 | 1 | 46.6 | 0 | 0.000 | 0.000 | 1.000 | 1.000 |
| 2 | 1 | 46.6 | 0.056 | 0.020 | 0.000 | 1.020 | 1.020 |
| 4 | 1 | 46.6 | 0.112 | 0.040 | 0.002 | 1.040 | 1.038 |
| 6 | 1 | 46.6 | 0.168 | 0.060 | 0.006 | 1.060 | 1.054 |
| 8 | 1 | 46.6 | 0.224 | 0.080 | 0.013 | 1.080 | 1.067 |
| 10 | 1 | 46.6 | 0.28 | 0.100 | 0.025 | 1.100 | 1.075 |
| 12 | 1 | 46.6 | 0.336 | 0.120 | 0.041 | 1.120 | 1.079 |
| 14 | 1 | 46.6 | 0.392 | 0.140 | 0.063 | 1.140 | 1.077 |
| 16 | 1 | 46.6 | 0.448 | 0.160 | 0.090 | 1.160 | 1.070 |
| 18 | 1 | 46.6 | 0.504 | 0.180 | 0.125 | 1.180 | 1.055 |
| 20 | 1 | 46.6 | 0.56 | 0.200 | 0.167 | 1.200 | 1.033 |
| 22 | 1 | 46.6 | 0.616 | 0.220 | 0.217 | 1.220 | 1.003 |
| 24 | 1 | 46.6 | 0.672 | 0.240 | 0.275 | 1.240 | 0.965 |
| 26 | 1 | 46.6 | 0.728 | 0.260 | 0.343 | 1.260 | 0.917 |
| 28 | 1 | 46.6 | 0.784 | 0.280 | 0.421 | 1.280 | 0.859 |
| 30 | 1 | 46.6 | 0.84 | 0.300 | 0.509 | 1.300 | 0.791 |
| 32 | 0.5 | 46.6 | 0.896 | 0.310 | 0.607 | 1.310 | 0.703 |
| 34 | 0.5 | 46.6 | 0.952 | 0.320 | 0.718 | 1.320 | 0.602 |
| 36 | 0.5 | 46.6 | 1.008 | 0.330 | 0.840 | 1.330 | 0.490 |
| 38 | 0.5 | 46.6 | 1.064 | 0.340 | 0.974 | 1.340 | 0.366 |
| 40 | 0.5 | 58.6 | 1.12 | 0.350 | 1.024 | 1.350 | 0.326 |
| 42 | 0.5 | 58.6 | 1.176 | 0.360 | 1.078 | 1.360 | 0.282 |
| 44 | 0.5 | 58.6 | 1.232 | 0.370 | 1.137 | 1.370 | 0.233 |
| 46 | 0.5 | 58.6 | 1.288 | 0.380 | 1.201 | 1.380 | 0.179 |
| 48 | 0.5 | 58.6 | 1.344 | 0.390 | 1.270 | 1.390 | 0.120 |
| 50 | 0.5 | 58.6 | 1.4 | 0.400 | 1.344 | 1.400 | 0.056 |

The software presented in this book are being developed by the author of the book with the collaboration of his students [3]. The software used for simulation of drip subunits is called as Drip System Simulation Programme (DSSP). This software was developed in Visual C++. This software does not need any installation process, and it can be copied on to any desktop and used. Another software namely dripmain.exe was

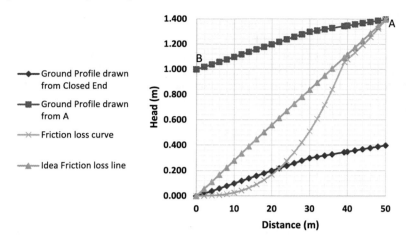

**Fig. 17** Friction loss curves for two diameter telescopic design problem in non-uniform slopes

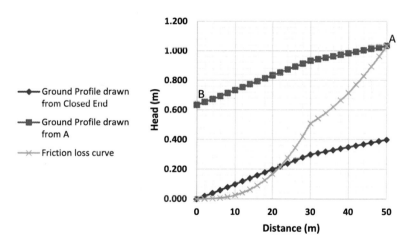

**Fig. 18** Improved friction loss curves for two diameter telescopic design problem in non-uniform slopes

developed in Visual Basic. The software is being upgraded, and the latest versions can be downloaded as and when new releases are made.

## Drip System Simulation Programme (DSSP)

This software simulates emitter discharges in a subunit with telescopic tapered laterals and submains in non-uniform slopes. Manufacturing variation of emitters is also accounted by assuming the variation as normally distributed. Lateral and submain can run on non-uniform slopes. Simulation can also be done for connecting sub main along with laterals. For tree crops, for each tree, many drippers each with different discharge rate and flow exponent are installed. These situations can also be very easily accounted in DSSP.

With DSSP many alternative designs can be evaluated for any input data with very good user-friendly data input interface and satisfactory designs can be stored in files and retrieved with ease.

## *Manufacturing Variation*

Due to manufacturing process, variations in diameter of flow passage of drip emitters occur and since the discharge coefficient 'k' depends on the diameter of flow passage, the k value of each dripper in a subunit is treated as a random variable and is assumed as normally distributed. The following equation is used to generate normally distributed random variate:

$$k = \mu + \sigma \left( \sum_{i=1}^{i=12} u_i - 6 \right) \tag{9}$$

where $k$ is the generated $k$ value, $\mu$ is the mean value of $k$ and $\sigma$ is the standard deviation of $k$ and $u_i$'s are random numbers generated from uniform distribution between 0 and 1.

## *Uniformity coefficient*

The uniformity of discharge in this software is represented using statistical uniformity which is defined as one minus coefficient of variation of discharge and following equation is used for finding the uniformity coefficient

$$\text{Uniformity} = \left( 1 - \sqrt{\frac{\sum_{i=1}^{i=n} q_i^2 - \bar{q}^2}{N}} \right) \times 100 \tag{10}$$

where $q_i$ is the discharge of emitter $i$, $\bar{q}$ is the average discharge of emitter and $N$ is the total number of emitters in the subunit.

### Barb Loss
Pressure head loss due to barbs from online emitters that extend through the pipe wall are called barb loss. The present version of the software calculates the barb loss which has significant error, and hence, it is cautioned that the software must not be used for accounting barb loss.

## *Screens of the DSSP*

Figure 19 shows the DSSP window, which consists of a menu, a toolbar and a status bar. The menu has six sections: "File", "View", "Design", "Format", "Window" and "Help". The menus are self-evident in their intent and are much similar to standard windows applications. The tool bar contains shortcuts for menu commands. The status bar at the bottom of the window displays help tip about the currently highlighted menu command.

The DSSP has a design wizard, which can be run by clicking "New" from the "File" menu. DSSP is capable of designing single lateral, single submain and full subunit. The subunit design wizard consists of eight steps. Each step displays a user interactive dialog box to enter the data.

An option for the first outlet spacing is provided for the submain, also for the laterals. This is to accommodate the situations, where the distance between the inlet and the first outlet is not equal to the spacing of other outlets.

The emitters are normally placed in two ways. One is, emitters are placed with respect to the plants. One or more number of emitters provided near the roots of each plant as shown in Fig. 20. Two is, emitters are placed individually as shown in Fig. 21. This is the case corresponding to inline drippers.

Figure 21 is the dialog box to enter data of emitters. Normally the discharge coefficient ($k$) and the flow exponent ($x$) values can be obtained from manufacturers catalogue. The $k$ and $x$ values of emitters can also be calculated from lab tests when they are not available. The data required to calculate the $k$ and $x$ values is discharges of emitter at two different pressures. Discharge at one pressure is enough if the $x$ value of emitter is known. To calculate $k$ and $x$ values from these data, the "Calculate" button is to be clicked (Fig. 22). Then the dialog box shown in Fig. 23 would appear

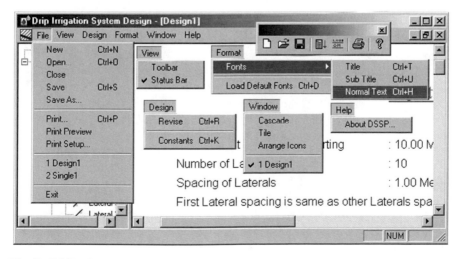

**Fig. 19** DSSP window

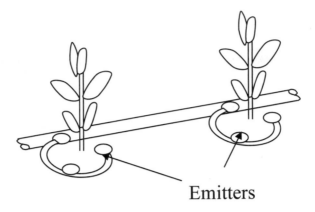

**Fig. 20**  Emitters placed with respect to plants

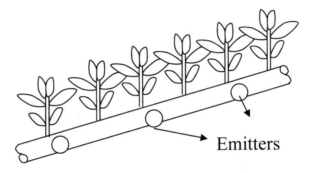

**Fig. 21**  Emitters placed independent of plants

on the screen. Then if the "Calculate" button in Fig. 23 is pressed the k and x values are calculated and automatically transferred to the dialog box in Fig. 22 when "OK" button is pressed.

The topographical data can be entered in three different types. "Uniform percentage slope" may be chosen if the slope along the submain and either side laterals are uniform for its entire length and is common to all the laterals in that side of the submain. "Non-uniform percentage slope-same for all Laterals" may be chosen if the slope is not uniform for entire length, still holding the common to all Laterals property as said in the previous case. In Fig. 24 a non-uniform slope lateral has been shown. In this case if slope data for one non-uniform sloped lateral is provided, the same slope is taken for all the laterals. "Fully non-uniform percentage slope" may be chosen when the slope data is different for each lateral.

The design result contains two splitters as shown in Fig. 25. The left one contains the headings such as "Input Data" and "Results" the "Result" has "Submain" and "Laterals" as subheadings. These appear as a tree view and can be expanded or collapsed. The right-side one displays the result for the currently selected option.

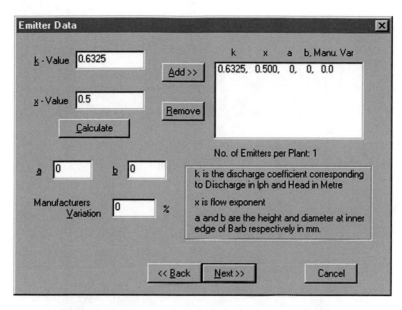

**Fig. 22** Emitter data dialog box

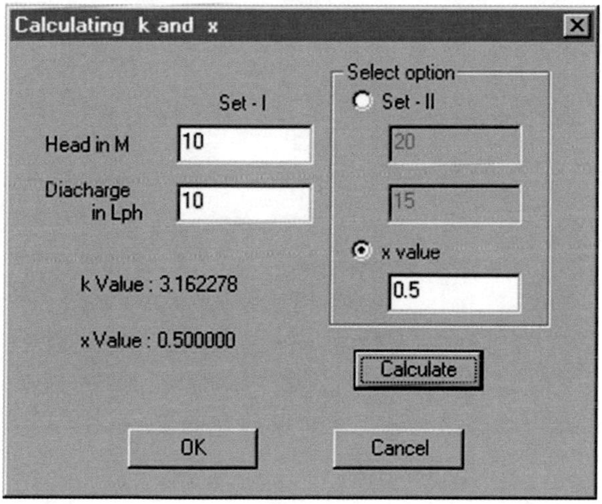

**Fig. 23** 'k' and 'x' calculation dialog box

Clicking on the "Input data" would show the input data, clicking on the "Results" would display the summarized results of the system.

Each design can be saved as a file by choosing "Save" from the "File" menu. The saved designs can be opened later by choosing "Open" from the "File" menu.

The emission uniformity is the only variable to decide whether a design is acceptable or not. For a good system the emission uniformity is expected to be above 95%.

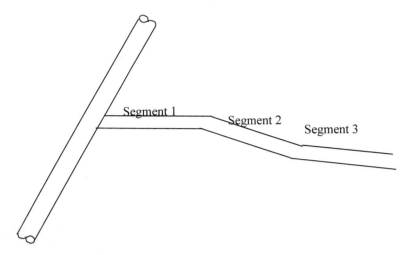

**Fig. 24** Laterals in non-uniform slopes divided into uniform slope segments

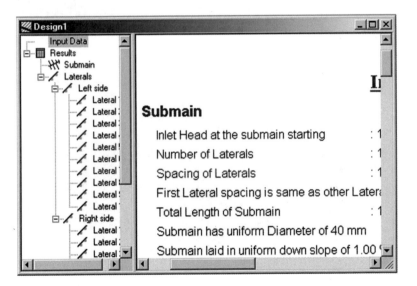

**Fig. 25** Design report view window

If the uniformity of the system is less than 95%, the design has to be revised in order to get the required uniformity. The revising can be done by choosing "Revise" from the "Design" menu. The input data can be modified and the change in the results can be evaluated. If any set of results is found satisfactory, that can be saved as a separate file.

If Hazen–Williams friction loss formula is used the values of $K = 9.58 \times 10^{-4}$, $m = 1.852$ and $n = 4.87$, for plastic pipes with diameter less than 120 mm and if the Darcy–Weisbach equation is used along with the Blasius equation, the values are $K = 7.89 \times 10^{-4}$, $m = 1.75$ and $n = 4.75$. In DSSP the constants can be changed by choosing "Constants" from the "Design" menu.

The report can also be printed directly from the program by choosing "Print" from the "File" menu. The style and colour of fonts in the report can also be changed by choosing the fonts from the format menu.

## Model Problem

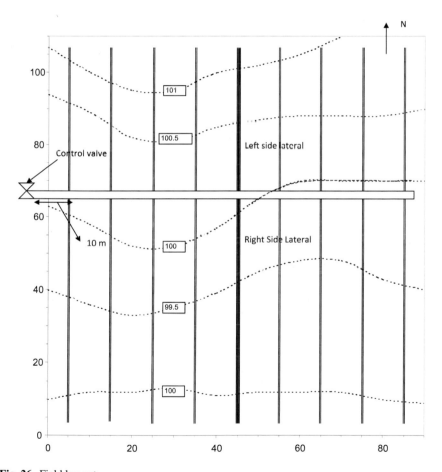

**Fig. 26** Field lay out

Figure 26 shows the field map with contours. The field data is furnished below.

Crop            : mango
Field size      : 110 m × 90 m
Plant spacing : 10 m × 10 m

Number of drippers/tree: 4

– 2 drippers of 10 l/h each

Dripper type : Vortex
Flow exponent($x$) = 0.4
Manufacturing variation = 2%

– 2 Microsprinklers of 30 l/h each

Flow exponent($x$) = 0.5
Manufacturing variation = 1%

Operating pressure    :    15 m head of water at the start of the submain.
Design the size of submain and lateral so that the uniformity of discharge is above 95%.

**Solution**
The orientation of submain (direction & location of the submain) must be decided first. The submain can be oriented either along east–west or north–south. In this case, it is better to orient east–west because the length along east–west is less. Along east–west, either along the Northern most row or southernmost row or somewhere in between has to be decided. Along the southern most row if the submain is located, the lateral would run in down slope between the 100 m contour and 99.5 m contour. Then after that the slope again rises. Therefore, the lateral has to run in the upslope for significantly longer length. This is also not recommended.

From northern most row, in the southern direction the gradient is gently sloping until 99.5 m contour. In this kind of situations, if the submain is laid in such a way that for shorter lengths the lateral runs upward and for longer lengths the lateral runs down slope the uniformity is higher and the pipe sizes are lower.

Let us assume that the submain is located as shown in the figure and we will design for this case.

# Design of Lateral

### Left-side lateral

Length   15 m

There will be 9 laterals on the left side. Let us select any one lateral for the design. Let us select an approximately mid lateral as shown in the figure.

From Fig. 26, we can interpolate from the contours and find out the reduced level at the starting point of the lateral. The lateral cuts the contour at two points. Therefore, we can assume two slope segments. One segment is from the start of lateral to the 100.5 m contour and another segment from 100.5 m contour to 101.0 m contour. For these slope segments, the slope is found out as follows:

1.   15 m–2.26%    upwards
2.   15 m–3.30%    upwards

For the remaining 5 m length of the lateral, the slope data need not be given. The slope will be extrapolated from the previous segment. If we want to give slope data for the next 5 m also, we can give. It is optional.

Number of plants       4
Spacing                      10 m
Head at inlet of lateral 15 m

Head at inlet of lateral is not known. However, head at the start of the subunit can be given as the head at the start of the lateral for the design purpose.

Spacing of the first plant   Half

Half means, the location of the first plant is half the distance of spacing of the plants.

Let us try 10 mm inside lateral (12 mm outside $\phi$) uniformly for the entire length.

1.   Double click the DSSP icon.
2.   Click on **file**. This will drop down a menu. Then click on **new**.
3.   Select "single pipe design" and click on ok.
4.   Select "design a lateral" and click on "next".
5.   Select "emitter spaced with respect to plants, one or more numbers provided for one plant" option and click on "next".
6.   Give data as below.

Number of plants        4
Spacing between plants    10 m
Pressure head at the inlet 15 m

Under first spacing heading, select "Half" option. Then click ok.

7.     Give the following data.

         Lateral diameter    10 mm. (Note: The unit should be properly selected)

         Laid to length       35 m

         Click "Add"

This will cause the data to be transferred to the right-side window. If you commit any error at this stage you can also use "remove" button to delete the unwanted data by highlighting the data not needed.

– Then click "next".

8.     For this screen, emitter data for each dripper of a plant should be provided. In the problem, only $x$ value of drippers are given and the discharge rates in $l/h$ are given. From this data, the $k$ value can be calculated. Therefore, click on "calculate".

         For head in metres under set—I, give 10 m head and for discharge in $l/h$ under set—I, give 10 $l/h$. Then select $x$ value option. Then enter 0.4 for the $x$ value. Click "calculate". The $k$ value will be calculated and shown in the screen. Then press "ok". Then the previous screen will be shown again. But in $k$ value field and $x$ value field, the calculated values will exist.

         For manufacturer's variations give 2. Then click "Add". This will transfer all the data entered to the big window in the right side. Once again press "Add". This will once again transfer all the data entered to the big window in the right side. "Add" is pressed 2 times because, 2 drippers of 10 l/h discharge are provided for each plant.

         Then click calculate again. Under set—I, give 10 for the head in m and give 30 for the discharge in $l/h$. Select $x$ value option. Enter 0.5 for $x$. Click on "calculate" Click ok. Once again, the previous window will appear. For manufacturers variation give 1. Then click "Add" 2 times because, this type of 2 microsprinklers are provided for each plant.

         Then click on "next". Then for "slope in %, "enter 2.26 and select upward. For length of the slope segment" enter 15. Then click on Add. Then the data will be transferred to the box on the right side. Then once again for "slope in %", enter 3.30 and select "upward" and for "length of the slope segment", enter 15 m. Then click on "Add".

         Now all your lateral data have been entered. Then click on "finish". Immediately the result would appear on the screen. The result will have all the input data. You can check whether your entered data and the data in the result are same. In the result, you can see the head at each emitter, the actual discharge rate of each emitter, total discharge to each plant.

         Also, you can see many relevant results, most importantly, the emission uniformity of the lateral. In this case it is above 95%. So, this design can be accepted.

Then click on "File". Click on "save as". Enter the file name as left1 (you can give any name). Also enter the directory in which it should be stored. Enter "drip" (you can store in any directory). Then click on "Save". The file is stored with the name left1.sds. The "sds" indicates that it is the file relating to single pipe design (either lateral or submain).

If you want to do redesign by changing the input data, you can once again click "file" and click "open". Then select the directory. In this case, it is "drip". Select file type. Select "sds" or "*. *". Then it will show the names of the files. Select the file name as left1. Let us assume that in this case, you do not intend to do redesign.

**Right-Side Lateral**
For the design of the right-side lateral, once again start from clicking on "File" and "New". The procedure is the same as the left-side lateral. The data for the right-side lateral is furnished below.

Length   65 m

Slope        3 segments

7.5 m   2.1% down
22.5 m 2.2% down
30 m     1.67% up

| No. of plants | 7 |
|---|---|
| Spacing | 10 m |
| Spacing of the first plant | Half |
| Head at the inlet of the lateral | 15 m |

Emitter data
  *Vortex drippers*

| No | : 2 |
|---|---|
| Discharge rate | : 10 1/h |
| x | : 0.4 |
| Manufacturing variation | : 2% |

  Microsprinklers

| No | 2 |
|---|---|
| Discharge rate | 30 1/h |
| x | 0.5 |
| Manufacturing variation | 1% |

## Lateral Diameter ($\phi$)

### Case-I
$\Phi$—10 mm, uniform throughout the length of 65 m. It can be observed from the results that the system cannot perform. Save the result in the file name right1.

### Case-II
Now the design can be revised by clicking on "design" and click on "revise". Then once again the data input windows will start appearing on the screen. Wherever you want to make changes, you can make changes. Otherwise, you just click on "next". For this case, only for the lateral diameter remove the existing data and enter 14 mm diameter for the length of 65 m.

The result for the changed data can be viewed immediately. The emission uniformity for this is 96.9%, which is more than 95%. Save this result by clicking on "file" and 'save as" to a file name right 2.

### Case-III
Now we can again redesign by incorporating multi-diameter segments in the lateral. Once again click on "design" and "revise". Then revise the data for the Lateral as follows after removing the existing data.

$$14 \, mm \, \phi \, for \, 35 \, m \, length$$
$$10 \, mm \, \phi \, for \, 30 \, m \, length$$

This data results in the emission uniformity of 94.89%, which is very near to 95%. Hence this design can be accepted. Save this result in a file name as right3.

## Design of Submain

### Case-I
For designing the submain once again click on "File" and "new". In the submain design both the laterals on right side and left side are taken as one lateral. Enter the following data.

> Number of laterals          9
> Spacing between laterals  10 m
> Pressure head at the inlet        15 m

Spacing of first lateral from the control valve and Pressure gauge   full

Submain diameter         —   47 mm for the length 90.00m

Discharge of one lateral   —   976.59 l/h

This value is obtained from the total discharge into the left-side lateral (file-left1) and the total discharge into the right-side lateral (file-right3) and then summed up.

Slope 0.25% upwards for 90.00 m    (from the map)

From the result, the pressure distribution uniformity can be seen that, it is 97.10%. This value is very high. So, the design can be revised. Save this file in the name sub1.

**Case-II**

Change the diameter data as below:

47 mm $\phi$ for 50 m length

37 mm $\phi$ for 20 m length

29 mm $\phi$ for 20 m length

The pressure distribution uniformity is 95.96%. Hence this design can be accepted and save this file in the name sub2.

# Design and Evaluation of the Subunit

We can examine, how the system would perform if the selected sizes of submain and lateral are used. For this once again start from clicking on "file" and "new". Select the design of subunit. Enter the following data.

Number of laterals        9

Spacing between laterals  10 m

Pressure head at the inlet 15 m

Submain location          Paired laterals(on both sides)

Spacing of first lateral    Full

**Submain data**

No. of segments 3

1.   47 mm $\phi$ for 50 m length
2.   37 mm $\phi$ for 20 m length
3.   29 mm $\phi$ for 20 m length

Select—Emitter spaced with respect to plants ….. option.

## Lateral Data

### Left-Side Laterals

No. of plants      4
Spacing between plants 10 m
First plant spacing      Half

### Lateral Diameter Data

### Left-Side Laterals

Number of segments1
10 mm $\phi$ for 35 m length

### Right-Side Laterals

No. of segments   2
14 mm $\phi$ for 35 m length
10 mm $\phi$ for 30 m length

### Emitter Data

| Emitter No | $k$ | $x$ | Manufacturing variation (%) |
|---|---|---|---|
| 1 | 3.9811 | 0.4 | 2.0 |
| 2 | 3.9811 | 0.4 | 2.0 |
| 3 | 9.4868 | 0.5 | 1.0 |
| 4 | 9.4868 | 0.5 | 1.0 |

### Slope Data

Select the option—Fully non-uniform percentage slope.
   Submain—0.25% up for 90.00 m.

| No | Left | Right |
|---|---|---|
| Lateral-1 | 1.67% up 20.00 m<br>3.33% up 15.00 m | 1.67% down 10 m<br>2.5% down 20 m<br>1.67% up 30 m |
| Lateral-2 | 1.67% up 12.5 m<br>3.33% up 15.0 m | 1.67% down 17.5 m<br>2.86% down 17.5 m<br>2 0.00% up 30 m |
| Lateral-3 | 1.67% up 10.00 m<br>3.33% up 15.00 m | 1.67% down 12 m<br>2.86% down 17.5 m<br>2.86% up 22.5 m |
| Lateral-4 | 1.85% up 15.00 m<br>2.85% up 17.50 m | 1.85% down 16 m<br>2.63% down 19 m<br>2.00% up 25 m |
| Lateral-5 | 2.20% up 15.00 m<br>3.33% up 15.00 m | 2.20% down 7.50 m<br>2.20% down 22.50 m<br>1.67% up 30.00 m |
| Lateral-6 | 2.86% up 17.5 m<br>3.3% up 15.0 m | 2.0% down 25.00 m<br>1.43% up 35.0 m |
| Lateral-7 | 2.86% up 35 m | 2.22% down 22.5 m<br>1.33% up 37.5 m |
| Lateral-8 | 2.86% up 17.5 m<br>1.8% up 17.5 m | 2.22% down 22.5 m<br>1.33% up 37.5 m |
| Lateral-9 | 2.5% up 20.0 m<br>1.5% up 15.0 m | 1.81% down 27.5 m<br>1.54% up 32.5 m |

The result would appear in two windows. In left-side window, you can see a small " + " mark. At the right side of ' + ', it would have been written as "Results". Above the "results", it would have been written as "input data". The input data" would have got highlighted with blue background. On the right-side window, the information of the highlighted line in the left-side window would appear.

If you click on the results in the left-side window, the results would appear on the right-side window. It can be seen that the emission uniformity of the entire system is 93.682%.

If you want to see the details of the results, you click on the " + " in the right-side window. The " + " will become "-". Under "result", you can see "submain" and "laterals". Before laterals, there is also a " + " sign. The " + " sign indicates that, there are some more sub headings under "laterals". Click on the " + " of laterals. Two subheadings namely "left side" and "right side" would appear. If you click on the " + " of the right side, all the lateral numbers on left side would appear. If you click on the " + " of the right side, all the lateral numbers in the right side would appear.

If you click on the "submain" in the left-side window, the consolidated results would appear in the right-side window. In the first column, the lateral number is provided. In the second column, the pressure head at the start of each lateral is given. In the third column, the discharge that would go into each lateral is given. In the fourth column, average discharge for each plant in each lateral is given. In the fifth column, the head loss due to friction in each lateral is provided. In the last column, the uniformity in each lateral is provided.

For the left-side laterals, the uniformity is more than 95%. But for the right-side laterals, the uniformity is less than 95% for many laterals. That is the reason why, the overall uniformity comes as 93.682%. Save this design in the file name as "subunit1". This will be saved in the type ".dds".

The design can once again be revised by changing the size of lateral for the right side alone as follows:

$$14 \text{ mm } \phi \text{ for } 45 \text{ m length}$$
$$10 \text{ mm } \phi \text{ for } 20 \text{ m length}$$

This results in the emission uniformity of 94.689% which is approximately 95%. Hence the design can be accepted. You can also examine, how in each lateral, each emitter discharges, by clicking on each lateral number in the left-side window.

It is to be noted that the preceding problem is to demonstrate the software to handle the more complicated situation. However, the software can handle the problem with very much simplified situations with land slope being horizontal or uniform slope and drippers without any manufacturing variation. For those situations, data entry will also be simple.

## Summary

For the design of telescopic pipes in non-uniform slopes, trial-and-error method is simple and easy to adopt. The difficulties in trial-and-error method are demonstrated with a numerical example. An optimal direct design method using polyplots that are adopted widely around the world is developed and demonstrated with numerical examples. Computer software for the design of drip irrigation subunits in non-uniform slopes was developed, and its use is demonstrated with a numerical example.

## Assessment

### Part A Choose the best option

1.  What is the best shape of energy gradient line in a telescopic submain to get maximum uniformity of discharge rate?

    a.  **Straight line**
    b.  Negative exponential

c. Positive exponential
d. Parabolic

2. Let us assume that it is possible to purchase any diameter of pipe from the market. If we use different diameters for each section for all the submain segments, we would get a straight pressure head curve for the telescopic submain. Say True or False (**True**)

3. In poly-plot method, when designing for downslopes, the maximum pressure head deviation is the _____ of absolute maximum positive and absolute negative deviation between the friction loss curve and the ground profile curve.

a. **sum**
b. difference
c. product
d. square

**Problems**

1. A submain must be designed for a length of 40 m. Laterals are placed at a spacing of 2 m with each lateral discharging at a rate of 700 l/h. The submain is laid on a down slope of 0.25% for the first 20 m from the start and on a down slope of 0.5% for the next 30 m. The allowable relative pressure head variation is 0.1. The average design pressure head is 10 m. Design a telescopic submain with two diameters of pipe. The diameters of pipe available in the market are 58.6, 46.6 and 37.2 mm.

# References and Further Reading

1. Studman, C. J. (1990). *Agricultural and horticultural engineering.* Butter Worths.
2. Wu, I. P., & Gitlin, H. M. (1979). Drip irrigation design on non-uniform slopes. *Journal of the Irrigation and Drainage Division, ASCE, 105*(3), 289–303.
3. Krishnan, M., & Ravikumar, V. (2002). A software for design of drip subunits with tapered pipes in non-uniform slopes. *Journal of Agriculture Engineering Indian Council of Agriculture Research, 39*(4). http://epubs.icar.org.in/ejournal/index.php/JAE/article/view/14143

# Design of Mains

## Introduction

Design methods for main pipes used in sprinkler as well as in drip irrigation are one and the same. In many simple situations, main pipe diameter is not designed by following rigorous engineering principles. The guidelines available in Table 1 are used as a rule of thumb. However, it is prudent to design main pipes because significant cost reduction is possible if proper main pipe design is done taking into account the real-field conditions.

Water hammer is sudden increase of pressure in the pipe due to sudden stoppage of water in the pipes due to pump trip or valve closure. Water hammer is an important aspect to be considered in design of main pipes.

## Water Hammer Management

Water hammer causes severe destruction if it is not understood properly, and suitable action is taken during design and installation of main pipes [1]. Water hammer usually occurs in pipes under two circumstances. One is during the pump trip when electric power supply is cut-off [2] (Fig. 1). When the pump suddenly stops, the water column in the delivery pipe gets separated as shown in Fig. 1 and vacuum gets formed in between the water column. If the velocity of flow is say 2 m/s, the length of vacuum would get formed for a height equal to 0.20 m ($v^2/2\ g$). Then the upper column would fall down due to gravity as well as suction with a velocity larger than 2 m/s. This kinetic energy is dissipated in a very short time. The time taken for dissipation of energy depends on the elastic properties of water and pipe and pump material. More the elasticity, more the time taken for energy dissipation and less the inertial/surge pressure increase. But unfortunately, the elasticity of the water and pipe material is so low that if proper care is not taken, water hammer causes breakage of pump and pipe [3].

© The Author(s), under exclusive license to Springer Nature Singapore Pte Ltd. 2023          353
V. Ravikumar, *Sprinkler and Drip Irrigation*,
https://doi.org/10.1007/978-981-19-2775-1_12

**Table 1** Guideline for
choosing main line diameter

| No. | Discharge range (l/s) | Recommended diameter (mm) |
|-----|----------------------|---------------------------|
| 1 | 0.07–0.13 | 20 |
| 2 | 0.13–0.25 | 25 |
| 3 | 0.25–0.50 | 32 |
| 4 | 0.50–1.0 | 40 |
| 5 | 1.00–1.80 | 50 |
| 6 | 1.80–3.60 | 63 |
| 7 | 3.60–5.75 | 75 |
| 8 | 5.75–8.50 | 90 |
| 9 | 8.50–15.00 | 110 |
| 10 | 15.00–20.00 | 140 |
| 11 | 20.00–30.00 | 160 |

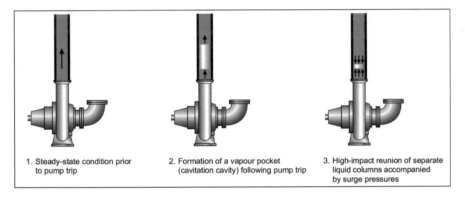

1. Steady-state condition prior to pump trip

2. Formation of a vapour pocket (cavitation cavity) following pump trip

3. High-impact reunion of separate liquid columns accompanied by surge pressures

**Fig. 1** Water hammer during pump trip (*Courtesy* KSB pumps)

During the water hammer at the pump delivery, an inertial pressure wave travels from the pump to the delivery pipe downstream. In order to manage this kind of situation, a combination of air/vacuum valve and a check valve is normally adopted (Fig. 2). If swing check valves are used, the returning water would close the check valve. The returning water bangs the valve every time, and this valve wears off in very less time. Therefore, specially designed surge check valves are used, and this valve closes even before the returning water reaches the check valve. These valves get closed by sensing the formation of vacuum. Since the valve is seated before the returning water reaches the valve, the hammer effect on the valve gets reduced.

The vacuum formed upstream of check valve gets released by provision of an air/vacuum valve between the pump and the check valve. During vacuum formation air is allowed inside the pipe. During the return of water, the air movement is restricted by another surge check valve placed along with the air valve (Fig. 3). This check

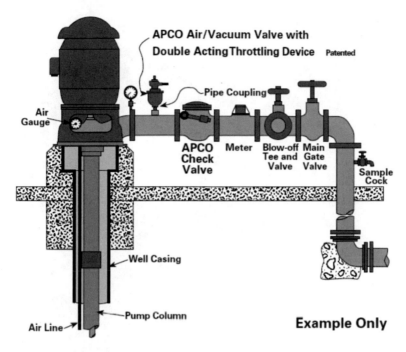

**APCO Air/Vacuum Valve with**
**Double Acting Throttling Device** Patented

Air Gauge

Pipe Coupling

APCO Check Valve

Meter

Blow-off Tee and Valve

Main Gate Valve

Sample Cock

Well Casing

Pump Column

Air Line

**Example Only**

**Fig. 2** Provision of air/vacuum valve and check valve for managing water hammer (www.dezurik. com)

valve restricts the movement of air to atmosphere, and this helps to reduce the water hammer effect by giving a cushion effect.

The second situation in which water hammer occurs is due to sudden closure of valves (Fig. 4). If there is a straight stretch of a long pipe and when a valve at the end of the stretch is closed, the moving water exerts inertial pressure on the valve. This inertial/surge pressure wave travels upstream and gets returned after reaching the other end. Simple solution for this problem is to close the valve slowly. But sometimes solenoid valves fail accidentally and shut down the flow causing water hammer. Provision of pressure relief valves at the upstream of each valve is also adopted to manage these situations. Provision of larger pipe diameter would decrease the velocity of flow.

The velocity in main pipes must be typically maintained below 1.5 m/s in order to reduce the possibilities of failure due to water hammer. This limit must be considered only when a pipe flow is stopped when flow happens. In open-end pipes this limit need not be imposed. Submain and lateral also can be considered as open ends because the flow can occur through drippers/sprinklers.

**Fig. 3** Air/vacuum valve
with additional surge check
valve (www.dezurik.com)

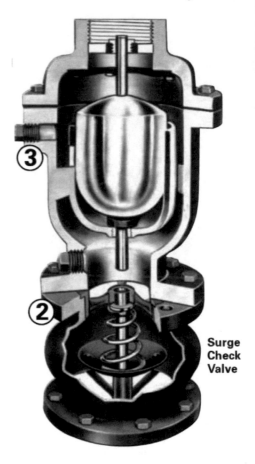

Surge
Check
Valve

## *Estimation of Surge Pressure Heads*

Let us consider a tank supplying water through a pipe, and a valve exists at the end
of the pipe (Fig. 5). The water in the pipe is brought to rest due to sudden closure of
the valve in the delivery pipe. Let $A$ be the cross-sectional area of pipe and $L$ be the
length of pipe. Let $V$ be the velocity of water.

As soon as the valve is closed, the water at the immediate vicinity of the valve
impinges on the valve and its kinetic energy is transferred to the valve and converted
into elastic energy. The elastic energy of the valve is again transferred to the water
causing pressure increase of water called as inertial pressure increase and, in this
case, specifically called as surge pressure increase. The portion of water that comes
to rest has a wave front, and it continues to move upstream until the tank. It is assumed
that the increased surge pressure on the upstream of wave front (valve side) remains
as such until the wave reaches the tank. Because of the pressure increase in the pipe
following things happen [4]:

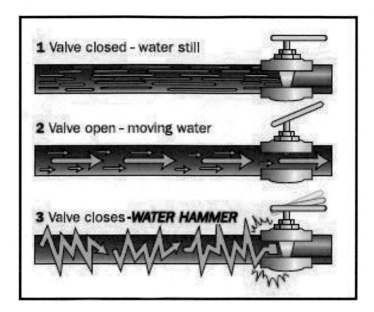

**Fig. 4** Water hammer due to sudden closure (Wenzhoue Weike Valve Co., Ltd.)

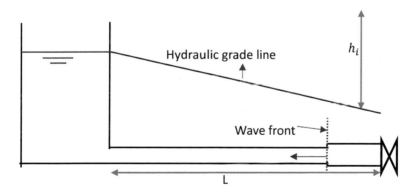

**Fig. 5** Surge wave front

- Decrease in the volume of water inside the pipe due to compression of water
- Increase in pipe diameter due to expansion of pipe in radial direction
- Increase in length of pipe (in this derivation, the tank end and valve end are assumed as constrained and increase in length is taken as zero)
- Due to the increase in diameter of pipe and compressibility of water, an additional volume is created in the pipe for filling additional volume of water in the pipe.

Let $\Delta t$ be the time taken for the wave front to travel from the valve to the tank, and in $\Delta t$ time all the water in the pipe comes to rest.

The mass of water stored in the pipe $(M) = \gamma LA/g$

where $\gamma$ is weight density of water.

By applying Newton's second law:

$$F = \frac{MV}{\Delta t}$$

where $F$ is the force exerted by the valve on the water.

$$F = \frac{\gamma AL}{g\Delta t}V$$

$$\frac{F}{A\gamma} = \frac{L}{\Delta t}\frac{V}{g}$$

$$\frac{P_i}{\gamma} = \frac{aV}{g} \tag{1}$$

where $P_i$ is called as surge pressure and $a = \frac{L}{\Delta t}$ is called as the velocity of surge wave. The velocity of the surge wave is approximately near the velocity of sound.

$$h_i = \frac{aV}{g} \tag{2}$$

where $h_i$ is surge pressure head.

It can be seen that the surge pressure head is a product of velocity of water and velocity of wave. If water and pipe are pure rigid bodies, the wave velocity will be equal to the velocity of sound. Due to the flexible nature of water and the pipe material, the velocity of the wave gets reduced.

Let us derive an expression to find out the velocity of the surge wave.

$$\Delta C = \Delta C_c + \Delta C_e \tag{3}$$

where $\Delta C$ is the additional volume created in the pipe $= VA\Delta t$;

$\Delta C_c$    is the additional volume created due to the compressibility of water;
$\Delta C_e$    is the additional volume created due to increase in diameter of pipe.

$$\Delta C_c = \frac{ALP_i}{K} \tag{4}$$

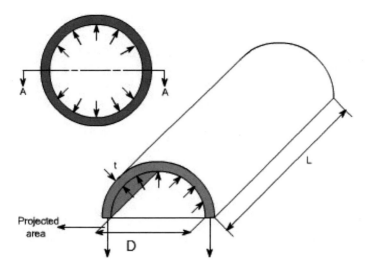

**Fig. 6** Illustration for hoop stress (campbell-sevey.com)

where $K$ is the bulk modulus of water. Let $dD$ be the increase in diameter due to expansion of the pipe. Then,

$$\Delta C_e = L\left(\frac{\pi}{4}((D + dD)^2 - D^2))\right)$$

Neglecting $dD^2$, following equation can be written:

$$\Delta C_e = \frac{\pi DLdD}{2} \tag{5}$$

Hoop stress in pipe wall is tangential to pipe circumference and is as follows (Fig. 6):

$$= \frac{P_i DL}{2t L} = \frac{P_i D}{2t} \tag{6}$$

where $D$ is the diameter of the pipe and $t$ is the thickness of the pipe.
Strain in the circumference of the pipe

$$= \frac{\pi(D + dD) - \pi D}{\pi D} = \frac{dD}{D}$$

where $dD$ is the increase in the diameter of the pipe. We know that Young's modulus ($E$) is ratio of stress to linear strain and is expressed as follows:

$$E = \frac{P_i D / 2t}{dD / D}$$

$$dD = \frac{P_i D^2}{2t E}$$

Substituting the preceding equation in Eq. 5, we get

$$\Delta C_e = \frac{\pi D L}{2} \frac{P_i D^2}{2t E} = \frac{A L D P_i}{t E} \tag{7}$$

Equation 3 can be written as follows by substituting Eqs. 4 and 7:

$$V = a P_i \left( \frac{1}{K} + \frac{D}{t E} \right) \tag{8}$$

Substituting $P_i = \gamma h_i$ in the preceding equation and rearranging, we get

$$a = \frac{V}{\gamma h_i} \frac{1}{\left( \frac{1}{K} + \frac{D}{t E} \right)}$$

Substitute for $h_i$ from Eq. 2 and $\gamma = \rho g$ in the preceding equation and rearrange as follows:

$$a = \sqrt{\frac{K / \rho}{\left( 1 + \frac{K D}{t E} \right)}} \tag{9}$$

It must be noted that the velocity of surge wave is independent of the velocity of flow as per this equation. Equations 1 and 9 are used together in estimating the surge pressure head.

Normally, when the solenoid valves are used the valve closing time must be greater than the time taken for the surge wave to come back again after getting reflected back from the tank.

$$t_c = \frac{2L}{a}$$

where $t_c$ is the minimum valve closure time.

**Problem**

Inside diameter of the pipe = 46 mm; thickness of the pipe = 2 mm; Young's modulus of PVC pipe = 3275 MPa; bulk modulus of water = $2.2 \times 10^9$ Pa. The pressure head before water hammer is 40 m. Find out the velocity of the surge wave and also the surge pressure head if the velocity of flow is 1.5 m/s and the moving water is suddenly stopped. Find also the total pressure head that could be developed?

Velocity of surge wave $(a) = \sqrt{\frac{K/\rho}{(1+\frac{KD}{tE})}} = 365.8$ m/s.

Recall that the velocity of the sound in air is 343 m/s.

Surge pressure head $(h_i) = \frac{aV}{g} = 55.93$ m

For the maximum allowable velocity of 1.5 m/s itself, the surge pressure head works out to 56 m.

Total pressure head $= 55.93 + 40 = 95.93$ m.

From the results of the preceding, we are able to gauge, how big the water hammer monster is. Mostly designing for total safety against water hammer is very costly, and hence proper management and prevention of water hammer must be resorted to.

Usually the fittings like bend, tee and couplers bear the brunt of water hammer. That is why it is conventional that the strength of fittings will be one level higher than the pipes. For instance, if pipes are 0.4 MPa, the fittings will be 0.6 MPa.

# Minimal Capital Method

It has been found from optimization analysis that maintaining a straight-line total energy gradient line from the water source to the outlets yields a near-minimum cost solution [5].

1.  In this method, the ground profile on which the main pipe is installed must be plotted as shown in Fig. 7.
2.  At all the submain starts which are operated simultaneously, draw vertical lines for a height equal to the pressure head required.
3.  Mark the pressure head available at the well on the vertical axis (point $x$)
4.  Draw line joining point $x$ to the last submain pressure head required point (point $d$ in the figure). This line indicates the energy gradient to be maintained along the mainline. This line should always have negative slope. If it is not possible to draw negative slope, the pressure head available at the control head should be increased. The pressure head required lines at each submain points (Aa, Bb, Cc, ….) all should be below the energy gradient line.
5.  If the energy gradient line cuts the lines Aa, Bb, Cc, then the energy gradient line should be drawn segment-wise as in Fig. 8. In this case, if points $d$ and $x$ are joined, it may cut the Cc line. In these situations, from the last point $d$, draw line dc. Then from the point $c$, see whether, if points $c$ and $x$ are joined, it will not cut lines Aa, Bb.
6.  If it cuts any vertical line, the procedure as used for points $d$ and $c$ should be used. Otherwise, the point cx can be joined. Now the required energy gradient line is xcd. Even when drawing the energy gradient line as segments the slope should always be a negative slope. In Fig. 9 if $c$ and $x$ are joined, it cuts Aa. In that case ca should be joined. It should be noted that cb should not be joined.

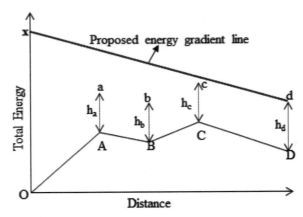

**Legend**

O – Pressure head available after the control head

A,B,C and D are submain start locations

$h_a$ ,$h_b$ ,$h_c$ ,$h_d$ are pressure head required at A,B,C,D respectively.

**Fig. 7** Plotting ground profile of main

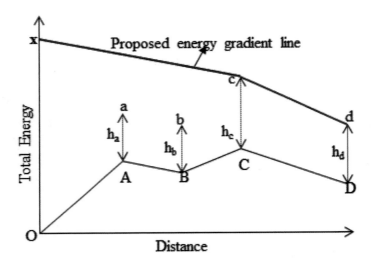

**Fig. 8** Correction of energy gradient line (Case-1)

It should also be observed that cb is infeasible. In this case xac is the required energy gradient line.

7. Find out the slope of the energy gradient line in each segment, starting from the last segment. This gives the allowable head loss due to friction per unit length

**Fig. 9** Correction of energy
gradient line (Case-2)

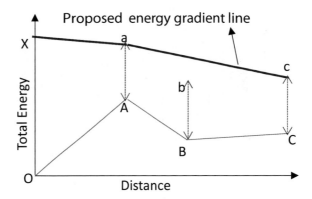

($\Delta h/L$). Then use the following equation to determine the diameter of the pipe for each segment [6].

When using this equation, the value of $Q$ should be carefully substituted for each segment.

$$\frac{\Delta h}{L} = \frac{7.89 \times 10^5 Q^{1.75}}{D^{4.75}} \qquad (10)$$

8.   From the computed diameter, select the nearest diameter of pipe available in the market.

**Problem**

In a DIS, pressure head available after the control head is 16 m. Four submains are to be operated simultaneously. The main runs on 1% upslope from the control head for 40 m, at which, the first submain start is located with 2 l/s discharge rate. Then the main runs for another 60 m on a level ground at which the second submain start is located with 1.5 l/s discharge rate. Then the submain runs for 70 m on a 3% upslope for 70 m at which the third submain is located with 1 l/s discharge rate. Then the submain runs for 50 m on a 4% downslope, at which the last submain is located with 1 l/s discharge rate. Pressure needed at the start of all the submains is 10 m. Design the size of mains at each section.

As per the procedure explained in the preceding sections, the ground profile, the pressure head required at each submain and the total energy along main are plotted as shown in Fig. 10. The total energy line has two slopes. From 0 to 170 m, it has slope ($\Delta h/l$) of 0.0206. From 170 to 220 m, it has a slope of 0.04. By using the preceding equation the design diameters obtained are as follows (Table 2):

A software called as Dripmain.exe in www.tnau.ac.in website can be used for doing this kind of computations. The software is capable of visually depicting the ground slope, friction loss curve and design pressure head for each outlet. It is also capable of providing recommended diameters based on the standard sizes available in the market. If the computer in which the software is installed has Visual Studio,

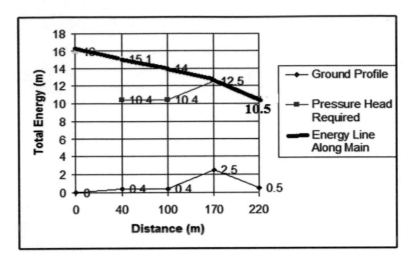

**Fig. 10** Ground slope, friction loss curve and design pressure head for a main design

**Table 2** Results of main design by minimal capital method

| Segment number | Discharge rate (l/s) | Design diameter (mm) | Nearest diameter available in market (mm) |
| --- | --- | --- | --- |
| 1 | 5.5 | 74 | 90 |
| 2 | 3.5 | 62 | 75 |
| 3 | 2 | 50 | 63 |
| 4 | 1 | 34 | 40 |

the software will work well. If the computer does not have Visual Studio installed, on double clicking the software icon, it will ask for few *.ocx files to be placed in c:\windows\system32. All these files are downloadable from Internet.

## Life Cycle Cost Analysis Method

The capital investment is done initially at the start of a project, and the operating cost occurs throughout the project life. For PVC pipes normally the project life is assumed as 20 years. The operating cost would also keep varying due to changes in cost of the power with respect to time. Therefore, bringing all the cost to a common time frame is a very big challenge in life cycle cost analysis. In the method presented here in this book, annualized costing is adopted. (Note: Fundamentals of financial analysis is provided in the Chap. 17 on Water Budgeting and Economic Analysis).

In the annualized cost method, the initial capital investment is assumed to be borrowed from a Public Financing Organization, and it is assumed that the borrowed money is repaid every year once.

Following formula is used for annualizing the capital investment [7]:

$$C_c = f_c \times I \qquad (11)$$

where

$C_c$ —annualized capital cost
$f_c$ —capital annualizing factor found out using Eq. 12.
$I$ —capital investment

$$f_c = \frac{r(1+r)^N}{(1+r)^N - 1} \qquad (12)$$

where $r$ is interest rate on the borrowed capital. If it is 8%, for an $N$ value of 20 years, $f_c$ works out to be 0.1019.

The operating cost incurred every year is brought to the first year time frame. In the method presented here, expenditure done throughout the first year is assumed to occur at the start of the year. The expenditures done towards operating cost in subsequent years are brought to the start of the first year. Following formula is used to bring the annual operating cost to the start of first year:

$$C_p = f_p \times V_A \qquad (13)$$

where $C_p$ is equivalent operating cost per year. $V_A$ is the operating cost for the first year, and $f_p$ is equivalent factor for bringing the annual expenditure done over $N$ years to the first-year time frame.

$$f_p = \frac{(1+e)^N - (1+r)^N}{(1+e) - (1+r)} \times \frac{r}{(1+r)^N - 1} \qquad (14)$$

where '$e$' is the percent of increase in power cost per year and $r$ and $N$ are as defined earlier. Generally, 5% price rise can be expected. A careful analysis in consultation with economists and past data would help in arriving at a reliable value. For the '$e$' value of 5% per year, $r = 8\%$ and $N = 20$ years, the value of $f_p$ works out to 1.4624.

**Table 3** Pumps cost

| Pump power (kW) | Cost (Rs.) |
|---|---|
| 0.75 | 7000 |
| 1.10 | 11,000 |
| 1.50 | 15,000 |
| 2.20 | 35,000 |
| 3.70 | 52,000 |
| 5.50 | 82,500 |
| 10 | 170,000 |

While substituting values for '$e$' and '$r$' in the preceding formulae, it must be in fraction.

**Problem**

Water must be carried to operate one submain operating at a distance of 500 m from the pond. There is no elevation difference between the water in the pond and the delivery valve of the submain. Pump as well as pipe diameter must be chosen for the system. The other relevant data is as follows:

Discharge rate—5 l/s.
The operating pressure head for the submain—15 m
Maximum allowable velocity in the main pipe—1.5 m/s
Daily operating hours of the submain—6 h
Cost of power—Rs. 5/(kw-h) at present and expected increase per year 5%
Interest on borrowed capital for agriculture—8%
Life time of the project—20 years
Assume an average efficiency of pump for 20 years—50%
Assume $f_p = 1.4624$ and $f_c = 0.1019$. In Table 3 data related to cost of pumps with respect to power and in Table 4 data related to pipe cost with respect to diameter of pipe is provided.

**Solution**

In Table 5 the assessed capital costs for five pipe diameters are presented.

**Table 4** Pipes cost

| No. | Pipe external diameter (mm) | Pipe internal diameter (mm) | Cost/m (Rs.) |
|---|---|---|---|
| 1 | 50 | 46.4 | 55 |
| 2 | 63 | 59.2 | 80 |
| 3 | 75 | 70.6 | 111 |
| 4 | 90 | 84.8 | 129 |
| 5 | 110 | 104 | 172 |
| 6 | 160 | 153.2 | 357 |

**Table 5** Capital cost assessment by life cycle cost method

| No. | Pipe int. dia. (mm) | Pipe cost (Rs./m) | Pipe cost (Rs.) | Friction loss (m) | Power (kW) | Selected pump power (kW) | Pump cost (Rs.) | Total cap cost (Rs.) | Equiv. ann. Cap. cost (Rs) |
|-----|---------------------|-------------------|-----------------|-------------------|------------|--------------------------|-----------------|----------------------|----------------------------|
| 1 | 46.4 | 55 | 27,500 | 80.0 | 9.32 | 10.0 | 170,000 | 197,500 | 20,116 |
| 2 | 59.2 | 80 | 40,000 | 25.2 | 3.94 | 5.5 | 82,500 | 122,500 | 12,477 |
| 3 | 70.6 | 111 | 55,500 | 10.9 | 2.54 | 3.7 | 52,000 | 107,500 | 10,949 |
| 4 | 84.8 | 129 | 64,500 | 4.6 | 1.92 | 2.2 | 35,000 | 99,500 | 10,134 |
| 5 | 104 | 172 | 86,000 | 1.7 | 1.64 | 2.2 | 35,000 | 121,000 | 12,324 |
| 6 | 153.2 | 357 | 178,500 | 0.3 | 1.50 | 1.5 | 15,000 | 193,500 | 19,708 |

**Table 6** Operating cost and total cost assessment by life cycle cost method

| No. | Ann. energy (kW) | Ann. energy cost (Rs.) | Equiv. ann. energy cost (Rs.) | Ann. total cost (Rs.) | Velocity of flow (m/s) |
|-----|------------------|------------------------|-------------------------------|-----------------------|------------------------|
| 1   | 21,900           | 109,500                | 160,132                       | 180,248               | **2.96**               |
| 2   | 12,045           | 60,225                 | 88,073                        | 100,550               | **1.82**               |
| 3   | 8103             | 40,515                 | 59,249                        | 70,198                | 1.28                   |
| 4   | 4818             | 24,090                 | 35,229                        | 45,363                | 0.89                   |
| 5   | 4818             | 24,090                 | 35,229                        | 47,553                | 0.59                   |
| 6   | 3285             | 16,425                 | 24,020                        | 43,728                | 0.27                   |

Friction loss for carrying 5 l/s for 500 m length is found using following formula:
$$\Delta h = \frac{7.89 \times 10^5 Q^{1.75}}{D^{4.75}} L$$
Total head to be developed by the pump $(H)$ = Frictional head + Submain operating head

Power is found out using following formula: $P = \frac{9810\left(\frac{Q}{1000}\right)H}{1000 \times \eta}$ kW

Total capital cost = Pump cost + Pipe cost

Equivalent annualized capital cost = $f_c \times$ Total capital cost

Annual energy consumption = $P_I \times 6 \times 365$ kW-h

where $P_I$ is the power of the installed pump.

Annual energy cost (Rs.) = Annual energy consumption $\times 5$

Equivalent annual energy cost = $f_p \times$ Annual energy cost

From the results, it can be seen that 90 mm external diameter (84.8 mm internal) can be selected as a suitable diameter for installation.

If we select 100 mm external diameter, annualized life cycle cost is very near to optimal cost and the velocity of flow is very less compared to the upper boundary velocity of 1.5 m/s. If the site conditions are very much prone for water hammer incidences, 100 mm diameter pipe selection can also be done.

But if we select 70 mm external diameter, annualized life cycle cost is significantly more and also the velocity of flow is very near to the upper boundary velocity of 1.5 m/s. Hence this selection may be avoided (Table 6).

# Summary

When a pump already exists in the farmer's field and if the available pressure head and discharge rate are known then the main line is designed based on minimizing the cost of the pipe alone. This method of designing is called as minimal capital method. When we design with the objective of minimizing only capital investment, we would select minimum diameter of pipe which can withstand the pressure head developed in the pipes. But this option may not be beneficial if time of operation is higher because the operational cost will be more due to spending of more power to overcome the frictional head losses. When we do life cycle cost analysis during design, then our design is more appropriate. When a new pump must be installed or pump may have already been installed with a variable speed drive, life cycle cost method is very useful in designing the main pipe. These facts are demonstrated with suitable numerical examples.

## Assessment

### Part A

### Say True or False for the following questions

1. If the velocity of flow is say 2 m/s, the length of vacuum would get formed in a vertical main pipe for a height equal to 0.20 m ($v^2/2$ g) when pump trips down. (True)
2. The velocity in main pipes must be typically maintained below 1.5 m/s in order to reduce the possibilities of failure due to water hammer. (True)
3. It is general practice to use the bends, tees and couplers with a greater strength than the pipe strength in order to manage water hammer. (True)

### Part B

### Brief Answers

1. What additional accessories are used for managing water hammer caused by vacuum formation?
2. Suggest a method for handling water hammer caused due to sudden closure of valves.
3. Discuss how formation of vacuum in main pipe causes water hammering.
4. Distinguish between an ordinary check valve and a surge check valve.
5. Discuss the advantages of life cycle cost analysis of main pipe design.
6. Write down the step-by-step procedure of main pipe design by minimal capital cost method.

## References and Further Reading

1. Wood, F. M. (1970). *History of water hammer* (C.E. Research Report No.65). Department of Civil Engineering, Queens University at Kingston, Ontario.
2. Anonymous. (2006). *Water hammer*. KSB. www.ksb.com
3. Jeppson, R. W., Flammer, G. H., & Watters, G. Z. (1972). *Experimental study of water hammer in buried PVC and Permastran pipes* (Reports Paper 293). https://digitalcommons.usu.edu/water_rep/293

4. Hanif Chaudhry, M. (2014). *Applied hydraulic transients*, 3rd edn. Springer
5. Wu, I. P. (1975). Design of drip irrigation mainlines. *Journal of irrigation & Drainage, ASCE, 101*(IR4), 265–278.
6. Ravikumar, V., Ranghaswami, M. V., Appavu, K., & Chellamuthu, S. (2011). *Microirrigation and irrigation pumps*. Kalyani Publishers.
7. Keller, J., & Bliesner, R. D. (1990). *Sprinkle and trickle irrigation*. Van Nostrand Reinhold.

# Fertigation

## Introduction

Fertigation is the process of applying fertilizers through irrigation water. When compared to conventional fertilizer application, the fertigation has many advantages. They are as follows:

- Nutrients can be maintained precisely throughout the root zone. Therefore, fertilizer waste is minimized and fertilizer use efficiency is very high.
- Total fertilizer demand can be divided into many doses according to the different phases of plant growth such as vegetative, flowering and fruit setting. This also results in higher efficiency.
- Microirrigation is amenable to automation, and hence fertigation can also be easily automated.

Fertigation also has some disadvantages. They are as follows:

- Many conventional phosphate fertilizers such as superphosphate and diammonium phosphate have very less solubility and tend to precipitate in the irrigation system equipment and tend to clog the system.
- At present special formulations of totally water-soluble fertilizers are available in the market, but they are very costly and the cost economics is not favourable to use such fertilizers for many crops.
- Sometimes when the quality of the irrigation water is saline, some fertilizers tend to get precipitated.
- The irrigation system accessories get corroded gradually.
- When more than one fertilizer is mixed, care should be taken not to mix the incompatible fertilizers.

For doing fertigation and proper operation and maintenance of drip irrigation systems, possessing a rudimentary knowledge in water chemistry is very much

© The Author(s), under exclusive license to Springer Nature Singapore Pte Ltd. 2023     371
V. Ravikumar, *Sprinkler and Drip Irrigation*,
https://doi.org/10.1007/978-981-19-2775-1_13

helpful. For proper fertigation, knowledge on fertilizer properties, crop characteristics and microirrigation system characteristics is essential. The interrelationships between them also must be properly understood. Hence this subject has become an active area of research to derive optimal way of fertigation scheduling. In this chapter, only a rudimentary concepts on fertilizer properties and plant nutrition are presented.

## Nutrient Elements

Nitrogen, phosphorous, potassium, calcium, magnesium and sulphur are called as macroelements as they are required in large quantities. Iron, zinc, manganese, copper, boron and molybdenum are called as microelements as they are required in small concentrations.

Nutrient elements are absorbed by plants only from aqueous solution. Chemically most of the fertilizers are salts. Similar to salts, fertilizers also get ionized in water. Positive ions are called as cations, and negative ions are called as anions.

Soil pH is one important factor based on which the nutrient availability varies significantly (Fig. 1). It can be seen from the figure that availability of most of the elements is better in the pH range of 6–7 (Fig. 2).

Generally, antagonism occurs between ions which possess the same kind of electrical charge. Surplus of one cation in soil solution diminishes the absorption of other cations. It is also true for anions. The most prominent known antagonism phenomena

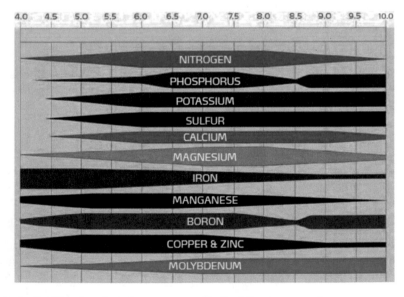

**Fig. 1** Availability of nutrient elements versus pH of soil (www.yara.com)

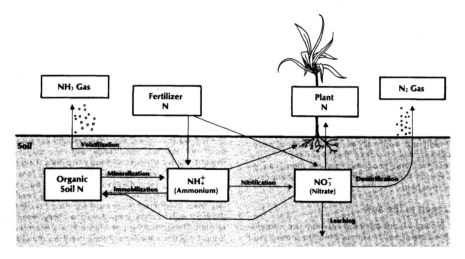

**Fig. 2**  Urea reaction in soil (http://www2.ca.uky.edu/agcomm/pubs/agr/agr105/agr105.htm)

are between potassium and magnesium, between ammonium and magnesium and between iron and manganese. When opposite kind of ions exists, synergy exists. Mutually one kind of anion helps the absorption of the other cations. Most known synergic phenomena exist between potassium and nitrate, ammonium and phosphate, magnesium and nitrate and zinc and nitrate.

In the following text, characteristics of microelements and macroelements are discussed.

## Nitrogen

Nitrogen is an important ingredient during the vegetative growth stage. Only if sufficient nitrogen is available for plants, they grow healthily. Nitrogen application beyond vegetative stage postpones flowering and also lodging of plants. Some of the nitrogen source of fertilizers are urea, ammonium nitrate and ammonium sulphate.

Urea is the cheapest source of nitrogen and commonly used. Urea is an organic fertilizer and highly soluble in water. Urea gets converted into ammonium in the soil by ammonification process. The ammonification is a two-stage process. The first stage is an enzymatic process, mediated by urease, a free enzyme existing in the soil. In this process, the urea is converted into ammonium carbonate $(NH_4)_2Co_3$. In the second stage, ammonium ion is produced. If the process of conversion of ammonium occurs in surface of the soil and if soil pH is high, instead of formation of ammonium ion, ammonia gas is produced and lost in the atmosphere. Ammonium ions get partially dissolved in soil solution. Since ammonium ions are positively charged and clay particles are negatively charged, they adsorb on clay particles. Therefore, ammonium ions do not move freely in the soil. Nitrification is the process in which soil bacteria oxidizes the ammonium into nitrate. The rate of oxidation depends on soil temperature, aeration and size of bacteria population. Volatilization of gaseous ammonia ($NH_3$) may occur when ammonium-containing fertilizers are spread on wet soil surface high in pH without any immediate cover.

When ammonium nitrate is applied through fertigation, how they get distributed in different sandy soils can be seen in Fig. 3. Since urea is electro-neutral, the distribution of urea when applied through drippers can be expected to be as that of the water distribution itself.

Urea is a very good source of nitrogen for using in fertigation because it is very well soluble in water. Ammonium nitrate and ammonium sulphate can also be used for application of nitrogen.

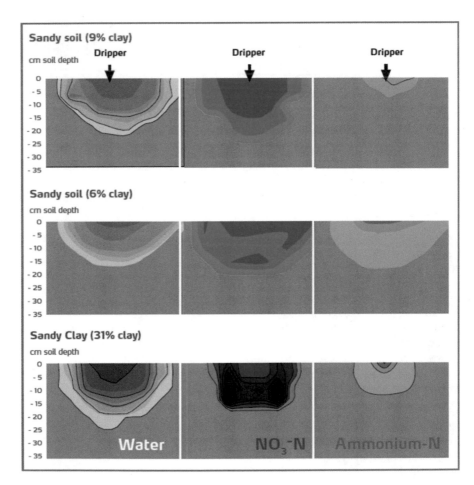

**Fig. 3** Relative mobility of ammonium and nitrate in sandy soils (www.yara.com)

## *Phosphorous*

Phosphorous is essential for cell division and encourages root development. It is also essential for the development of seeds and fruits. The soil pH has a significant influence over the availability of Phosphate ions in the soil solution. The optimum soil solution pH for Phosphorous utilization by crops is between 5 and 7. Generally it is not recommended to apply phosphorous fertilizers through fertigation, because the relative movement of phosphorus fertilizer is very low (Fig. 4). When superphosphate is used as Phosphorous source, it is better to apply to the soil. Diammonium phosphate (DAP) fertilizer can also be used as phosphorous source. When DAP is used in fertigation, it is better to soak the fertilizers in water over night and filter the solution and fertigated. If the water contains calcium salts, formation of calcium phosphate would cause significant clogging of filters and drippers.

Monoammonium phosphate and ammonium polyphosphate are highly soluble and costlier. They can also be used as phosphorous sources.

## *Potassium*

Normally soils are rich in potassium. But the availability of potassium from the soil is limited because the form of potassium is not in available form. As clay content increases, the potassium adsorption by clay particles increases (Fig. 5). Micaceous clay holds up more potassium and makes it unavailable for plant than kaolinite clay. But, the adsorbed K in micaceous clay would also be available during weathering process. Weathering to the extent of releasing K can occur within few months itself.

Muriate of potash (MOP) is a cheap source of potassium. The chemical name of MOP is potassium chloride. MOP is available in two forms. One is red potash, and another is white potash. White potash is more soluble in water than red potash. Fertigation of MOP can be done after overnight soaking and filtering. Some crops are sensitive to chlorine. For those crops, instead of using MOP, potassium nitrate or potassium sulphate can be used as a potassium source.

## *Calcium, Magnesium and Sulphur*

Generally, crops need more calcium. But the availability of calcium from soil and water is sufficient for crops. Sulphur is also available in sufficient quantities in soils. In sandy soils and in high rainfall areas, sulphur deficiency is possible. In such situations ammonium sulphate can be used in fertigation or superphosphate can be used as soil application.

Calcium, magnesium and sulphur are all highly mobile elements (Fig. 6). Proper

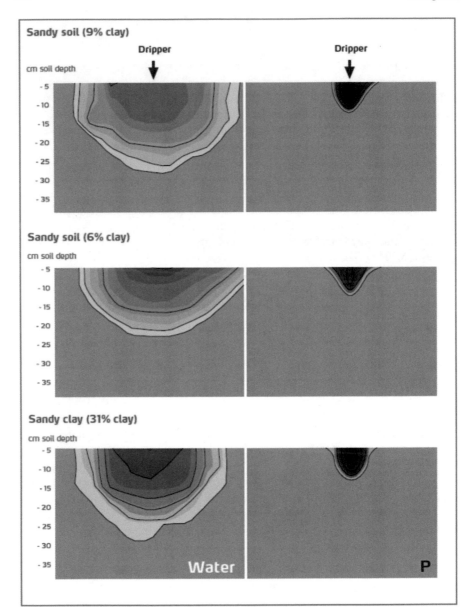

**Fig. 4** Relative movement of phosphorus in sandy soils (www.yara.com)

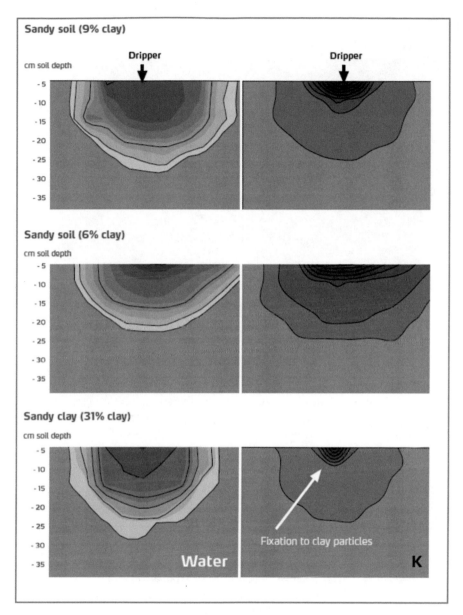

**Fig. 5** Potassium mobility in sandy soils (www.yara.com)

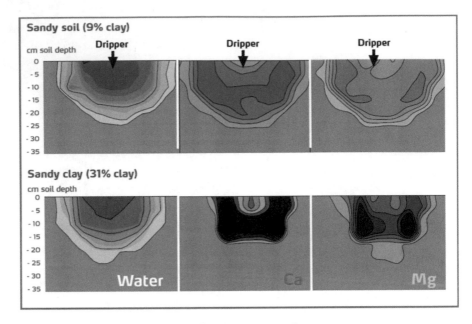

**Fig. 6** Mobility of calcium and magnesium (www.yara.com)

balance of calcium, magnesium and potassium is necessary to maintain good nutrient balance in the wetted zone.

In case of calcium deficiency, calcium ammonium nitrate can be used for fertigation. Calcium and nitrate are highly mobile and care must be taken to retain them in the wetted bulb. During flushing operation after fertigation or during frequent irrigation, there is a possibility of leaching (Fig. 7) and proper care is needed in such occasions.

## *Micronutrients*

Iron, manganese, copper and zinc are needed for crops in very small quantities. In many cases, they exist in adequate amounts in soil, but adverse soil conditions make them unavailable. The adverse conditions are too high or too low pH, high calcium and bicarbonate, water logging and poor aeration. If deficiency is due to precipitation from soil solution, application of chelates can be beneficial. The chelates are complex organic compounds in which metal ion is isolated from other ions in soil solution by the electric charge of organic molecule. Chelates are sensitive to prolonged exposure to light as well as extreme pH levels and heat. Both soil and foliar applications are practised with chelates. Chelate application is normally done for alkaline pH and when phosphate is significantly higher.

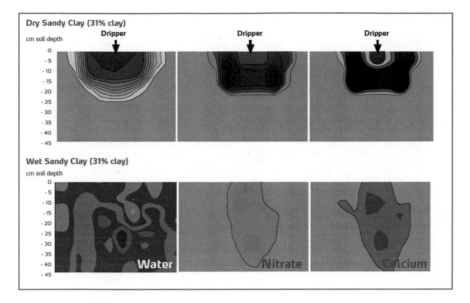

**Fig. 7** Calcium and nitrate mobility in dry and wet soils (www.yara.com)

Foliar sprays of chelates cause less leaf scorching. Foliar application of micronutrients is normally practised when visual symptoms of deficiency are observed in plants. Foliar spray is generally done during early morning or late afternoon. This practice is reported to have less leaf burning problem due to foliar spray. Boron and molybdenum are rarely needed to be supplied through fertilizers.

## Nutrients in Fertilizers

All the fertilizers are available in the form of salts. The per cent weight of element N in fertilizer is the nutrient content of nitrogen. The per cent weight of $P_2O_5$ molecules for phosphorous, $K_2O$ molecules for potassium, sulphur atoms for sulphur and MgO molecules for magnesium are considered as fertilizer contents for each element. Table 1 provides the nutrient content of different fertilizers.

## Fertilizers and Irrigation Water

The solubility property of fertilizers is very important for fertigation. Table 2 provides the solubility of the commonly used fertilizers. Solubility increases with increase in temperature, and solubility of phosphorous and potassium fertilizers is generally low.

**Table 1** Nutrient content of different fertilizers

| No | Fertilizer | N (%) | P$_2$O$_5$ (%) | K$_2$O (%) | S (%) | MgO (%) |
|----|------------|-------|----------------|------------|-------|---------|
| 1 | Ammonium nitrate | 33 | | | | |
| 2 | Calcium ammonium nitrate | 26 | | | | |
| 3 | Ammonium sulphate | 21 | – | – | 24 | |
| 4 | Calcium nitrate | 15 | | | | |
| 5 | Sodium nitrate | 16 | | | | |
| 6 | Urea | 46 | | | | |
| 7 | Ammonium chloride | 25 | | | | |
| 8 | Potassium nitrate | 13 | | 46 | | |
| 9 | Monoammonium phosphate | 12 | 61 | | | |
| 10 | Diammonium phosphate | 18 | 46 | | | |
| 11 | Ammonium phosphate sulphate | 16 20 | 20 20 | | | |
| 12 | Ammonium phosphate nitrate | 20 | 17 | | | |
| 13 | Potassium chloride (MOP) | | | 60 | | |
| 14 | Potassium sulphate | | | 50 | 18 | |
| 15 | Potassium magnesium sulphate | | | 26 | 15 | 30 |

**Table 2** Solubility of different fertilizers

| Fertilizer | Temperature | | | | |
|------------|-------|-------|-------|-------|-------|
| | 0 °C | 10 °C | 20 °C | 30 °C | 40 °C |
| Urea | 40 | 45 | 51 | 57 | 62 |
| Ammonium nitrate | 54 | 61 | 66 | 70 | 74 |
| Ammonium sulphate | 41 | 42 | 43 | 44 | 45 |
| Potassium sulphate | 6 | 8 | 9 | 11 | 13 |
| Potassium chloride | 22 | 23 | 25 | 27 | 28 |
| Potassium nitrate | 11 | 17 | 24 | 31 | 39 |
| Diammonium phosphate | 23 | 27 | 29 | 30 | 31 |
| Monoammonium phosphate | 15 | 18 | 21 | 24 | 26 |

Irrigation water contains various chemical elements. Some of them may interact with dissolved fertilizers causing unwanted results.

If water contains high levels of calcium, magnesium and bicarbonate ions, phosphate-containing fertilizers precipitate calcium and magnesium phosphates. Acidification of irrigation water may reduce the detrimental effect. Polyphosphate fertilizers react with calcium and magnesium ions producing gel suspensions and clog the filters and drippers. If water is rich in Ca, precipitation of gypsum would occur with sulphate-containing fertilizers. Solubility of gypsum gets decreased as

**Table 3** Compatibility chart among different fertilizers

| | Urea | Ammonium Nitrate | Ammonium | Calcium Nitrate | Potassium Nitrate | Potassium Chloride | Potassium Sulphate | Ammonium Phosphate | Fe,Zn, Cu, Mn Sulphate | Fe, Zn, Cu, Mn Chelate | Magnesium Sulphate | Phosphoric Acid | Sulphuric Acid | Nitric Acid |
|---|---|---|---|---|---|---|---|---|---|---|---|---|---|---|
| Urea | | | | | | | | | | | | | | |
| Ammonium Nitrate | | | | | | | | | | | | | | |
| Ammonium Sulphate | | | | | | | | | | | | | | |
| Calcium Nitrate | | | ■ | | | | | | | | | | | |
| Potassium Nitrate | | | | | | | | | | | | | | |
| Potassium Chloride | | | | | | | | | | | | | | |
| Potssium Sulphate | | | | ■ | | | | | | | | | | |
| Ammonium Phosphate | | | | | | | | | | | | | | |
| Fe, Zn, Cu, Mn Sulphate | | | | | | | | ■ | | | | | | |
| Fe, Zn, Cu, Mn Chelate | | | | | | | | | | | | | | |
| Magnesium Sulphate | | | | ■ | | | | ■ | | · | | | | |
| Phosphoric Acid | | | | | | | | | | | | | | |
| Sulfuric Acid | | | | ■ | | | | | | | | | | |
| Nitric acid | | | | | | | | | ■ | | | | | |

Legend: ■ Imcompatible   ▨ Reduced Solubility   ☐ Compatible

temperature increases. Therefore during summer this problem is often encountered. Urea precipitates lime if water contains more calcium and bicarbonate.

When two or more fertilizers are used simultaneously, precipitates may get formed. Table 3 provides data about compatibility between different fertilizers. For any specific irrigation water, before fertigation is attempted, the fertilizer should be dissolved in the recommended concentration and left for sometime. If any precipitate occurs, that fertilizer combinations should not be adopted (Fig. 8).

## *Concentrations of Fertilizers Solution*

The maximum concentrations of fertilizer solutions are limited by the solubility of the fertilizers. But the stock solutions are to be diluted so that finally when the fertilizer mixed irrigation water reaches the plant, the concentration has to fall within the limits as given in Table 4 for desirable results. The concentrations above the maximum may be injurious to plants.

## Water Soluble Solid and Liquid Fertilizers

In order to overcome some disadvantages of the conventional fertilizers and to produce the yield of any crop to its fullest potential, water-soluble solid and liquid

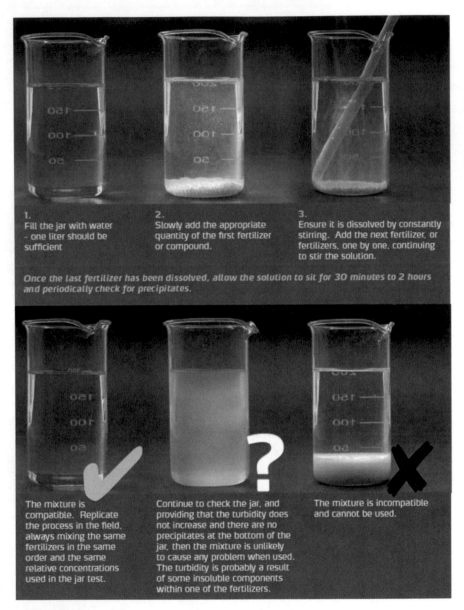

**Fig. 8** Test for compatibility of fertilizer mix (www.yara.com)

fertilizers have been developed. Some of the fertilizers available in the market are as follows:

- Urea–ammonium–nitrate (UAN)
- All 19 (19N-19P-19K)

**Table 4** Acceptable concentration of different nutrients [1]

| No. | Nutrients | Acceptable limits of concentration (ppm) | Average acceptable concentration (ppm) |
|---|---|---|---|
| 1 | Nitrogen | 150–1000 | 250 |
| 2 | Phosphorus | 50–100 | 80 |
| 3 | Potassium | 100–400 | 300 |
| 4 | Calcium | 100–500 | 200 |
| 5 | Magnesium | 50–100 | 75 |
| 6 | Sulphur | 200–1000 | 400 |
| 7 | Copper | 0.1–0.5 | 0.05 |
| 8 | Boron | 0.5–5.0 | 1.0 |
| 9 | Iron | 2.0–10 | 5.0 |
| 10 | Manganese | 0.5–5.0 | 2.0 |
| 11 | Molybdenum | 0.01–0.05 | 0.02 |
| 12 | Zinc | 0.5–1.0 | 0.5 |
| 13 | Sodium | 20–100 | 50 |
| 14 | Carbonates | 20–100 | 60 |
| 15 | Sulphate | 200–300 | 250 |
| 16 | Chloride | 50–100 | 70 |

- All 19 with micronutrients
- Multi-K (13N-46K)
- Multi-K with micronutrients.

They are very costly, and use of them is advised only when the economics works out in favour of using those fertilizers. A thumb rule of using 50% of water-soluble fertilizers and 50% of conventional fertilizers is followed in many places in India.

**Problem**

Tomato crop needs nitrogen, phosphorous and potassium in the ratio of 200:250:250 kg/ha. If the water-soluble fertilizers NPK 13:40:13 and potassium sulphate (0:0:50) are used to meet a 50% of the phosphorous and potassium requirement and urea is utilized for meeting the remaining nitrogen requirement, find out the quantity of respective fertilizers needed?

**Solution**

50% recommended Phosphorous     $= 250/2 = 125$ kg

DAP needed for 125 kg Phosphorous $= (100/46) \times 125 = 271.75$ kg

Nitrogen in 271.75 kg DAP     $= (18/100) \times 271.75 = 48.91$ kg

Total Nitrogen needed     $= 200$ kg

Apart from DAP, Nitrogen to be given using other fertilizers

$$= 200 - 48.91 = 151.1 \text{ kg}$$

50% recommended potassium     $= 250/2 = 125$ kg

MOP needed for 125 kg Potassium     $= (100/60) \times 125 = 208.33$ kg

50% recommended Phosphorous     $= 250/2 = 125$ kg

NPK 13 : 40 : 13 fertilizer needed for 125 kg Phosphorous

$$= (125/40) \times 100 = 312.50 \text{ kg}$$

Nitrogen in 312.5 kg of NPK 13 : 40 : 13 fertilizer

$$= (13/100) \times 312.5 = 40.625 \text{ kg}$$

Potassium in 312.5 kg of NPK 13 : 40 : 13 fertilizer

$$= (13/100) \times 312.5 = 40.625 \text{ kg}$$

Potassium to be given through Potassium Sulphate

$$= 250 - 125 = 125 \text{ kg}$$

Potassium available from 40.625 kg of NPK 13 : 40 : 13

$$= 84.375 \text{ kg}$$

NPK 0 : 0 : 50 fertilizer needed for 84.375 kg of Potassium

$$= (100/50) \times 84.375 = 168.75 \text{ kg}$$

Nitrogen to be given through Urea     $= 200 - 48.91 - 40.625 = 110.465$ kg

Urea needed for 110.465 kg Nitrogen   $= (100/46) \times 110.465 = 240$ kg

**Fertilizer requirement**

    Urea:            240 kg
    DAP:             271.75 kg
    MOP:             208.33 kg
    NPK: 13 : 40 : 13 312.50 kg
    NPK: 0 : 0 : 50    168.72 kg
    (Table 5)

# Fertilizer Injection Times

When we opt for larger fertigation intervals, the amount of fertilizer needed to be applied in a single fertigation event will be more. In such a situation, if more quantity

**Table 5** Recommended dose of fertilizers for different crops

| Sl. No. | Crop | N (kg/ha) | $P_2O_5$ (kg/ha) | $K_2O$(kg/ha) |
|---------|------|-----------|------------------|----------------|
| 1 | Arecanut | 132 | 52 | 198 |
| 2 | Banana | 275 | 75 | 825 |
| 3 | Brinjal | 200 | 150 | 100 |
| 4 | Coconut | 98 | 56 | 210 |
| 5 | Chilli | 120 | 60 | 30 |
| 6 | Gloriosa | 100 | 50 | 75 |
| 7 | Gourds | 200 | 100 | 100 |
| 8 | Grapes | 400 | 360 | 1040 |
| 9 | Maize | 135 | 62.5 | 50 |
| 10 | Onion | 60 | 60 | 30 |
| 11 | Sugarcane | 275 | 63 | 133 |
| 12 | Tomato | 200 | 250 | 250 |
| 13 | Tapioca | 90 | 90 | 240 |

of fertilizers is injected in a shorter duration of time, the concentration of the nutrients in the irrigation water will be more than the acceptable concentration levels. Therefore, it is better to increase the fertigation frequency during the early stages of the crop, because the withstanding ability of young plants against higher fertilizer concentration in soil solution is lower.

### Problem
In a drip irrigation system for sugarcane of area 2300 m², the total number of drippers in the system is 2625. Each dripper discharges at a rate of 2 l/h. 12 kg of urea is applied for each fertigation. A venturi in the system sucks at a rate of 25 l/h. It has been decided to limit the concentration of nitrogen in irrigation water below 500 mg/l. Find out the duration of fertigation and the volume of stock solution to be prepared.

### Solution
Total discharge rate of the system = 2526 × 2 l/h = 5250 l/h
Nitrogen in 12 kg of Urea         = (46/100) × 12 = 5.52 kg

Let y be volume of water needed to go in the drip irrigation system with 500 ppm of nitrogen in the flow.

$$500 = (5.52 \times 1000 \times 1000)/y \quad \text{mg/l}$$
$$y = 11,040 \text{ litres}$$

Duration of fertigation = 11,040/5250 h = 2.1 h

Volume of stock solution needed = Venturi suction rate × Duration of irrigation
$$= 25 \times 2.1 = 52.6 \text{ litres}$$

Generally, it is recommended that only irrigation water is fed through the system for a half an hour to one hour. Then the fertigation is done for a duration typically around one to three hours. Then flushing of pipes should be done at least for half an hour. Otherwise, precipitates get formed in the dripper passages and cause clogging [2].

## Fertigation Scheduling

Growth curve nutrition approach is the general method of fertigation scheduling method. For this method, nutrient uptake curves for every crop are needed to be collected [3, 4]. Figures 9 and 10 show a sample curves for potato and citrus. How much nutrient a plant needs is noted from the curve and that amount is supplied to the plant. Fertigation interval may equal irrigation interval, or it may also be longer than irrigation interval like weekly or fortnightly.

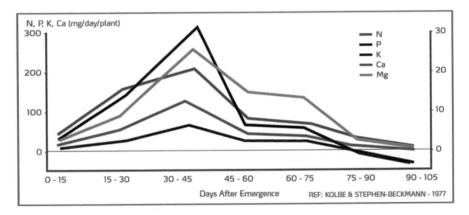

**Fig. 9** A sample nutrient uptake curve for potato (www.yara.com)

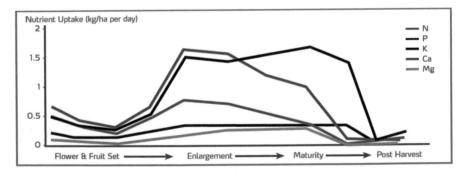

**Fig. 10** A sample nutrient uptake curve for citrus (www.yara.com)

**Table 6** Macronutrients leaf analysis at mid-fruit development for pome fruit (www.yara.com)

| Nutrient (%) | Deficient | Marginal | Adequate | High | Excessive |
|---|---|---|---|---|---|
| N | < 2.00 | 2.00–2.25 | 2.25–2.70 | 2.70–2.85 | > 2.85 |
| P | < 0.14 | 0.14–0.16 | 0.16–0.28 | 0.28–0.30 | > 0.30 |
| K | < 1.00 | 1.00–1.20 | 1.20–1.85 | 1.85–2.00 | > 2.00 |
| Ca | < 1.04 | 1.04–1.20 | 1.20–2.00 | 2.00–2.20 | > 2.20 |
| Mg | < 0.15 | 0.15–0.19 | 0.19–0.36 | 0.36–0.39 | > 0.39 |
| S | < 0.10 | 0.10–0.14 | 0.14–0.18 | 0.18–0.22 | > 0.39 |

These curves provide only plant uptake but the efficiency of uptake for each nutrient would vary. So, the farmer must fertigate with little extra dose. This knowledge usually comes with experience of operation in fertigation. During the fertigation scheduling process, leaf analysis is done and nutrient deficiency if any is estimated. If any nutrient is deficient, then that nutrient application is increased. A sample guideline to gauge the nutrient sufficiency by leaf analysis for pome fruit is shown in Table 6.

# Summary

Fertigation scheduling is a very wide subject, and most of the things discussed in this chapter are summarized from researches and experiences. While choosing a fertilizer for fertigation, cost of fertilizer, interaction of the fertilizer with water and soil and interaction with accompanying fertilizer must be carefully studied to have a problem-free fertigation. Fertigation scheduling must be done with a guidance of plant nutrient uptake curves and frequent leaf analysis.

# Assessment

## Part A

1.  Choose the incorrect statement

    a.  Cations are positively charged ions
    b.  Anions are negatively charged ions
    c.  Surplus of one cation in soil solution diminishes the absorption of other cations
    d.  **Cations are negatively charged ions**

2.  Choose the incorrect statement

    a.  Cations are positively charged ions
    b.  Anions are negatively charged ions
    c.  Surplus of one cation in soil solution diminishes the absorption of other cations
    d.  **Anions are negatively charged ions**

3.  _____application beyond vegetative stage postpones flowering and also lodging of plants.

    **a. Nitrogen**  b. Phosphorus  c. Potash  d. Micronutrients

4.  Phosphorous is essential for cell division and encourages _____development.

    **a. Root**  b. Flower  c. Leaf  d. Trunk

5.  _____ is micronutrient

    **a. Iron**  b. Nitrogen  c. Phosphorus  d. Potash

6.  _____of irrigation water increases the solubility of fertilizers

    a. **Acidification**  b. Cooling  d. Blending  d. Colouring

7.  It is better to increase the fertigation frequency during the _____of the crop.

    a. **Early stages**  b. Mid-stage  c. Harvesting stage  d. None of these

8.  Potassium nitrate is a _____soluble fertilizer

    **a. Water**  b. Acid  c. Base  d. None of these

9.  Monoammonium phosphate is a ——-soluble fertilizer

    **a. Water**  b. Acid  c. Base  d. None of these

10. Most of the elements are better absorbed by the plants in the pH range of six to seven. Say True or False (True)
11. Volatilization of gaseous ammonia ($NH_3$) may occur when ammonium-containing fertilizers are spread on wet soil surface high in pH without any immediate cover. Say True or False (True)
12. In sandy soils and in high rainfall areas, sulphur deficiency does not occur. Say True or False (False)
13. Calcium and nitrate tend to leach down while flushing operation after fertigation. Say True or False (True)
14. Too high or too low pH, high calcium and bicarbonate, water logging and poor aeration may cause micronutrient deficiency. Say True or False (True)
15. Soil applications of chelates should be done when soil pH is alkaline or if phosphate level in the soil is high. Say True or False (True)

**Part B**
**Short Answers**

1. What are the recommended potassium fertilizers for fertigating chlorine-sensitive crops?
2. Which are the situations in which sulphur deficiency is possible?
3. What is the relationship between solubility of fertilizers and temperature?
4. Why is the concentration of fertilizers in irrigation water important when planning fertigation?

**Part C**
**Brief Answers**

1. Write down at least four micronutrients needed for crop growth?
2. What are chelates? What situation warrants for application of chelates?
3. Write down the names of 6 macronutrients for plants?
4. How to check the compatibility of different fertilizers during fertigation?
5. Write a short note on deciding fertilizer injection times?
6. What atoms/molecules are considered for nutrient calculation for the following elements?
7. Write a short note on fertilizers and irrigation water reactions?

# References and Further Reading

1. Patel, N. (2017). *Precision farming development centres research findings on fertigation techniques*. Department of Agriculture, Cooperation & Farmers Welfare Ministry of Agriculture & Farmers Welfare Government of India.
2. Nathan, R. (2005). *Fertigation, extension service, irrigation and soil field service*. State of Israel.
3. Lamm, F. R., Ayars, J. E., & Nakayama, F. S. (2006). *Microirrigation for crop production-design, operation, and management*. eBook ISBN: 9780080465814.
4. Incrocci, L., Massa, D., & Pardossi, A. (2017). New trends in the fertigation management of irrigated vegetable crops. *Horticulturae*. https://doi.org/10.3390/horticulturae3020037

# Maintenance and Operation of Drip Irrigation Systems

## Introduction

Maintenance problems in drip irrigation systems primarily arise due to dissolved salts, algae and incompatible mixture of fertilizers during fertigation. Dripper plugging also occurs when very fine soil sediment particles settle down due to sluggish flow in the dripper passage. Prior knowledge of possible chemical reaction of water with salts and fertilizers and bio-chemical reaction of water with algae would be very helpful in understanding the mechanism behind plugging of emitters. A water test is a necessity to chart out plans for operation and maintenance. A sample water test would help classify the plugging potential of water, and the parameters those are tested normally include total dissolved salts (TDS), pH, electrical conductivity, iron, manganese, calcium carbonate and bacteria [1]. Classification guideline for irrigation water is available in Chap. 7, namely drip irrigation components.

Drip irrigation operation specific to drip irrigation is also discussed here.

## Water Chemical Reactions

Water is called as a universal solvent because it can dissolve any substance to certain extent. The proportion of dissolved gases such as oxygen, carbon dioxide and hydrogen sulphide significantly affects the dissolving power of the water.

Studying the role of carbon dioxide in water is very important. Carbon dioxide finds place into the water during rain itself as the rain passes through atmosphere. When rain water falls on ground, it passes through organic matter which enriches carbon content in water. Finally the water becomes a weak carbonic acid. The pressure deep underground is very much conducive for the weak carbonic acid to dissolve calcium carbonate. In the water, calcium carbonate is in the form of calcium bicarbonate. This calcium bicarbonate in underground water gets precipitated when it is passing through pipes and drippers on the surface of the ground (Fig. 1). The pressure

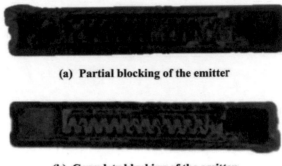

**(a) Partial blocking of the emitter**

**(b) Complete blocking of the emitter**

**Fig. 1** Salt encrustation in inline emitter [2]

and temperature on the surface of the ground are conducive for precipitation in the form of calcium carbonate. Higher pH and temperature are conducive for this kind of precipitation.

In airlift pumps, air is passed through water with a high pressure. At a higher pressure, the proportion of carbon dioxide dissolved gets increased significantly higher. Figure 2 shows the extent of deposition of calcium and magnesium salts in a pipe operated with airlift pump.

Iron is available in soil as oxides, and when water passes through the soil it gets converted into ferrous bicarbonate. This ferrous bicarbonate when exposed to air would precipitate as ferric hydroids, and this precipitate has a very high potential to plug the drippers. Mostly if iron is available in water, we can also expect certain amount of manganese also in water at lower concentrations.

Hydrogen sulphide is also a plugging type of salt; but when it is present with calcium carbonate, precipitation of calcium carbonate gets reduced. This is because

**Fig. 2** Salt encrustation in pipes of airlift pumps (https://agritech.tnau.ac.in)

presence of hydrogen sulphide increases acidity. Presence of hydrogen sulphide with iron causes precipitation of iron sulphate which is also a plugging agent.

## Bio-chemical Reactions with Water

Algae and bacteria growth is inevitable when water source is surface water. Sand filters normally used while using surface water cannot filter the spores of the algae. When water comes out from the dripper, water velocity is very low and also the nutrients needed for algae growth are available in water itself from the fertilizers added with irrigation water. Additionally sunlight is also available at the exit of the dripper. Therefore, the growth of algae and plugging of the emitter by algae invariably happens when surface water is used.

Due to bacterial action precipitation of sulphur, manganese and iron also would happen. Certain bacteria oxidize hydrogen sulphide and produce elemental sulphur slimes.

## Reaction of Fertilizers With Water

When phosphate fertilizers are used with the water containing calcium, magnesium and bicarbonates, precipitate of calcium and magnesium would form. Acidification of irrigation water would reduce this problem.

When polyphosphate fertilizers are used with water containing calcium and magnesium, gel-like suspensions are formed. These gel-like suspensions would clog filters as well as drippers.

Calcium sulphate is precipitated if water containing calcium is used when sulphate fertilizers are used. This problem would be more at higher temperatures. During fertigation of urea with calcium bicarbonate-rich water, calcium oxide (lime) precipitation is possible.

When water contains calcium, fertigation with phosphoric acid would cause precipitation of calcium phosphate. Iron concentration above 0.1 ppm in water would produce precipitates if calcium or phosphate fertilizers are used.

It is very difficult to know in advance about the nature of chemical reaction that would occur for the set of fertilizers used in fertigation. Hence conduct of jar test is recommended before doing fertigation. Details of conducting a jar test are discussed in the preceding chapter.

## Acid Treatment

Hydrochloric acid, sulphuric acid and phosphoric acid are commonly used in acid treatment. Among these three acids, hydrochloric acid is very popular because it is less costly and safer to use. Sulphuric acid is efficient in removing very thick scales, but it is highly toxic. More safety precautions are necessary with sulphuric acid. Phosphoric acid can also be used, and it is also a good source of phosphorus. Nitric acid can also be used. If calcium content in water is more than 50 ppm, calcium phosphate tends to precipitate.

Care must be exercised when doing acid treatment. Protective gloves and goggles are essential (Fig. 3). When diluting acid, water must be taken in a container and acid must be poured in the water in container. Taking acid in a container and pouring water into the container would splash the acid out of the container and would cause an accident.

### *Acid Treatment for Prevention of Carbonate scales*

Acid treatment can be continuously done to prevent the scale formation due to carbonate salts in the water. If these salts plug the emitters and pipe, reclamation by acid treatment becomes difficult. It is because the emitters will be discharging the water, and it takes more time for removal of the salts. Hence, if the carbonate content is more than a critical level continuous treatment with acid for all irrigations is recommended.

The best way to find out the acid dose for any irrigation water is to titrate a specific volume of irrigation water to bring its pH to 7 (Fig. 4). It must be kept in mind that the pH of irrigation water does not remain constant throughout the year. During rainy reason, pH normally lower during summer, pH is higher. When acid treatment is

**Fig. 3** Protective wear for acid treatment (https://wik ihow.com)

**Titration curve of water with chloric acid (33%)**

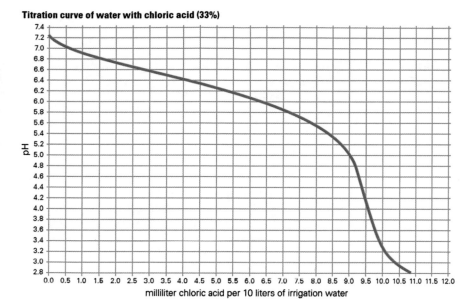

**Fig. 4** Typical titration curve depicting relationship between pH and HCl addition

implemented in the field, always test the pH of irrigation water by pH paper after acidification.

Following procedure can also be used for estimating the quantity of acid needed if concentrations of Ca and Mg are known in mg/l unit. This estimation assumes that most of the Ca and Mg in a water sample is a result of dissolved Ca and Mg carbonates. If the Ca and Mg concentration is known in mg/l, they can be converted into meq/l as follows:

$$Ca\left(\frac{meq}{l}\right) = \frac{Ca\left(\frac{mg}{l}\right)}{\text{Equivalent weight of Ca}} = 0.05Ca\left(\frac{mg}{l}\right)$$

$$Mg\left(\frac{meq}{l}\right) = \frac{Mg\left(\frac{mg}{l}\right)}{\text{Equivalent weight of Mg}} = 0.083Mg\left(\frac{mg}{l}\right)$$

Total concentration of Ca and Mg is the sum of concentrations Ca and Mg found out using preceding equations. Chemistry basics needed to understand these equations are introduced in the basic section chapter, namely salinity.

If the water test does not show Ca and Mg concentrations but shows only hardness in mg/l of $CaCO_3$, dividing the hardness by 50 provides a good estimate of bases in unit of meq/l. For neutralizing 80% of bases of Ca and Mg carbonates, 23 ml of

$H_2SO_4$ (93%) or 78 ml of HCl (32%) or 18 ml of $H_3PO_4$ (85%) is needed per 1000 L of water per meq.

**Problem**
For a $CaCo_3$ hardness of 200 mg/l in the irrigation water, find out the HCl requirement per thousand litres of irrigation water to neutralize 80% of its bases.

**Solution**
meq/L of bases = 200/50 = 4 meq/L.

HCl needed = 4 × 78 = 312 ml per 1000 L of irrigation water.

A water quality analysis usually lists electrical conductivity in millimhos per centimetre (mmho/cm). (One mmho/cm is also equivalent to 1 desiSiemens per metre (dS/m)) To convert from dS/m approximately to ppm unit or mg/l unit, multiply by 640. For example, if the electric conductivity metre is 1 dS/m, then dissolved solids are converted as 640 ppm.


## *Acid Treatment for Clogged Systems*

When systems get either partially clogged or fully clogged, it becomes difficult to reclaim the system by acid treatment. The pH of the water should be brought down to the range of 4 to 2. It should be borne in mind that as the pH is reduced more and more, corrosion of various components of irrigation system would also occur. So, the pH reduction should be planned according to the requirement. Once the pH to be maintained for acid treatment is decided, then the time of operation should be estimated. This can be done by placing a small piece of affected lateral inside the water of the specified pH for a duration so that the scales get removed satisfactorily. For this duration, the acid treatment should be done in the field.

It is economical to roll the inline laterals and place inside the acidified water (Fig. 5). The pH of the acidified water may be maintained around 2. The time duration for which immersion is done may be for few days.

Usually the emitter plugging starts at the farther end of laterals. Sometimes, it may be useful to identify the partially plugged emitters visually, and those drippers have to be marked. Then the water supply must be stopped, and acidified water must be taken in a container as shown in Fig. 6. The laterals must be immersed with a wooden fork for some time as shown in Fig. 6. Very high concentration of acidified water may be used in this case because the solution is not wasted and reused. When this type plugging removal is done, water must not flow in the laterals.

Mostly application of acid is done for half an hour at the end of irrigation by injecting acid into the system through venturi. Generally, it is recommended that 1–2 L of HCl is to be injected for every 1000 L of water. The water in the pipes should be left as such for overnight, and next day the system should be flushed. During flushing first main lines should be flushed and then laterals. Otherwise, there will be possibility of emitters getting clogged with the removed scales.

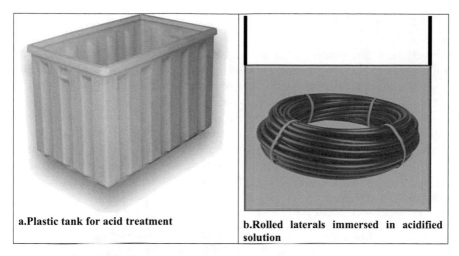

**Fig. 5** Lateral rolls placed inside acidified water

a.Identification of plugged Locations in Inline          b. Immerse Inline lateral in Acid Solution
Laterals

**Fig. 6** Acid treatment in case of partial plugging

Let us assume that the drippers are discharging at an average rate of 1.5 l/h and 2000 drippers are simultaneously operating. Then the total discharge rate of the system is 3500 l/h. If the acid treatment is done for half an hour, the total discharge from the system would be 1750 L. If we decide to inject 1 L of acid per 1000 L of water, the amount of HCl needed is 1.75 L.

Let us assume that venturi sucks at a rate 50 l/h. So, for an half an hour, it can suck 25 L. Therefore, dilute 1.75 L of acid into around 25 L and place under the venturi.

## Biological Plugging Removal

For surface water sources, water would contain organic matter due to algae and bacterial slime (Fig. 7). Within few days of operation of drip system, with these types of water, plugging of emitters due to algae growth can be expected (Fig. 8). Chlorination is the popular option for handling biological clogging.

Bleaching powder ($CaOCl_2$) commercially available for domestic use contains only 20% chlorine, whereas the calcium hypochlorite $Ca(OCl_2)$ would contain 65% chlorine. Bleaching powder and calcium hypochlorite are solids. If water is alkaline, bleaching powder as well calcium hypochlorite do not dissolve well in water, and in such situation, sodium hypochlorite solution is used. Sodium hypochlorite solution contains 10% chlorine.

**Fig. 7** Algae in open water sources

shutterstock.com · 1465365128

**Fig. 8** Plugging due to algae https://doi.org/10.1590/ S0103-90162008000100001

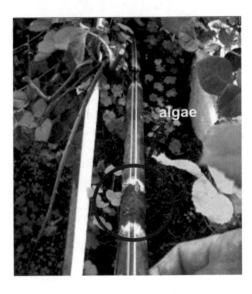

**Fig. 9** pH versus HOCL and
OCL concentration

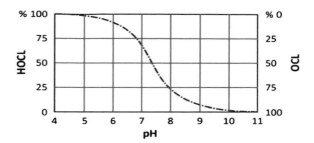

Sometimes for alkaline water acidification is done with hydrochloric acid. When chlorine compounds are added to the water they produce hypochlorous acid. Hypochlorous acid is 60 times more powerful oxidizing capacity than hypochlorite. This hypochlorous acid may further ionize into hydrogen ion and hypochlorite ion. This chemical reaction is depicted below:

$$HOCl \leftrightarrow H^+ + OCl^-$$

Both hypochlorous acid and hypochlorite ion are oxidizing agents. But they differ in oxidizing ability. Hypochlorous acid is 60 times more powerful oxidizing capacity than hypochlorite. Note that the preceding reaction is reversible, and when pH is low the reaction proceeds towards combining hypochlorite ion and hydrogen ion to form hypochlorous acid. Therefore, reduction of pH during chlorination helps in increasing the capacity of oxidization.

From Fig. 9, it can be seen that at pH 7.5, the HOCl concentration is around 45%, and at pH 6.5, the HOCl concentration increases to around 90%. It can also be noted that reducing pH below 6.5 does not have much effect.

If acid and chlorine are injected at the same time, it should be done at two different injection points. Either two venturies in series can be fixed or venturi with two inlets can be used for this purpose (Fig. 10). Mixing acid and liquid chlorine together will produce highly toxic chlorine gas. Never store acids and chlorine together.

There are three methods of chlorine application:

1. Continuous treatment
2. Super chlorination
3. Intermittent treatment.

## Continuous Treatment

It is the best and simplest method. In the long run, it provides good results as this method continually removes the organic growth, and the system is all the time clean from organic material. The required concentration varies according to the water quality. Normal injection rate is 1 ppm and to begin with 5–10 ppm can be adopted.

**Fig. 10** Venturi with two inlets

## *Intermittent Treatment*

When less organic matter exists in the water, 10–20 ppm injection for 30 min at the end of irrigation can be done. Chlorine is left in pipes till next irrigation. This is called as intermittent method.

## *Super Chlorination*

Without chlorination if the irrigation system is operated for some time, there is a possibility of high accumulation of organic matter. This may completely clog the system. In such situations, 500 ppm is injected, and the system is shut down and left for up to 24 h. Then the system is flushed step by step. First main line, then submains and laterals. Flushing first through laterals would cause emitter plugging. Superchlorination may damage sensitive plants and young plants. Irrigation system components would also get corroded to some extent.

Exactly finding the correct dosage of chlorine is difficult because the chlorine demand of each and every water is different. Use a trial-and-error method to achieve a specific free residual chlorine concentration. At least 0.5 ppm of free chlorine at the end of laterals ensures that all the organic matter has been destructed. Use a field test kit to measure chlorine concentration from water obtained from drippers at the farthest end.

The sodium hypochlorite has limited stability. After preparing stock solution with sodium hypochlorite, it must be injected into the system immediately. Stock solutions with higher chlorine concentration would release chlorine faster, and the problem aggravates when the stock solution is placed under direct sun radiation or in the higher temperature ambience. Hence covering the stock solution container with wet cloth or shade is recommended.

Following formula can be used for estimating the total quantity of chemical (Wc) in grams needed:

$$W_c = (QT\ C_n)/(10 \times \%C_c)$$

where

$Q$    = discharge rate of pump litres per hour
$T$    = chemical injection time (hours)
$C_n$  = concentration of the chemical needed in irrigation water (ppm)
$C_C$  = per cent concentration of chemical used

## Problem

- Pump discharge—2 l/s
- Concentration required ($C_c$)—5 ppm
- Injection time ($T$)—3 h
- Calculate sodium hypochlorite—10% chlorine.

## Solution

Total volume of water ($Q$) = 2 × 3600= 7200 l/h

By preceding Eqn. sodium hypochlorite required = 1080 gm

Recently, use of hydrogen peroxide is recommended for controlling algae growth in dripper passages. For saline water, effect of hydrogen peroxide is better. It is much friendlier to environment than chlorine compounds. Hydrogen peroxide is available in 2 concentrations such as 35 and 50%. The recommended dose is 300–500 g per 1000 L of water.

## *Algae Control in Ponds and Wells*

Most simple way of reduction of algae growth in wells is to obstruct direct sunlight falling on the surface of water in wells directly by providing a cover. The algae growth occurs primarily on the surface of the water. Hence application of oxidizing agent like bleaching powder on the surface of the water is recommended. Bleaching powder that has to be sprinkled over the surface of water at constant intervals would help prevent the algae growth in ponds and wells. Mixing of bleaching powder for the full depth of well and pond need not be done, and it is wasteful effort. It is very important to check the residual chlorine level by simple test kits available in the market.

Algae growth control in ponds and open wells is also possible with copper sulphate. When the water is used for rearing animals and domestic use copper sulphate

**Table 1** Copper sulphate (CuSO$_4$) levels safe for fish

| Alkalinity value (CaCO$_3$, mg/l) | Addition of copper sulphate |
|---|---|
| Below 40 | Do not use |
| 40–60 | 0.4 gm/m$^3$ of water |
| 60–100 | 0.5 gm/m$^3$ of water |
| Over 100 | 1 gm/m$^3$ of water |

Dupree and Huner [3]

must be avoided. Highly porous cloth like gunny bag is used to make small bundle, and the bundle is made to float using inflated tubes filled with air. The recommended concentration of copper sulphate is 0.5–2 ppm. If fish is grown in ponds, concentration of copper sulphate must be within critical levels (see Table 1 for safe levels).

## *Remedy for Iron*

Iron problem is the most difficult to handle. Many a times famers get up while handling the iron contaminated water. Plugged emitters can be cleaned by soaking with citric acid (0.5–1.0%) solution for 24–48 h. For controlling the iron scales, 1 mg/l of chlorine or 1% citric acid can be injected with all the irrigation. Iron consumes about 0.7 ppm chlorine for each ppm of iron.

Stabilization is a process of sequestering iron by chelating with the chemicals like poly phosphates and phosphonates, and this process helps in preventing conversion of ferrous into ferric form. Stabilization process is widely adopted for iron management in irrigation water.

It is to be recalled that even 1.5 mg/l of iron would cause severe plugging problems. In the event of larger iron concentrations, water is to be pumped to a surface tank and aerated sufficiently so that the dissolved bicarbonate gets oxidized and precipitated into ferric forms (Fig. 11). A media filter is also used additionally to filter the ferric precipitates.

Usually when iron is present in water a small quantity of manganese would also be present in water. All the treatment processes discussed in for iron management are useful for manganese management as well as hydrogen sulphide. Hydrogen sulphide consumes about 2 ppm chlorine for each ppm of sulphide.

## Regular Maintenance of Accessories

For successful maintenance of drip irrigation systems, maintenance of various devices is very important. Measurement of pressure at different locations like inlet and outlet of filters and laterals is very useful way ascertaining the goodness of working of system (Fig. 12).

**Fig. 11** Iron precipitate in open ponds

**Fig. 12** Pressure measurement in lateral

## Filters

Examine screens and seams of filters for damages. The maximum allowable pressure differential between inlet and outlet of the filter is 5 m. It is always better to flush the filters before each irrigation.

## Media Filters

Back washing of sand filter is to be done periodically. The discharge rate during back washing must be such that the media do not escape out with flowing water. After back washing, bleaching powder is sprinkled on the water standing on the surface of

media and allowed to soak for some time. Then the final washing of oxidized algae
and organic matter is done.

## Flushing Main and Laterals

During installation process, before connecting laterals flush main and submain.
During failure or burst of the main lines, flush dirt from lines for at least 10 min.
Filtration does not prevent very small particles and water soluble matter from going
into the pipe network. A layer of sediment is deposited on the inner walls of the pipe
and dripper passages, and accumulation occurs more at the end of laterals. Therefore
flush the laterals after 2–4 weeks. As soon as lateral end is opened for flushing dirty
water flows out initially, and after few some time, dirt-free water would come out
(Figs. 13 and 14). Then the lateral can be closed.

A minimum flow velocity of 0.5 m/s is generally recommended for flushing lateral
lines. Figure 15 provides useful data about flow velocity versus throw distance of jet
for pipe heights of 0.3, 0.5 and 1 m.

## Handling Rats and Squirrel Problems

Rats and squirrel bite the laterals. This problem is very difficult to solve. One way is
directly controlling rats by chemicals. These animals generally damage the pipes to
meet their water needs. Some of the Indian farmers practise the use of cow's urine or
neem oil through fertigation system to repel the rats and squirrels. The basis is that
every animal dislikes the urine smell of the other. The sour taste of neem oil is not

**Fig. 14** Dirt-free water after sometime during flushing www.netafim.com

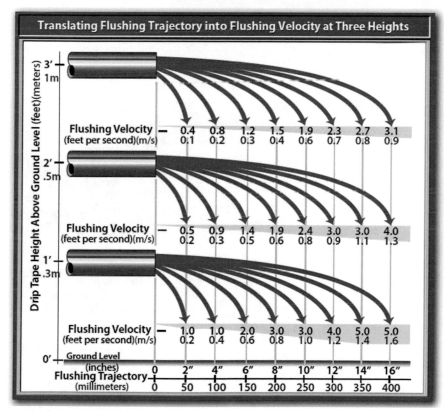

**Fig. 15** Jet throw distance versus velocity of flow for three heights of pipe (www.netafim.com)

liked by them. Results of intensive studies are not available in this regard. The drip lateral production companies claim that they produce laterals with a rat repellents.

## Operation of Drip Irrigation System

The discussions made in the chapter on operation of sprinkler irrigation system are all applicable to drip irrigation system also. However, some points which are specific to drip irrigation are discussed here.

### *Minimum Irrigation Duration*

In row crops, the minimum duration of irrigation is the time taken for getting a continuous wetted strip. In Fig. 16, we can see the wetted circles sometime after irrigation is started. In Fig. 17, we can see the formation of wetted strip after the adjacent wetted circles overlap.

**Fig. 16** Wetted circles before overlapping

**Fig. 17** Continuous wetted strip

**Fig. 18** Wetted strip for two crop rows

**Fig. 19** One lateral for four crop rows

For finding the minimum duration of irrigation, the initial soil moisture before starting the irrigation must be approximately equal to the soil moisture level at which every irrigation is done. During irrigation, note down the time taken for the wetted circles to join together throughout the field, and this is the minimum duration of irrigation.

If more than one crop rows are planted for a lateral, find out the time taken for the width of wetted strip to reach all the rows of crop sufficiently (Figs. 18 and 19). This duration is the minimum duration needed to be irrigated. Also note down the depth of soil wetting after irrigation is done for the minimum duration of irrigation.

## Maximum Duration of Irrigation

Data of effective depth of roots is needed to find out the maximum duration of irrigation. This data can be easily obtained by cutting-open the soil of a fully developed crop and measuring the root depth. Published data for different crops is also presented in the chapter, namely operation and maintenance of sprinkler irrigation systems.

For finding the maximum duration of irrigation, a trench as in Fig. 18 must be dug in the field at the end of any lateral. The location of the last dripper in the lateral must be at half of normal spacing of dripper from the wall of the trench near the lateral (Fig. 20). It is very important to adhere to this norm of placing the last dripper

**Fig. 20** Pit for finding
maximum duration of
irrigation

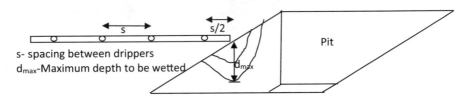

s- spacing between drippers
$d_{max}$-Maximum depth to be wetted

**Fig. 21** Configuration of dripper and pit

as shown in Fig. 21. If erroneously the dripper location is brought near the wall of
the trench, error involved would be very high. The irrigation system must be allowed
to operate, the depth of wetting of soil must be noted for every half an hour, and
the experiment must be ended up by noting down the time taken for wetting the
maximum expected rooting depth. Maximum duration of irrigation is the time taken
to wet the maximum rooting depth.

## Irrigation for Germination

First few irrigations for both drip and sprinklers must be significantly higher. Irrigation must drench the soil to a higher level. Only then germination will be better.
Even if nursery plants are planted, application of higher irrigation amount will be
better during initial stages. Evapotranspiration (ET) need not be considered while
irrigating for the first two or three irrigations.

## *Irrigation Scheduling*

After initial crop establishment is achieved, the irrigation scheduling based on crop ET may be initiated. Methods for estimating ET are presented as a chapter, namely evapotranspiration for microirrigated crops. Nowadays, NETAFIM application software provides reference crop ET for any place on daily basis with free of cost through Internet.

To start with, a suitable irrigation interval must be decided by the farmer. There is a general feeling existing among the farmers that for drip irrigation, the irrigation interval must be daily. Many a times, adoption of daily drip irrigation has been a cause for the decreased yield because of the possibility of over irrigation and subsequent leaching of nutrients below the root zone.

We will see few guidelines for deciding the irrigation interval. For the following data, we will work out the irrigation interval during the initial stage and also for the developed stage.

**Data**

| | |
|---|---|
| Crop | $=$ Sugarcane |
| Crop factor $(K_c)$ | $= 0.4$ |
| Dripper discharge $(q)$ | $= 4$ l/h |
| Dripper spacing $(s_p)$ | $= 60$ cm |
| Lateral spacing $(s_l)$ | $= 150$ cm |
| Minimum irrigation duration (t) found by field experimentation | $= 1$ h |
| Average reference crop evapotranspiration during initial stage $(ET_o)$ | $= 0.4$ cm/d |

**Solution**

$$\text{Minimum depth of water applied} = \frac{q(l/h) \times t(h)}{S_l \times S_p \, \text{cm}^2} \times 1000 \, \text{cm}^3/l$$
$$= \frac{4 \times 1 \times 1000}{150 \times 60}$$
$$= 0.44 \, \text{cm}$$

$$\text{Minimum irrigation interval} = \frac{\text{Minimum depth of water applied (cm)}}{K_c ET_o \, (\text{cm/d})}$$
$$= \frac{0.44}{0.4 \times 0.4}$$
$$= 2.78 \, \text{days}$$

Hence, minimum irrigation interval of 2 days can be adopted. It means that irrigation interval below 2 days would be a wasteful affair.

As crop grows, the depth of soil wetted must be increased because of root growth. Let us assume that depth of root zone gets increased to 40 cm and the duration of irrigation needed to wet 40 cm depth be 4 h. Let the average reference crop evapotranspiration during this period be 0.6 cm/d and the crop factor ($K_c$) be 1.25.

Now, we will estimate the irrigation interval for the developed stage:

Following data does not change: $q = 4$ l/h, $S_1 = 150$ cm and $S_p = 60$ cm

Depth of water applied $= \frac{q(l/h) \times t(h)}{S_l \times S_p \, \text{cm}^2} \times 1000 \, \text{cm}^3/l = \frac{4 \times 4 \times 1000}{150 \times 60} = 1.78$ cm

Irrigation interval $= \frac{\text{Depth of water applied (cm)}}{K_c \text{ETo(cm/d)}} = \frac{1.78}{1.25 \times 0.6} = 2.37$ days

The irrigation interval adopted may be 2 or 3 days. It must be kept in mind that all these calculations are based on our calculations. Practically, the farmer must look at the plant symptoms, and if the plant shows wilt symptoms, the irrigation interval must be reduced.

Let us assume that the farmer decides to adopt 2 days irrigation interval. For every irrigation the farmer need not operate the irrigation system for 4 h.

When irrigation is done, the time duration of irrigation may be adjusted based on the real-time $ET_o$ values obtained from Internet or by necessary computation from meteorological data. Application efficiency for drip irrigation is usually taken as 0.9 and for sprinkler irrigation 0.8. Let us assume that the sum of $ET_o$ values for 2 days after a preceding irrigation is 1.6 cm.

Then the duration of forthcoming irrigation can be calculated as follows:

Depth of water due to evapotranspiration $= K_c \, ET_o$.

$= 1.25 \times 1.6 = 2$ cm.

$$\text{Rate of irrigation} = \frac{q\left(\frac{l}{h}\right)}{S_l \times S_p(\text{cm}^2)} 1000 \, \text{cm}^3/l$$

$$= \frac{4\left(\frac{l}{h}\right)}{150 \times 60 \, (\text{cm}^2)} 1000 \, \text{cm}^3/l = 4.55 \, \text{h}$$

$$= 0.44 \, \text{cm/h}$$

$$\text{Duration of irrigation} = \frac{\text{Depth of water evapotranspirated}}{\text{Rate of irrigation}}$$

$$= \frac{2 \, \text{cm}}{0.44 \, \text{cm/h}}$$

$$= 4.55 \text{h}$$

Note that the irrigation duration has got increased from 4 to 4.55 h. Similarly if the sum of two days $ET_o$ becomes a reduced quantity, accordingly, the irrigation duration can also be reduced.

Adjusting the duration of irrigation every time using the $ET_o$ data has been found to be very efficient in managing irrigation water and also nutrients in root zone.

When irrigation is done in saline lands the quantity of irrigation water supplied must be larger than the ET of crop. The irrigation scheduling calculations considering salinity also are similar to sprinkler irrigation, and it is presented in Chapter number 6.

## Summary

Maintenance of drip irrigation system must be based on a water test. Emitter plugging due to salt deposition is treated by acid application. Emitter plugging due to algae growth has been so far controlled by chlorination. Hydrogen peroxide is recommended for chlorination nowadays as it is more environment-friendly. Real-time irrigation scheduling can be improved by estimating evapotranspiration using Internet application. Optimal management of water and nutrients in root zone can be done well with a simple method of digging a very small pit at the end of any one lateral and observing the wetting front as well as root development.

## Assessment

### Part A

1. The presence of _____ can be easily known if the water has rotten egg smell.

   a. **hydrogen sulphide**
   b. iron
   c. carbonate
   d. bicarbonate

2. Water absorbs some _____ from the air, but larger quantities are absorbed from decaying organic matter as water passes through the soil.

   a. **$CO_2$**
   b. iron
   c. aluminium
   d. oxygen.

3. In the wells with _____ pumps, since the water is lifted by supplying air to the well water, the amount of $CO_2$ dissolved is very high. Therefore, the amount of calcium bicarbonate dissolved from the subsurface formations is also high.

   a. **airlift**
   b. jet
   c. submersible
   d. turbine

4. Iron is usually dissolved in water in the form of ferrous oxides. When exposed to air, soluble ferrous bicarbonate oxidizes to the ——— ferric hydroids and precipitates.

    a. **insoluble**
    b. soluble
    c. white
    d. Mono

5. In water with calcium, magnesium and bicarbonates, if phosphate-containing fertilizers are mixed, calcium and magnesium phosphates are formed. By_____, this effect can be reduced.

    a. **reducing pH**
    b. increasing pH
    c. neutralizing
    d. none of these

6. Which among the following is more powerful oxidizing agent?

    a. **Hypochlorous acid**
    b. Hypochlorite
    c. Water
    d. Urea solution

## Part B

1. How much free chlorine is available in sodium hypochlorite?
2. What is chlorination? What are the chemical compounds used for chlorination?
3. What do you understand from the term free chlorine? What is the relationship between pH of water and free chlorine?
4. Discuss briefly about three chlorination methods, namely continuous treatment and intermittent treatment.
5. Discuss briefly about the superchlorination method.
6. What are the chemicals used for acid treatment? Discuss about acid treatment for prevention of carbonate scales. Discuss about acid treatment for clogged systems?

## Problem

1. Develop a suitable irrigation schedule for the following data:

| Crop | = Banana |
|---|---|
| Crop factor ($K_c$) | = 1.2 |
| Dripper discharge ($q$) | = 4 l/h |
| Dripper spacing ($s_p$) | = 30 cm |
| Lateral spacing ($s_l$) | = 180 cm |

(continued)

(continued)

| Minimum irrigation duration ($t$) found by field experimentation | = 2 h |
| Reference crop evapotranspiration ($ET_o$) | = 0.6 cm/d |

# References and Further Reading

1. Maintenance Guide for Micro irrigation systems in the Southern Region, By the Irrigation Water Management team, A partnership of USDA-CSREES and Land Grant Colleges and Universities, USA.
2. Lili, Z., Peiling Y., Wengang Z., Yunkai L., Yu L. (2021). Effects of water salinity on emitter clogging in surface drip irrigation systems. *Irrigation Science* 39(1). https://doi.org/10.1007/s00271-020-00690-3
3. Dupree, H. K., & Huner, J. V. (1984). *The status of warm water fish farming and progress in fish farming research* (pp. 165–176). Washington, D.C., U.S. Fish and Wildlife Service.

# Evapotranspiration for Microirrigated Crops

## Introduction

Water applied to the soil through rain or by irrigation is again sent back to atmosphere by two processes, namely evaporation and transpiration. The water vapour escaping from soil surface is termed as evaporation. Water extracted by plants from soil also escapes to the atmosphere from plant leaves which is called as transpiration. Estimation of evaporation and transpiration separately is very difficult, and it is conventional to combine both evaporation and transpiration together as evapotranspiration (ET). However, calculation methods have been developed to calculate evaporative and transpirative components which are very much useful for precise irrigation methods like sprinkler and drip irrigation (Fig. 1).

## Reference Crop Evapotranspiration

A term most commonly used in evapotranspiration (ET) literature is reference crop ET. Previously another term, namely potential ET, was used. Nowadays, reference crop ET is recommended to be used instead of potential ET.

Reference crop ET is the ET occurring from a hypothetical grass reference with a crop height of 0.12 m, surface resistance of 70 s m$^{-1}$ and an albedo of 0.23 m. The reference surface closely resembles an extensive surface of green, well-watered grass of uniform height, actively growing and completely shading the ground. The surface resistance of 70 s m$^{-1}$ corresponds to a moderately dry soil surface resulting from an irrigation of weekly frequency.

The ET demand of atmosphere is due to the factors such as solar radiation, extraterrestrial radiation, soil heat flux and aerodynamic resistance of the soil.

A method, namely Penman-Monteith approach, is the standard method for calculating the reference crop ET. Details of the method are not discussed here, and the reader is advised to refer to Allen et al. [1]. Standard free to download software are

© The Author(s), under exclusive license to Springer Nature Singapore Pte Ltd. 2023
V. Ravikumar, *Sprinkler and Drip Irrigation*,
https://doi.org/10.1007/978-981-19-2775-1_15

**Fig. 1** Evapotranspiration
(Wikipedia.org)

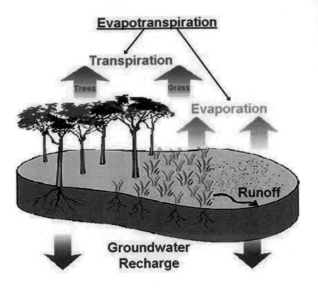

available in public domain for calculating reference crop ET (e.g. http://www.fao.
org/land-water/databases-and-software/eto-calculator/en/).

The present state-of-the-art remote sensing technology and cell phone technology
are so advanced and so it is now possible to get the reference crop evapotranspiration
data for any location easily. This facility is made available by a renowned irrigation
firm, namely Netafim. One can download this software application from Netafim
website. Working with this application is very simple. The software application itself
would find the user's latitude and longitude using global positioning system (GPS) in
the cell phone. The user must just confirm his location on the Google Map by clicking
on the Google Map corresponding to the user's field location. The application would
provide actual $ET_o$ data for all the previous days and forecasted $ET_o$ for one week
(Fig. 2).

Estimation of evaporation from a pool of water of a standard pan is one easy way
of obtaining data for reference crop ET. Data from pan evaporimeter continues to be
recommended in India due to its simplicity. Some of the reasons that devalue the data
obtained from the pan evaporimeter are as follows: pan evaporation occurs during
night also, whereas plant transpiration becomes near zero at night. Painting of the
metal and water level in the metal affect the quality of data significantly.

Reference crop ET is obtained by multiplying the pan evaporation depth ($E_p$) by
a factor known as pan coefficient ($K_p$).

$$ET_o = K_p \times E_p \qquad (1)$$

The pan coefficient for Class A pan based on different siting and local climatic
conditions is provided in Table 1 (Fig. 3).

**Fig. 2** Screen shot of
Netafim application software
for knowing reference crop
ET

# ETO Yesterday

**Coimbatore, India**
(Current Location)

3 mm

## Weather Today

5 mm
**RAINFALL**

2 m/sec
**WIND SPEED**

82 %
**HUMIDITY**

29 °C 22 °C
**TEMPERATURE**

## Evapotranspiration Estimation by Climatological Data

Two methods, namely single crop coefficient approach and dual crop coefficient, are very much useful in estimating ET.

### *Single Crop Coefficient Approach*

In the single crop coefficient approach, ET of a crop ($ET_c$) is estimated using following equation:

$$ET_c = K_c \times ET_o \tag{2}$$

where $K_c$ is called as crop coefficient and $ET_o$ is reference crop ET. Most of the effects of the various weather conditions are incorporated into the $ET_o$ estimate. Therefore, as $ET_o$ represents an index of climatic demand, $K_c$ varies predominately with the specific crop characteristics and only to a limited extent with climate. This enables the transfer of standard values for $K_c$ between locations and between climates. This has been a primary reason for the global acceptance and usefulness of the crop coefficient approach and the $K_c$ factors developed in past studies [1].

For sprinkler irrigation method, where there is no partial wetting of the land, the single crop coefficient approach is sufficient. But accuracy of determination of

**Table 1** Pan coefficients for Class A pan for different pan siting and environment conditions [2]

| Class A pan | Case A: pan placed in short green cropped | | | | Case B: pan placed in dry fallow area | | | |
|---|---|---|---|---|---|---|---|---|
| RH mean (%) | Windward side distance of green crop (m) | Low < 40 | Medium 40–70 | High > 70 | Windward side distance of dry fallow (m) | Low < 40 | Medium 40–70 | High > 70 |
| Wind speed (m s$^{-1}$) | | | | | | | | |
| Light < 2 | 1 | 55 | 65 | 75 | 1 | 7 | 8 | 85 |
| | 10 | 65 | 75 | 85 | 10 | 6 | 7 | 8 |
| | 100 | 7 | 8 | 85 | 100 | 55 | 65 | 75 |
| | 1000 | 75 | 85 | 85 | 1000 | 5 | 6 | 7 |
| Moderate 2–5 | 1 | 5 | 6 | 65 | 1 | 65 | 75 | 8 |
| | 10 | 6 | 7 | 75 | 10 | 55 | 65 | 7 |
| | 100 | 65 | 75 | 8 | 100 | 5 | 6 | 65 |
| | 1000 | 7 | 8 | 8 | 1000 | 45 | 55 | 6 |
| Strong 5–8 | 1 | 45 | 5 | 6 | 1 | 6 | 65 | 7 |
| | 10 | 55 | 6 | 65 | 10 | 5 | 55 | 65 |
| | 100 | 6 | 65 | 7 | 100 | 45 | 5 | 6 |
| | 1000 | 65 | 7 | 75 | 1000 | 4 | 45 | 55 |
| Very strong > 8 | 1 | 4 | 45 | 5 | 1 | 5 | 6 | 65 |
| | 10 | 45 | 55 | 6 | 10 | 45 | 5 | 55 |
| | 100 | 5 | 6 | 65 | 100 | 4 | 45 | 5 |
| | 1000 | 55 | 6 | 65 | 1000 | 35 | 4 | 45 |

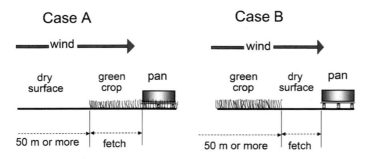

**Fig. 3** Two cases of evaporation pan siting and their environment [1]

ET gets improved when we correct the $K_c$ values depending on the frequency of irrigation.

## Single Crop Coefficient Approach

The crop coefficients for different crop growth stages would vary. Table 2 indicates the crop coefficient values for different growth stages. The $K_c$ values of Table 2 are typical values expected for subhumid climate with average daytime minimum relative humidity 45% and having calm to moderate wind speeds averaging 2 m/s.

Length of each stage of crop development is subjective, and initial period is the period at which the transpiration component is minimal and soil evaporative component is significant. Indicative data on the length of crop stages for different crop is available in Allen et al. [1].

The initial stage runs from planting date to the date at which the plant cover reaches 10%. The development stage runs from 10% plant cover to the full foliage development or initiation of flowering. The mid-season stage runs until the start of maturity. Yellowing of leaves starts to occur when the maturity starts to set in.

It should be observed that with only three values of crop coefficients, how four stages of crop growth are represented (Fig. 4). During the development stage, the crop coefficient is assumed to increase linearly from $K_{cini}$ to $K_{cmid}$. Similarly, during late season, the crop coefficient is assumed to decrease linearly from $K_{cmid}$ to $K_{cend}$.

$K_{cini}$ primarily depends on evaporation from soil, and it does not depend much on the crop. Evaporation from soil depends on soil type, frequency of wetting and fraction of area wetted.

Following formula is used to convert the depth of water application considering the entire area of the field ($I$) to the depth of wetting in the wetted area alone ($I_w$).

$$I_W = \frac{I}{f_W} \tag{3}$$

where $f_w$ is fraction of the area wetted (Fig. 5).

**Table 2** Crop coefficients of selected vegetables, fruits and other crops

| S. No. | Name of crops and fruits | Crop coefficient | | |
|---|---|---|---|---|
| | | Kc initial | Kc middle | Kc maturity |
| | *Vegetables crop* | | | |
| 1 | Broccoli | 0.7 | 1.05 | 0.95 |
| 2 | Cabbage | 0.7 | 1.05 | 0.95 |
| 3 | Carrots | 0.7 | 1.05 | 0.95 |
| 4 | Cauliflower | 0.7 | 1.05 | 0.95 |
| 5 | Garlic | 0.7 | 1.00 | 0.70 |
| 6 | Lettuce | 0.7 | 1.00 | 0.95 |
| 7 | a. Onion (dry) | 0.7 | 1.05 | 0.75 |
| | b. Onion (green) | 0.7 | 1.00 | 1.00 |
| | c. Onion (seed) | 0.7 | 1.05 | 0.80 |
| 8 | Spinach | 0.7 | 1.00 | 0.95 |
| 9 | Radish | 0.7 | 0.90 | 0.85 |
| 10 | Brinjal | 0.6 | 1.05 | 0.90 |
| 11 | Tomato | 0.6 | 1.052 | 0.70–0.90 |
| 12 | Cucumber | | | |
| | Fresh market | 0.60 | 1.002 | 0.75 |
| | Machine harvest | 0.5 | 1.00 | 0.90 |
| 13 | Pumpkin | 0.5 | 1.00 | 0.80 |
| 14 | Winter squash | 0.5 | 1.00 | 0.80 |
| 15 | Squash | 0.5 | 0.95 | 0.75 |
| 16 | Sweet melons | 0.5 | 1.05 | 0.75 |
| 17 | Beet | 0.5 | 1.05 | 0.95 |
| 18 | Potato | 0.5 | 1.15 | 0.754 |
| 19 | Sweet potato | 0.5 | 1.15 | 0.65 |
| 20 | Sugar beet | 0.35 | 1.20 | 0.705 |
| 21 | Beans | | | |
| | Green | 0.5 | 1.052 | 0.90 |
| | Dry | 0.4 | 1.152 | 0.35 |
| | *Fruits crop* | | | |
| 22 | Strawberries | 0.4 | 0.85 | 0.75 |
| 23 | Banana | | | |
| | First year | 0.50 | 1.10 | 1.00 |
| | Second year | 1.00 | 1.20 | 1.10 |
| 24 | Coffee | | | |

(continued)

**Table 2** (continued)

| S. No. | Name of crops and fruits | Crop coefficient | | |
|--------|--------------------------|------------|-----------|-------------|
| | | Kc initial | Kc middle | Kc maturity |
| | – Bare ground cover | 0.9 | 0.95 | 0.95 |
| | – With weeds | 1.05 | 1.1 | 1.1 |
| 25 | Date palm | 0.90 | 0.95 | 0.95 |
| 26 | Pineapple | | | |
| | Bare soil | 0.50 | 0.30 | 0.30 |
| | With grass cover | 0.50 | 0.50 | 0.50 |
| 27 | Berried (bushes) | 0.30 | 1.05 | 0.50 |
| 28 | Grapes | | | |
| | Raining | 0.30 | 0.85 | 0.45 |
| | Wine | 0.30 | 0.70 | 0.45 |
| 29 | Apples, cherries, pears | | | |
| | No ground cover, killing frost | 0.45 | 0.95 | 0.7018 |
| | No ground cover, no frosts | 0.6 | 0.95 | 0.7518 |
| | Active ground cover, killing frosts | 0.5 | 1.2 | 0.9518 |
| | Active ground over, no frosts | 0.8 | 1.2 | 0.8518 |
| 30 | Apricots, peaches | | | |
| | No ground cover, killing frost | 0.45 | 0.90 | 0.6518 |
| | No ground cover, no frost | 0.55 | 0.90 | 0.6518 |
| | Active ground cover, killing frosts | 0.50 | 1.15 | 0.9018 |
| | Active ground over, no frosts | 0.80 | 1.15 | 0.8518 |
| 31 | Citrus (no ground cover) | | | |
| | 70% canopy | 0.70 | 0.65 | 0.70 |
| | 50% canopy | 0.65 | 0.60 | 0.65 |
| | 20% canopy | 0.50 | 0.45 | 0.55 |
| 32 | Citrus (with active ground cover) | | | |
| | 70% canopy | 0.75 | 0.70 | 0.75 |
| | 50% canopy | 0.80 | 0.80 | 0.80 |
| | 20% canopy | 0.85 | 0.85 | 0.85 |
| 33 | Watermelon | 0.4 | 1.00 | 0.75 |
| 34 | Mushmelon | 0.5 | 0.85 | 0.60 |
| 35 | Kiwi | 0.4 | 1.05 | 1.05 |
| | *Other crops* | | | |
| 36 | Pulses | 0.4 | 1.152 | 0.35 |

(continued)

**Table 2** (continued)

| S. No. | Name of crops and fruits | Crop coefficient | | |
|---|---|---|---|---|
| | | Kc initial | Kc middle | Kc maturity |
| 37 | Soybeans | 0.4 | 1.15 | 0.50 |
| 38 | Cotton | 0.15 | 1.10–1.15 | 0.50–0.40 |
| 39 | Sugarcane | 0.4 | 1.25 | 0.75 |
| 40 | Groundnut (peanut) | 0.4 | 1.15 | 0.6 |
| 41 | Chick pea | 0.4 | 1.00 | 0.35 |
| 42 | Green gram and cowpea | 0.4 | 1.05 | 0.6–0.356 |
| 43 | Lentil | 0.4 | 0.3 | 0.5 |
| 44 | Peas | | | |
| | – Fresh | 0.5 | 1.152 | 1.1 |
| | – Dry/seed | 0.5 | 1.15 | 0.3 |
| 45 | Mint | 0.6 | 1.15 | 1.1 |
| 46 | Safflower | 0.35 | 1.0–1.159 | 0.25 |
| 47 | Sesame | 0.35 | 1.1 | 0.25 |
| 48 | Sunflower | 0.35 | 1.0–1.159 | 0.35 |

Adapted from Allen et al. [1]

**Fig. 4** Crop coefficient curve [1]

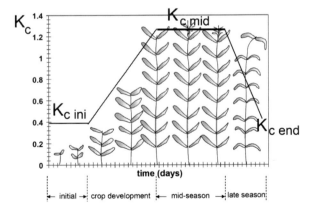

If the depth of application ($I_w$) is below 10 mm, Fig. 6 must be used for finding out the value of $K_{cini}$. If the $I_w$ values are above 40 mm, the $K_{cini}$ must be found out using Fig. 7a or Fig. 7 which depends on type of soil. If the depth of application is between 10 and 40 mm, following interpolation formula may be used to get $K_{cini}$.

$$K_{cini} = K_{cini\ (Fig.\ 5)} + \frac{I_W - 10}{40 - 10}\left(K_{cini\ (Fig.\ 6)} - K_{cini\ (Fig.\ 5)}\right) \qquad (4)$$

**Fig. 5** Partial wetting by irrigation [1]

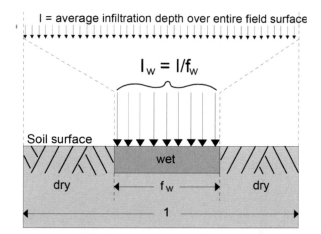

**Fig. 6** $K_{cini}$ related to interval between irrigations for wetting from 3 to 10 mm for all types of soil [1]

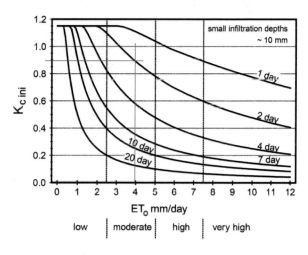

**Fig. 7 a** $K_{cini}$ related to interval between irrigations for wetting above 40 mm for coarse textured soil [1]. **b** $K_{cini}$ related to interval between irrigations for wetting above 40 mm for medium and fine textured soil [1]

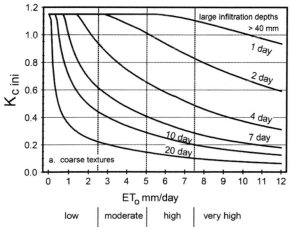

**Fig. 7** (continued)

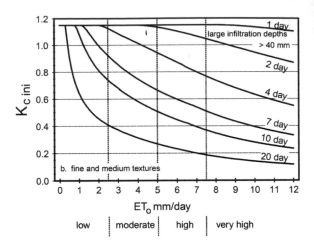

b. fine and medium textures

The $K_{cini}$ values thus obtained should also be adjusted for the partial wetting by multiplying the $K_{cini}$ by the fraction of area wetted. It must be noted that $K_{cini}$ values given in Table 2 are indicative values for conventional irrigation methods. It is suggested to use the preceding correction procedures for sprinkler and drip irrigation.

Normally, $K_{cmid}$ and $K_{cend}$ values do not need any adjustment for the changes in frequency of irrigation and the fraction of area wetted. But $K_{cmid}$ and $K_{cend}$ values are to be adjusted for relative humidity and wind velocity using the following formula:

$$\text{Adj.}\,K_{cmid} \text{ or } K_{cend} = K_{cmid} \text{or} K_{cend} + (0.04(u_2 - 2) - 0.004(\text{RH}_{min} - 45))\left(\frac{h}{3}\right)^{0.33}$$

$$(5)$$

where $u_2$ is the wind velocity (m/s).

RH$_{min}$ is the minimum relative humidity (%).

'$h$' is the mean plant height (m).

After adjusting the $K_c$ values in Table 2, $K_c$ values for the entire crop season have to be constructed as shown in Fig. 4. From these $K_c$ values, daily ET can be found out.

### Exercise

Sugarcane crop is drip irrigated with a wetted fraction of 0.5 and an irrigation depth of 7.5 mm with 2 days of irrigation interval. Reference crop ET is 4 mm. Find out the $K_{cini}$ for initial stage and estimate the ET.

### Solution

From Fig. 6, for ET$_o$ of 4 mm/day and irrigation interval of 2 days, value of $K_{cini}$ is 0.9.

Since the wetted fraction is 0.5, the adjusted $K_{cini} = K_{cini} \times f_w = 0.9 \times 0.5 = 0.45$.

$$ET = K_{cini} \times ET_o = 0.45 \times 4 = 1.8 \text{ mm/d}.$$

**Exercise**

The $K_{cmid}$ value from Table 2 is 1.25. Adjust the $K_{cmid}$ for the local climate crop specific conditions for the following data. Wind velocity is 4 m/s; minimum relative humidity is 50%, and height of sugarcane crop is 1.5 m.

**Solution**

$$\text{Adjusted} K_{cmid} = K_{cmid} + (0.04(u_2 - 2) - 0.004(\text{RH}_{min} - 45))\left(\frac{h}{3}\right)^{0.33}$$

$$\text{Adjusted} K_{cmid} = 1.25 + (0.04(4 - 2) - 0.004(50 - 45))(1.5)^{0.33} = 1.30$$

There is a common belief that ET gets decreased for drip irrigated crop when compared to conventional flood irrigation methods. Hence, there is a tendency to underestimate the ET by multiplying with wetted area fraction especially in India [3]. But research results have shown that for fully grown-up crops, the ET is not lower compared to conventional irrigation methods. It is sometimes even higher due to increased occurrence of evaporation as surface of soil remains wetted always [1].

Therefore, it is recommended that when single crop coefficient approach is used for drip irrigated crops, $K_c$ values need not be adjusted at all for the crop stages other than initial stages. By doing so, only in crop development stage, there is some possibility of error coming in. Best way is to adopt dual crop coefficient approach presented in the succeeding sections.

## Dual Crop Coefficient Approach

In the dual crop coefficient approach, ET of a crop ($ET_c$) is estimated using the following equation:

$$ET_c = (K_{cb} + K_e) \times ET_o \qquad (6)$$

where $K_{cb}$ is called as basal crop coefficient and $K_e$ is called as evaporation coefficient. $K_{cb}$ is defined as a ratio of $ET_C$ to $ET_o$ when soil evaporation is zero and soil surface is dry and transpiration component is fully met by sufficient water available in root zone below soil surface. $K_{cb}$ component also includes the diffusive evaporative water movement from the root zone below the surface. $K_{cb}$ values can be estimated from the $K_c$ values in Table 2 by adopting the conversion guidelines available in Table 3. The $K_{cb}$ values thus obtained may also be adjusted for climate using Eq. 5.

The soil evaporation coefficient ($K_e$) describes evaporation component from soil surface only. If a soil is wet following a rain or irrigation, the value of $K_e$ may be

**Table 3** General guidelines to derive $K_{cb}$ from the $K_c$ values listed in Table 2 [1]

| Growth stage | Ground condition, irrigation and cultural practices | $K_{cb}$ |
|---|---|---|
| Initial | Annual crop—(nearly) bare soil surface<br>Perennial crop—(nearly) bare soil surface | 0.15<br>0.15–0.20 |
| | Grasses, brush and trees—killing frost | 0.30–0.40 |
| | Perennial crop—some ground cover or leaf cover infrequently irrigated (olives, palm trees, fruit trees, …) | $K_{cini}$ (Table 2)—0.1 |
| | Frequently irrigated (garden-type vegetables, …) | $K_{cini}$ (Table 2)—0.2 |
| Mid-season | Ground cover more than 80%<br>Ground cover less than 80% (vegetables) | $K_{cmid}$ (Table 2)—0.05<br>$K_{cmid}$ (Table 2)—0.10 |
| At the end of season | Infrequently irrigated or wetted during late season<br>Frequently irrigated or wetted during late season | $K_{cend}$—0.05<br>$K_{cend}$—0.1 |

large. The sum of $K_e$ and $K_{cb}$ can never exceed a maximum value ($K_{cmax}$) determined by the energy available for ET. As a soil surface becomes drier, $K_e$ becomes smaller and falls to zero when no water is left for evaporation. Estimation of $K_e$ requires a daily water balance computation for calculation of soil water content remaining in upper top soil.

When the top soil dries out, less water is available for evaporation from the surface soil, and a reduction in evaporation begins to occur in proportion to the amount of water remaining in the surface soil layer. Following equation is used for finding the coefficient for estimation of soil evaporation ($K_e$).

$$K_e = \text{Min}(K_r(K_{cmax} - K_{cb}), f_{ew} \times K_{cmax}) \tag{7}$$

where $K_{cmax}$ is the maximum $K_c$ value limited by the energy available at the surface, $K_r$ is a dimensionless reduction coefficient depending on the cumulative depth of water depleted from the top soil, and $f_{ew}$ is fraction of the soil which is both exposed as well as wetted.

In order to use the preceding equation, values of $f_{ew}$ and also $K_{cmax}$ on each day should be found out. The value of $f_{ew}$ can be found out using following equation for conventional and sprinkler irrigation (Fig. 8).

$$f_{ew} = \text{Min}((1 - f_c), f_w) \tag{8}$$

where $f_c$ is fraction of area covered by vegetation and $f_w$ is fraction of area wetted. For drip irrigation, following equation is used:

$$f_{ew} = f_w \left(1 - \frac{2f_c}{3}\right) \tag{9}$$

It is better to estimate the values of $f_{ew}$, $f_w$ and $f_c$ in the field, instead of using Eqs. 8 and 9.

**Fig. 8** Fraction of foliage cover, wetted area and exposed and wetted fraction [1]

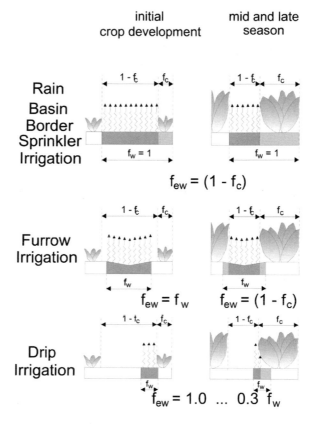

For high-frequency irrigation, $K_{cmax}$ can be found out using the following equation:

$$K_{cmax} = \text{Max}\left(\begin{cases} 1.1 + [0.04(u_2 - 2) - 0.004(RH_{min} - 45)]\left(\dfrac{h}{3}\right)^{0.33} \end{cases}, \\ \{K_{cb} + 0.05\} \right) \quad (10)$$

The soil evaporation from the exposed soil can be assumed to take place in two stages (Fig. 9). One is energy limiting stage, and another is falling rate stage. When soil is wet, evaporation reduction coefficient is one. When water content in upper soil layer is limiting, $K_r$ decreases and becomes zero when total amount of water that can be evaporated from the top soil is depleted completely. In order to find out the value of $K_r$, first the total evaporable water in mm (TEW) from the soil surface is found out using the following equation:

$$TEW = 1000(\theta_{fc} - 0.5\theta_{wp})Z_e \quad (11)$$

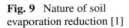

**Fig. 9** Nature of soil
evaporation reduction [1]

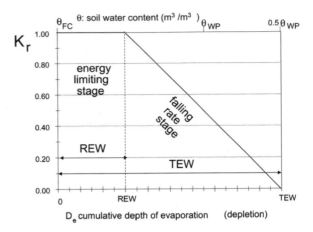

where $\theta_{\text{fc}}$ is soil water content at the field capacity (volumetric), $\theta_{\text{fp}}$ is the soil water content at wilting point (volumetric), and $Z_e$ is the depth of the soil layer that is subject to evaporation. Usually $Z_e$ ranges from 0.10 m to 0.15 m (Table 4).

Following equation is used for finding the value of $K_r$:

$$K_r = \frac{\text{TEW} - D_{e,i-1}}{\text{TEW} - \text{REW}} \quad \text{for } D_{e,i-1} > \text{REW} \tag{12}$$

where $D_{e,\,i-1}$ is cumulative depth of evaporation from the soil surface layer till the previous day (mm) and REW is readily evaporable water which is the total depth of evaporation possible till the start of energy limiting stage.

**Table 4** Typical soil water characteristics of different soil types [1]

| Soil type (USA soil texture classification) | Soil water characteristics | | | Evaporation parameters | |
|---|---|---|---|---|---|
| | $\theta_{FC}$ m$^3$/m$^3$ | $\theta$wp m$^3$/m$^3$ | $\theta_{FC} - \theta$wp m$^3$/m$^3$ | Amount of water that can be depleted by evaporation | |
| | | | | Stage 1 REW mm | Stages 1 and 2 TEW ($Z_e$ = 0.10 m) mm |
| Sand | 0.07–0.17 | 0.02–0.07 | 0.05–0.11 | 2–7 | 6–12 |
| Loamy sand | 0.11–0.19 | 0.03–0.10 | 0.06–0.12 | 4–8 | 9–14 |
| Sandy loam | 0.18–0.28 | 0.06–0.16 | 0.11–0.15 | 6–10 | 15–20 |
| Loam | 0.20–0.30 | 0.07–0.17 | 0.13–0.18 | 8–10 | 16–22 |
| Silt loam | 0.22–0.36 | 0.09–0.21 | 0.13–0.19 | 8–11 | 18–25 |
| Silt | 0.28–0.36 | 0.12–0.22 | 0.16–0.20 | 8–11 | 22–26 |
| Silt clay loam | 0.30–0.37 | 0.17–0.24 | 0.13–0.18 | 8–11 | 22–27 |
| Silty clay | 0.30–0.42 | 0.17–0.29 | 0.13–0.19 | 8–12 | 22–28 |
| Clay | 0.32–0.40 | 0.20–0.24 | 0.12–0.20 | 8–12 | 22–29 |

The water balance equation for the evaporating soil layer can be written as follows (Fig. 10):

$$D_{e,i} = D_{e,i-1} - P_i - \frac{I_i}{f_w} + \frac{E_i}{f_{ew}} + T_{ew} + DP_{e,i} \qquad (13)$$

where

$D_{e,i-1}$ cumulative depth of evaporation following complete wetting from the exposed and wetted fraction of the topsoil at the end of the day 'i-1' [mm];

$D_{e,i}$ cumulative depth of evaporation (depletion) following complete wetting at the end of the day 'i' [mm];

$P_i$ precipitation on day 'i' [mm];

$I_i$ irrigation depth on day 'i' that infiltrates in the soil [mm];

$E_i$ evaporation on day i (i.e. $Ei = K_eET_o$) [mm];

$T_{ew,I}$ depth of transpiration from the exposed and wetted fraction of the soil surface layer on day 'i' [mm], and it can be usually assumed as zero;

$DP_{e,i}$ deep percolation loss from the topsoil layer on day 'i' if soil water content exceeds field capacity [mm].

$f_w$ fraction of soil surface wetted by irrigation, $f_{ew}$ exposed and wetted soil fraction.

Following heavy rain or irrigation, the soil water content in the topsoil ($Z_e$ layer) might exceed field capacity. However, it is assumed that the soil water content is nearly at field capacity immediately following a complete wetting event, so that

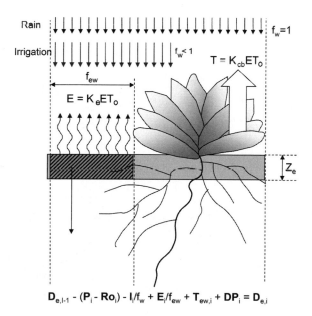

**Fig. 10** Water balance of the top soil layer [1]

$$D_{e,I-1} - (P_i - Ro_i) - I_i/f_w + E_i/f_{ew} + T_{ew,i} + DP_i = D_{e,i}$$

the depletion $D_{e,i}$ in Eq. 13 is zero. Following heavy rain or irrigation, downward drainage (percolation) of water from the topsoil layer is calculated as follows:

$$DP_{e,i} = P_i + \frac{I_i}{f_w} - D_{e,i-1} \geq 0 \qquad (14)$$

As long as the soil water content in the evaporation layer is below field capacity (i.e. $D_{e,i} > 0$), the soil will not drain and $DP_{e,i} = 0$.

### Exercise
Following is the data for sugarcane crop grown in sandy clay loam soil:

Field capacity—0.22.
Permanent wilting point—0.14.
Readily evaporable water—8 mm.
Total evaporable water—15 mm.
Fraction wetted—0.5
Depth of evaporating soil layer $Z_e$—0.10 m.
Find out the ET for the 50 days of data in Table 5.

### Solution
The relevant equations are substituted, and results are presented in Table 6. Detailed computations are also presented for 15th day for a complete understanding.

### Column 2

$$f_{ew} = f_w \left( 1 - \frac{2 f_c}{3} \right)$$

$$f_{ew} = 0.5 \left( 1 - \frac{2 \times 0.05}{3} \right) = 0.483$$

### Column 3

$I/f_w = 0/0.5 = 0.0$

### Column 4

First part of

$$K_{cmax} = 1.1 + [0.04(u_2 - 2) - 0.004(RH_{min} - 45)] \left( \frac{h}{3} \right)^{0.33}$$

$$= 1.1 + [0.04(3.47 - 2) - 0.004(41 - 45)] \left( \frac{0.09}{3} \right)^{0.33} = 1.22$$

Second part of $K_{cmax} = K_{cb} + 0.05 = 0.04 + 0.05 = 0.09$

$$K_{cmax} = \text{Maximum}(1.22, 0.09) = 1.22$$

**Table 5** Data for computation of $ET_0$

| Days after planting | Plant height (m) | Wind velocity (m/s) | Min. rel. hum. (%) | $P_i$ (mm) | $K_{cb}$ | $ET_0$ (mm) | Irrigation (mm) | $f_c$ | $(1 - f_c)$ |
|---|---|---|---|---|---|---|---|---|---|
| 1 | 0.00 | 3.14 | 54.00 | 0.0 | 0.00 | 4.4 | 7.50 | 0.00 | 1.00 |
| 2 | 0.00 | 4.08 | 60.00 | 3.5 | 0.00 | 4.9 | 0.00 | 0.00 | 1.00 |
| 3 | 0.00 | 3.94 | 62.00 | 14.0 | 0.00 | 3.9 | 0.00 | 0.00 | 1.00 |
| 4 | 0.00 | 2.86 | 66.00 | 12.0 | 0.00 | 3.8 | 0.00 | 0.00 | 1.00 |
| 5 | 0.00 | 1.92 | 57.00 | 18.0 | 0.00 | 3.2 | 0.00 | 0.00 | 1.00 |
| 6 | 0.00 | 1.22 | 54.00 | 15.6 | 0.00 | 3.9 | 0.00 | 0.00 | 1.00 |
| 7 | 0.00 | 1.50 | 51.00 | 6.5 | 0.00 | 4.1 | 0.00 | 0.00 | 1.00 |
| 8 | 0.00 | 1.42 | 40.00 | 0.8 | 0.00 | 4.3 | 0.00 | 0.00 | 1.00 |
| 9 | 0.00 | 1.86 | 45.00 | 1.0 | 0.00 | 4.5 | 7.50 | 0.00 | 1.00 |
| 10 | 0.00 | 4.14 | 50.00 | 4.8 | 0.00 | 6.4 | 0.00 | 0.00 | 1.00 |
| 11 | 0.02 | 4.36 | 42.00 | 0.0 | 0.01 | 6.7 | 0.00 | 0.01 | 0.99 |
| 12 | 0.04 | 3.94 | 50.00 | 1.0 | 0.02 | 6.3 | 7.50 | 0.02 | 0.98 |
| 13 | 0.05 | 3.42 | 49.00 | 0.8 | 0.02 | 5.3 | 0.00 | 0.03 | 0.97 |
| 14 | 0.07 | 4.03 | 47.00 | 0.0 | 0.03 | 5.6 | 7.50 | 0.04 | 0.96 |
| **15** | **0.09** | **3.47** | **41.00** | **0.0** | **0.04** | **5.5** | **0.00** | **0.05** | **0.95** |
| 16 | 0.11 | 3.61 | 53.0 | 0.0 | 0.05 | 5.6 | 7.50 | 0.06 | 0.94 |
| 17 | 0.13 | 4.50 | 76.5 | 2.2 | 0.05 | 4.4 | 0.00 | 0.07 | 0.93 |
| 18 | 0.14 | 3.53 | 30.5 | 2.5 | 0.06 | 2.7 | 0.00 | 0.08 | 0.92 |
| 19 | 0.16 | 3.42 | 83.0 | 26.8 | 0.07 | 2.8 | 0.00 | 0.09 | 0.91 |
| 20 | 0.18 | 3.06 | 88.0 | 17.6 | 0.08 | 2.2 | 0.00 | 0.10 | 0.90 |
| 21 | 0.20 | 2.33 | 74.0 | 8.4 | 0.08 | 2.9 | 0.00 | 0.11 | 0.89 |

(continued)

**Table 5** (continued)

| Days after planting | Plant height (m) | Wind velocity (m/s) | Min. rel. hum. (%) | $P_i$ (mm) | $K_{cb}$ | $ET_0$ (mm) | Irrigation (mm) | $f_c$ | $(1 - f_c)$ |
|---|---|---|---|---|---|---|---|---|---|
| 22 | 0.22 | 2.22 | 69.5 | 1.0 | 0.09 | 3.1 | 0.00 | 0.12 | 0.88 |
| 23 | 0.23 | 3.36 | 70.5 | 0.8 | 0.10 | 4.0 | 7.50 | 0.13 | 0.87 |
| 24 | 0.25 | 4.39 | 67.0 | 0.4 | 0.11 | 5.0 | 0.00 | 0.14 | 0.86 |
| 25 | 0.27 | 4.67 | 57.0 | 0.0 | 0.11 | 6.2 | 7.50 | 0.15 | 0.85 |
| 26 | 0.29 | 3.83 | 64.5 | 0.0 | 0.12 | 5.5 | 0.00 | 0.16 | 0.84 |
| 27 | 0.31 | 2.72 | 70.0 | 0.0 | 0.13 | 4.8 | 7.50 | 0.17 | 0.83 |
| 28 | 0.32 | 1.36 | 71.5 | 0.0 | 0.14 | 3.9 | 0.00 | 0.18 | 0.82 |
| 29 | 0.34 | 1.39 | 69.0 | 0.0 | 0.14 | 4.2 | 0.00 | 0.19 | 0.81 |
| 30 | 0.36 | 1.33 | 71.5 | 0.0 | 0.15 | 4.1 | 7.50 | 0.20 | 0.80 |
| 31 | 0.38 | 3.83 | 64.5 | 0.0 | 0.16 | 5.5 | 0.00 | 0.21 | 0.79 |
| 32 | 0.40 | 2.97 | 70.5 | 0.0 | 0.17 | 4.7 | 0.00 | 0.22 | 0.78 |
| 33 | 0.41 | 1.44 | 76.0 | 0.0 | 0.19 | 3.5 | 0.00 | 0.23 | 0.77 |
| 34 | 0.43 | 1.22 | 70.5 | 0.0 | 0.20 | 3.2 | 0.00 | 0.24 | 0.76 |
| 35 | 0.45 | 1.97 | 71.5 | 0.0 | 0.21 | 3.7 | 7.50 | 0.25 | 1.00 |
| 36 | 0.47 | 2.72 | 58.0 | 0.0 | 0.22 | 5.2 | 0.00 | 0.26 | 1.00 |
| 37 | 0.49 | 2.11 | 68.5 | 0.0 | 0.23 | 4.7 | 0.00 | 0.27 | 1.00 |
| 38 | 0.51 | 2.08 | 62.5 | 0.0 | 0.24 | 5.1 | 0.00 | 0.28 | 1.00 |
| 39 | 0.53 | 3.92 | 57.0 | 0.0 | 0.26 | 5.7 | 0.00 | 0.29 | 1.00 |
| 40 | 0.55 | 2.89 | 68.5 | 0.0 | 0.27 | 4.7 | 0.00 | 0.30 | 1.00 |
| 41 | 0.57 | 1.92 | 66.5 | 0.0 | 0.28 | 4.7 | 0.00 | 0.31 | 1.00 |
| 42 | 0.59 | 1.94 | 68.5 | 0.0 | 0.29 | 4.8 | 0.00 | 0.32 | 1.00 |

(continued)

**Table 5** (continued)

| Days after planting | Plant height (m) | Wind velocity (m/s) | Min. rel. hum. (%) | $P_i$ (mm) | $K_{cb}$ | $ET_0$ (mm) | Irrigation (mm) | $f_c$ | $(1 - f_c)$ |
|---|---|---|---|---|---|---|---|---|---|
| 43 | 0.62 | 1.78 | 62.0 | 0.0 | 0.30 | 4.6 | 7.50 | 0.33 | 1.00 |
| 44 | 0.64 | 1.67 | 66.5 | 0.0 | 0.31 | 4.7 | 0.00 | 0.34 | 1.00 |
| 45 | 0.66 | 2.03 | 68.5 | 0.0 | 0.33 | 4.7 | 0.00 | 0.35 | 0.99 |
| 46 | 0.68 | 1.25 | 71.0 | 1.0 | 0.34 | 3.3 | 7.50 | 0.36 | 0.98 |
| 47 | 0.70 | 1.58 | 87.5 | 0.2 | 0.35 | 3.2 | 0.00 | 0.37 | 0.97 |
| 48 | 0.72 | 1.17 | 76.0 | 12.0 | 0.36 | 3.5 | 7.50 | 0.38 | 0.96 |
| 49 | 0.74 | 1.39 | 80.5 | 1.6 | 0.37 | 3.7 | 0.00 | 0.39 | 0.95 |
| 50 | 0.76 | 0.86 | 73.5 | 0.5 | 0.38 | 2.4 | 7.50 | 0.40 | 0.94 |

**Table 6** Estimation of ET for data in Table 5

| (1) | (2) | (3) | (4) | (5) | (6) | (7) | (8) | (9) | (10) | (11) | (12) | (13) |
|---|---|---|---|---|---|---|---|---|---|---|---|---|
| Days | $f_{ew}$ | I/ $f_w$ (mm) | $K_{cmax}$ | $D_{e,i}$ Start (mm) | $K_r$ | $K_e$ | $K_e \times$ ETo/ $f_{ew}$ (mm) | $DPe_i$ (mm) | $D_{e,i}$ End (mm) | E (mm) | $K_c$ | ET (mm) |
| 1 | 0.50 | 15.00 | 1.11 | 0.00 | 1.00 | 0.55 | 4.87 | 15.00 | 4.87 | 2.44 | 0.55 | 2.44 |
| 2 | 0.50 | 0.00 | 1.12 | 1.37 | 1.00 | 0.56 | 5.49 | 0.00 | 6.86 | 2.75 | 0.56 | 2.75 |
| 3 | 0.50 | 0.00 | 1.11 | 0.00 | 1.00 | 0.55 | 4.33 | 7.14 | 4.33 | 2.16 | 0.55 | 2.16 |
| 4 | 0.50 | 0.00 | 1.05 | 0.00 | 1.00 | 0.52 | 3.97 | 7.67 | 3.97 | 1.99 | 0.52 | 1.99 |
| 5 | 0.50 | 0.00 | 1.04 | 0.00 | 1.00 | 0.52 | 3.33 | 14.03 | 3.33 | 1.66 | 0.52 | 1.66 |
| 6 | 0.50 | 0.00 | 1.02 | 0.00 | 1.00 | 0.51 | 3.97 | 12.27 | 3.97 | 1.98 | 0.51 | 1.98 |
| 7 | 0.50 | 0.00 | 1.04 | 0.00 | 1.00 | 0.52 | 4.28 | 2.53 | 4.28 | 2.14 | 0.52 | 2.14 |
| 8 | 0.50 | 0.00 | 1.10 | 3.48 | 1.00 | 0.55 | 4.71 | 0.00 | 8.19 | 2.36 | 0.55 | 2.36 |
| 9 | 0.50 | 15.00 | 1.09 | 0.00 | 1.00 | 0.55 | 4.92 | 7.81 | 4.92 | 2.46 | 0.55 | 2.46 |
| 10 | 0.50 | 0.00 | 1.19 | 0.12 | 1.00 | 0.60 | 7.64 | 0.00 | 7.76 | 3.82 | 0.60 | 3.82 |
| 11 | 0.50 | 0.00 | 1.26 | 7.76 | 1.00 | 0.62 | 8.42 | 0.00 | 15.00 | 4.18 | 0.63 | 4.23 |
| 12 | 0.49 | 15.00 | 1.19 | 0.00 | 1.00 | 0.59 | 7.48 | 1.00 | 7.48 | 3.69 | 0.60 | 3.79 |
| 13 | 0.49 | 0.00 | 1.16 | 6.68 | 1.00 | 0.57 | 6.16 | 0.00 | 12.85 | 3.02 | 0.59 | 3.14 |
| 14 | 0.49 | 15.00 | 1.22 | 0.00 | 1.00 | 0.59 | 6.81 | 2.15 | 6.81 | 3.31 | 0.62 | 3.48 |
| **15** | **0.48** | **0.00** | **1.22** | **6.81** | **1.00** | **0.59** | **6.72** | **0.00** | **13.53** | **3.25** | **0.63** | **3.45** |
| 16 | 0.48 | 15.00 | 1.09 | 0.00 | 1.00 | 0.52 | 6.09 | 1.47 | 6.09 | 2.92 | 0.57 | 3.18 |
| 17 | 0.48 | 0.00 | 1.06 | 3.89 | 1.00 | 0.50 | 4.65 | 0.00 | 8.54 | 2.22 | 0.56 | 2.45 |
| 18 | 0.47 | 0.00 | 0.96 | 6.04 | 1.00 | 0.46 | 2.60 | 0.00 | 8.63 | 1.23 | 0.52 | 1.39 |
| 19 | 0.47 | 0.00 | 0.93 | 0.00 | 1.00 | 0.44 | 2.62 | 18.17 | 2.62 | 1.23 | 0.51 | 1.42 |
| 20 | 0.47 | 0.00 | 0.87 | 0.00 | 1.00 | 0.41 | 1.92 | 14.98 | 1.92 | 0.89 | 0.48 | 1.06 |
| 21 | 0.46 | 0.00 | 0.92 | 0.00 | 1.00 | 0.42 | 2.66 | 6.48 | 2.66 | 1.23 | 0.51 | 1.47 |
| 22 | 0.46 | 0.00 | 0.94 | 1.66 | 1.00 | 0.43 | 2.91 | 0.00 | 4.56 | 1.34 | 0.52 | 1.62 |
| 23 | 0.46 | 15.00 | 1.01 | 0.00 | 1.00 | 0.46 | 4.05 | 11.24 | 4.05 | 1.85 | 0.56 | 2.24 |
| 24 | 0.45 | 0.00 | 1.11 | 3.65 | 1.00 | 0.51 | 5.57 | 0.00 | 9.22 | 2.53 | 0.61 | 3.05 |
| 25 | 0.45 | 15.00 | 1.21 | 0.00 | 1.00 | 0.54 | 7.51 | 5.78 | 7.51 | 3.38 | 0.66 | 4.08 |
| 26 | 0.45 | 0.00 | 1.09 | 7.51 | 1.00 | 0.49 | 6.00 | 0.00 | 13.51 | 2.68 | 0.61 | 3.34 |
| 27 | 0.44 | 15.00 | 0.96 | 0.00 | 1.00 | 0.43 | 4.62 | 1.49 | 4.62 | 2.05 | 0.55 | 2.66 |
| 28 | 0.44 | 0.00 | 0.84 | 4.62 | 1.00 | 0.37 | 3.29 | 0.00 | 7.91 | 1.45 | 0.51 | 1.97 |
| 29 | 0.44 | 0.00 | 0.86 | 7.91 | 1.00 | 0.38 | 3.62 | 0.00 | 11.53 | 1.58 | 0.52 | 2.18 |
| 30 | 0.43 | 15.00 | 0.84 | 0.00 | 1.00 | 0.36 | 3.42 | 3.47 | 3.42 | 1.48 | 0.51 | 2.10 |
| 31 | 0.43 | 0.00 | 1.09 | 3.42 | 1.00 | 0.47 | 6.00 | 0.00 | 9.42 | 2.58 | 0.63 | 3.47 |
| 32 | 0.43 | 0.00 | 0.97 | 9.42 | 0.80 | 0.41 | 4.57 | 0.00 | 13.99 | 1.95 | 0.59 | 2.76 |
| 33 | 0.42 | 0.00 | 0.80 | 13.99 | 0.14 | 0.09 | 0.73 | 0.00 | 14.72 | 0.31 | 0.27 | 0.96 |

(continued)

**Table 6** (continued)

| (1) | (2) | (3) | (4) | (5) | (6) | (7) | (8) | (9) | (10) | (11) | (12) | (13) |
|-----|-----|-----|-----|-----|-----|-----|-----|-----|------|------|------|------|
| 34 | 0.42 | 0.00 | 0.82 | 14.72 | 0.04 | 0.02 | 0.19 | 0.00 | 14.91 | 0.08 | 0.22 | 0.71 |
| 35 | 0.50 | 15.00 | 1.11 | 0.00 | 1.00 | 0.55 | 4.87 | 15.00 | 4.87 | 2.44 | 0.55 | 2.44 |
| 36 | 0.50 | 0.00 | 1.12 | 1.37 | 1.00 | 0.56 | 5.49 | 0.00 | 6.86 | 2.75 | 0.56 | 2.75 |
| 37 | 0.50 | 0.00 | 1.11 | 0.00 | 1.00 | 0.55 | 4.33 | 7.14 | 4.33 | 2.16 | 0.55 | 2.16 |
| 38 | 0.50 | 0.00 | 1.05 | 0.00 | 1.00 | 0.52 | 3.97 | 7.67 | 3.97 | 1.99 | 0.52 | 1.99 |
| 39 | 0.50 | 0.00 | 1.04 | 0.00 | 1.00 | 0.52 | 3.33 | 14.03 | 3.33 | 1.66 | 0.52 | 1.66 |
| 40 | 0.50 | 0.00 | 1.02 | 0.00 | 1.00 | 0.51 | 3.97 | 12.27 | 3.97 | 1.98 | 0.51 | 1.98 |
| 41 | 0.50 | 0.00 | 1.04 | 0.00 | 1.00 | 0.52 | 4.28 | 2.53 | 4.28 | 2.14 | 0.52 | 2.14 |
| 42 | 0.50 | 0.00 | 1.10 | 3.48 | 1.00 | 0.55 | 4.71 | 0.00 | 8.19 | 2.36 | 0.55 | 2.36 |
| 43 | 0.50 | 15.00 | 1.09 | 0.00 | 1.00 | 0.55 | 4.92 | 7.81 | 4.92 | 2.46 | 0.55 | 2.46 |
| 44 | 0.50 | 0.00 | 1.19 | 0.12 | 1.00 | 0.60 | 7.64 | 0.00 | 7.76 | 3.82 | 0.60 | 3.82 |
| 45 | 0.50 | 0.00 | 1.26 | 7.76 | 1.00 | 0.62 | 8.42 | 0.00 | 15.00 | 4.18 | 0.63 | 4.23 |
| 46 | 0.49 | 15.00 | 1.19 | 0.00 | 1.00 | 0.59 | 7.48 | 1.00 | 7.48 | 3.69 | 0.60 | 3.79 |
| 47 | 0.49 | 0.00 | 1.16 | 6.68 | 1.00 | 0.57 | 6.16 | 0.00 | 12.85 | 3.02 | 0.59 | 3.14 |
| 48 | 0.49 | 15.00 | 1.22 | 0.00 | 1.00 | 0.59 | 6.81 | 2.15 | 6.81 | 3.31 | 0.62 | 3.48 |
| 49 | 0.48 | 0.00 | 1.22 | 6.81 | 1.00 | 0.59 | 6.72 | 0.00 | 13.53 | 3.25 | 0.63 | 3.45 |
| 50 | 0.48 | 15.00 | 1.09 | 0.00 | 1.00 | 0.52 | 6.09 | 1.47 | 6.09 | 2.92 | 0.57 | 3.18 |

## Column 5

Let us assume precipitation and irrigation occur at the start of the day.

Maximum$\{(D_{e,i}$ of previous day- $I/f_w$- $P_i$), 0$\}$ = Maximum$\{(6.81–0-0), 0\}$ = 6.81 mm.

## Column 6

Only if $D_{e,i-1}$ –start > REW, reduction factor must be worked out using $K_r = \frac{TEW - D_{e,i-1}}{TEW - REW}$ for $D_{e,i-1}$ > REW.

If $D_{e,i}$ –start is less than or equal to REW, then $K_r = 1$. In this case, $K_r = 1$.

## Column 7

$$K_e = Min(K_r(K_{cmax} - K_{cb}), f_{ew} \times K_{cmax})$$

$$K_e = Min(1(1.22 - 0.04), 0.48 \times 1.22) = Min(1.18, 0.59) = 0.59$$

## Column 8

$$\text{Evaporation from the exposed wetted surface alone} = K_e \times ET_o/f_{ew}$$
$$= 0.59 \times 5.5/0.483$$

$$= 6.72 \, \text{mm}$$

**Column 9**

$$DP_{e,i} = (0 + 0 - 6.81) \le 0)\,) = 0 \, \text{mm}$$

**Column 10**

$$D_{e,i} = D_{e,i-1} - P_i - \frac{I_i}{f_w} + \frac{E_i}{f_{ew}} + T_{ew} + DP_{e,i}$$

$$D_{e,i} = 6.81 - 0 - 0 + 6.72 + 0 + 0 = 13.53 \, \text{mm}$$

**Column 11**

$$\text{Mean Evaporation over entire area} = K_e \times \text{ET}_o$$
$$= 0.59 \times 5.5 = 3.25 \, \text{mm}$$

**Column 12**

$$K_c = K_{cb} + K_e = 0.04 + 0.59 = 0.63$$

**Column 13**

$$\text{ET} = K_c \times \text{ET}_o = 0.63 \times 5.5 = 3.45 \, \text{mm}$$

## Summary

Reference crop ET (ET$_o$) is a standard method to estimate the evaporating power of atmosphere. ET$_o$ is estimated for a synthetic reference grass surface which is assumed to be well watered with a weekly irrigation frequency. Real-time ET$_o$ data for any place is freely available in Internet. Pan evaporimeter is widely adopted method to estimate ET$_o$ in India due to its simplicity even though this method suffers from certain disadvantages. ET of any crop is obtained by multiplying ET$_o$ with a crop factor ($K_c$). Most of the effects of the various weather conditions are incorporated into the ET$_o$ estimate. Therefore, as ET$_o$ represents an index of climatic demand, $K_c$ varies predominately with the specific crop characteristics and only to a limited extent with climate. This enables the transfer of standard values for $K_c$ between locations and between climates. For drip irrigated crops, ET is estimated more accurately by partitioning $K_c$ into evaporation coefficient and transpiration coefficient. ET of fully grown drip irrigated crops is believed to be significantly less than the conventional irrigated crops. But the research results have proved that there is no significant reduction in ET, and it is found to be sometimes higher also.

## Assessment

### Part A. Say True or False

1. Potential ET is obsolete, and reference crop ET is recommended to be used. (True)
2. Penman-Monteith approach of determining reference crop ET is the current standard. (True)
3. Reference crop ET for any place is available through an application software that can work in cell phones. (True
4. Estimation of reference crop ET by pan evaporimeter is more accurate than Penman-Montieth method. (False)
5. ET of drip irrigated fully grown crops is generally lower than when sprinkler irrigation is used to irrigate the same crop. (False)
6. Crop coefficients for estimating crop ET developed at one climatic conditions with a set of cultivation and irrigation practice is transferable to any other climatic conditions cultivated with the same set of practices. (True)
7. Dual crop coefficient approach of determining crop ET is more accurate method for any kind of irrigation practice. (True)
8. The crop coefficient for initial stage of crop is more dependent on the frequency of irrigation and fraction of area wetted than the type of crop grown. (True)
9. When a crop coefficient developed in one climatic condition is used in another climatic zones, the crop coefficient can be converted based on the data such as wind velocity, minimum relative humidity and crop height. (True)
10. $K_{cmax}$ is the maximum $K_c$ value limited by the energy available at the land surface. (True)

### Part B. Short Answer Questions

1. Define reference crop ET?
2. Define basal crop coefficient?
3. The $K_{cmid}$ value from Table 2 is 1.10. Adjust the $K_{cmid}$ for the local climate crop specific conditions for the following data. Wind velocity is 10 m/s; minimum relative humidity is 30%; and height of the crop is 0.5 m.
4. The evaporation coefficient is 0.2, basal crop coefficient is 0.9, and reference crop ET is 8 mm. Find out the crop ET.
5. What is the data needed to estimate total evaporable water from a soil?

# References and Further Reading

1. Allen, R. G., Pereira, L. S., Raes, D., & Smith, M. (1998). *Crop evapotranspiration. Guidelines for computing crop water requirements*. FAO irrigation and Drainage Paper No. 56, FAO, Rome, Italy, p. 300.
2. Doorenbos, J., & Pruitt, W.O. (1977). *Guidelines for predicting crop water requirements*. Food and Agriculture Organisation of the United Nations, FAO Irrigation and Drainage Paper 24, Rome. https://www.facebook.com/Netafim/videos/2706130609711081/
3. Anonymous. (2010). *National Mission on Micro irrigation, Operational Guidelines*. Ministry of Agriculture, Government of India.

# Rhizosphere Modelling

## Introduction

In this book the subject of rhizosphere modelling is introduced to help the reader develop skills to improve irrigation and fertigation management. The rhizosphere modelling hinges on solving two partial differential equations, namely Richard's equation and convection–dispersion equation. Richard's equation represents the processes involved in flow of water in variably saturated root zone. Convection–dispersion equation represents the processes involved in flow of nutrients in the root zone. Both of them are highly nonlinear equations, and computer software are used for solving them.

Though the rhizosphere modelling has significant advantages, practising engineering professionals were daunted to solve the rhizosphere modelling equations for which computer coding was inevitable. As of today, many user-friendly free software have come up to aid in handling this task. Hydrus is one among many softwares which is very useful in solving these equations. A big challenge for the rhizosphere modeller is collecting field data to use in software packages and estimating various parameters to be used in the models. This chapter would help modellers do this job very easily so that significant potential of rhizosphere modelling could be harnessed.

## Derivation of Richards Equation

Flow in variably saturated state can be expressed by the Darcy–Buckingham equation already discussed in Basic section. If the flow studied is a steady-state flow and the volumetric water content in soils ($\theta$) does not vary with respect to time, the Darcy–Buckingham equation is sufficient to represent the flow process. But in real-field situations, unsteady flow occurs under variable volumetric content over space and time. It necessitates the derivation of partial differential equation by using the concept of continuity along with the Darcy–Buckingham equation.

© The Author(s), under exclusive license to Springer Nature Singapore Pte Ltd. 2023    439
V. Ravikumar, *Sprinkler and Drip Irrigation*,
https://doi.org/10.1007/978-981-19-2775-1_16

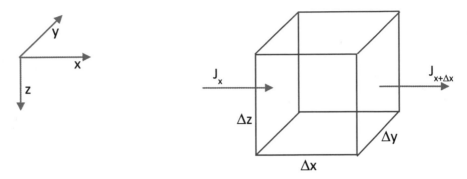

**Fig. 1** Infinitesimal control volume for derivation of Richard's equation

Whenever any continuity equation is derived, we must pay attention to the direction of reference axis. In this case, x is positive towards right from the origin, y is positive towards the direction going into the paper and z is positive downwards from the origin (Fig. 1).

Let us assume that volumetric water content ($\theta$) is a function of space and time. Let us consider a representative elementary volume of soil (REV) whose dimensions are $\Delta x$, $\Delta y$ and $\Delta z$ in X-, Y- and Z-directions, respectively, as shown in Fig. 1. Let $J_x$ be the flux (flow rate/unit area) entering the REV in X-direction and $J_{x+\Delta x}$ be the flux leaving from REV in X-direction.

By using Taylor series expansion and truncating the second-order terms, we can write as follows:

$$J_{x+\Delta x} = J_x + \frac{\partial J_x}{\partial X} \Delta x \tag{1}$$

$$\text{Net flux into REV in } X\text{-direction} = (J_x - J_{x+\Delta x})\Delta y \, \Delta z = -\frac{\partial J_x}{\partial X} \Delta x \, \Delta y \, \Delta z \tag{2}$$

By following the preceding logic for Y- and Z-directions, we can write the following equation:

$$\text{Net flux into REV in all three directions} = -\left(\frac{\partial J_x}{\partial X} + \frac{\partial J_x}{\partial Y} + \frac{\partial J_x}{\partial Z}\right)\Delta x \, \Delta y \, \Delta z \tag{3}$$

By conservation of mass, the rate of change of water content in REV must be equal to the net flux into the REV. Therefore,

$$-\left(\frac{\partial J_x}{\partial X} + \frac{\partial J_x}{\partial Y} + \frac{\partial J_x}{\partial Z}\right)\Delta x \Delta y \Delta z = \frac{\partial \theta}{\partial t} \Delta x \Delta y \Delta z \tag{4}$$

By Darcy–Buckinghum equation we can write the following equations:

$$J_x = -K(h)\frac{\partial H}{\partial X} \tag{5}$$

$$J_y = -K(h)\frac{\partial H}{\partial Y} \tag{6}$$

$$J_z = -K(h)\frac{\partial H}{\partial Z} \tag{7}$$

where $K(h)$ is the unsaturated hydraulic conductivity which depends on the capillary pressure head '$h$' of soil water. '$H$' is called as hydraulic head which is as follows:

$$H = h - z \tag{8}$$

where '$z$' is called as gravitational energy head. Usually, the reference axes adopted is such that direction $Z$ is positive upwards. In such situations, we are able to derive work when a body is brought to origin from any positive elevation. Hence $H$ is expressed as $h + z$. In our case positive values of elevation indicates that the body is lower than the origin and work needs to be done on the object to bring to the datum. Hence, in our case H is to be expressed as equal to $h - z$.

It is to be noted that the soil is assumed as isotropic in the preceding set of equations Eqs. 5 to 7. By substituting Eqs. (5) to (8) in Eq. (4), we get as follows:

$$\frac{\partial}{\partial X}\left(K(h)\frac{\partial h}{\partial X}\right) + \frac{\partial}{\partial Y}\left(K(h)\frac{\partial h}{\partial Y}\right) + \frac{\partial}{\partial Z}\left(K(h)(\frac{\partial h}{\partial Z} - 1)\right) = \frac{\partial \theta}{\partial t} \tag{9}$$

Equation 9 is popularly called as Richard's equation, and there is a big debate on the name of the equation, and it is also called as Richardson equation. We will follow the convention of naming it as Richard's equation.

It is to be noted that there are two dependent variables '$\theta$' and '$h$' in Eq. 9, and this form of equation is called as mixed form. Since '$\theta$' and '$h$' can be related by functional relationship, it is possible to express Eq. 9 either in terms of only '$\theta$' or in terms of '$h$'.

$\frac{\partial \theta}{\partial t}$ can be expressed as follows:

$$\frac{\partial \theta}{\partial t} = \frac{\partial \theta}{\partial h}\frac{\partial h}{\partial t} = \frac{d\theta}{dh}\frac{\partial h}{\partial t} = C_w \frac{\partial h}{\partial t} \tag{10}$$

where $C_w$ is called as hydraulic capacity function. By substituting Eq. 10 in Eq. 9, we get as follows:

$$\frac{\partial}{\partial X}\left(K(h)\frac{\partial h}{\partial X}\right) + \frac{\partial}{\partial Y}\left(K(h)\frac{\partial h}{\partial Y}\right) + \frac{\partial}{\partial Z}\left(K(h)\left(\frac{\partial h}{\partial Z} - 1\right)\right) = C_w\frac{\partial h}{\partial t} \quad (11)$$

Note that the preceding equation is in terms of 'h' alone.

Similarly we can substitute $\frac{\partial h}{\partial X} = \frac{dh}{d\theta}\frac{\partial \theta}{\partial X}, \frac{\partial h}{\partial y} = \frac{dh}{d\theta}\frac{\partial \theta}{\partial y}$ and $\frac{\partial h}{\partial Z} = \frac{dh}{d\theta}\frac{\partial \theta}{\partial Z}$ in Eq. 9, we get as follows:

$$\frac{\partial}{\partial X}\left(D(\theta)\frac{\partial \theta}{\partial X}\right) + \frac{\partial}{\partial Y}\left(D(\theta)\frac{\partial \theta}{\partial Y}\right) + \frac{\partial}{\partial Z}\left(D(\theta)\frac{\partial \theta}{\partial Z}\right) - \frac{\partial K(\theta)}{\partial Z} = \frac{\partial \theta}{\partial t} \quad (12)$$

where $D(\theta)$ is called as diffusivity which is equal to $K(\theta)\frac{dh}{d\theta}$. The term diffusivity does not have any physical meaning, and it has been termed like this because by introducing this substitution, we get an equation similar to the popular thermal diffusivity equation for which analytical solutions are available. Note that the preceding equation is in terms of '$\theta$' alone.

Equation 12 cannot be used if the flow field contains saturated flow, whereas Eqs. 9 and 11 can be used even if the flow field contains both saturated and unsaturated flows. While solving the mixed form (Eq. 9), mass gets balanced better than when the head form (Eq. 11) is used.

The Richard's equation can be solved if we specify the initial conditions as well as boundary conditions for any specific problem. Initial condition may be specified in terms of initial soil water content in the flow domain or initial soil water pressure heads.

$$\text{Either } \theta(x, y, z, 0) = \theta \text{ or } h(x, y, z, 0) = h \quad (13)$$

$\theta(x, y, z, 0)$ is water content at zero time, and $h(x, y, z, 0)$ is soil water pressure head at zero time. This data often comes from field data collection. It is generally recommended to provide initial condition in terms of pressure head rather than water content because erroneous initial condition data is more possible in water content data.

## Boundary Conditions

There are two types of boundary conditions with which the Richard's equation is usually solved. One is Dirichlet boundary condition, and another is Neumann boundary condition.

If at boundaries, soil water pressure heads are known for different time periods, those boundary conditions are called as Dirichlet boundary condition (Type 1). It is mathematically expressed as follows:

$$h(x_b, y_b, z_b, t) = h_b \quad (14)$$

At the boundary node $(x_b, y_b, z_b)$ at time '$t$', the head to be specified is denoted as $h_b$. For instance, when simulating an infiltration experiment, the depth of water standing on the ground is known and that data is used when specifying the surface soil boundary.

If at boundaries, flux across boundaries are known for different time periods, those boundary conditions are known as Neumann boundary condition (Type 2). It is mathematically expressed as follows:

$$J(x_b, y_b, z_b, t) = J_b \tag{15}$$

At the boundary node $(x_b, y_b, z_b)$ at time '$t$', the flux to be specified is denoted as $J_b$. For instance, when flow from a pressure compensating subsurface dripper is simulated, the flux from the dripper boundary can be specified as a Neumann boundary.

Sometimes at boundaries like free drainage boundary often encountered in deep wetted profiles with deep water tables, unit hydraulic gradient exists. Across these boundaries, the flux is equal to the hydraulic conductivity and this boundary condition may be mathematically expressed as follows:

$$J(x_b, y_b, z_b, t) = -K(h) \tag{16}$$

When simulating the root zone with infiltration from rainfall or evaporation from the soil surface, the flux passing across soil surface is governed by the soil pressure head status of the soil surface. In such situations, the boundary condition is mathematically expressed as follows:

$$J(x_b, y_b, z_b, t) \leq J_b^{\max} \tag{17}$$

where $J_b^{\max}$ is the maximum allowable flux of the soil surface. When we conduct leaching test in soil columns, we encounter a seepage face at the bottom of the column. Bottom seepage face boundary is exposed to atmosphere. In these situations, we cannot fix up the priory boundary condition. When solving the equation for the bottom boundary, we must examine the pressure head of the bottom boundary. If it is greater than zero, the pressure head must be forced to zero. This is mathematically expressed as follows:

$$h(x_b, y_b, z_b, t) = 0 \text{ when computed } h(x_b, y_b, z_b, t) \geq 0 \tag{18}$$

If it is less than zero, the pressure head is left as equal to computed pressure head. This is mathematically expressed as follows:

$$J(x_b, y_b, z_b, t) = 0 \text{ when computed } h(x_b, y_b, z_b, t) \leq 0 \tag{19}$$

Finite difference and finite element methods are often used in solving the Richard's equation with necessary initial and boundary condition. The details of numerical solving procedures are not presented in this book. However, a brief introduction on finite difference methods (FDM) is provided for the readers in Appendix II of this chapter.

## Unsaturated Hydraulic Conductivity Parameters

Nowadays, software like Hydrus is very much used to solve the Richards equation. The results obtained while using such software relies primarily on feeding in quality data. One essential data needed is the unsaturated hydraulic conductivity characteristics of soil.

One set of equations often used is describing the unsaturated retention and hydraulic conductivity functions is the van Genuchten–Mualem model. These equations are as follows:

$$K(h) = K_s S_e^l \left(1 - \left(1 - S_e^{1/m}\right)^m\right)^2 \tag{20}$$

$$h = -\frac{1}{\alpha}\left[\left(\frac{\theta_s - \theta_r}{\theta - \theta_r}\right)^{\frac{1}{m}} - 1\right]^{\frac{1}{n}} \text{ for } \theta < \theta_s \tag{21}$$

$$K(h) = K_s \text{ for } \theta = \theta_s \tag{22}$$

$$S_e = \frac{\theta - \theta_r}{\theta_s - \theta_r} \text{ or } S_e = \frac{1}{\left(1 + (-\alpha h)^n\right)^m} \tag{23}$$

$$m = 1 - \frac{1}{n}; \ n > 1 \tag{24}$$

where $\theta$ is volumetric water content (cm$^3$/cm$^3$), $\theta_r$ and $\theta_s$ denote residual and saturated water contents (cm$^3$/cm$^3$), respectively; $S_e$ is the effective saturation, $K_s$ is the saturated hydraulic conductivity (cm/d), $\alpha$ is the inverse of air-entry value or (bubbling pressure (1/cm), $n$ is the pore size distribution index, and $l$ is the pore connectivity parameter. '$l$' is conventionally taken as 0.5

The first-level approximation adopted in knowing hydraulic conductivity data is based on the textural classification of soil as in Table 1 [1]. To generate a much representative data of unsaturated hydraulic conductivity parameters, inverse modelling can be done with Hydrus-1D by conducting a simple double-ring infiltrometer test.

Another better method is to collect simultaneously water content and suction pressure head data for soil. Pressure plate apparatus can be used to get this data from the laboratory. Better than this is installation of soil suction measurement equipment

**Table 1** Soil hydraulic parameters for van Genuchten unsaturated hydraulic conductivity parameters

| Textural class | $\theta_r$ $(L^3\ L^{-3})$ | $\theta_s$ $(L^3\ L^{-3})$ | $\alpha$ $(L^3\ L^{-3})$ | $n$ $(-)$ | $Ks$ (cm d$^{-1}$) |
|---|---|---|---|---|---|
| Sand | 0.053 | 0.375 | 0.035 | 3.18 | 643 |
| Loamy sand | 0.049 | 0.390 | 0.035 | 1.75 | 105 |
| Sandy loam | 0.039 | 0.387 | 0.027 | 1.45 | 38.2 |
| Loam | 0.061 | 0.399 | 0.011 | 1.47 | 12.0 |
| Silt | 0.050 | 0.489 | 0.007 | 1.68 | 43.7 |
| Silt loam | 0.065 | 0.439 | 0.005 | 1.66 | 18.3 |
| Sandy clay loam | 0.063 | 0.384 | 0.021 | 1.33 | 13.2 |
| Clay loam | 0.079 | 0.442 | 0.016 | 1.41 | 8.18 |
| Silty clay loam | 0.090 | 0.482 | 0.008 | 1.52 | 11.1 |
| Sandy clay | 0.117 | 0.385 | 0.033 | 1.21 | 11.4 |
| Silty clay | 0.111 | 0.481 | 0.016 | 1.32 | 9.61 |
| Clay | 0.098 | 0.459 | 0.015 | 1.25 | 14.8 |

like tensiometer and any volumetric water content measurement equipment like time domain reflectometry sensor in soil at the same place. Figure 2 shows such an installation in field. These types of installation must be done in field without any crop, and surface of soil must be covered with black polythene sheet. After installation of probes uniform watering of soil profile must be done by ponding water over the surface of soil. When infiltration is about to get completed, the surface pond must disappear at the same instant on the surface. This is to ensure uniform watering of the soil profile. Nowadays, integrated tensiometer and TDR sensor are also available (Fig. 3). If they are installed, observations can be taken up even without uniform watering of soil profile and covering the surface with black polythene sheet.

After the observations are taken up, any suitable relationship like van Genuchten–Mualem model can be fitted by parameter optimization methods.

**Fig. 2** Tensiometer and TDR at the same depth in a crop less field surface covered with plastic sheet [18]

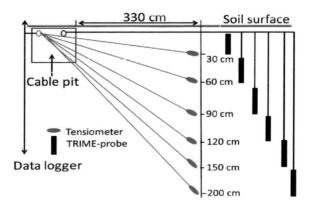

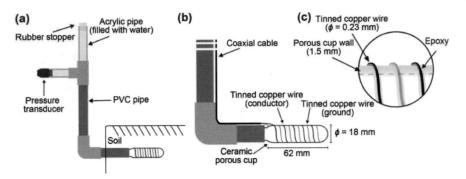

**Fig. 3** Integrated tensiometer with TDR [2]

## Parameter Optimization

Parameter optimization method is explained by taking the case of fitting van Genuchten–Mualem relationship for field observations of water content and suction pressure head data.

Let us assume that we want to optimize the parameters $\theta_r$, $\theta_s$, $\alpha$ and $n$ of the van Genuchten–Mualem relationship. Let $B_{opt}$ be optimized set of values for $\theta_r$, $\theta_s$, $\alpha$ and n which is expressed as follows:

$$B^{opt} = \{\theta_r, \theta_s, \alpha, n\} \tag{25}$$

The objective function may be minimization of sum of squared deviations between observed and estimated water content which is defined as follows:

$$B^{opt} = \min_{j} \ Dev[B_j] = \sum_{i=1}^{n} \left( W_i \left( \theta_i^{obs} - \widehat{\theta}_i[B_j] \right) \right)^2 \tag{26}$$

where '$i$' is index for data points and '$n$' is total number of data points; $\theta_i^{obs}$ is observed water content and $\widehat{\theta}_i[B_j]$ is the estimated water content for the parameter set $B_j$; $W_i$ is user defined weight for observation number '$i$'. $Dev[B_j]$ is the sum of squared deviations for the parameter set '$j$'. If all the data points are to be treated equal, then W takes a value of 1. Sometimes, when we feel that some points be given lower importance, then for those points relative weights may be given with values less than one.

Fitting a water content-pressure head relationship using the objective functions as in Eq. 26 is very easy by using a free software, namely RETC [21]. A sample data and fitted van Genuchten unsaturated hydraulic conductivity parameters using an Internet source is shown in Fig. 4 (http://purl.org/net/swrc/).

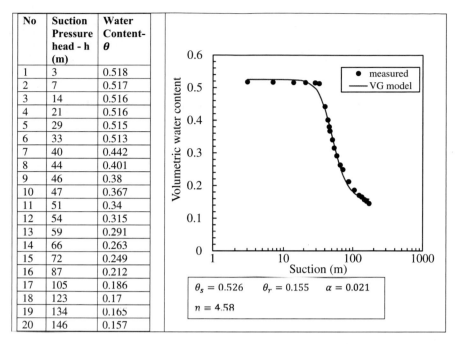

| No | Suction Pressure head - h (m) | Water Content- $\theta$ |
|----|----|----|
| 1 | 3 | 0.518 |
| 2 | 7 | 0.517 |
| 3 | 14 | 0.516 |
| 4 | 21 | 0.516 |
| 5 | 29 | 0.515 |
| 6 | 33 | 0.513 |
| 7 | 40 | 0.442 |
| 8 | 44 | 0.401 |
| 9 | 46 | 0.38 |
| 10 | 47 | 0.367 |
| 11 | 51 | 0.34 |
| 12 | 54 | 0.315 |
| 13 | 59 | 0.291 |
| 14 | 66 | 0.263 |
| 15 | 72 | 0.249 |
| 16 | 87 | 0.212 |
| 17 | 105 | 0.186 |
| 18 | 123 | 0.17 |
| 19 | 134 | 0.165 |
| 20 | 146 | 0.157 |

$\theta_s = 0.526 \quad \theta_r = 0.155 \quad \alpha = 0.021$

$n = 4.58$

**Fig. 4** Sample data set and fitting unsaturated hydraulic conductivity parameters

**Fig. 5** Double-ring infiltrometer test. Image: environmentalbiophysics.org

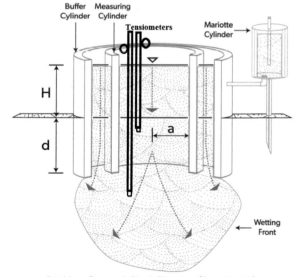

Double or Concentric Ring Infiltrometer (Cross Section)

A very simple way to get saturated hydraulic conductivity is to conduct a double-ring infiltrometer test (Fig. 5). The steady infiltration rate value we get at the end of the test can be taken as the saturated hydraulic conductivity value.

## Inverse Modelling with a Double-Ring Infiltrometer Test

Parameter estimation using inverse modelling option in Hydrus-1D is a relatively easy option. A double-ring infiltrometer field test observations can be used to do inverse modelling. Flow in the inner ring of a double-ring infiltrometer test is a one-dimensional vertical downward flow, and hence, one-dimensional Richard's equation is used for analysis which is as below:

$$\frac{\partial}{\partial Z}\left(K(h)(\frac{\partial h}{\partial Z} - 1)\right) = \frac{\partial \theta}{\partial t} \tag{27}$$

Before the test is done in a field, the field must be wetted uniformly and at least two tensiometers must be installed at the site at two different depths as shown in Fig. 5. These tensiometers record soil water pressure head just before the start of the experiment. Initial condition data is very essential for running the model. The depth of standing water at different times as well as the depths of water infiltrated into the soil for different time periods have to be observed. The experiment need not be done until the steady infiltration rate is reached. However, doing the experiment until steady infiltration rate is reached is very useful.

The boundary condition for the top surface of the soil to be selected in Hydrus-1D is variable pressure head. Simulation may be done for a depth of soil at which we may expect unit hydraulic gradient to get formed. We can expect approximately unit hydraulic gradient at a depth around 1 m. The bottom boundary condition in Hydrus-1D can be selected as free drainage boundary.

In inverse modelling, the inverse equation is said to be well-posed, if we get global optimum solution. If the equation is said to be ill-posed, we are not guaranteed to get global optimum solution. The inverse equation developed in this case is ill-posed. In such conditions, the number of parameters that must be found out using inverse modelling must be decided with care. While it is possible to optimize five parameters, namely $K_s$, $\theta_s$, $\theta_r$, $\alpha$ and $n$ simultaneously with this method, it is not advisable to optimize all the parameters. Field determination of $K_s$ and laboratory estimation of $\theta_s$ using the soil cores taken from the field is recommended.

A complete set of data and the procedure of using Hydrus-1D is provided in Appendix I of this chapter.

**Fig. 6** Water balance in the root zone

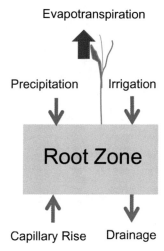

Evapotranspiration

Precipitation    Irrigation

Root Zone

Capillary Rise    Drainage

# Soil Water Balance Methods for Estimation of Evapotranspiration

Soil water balance method is emerging as a reliable method for estimating evapotranspiration (ET) due to the recent advancements in measurement of soil moisture and tension equipment and their availability at an affordable cost. Therefore nowadays, ET for any time period is estimated using water balance method by finding change in soil water content between sampling time periods, rainfall, capillary rise from water table, deep percolation and surface runoff occurring during the time period (Fig. 6).

## *Water Balance Models*

Three types of water balance models are presented here. They are as follows:

(a)    Single cell–single step
(b)    Single cell–multiple step
(c)    Multiple cell–multiple step.

## *Single Cell–Single Step Model*

In this method, entire root zone is treated as single cell and soil moisture is sensed at a location where spatial average soil moisture is expected (Fig. 7).

$$\mathrm{ET}_t = I_t - (\theta_{t+1} - \theta_t)z - \mathrm{Dr}_t \tag{28}$$

**Fig. 7** Illustration for single
cell–single step model

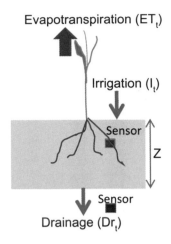

where

ET$_t$    is evapotranspiration during a period '$t$';

'$t$'    is called as time step and usually daily time steps are conventional;

$I_t$    is the depth of irrigation done during period $t$;

$\theta_t$    are water contents in root zone at the start of time step $t$ and $t+1$, respectively;

$z$    is depth of root zone;

Dr$_t$    is the depth of water drained below the root zone.

In this method, one of the most difficult variables to be estimated is drainage below root zone. In case of underground water table being at sufficiently deeper depths, unit hydraulic gradient may be assumed. Installation of one sensor below the root zone and recording the soil moisture can improve the estimate of drainage. For the assumption of unit hydraulic gradient Dr$_t$ is numerically equal to the unsaturated hydraulic conductivity $K(\theta)$. A suitable model relating unsaturated hydraulic conductivity to the water content must be used to estimate $K(\theta)$.

Campbell [3] equation is easy to use while using $K(\theta)$. The Campbell equation is as follows:

$$K(\theta) = K_s \left(\frac{\theta}{\theta_s}\right)^{2b+3} \tag{29}$$

where

$K_s$    is the saturated hydraulic conductivity;

$K(\theta)$    is the unsaturated hydraulic conductivity at water content $\theta$;

$\theta_s$    is the saturated water content and

$b$    is a fitting parameter and guide line values are provided in Table 2.

Gardner equation as given below (discussed in basic section) can also be used if the soil moisture is estimated in suction head

$$K(h) = K_s e^{\alpha h} \tag{30}$$

If there is rainfall on the day of estimation in small amounts, its amount can be added with depth of irrigation. If rainfall is high, runoff and deep percolation estimation is impossible with this method. Hence ET is not estimated on the days of higher rainfall with this method.

## Problem

In a silty loam soil with saturated water content of 0.35, irrigation depth of 2 cm through sprinkler irrigation was given on a day. At the start of the day the moisture content in the root zone was 0.28, and after a day, the moisture content in the root zone is 0.30. The root zone depth is 60 cm. Soil moisture content just below the root zone remains constant at 0.25 during the day. Find out the ET?

If we assume unit hydraulic gradient below the root zone, the drainage rate will be equal to unsaturated hydraulic conductivity.

From Table 2, we can get $K_s = 0.68$ cm/h (i.e. $= 16.32$ cm/d) and Campbell fitting parameter b $= 4.7$

$$K(\theta) = 16.32 \left(\frac{0.25}{0.35}\right)^{2 \times 4.7 + 3} = 0.251 \frac{cm}{d} = Dr_t$$

$$ET_t = 2 - (0.30 - 0.28)60 - 0.251 = 0.549 \text{ cm}$$

Table 2 Campbell fitting parameter values for different soil textures [20]

| Texture | Silt | Clay | Saturated hydraulic conductivity (cm/h) | Campbell fitting parameter '$b$' |
|---|---|---|---|---|
| Sand | 0.05 | 0.03 | 21 | 1.7 |
| Loamy sand | 0.12 | 0.07 | 6.1 | 2.1 |
| Sandy loam | 0.25 | 0.10 | 2.6 | 3.1 |
| Loam | 0.40 | 0.18 | 1.3 | 4.5 |
| Silty loam | 0.65 | 0.15 | 0.68 | 4.7 |
| Sandy clay loam | 0.13 | 0.27 | 0.43 | 4.0 |
| Clay loam | 0.34 | 0.34 | 0.23 | 5.2 |
| Silty clay loam | 0.58 | 0.33 | 0.15 | 6.6 |
| Sandy clay | 0.07 | 0.40 | 0.12 | 6.0 |
| Silty clay | 0.45 | 0.45 | 0.09 | 7.9 |
| Clay | 0.20 | 0.60 | 0.06 | 7.6 |

## *Single Cell–Multiple Step Model*

In this model, soil moisture sensing is done in root zone at smaller time intervals like minutes or hours. The number of sensors used in root zone and also below root zone may be one or more. Accounting rainfall to a certain extent in this model is possible. As long as the infiltration rate at any instant is greater than rainfall rate, the method can be adopted. When the rainfall rate is greater than infiltration rate, surface runoff occurs. During such a situation, only effective rainfall infiltrating into soil must be used for computation. Often at point locations, estimation of effective rainfall becomes very difficult and error in computation of ET gets introduced. The water balance equation used in this model is as follows:

$$\mathrm{ET}_t = I_t + P_t - (\overline{\theta}_{t+1} - \overline{\theta}_t)z - \mathrm{Dr}_t \tag{31}$$

where $P_t$ is rainfall during time step '$t$' and other terms are as in Eq. 28.

ET is estimated by summing up over any larger interval. For instance, if time step followed in Eq. 31 is an hour, ET for one day is computed as follows:

$$\mathrm{ET\ for\ a\ day} = \sum_{t=1}^{t=24} \mathrm{ET}_t \tag{32}$$

Single cell–multiple step model was implemented for a real-field situation, and the following discussion is done with reference to the case studied [4].

The case study is for sugarcane crop grown in sandy clay loam soil under drip irrigation. Figure 8 shows the configuration of plant rows and drip lateral adopted. One vertical boundary in Fig. 9 is below the dripper across which apparently flow would not occur. But in reality due to the difference in root water uptake on both the sides of the boundary, there may be little flow across the boundary. Hence, practically

**Fig. 8** Area irrigated by one dripper [4]

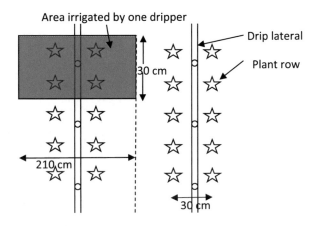

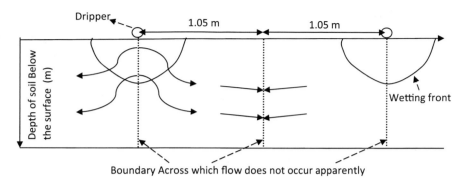

**Fig. 9** Vertical two dimension of the root zone [4]

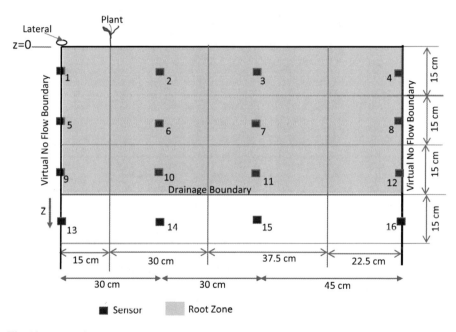

**Fig. 10** Soil moisture sensors, dripper and plant

this may be treated as virtual no-flow boundary. There is another no-flow boundary exactly mid-way between two adjacent laterals. Identification of these two boundaries is useful in placement of soil moisture sensors. If sensors are placed on these no-flow boundaries, it is possible to capture the locations of maximum and minimum moisture content locations.

In order to account for the variability of soil water content variations within the root zone, the soil moisture contents are logged at 16 locations within the root zone and the average soil moisture content is found out (Fig. 10). The observations of soil water content have been recorded at 15 min interval.

A sample data set in Table 3 shows the logged soil moisture content at 15 min interval for one day at all the 16 sensors.

## Drainage from Root Zone and Capillary Rise to Root Zone

The flow across drainage boundary (Fig. 10) occurring from the subsurface to root zone upwards and from the root zone to the subsurface zone downwards is assumed as one dimensional in vertical direction. Darcy–Buckingham equation used for one-dimensional flow with positive downward z-axis is as follows:

$$\text{Dr} = -K(h)\left(\frac{\Delta h}{\Delta z} - 1\right) \tag{33}$$

where Dr is the drainage rate (L/T) in one-dimensional vertical flow occurring through the nodes in which sensors 9 & 13, 10 & 14, 11 & 15 and 12 & 16 are placed. $K(h)$ is the unsaturated hydraulic conductivity at the pressure head '$h$' and '$z$' is the distance in vertical direction. Note that '$K$' is a function of '$h$'. Hence, while calculating drainage between any two points for a period, the water contents at the start and end of the period for both the points must be averaged. In this numerical example, the water contents at the end of the period are not considered and only the water contents at the start of the period for both the points (upstream and downstream) is considered for calculating $K(h)$ for simplicity in calculation.

The analytical model describing the unsaturated retention and hydraulic conductivity functions adopted in this work is the van Genuchten–Mualem model.

The unsaturated hydraulic conductivity parameter values used for the sandy clay loam soil in the study area are as follows:

$$\theta_x = 0.063, \theta_s = 0.384, \alpha = 0.021 \text{ cm}^{-1}, n = 1.33 \text{ and } K_s = 13.2 \text{ cm/d}$$

From Fig. 10, we can see that the width of cell represented by senor 13 is 15 cm, sensor 14 is 30 cm, sensor 15 is 37.5 cm and sensor 16 is 22.5 cm, respectively. Therefore average rate of drainage below the root zone is estimated by spatial averaging technique:

$$\text{Dr}_{\text{average}} = 0.143 \text{ Dr}_{9\&13} + 0.286\text{Dr}_{10\&14} + 0.357\text{Dr}_{11\&15} + 0.214\text{Dr}_{12\&16} \tag{34}$$

The coefficients of the preceding equation such as 0.143, 0.286, 0.357 and 0.214 are obtained by dividing the representative cell widths by total width of 105 cm.

**Table 3** Water content in 16 sensors on a day at 15 min intervals

| Time | Water content at sensor 1 | Water content at sensor 2 | Water content at sensor 3 | Water content at sensor 4 | Water content at sensor 5 | Water content at sensor 6 | Water content at sensor 7 | Water content at sensor 3 | Water content at sensor 9 | Water content at sensor 10 | Water content at sensor 11 | Water content at sensor 12 | Water content at sensor 13 | Water content at sensor 14 | Water content at sensor 15 | Water content at sensor 16 |
|---|---|---|---|---|---|---|---|---|---|---|---|---|---|---|---|---|
| 8:00 | 0.27488 | 0.22248 | 0.21036 | 0.16174 | 0.17938 | 0.25637 | 0.25212 | 0.17122 | 0.21998 | 0.19401 | 0.19456 | 0.18950 | 0.18990 | 0.15919 | 0.18300 | 0.18862 |
| 8:15 | 0.27444 | 0.22237 | 0.21036 | 0.16170 | 0.17934 | 0.25635 | 0.26197 | 0.17115 | 0.22000 | 0.19392 | 0.19457 | 0.18953 | 0.18990 | 0.15917 | 0.18300 | 0.18861 |
| 8:30 | 0.27405 | 0.22240 | 0.21037 | 0.16184 | 0.17936 | 0.25627 | 0.26198 | 0.17119 | 0.22003 | 0.19402 | 0.19457 | 0.18945 | 0.18991 | 0.15904 | 0.18300 | 0.18871 |
| 8:45 | 0.27338 | 0.22239 | 0.21036 | 0.16180 | 0.17941 | 0.25618 | 0.26192 | 0.17105 | 0.21997 | 0.19394 | 0.19456 | 0.18944 | 0.18987 | 0.15910 | 0.18297 | 0.18858 |
| 9:00 | 0.27244 | 0.22242 | 0.21036 | 0.16182 | 0.17935 | 0.25614 | 0.26193 | 0.17108 | 0.21993 | 0.19389 | 0.19456 | 0.18938 | 0.18984 | 0.15916 | 0.18293 | 0.18856 |
| 9:15 | 0.27124 | 0.22239 | 0.21036 | 0.16182 | 0.17923 | 0.25604 | 0.26190 | 0.17099 | 0.21985 | 0.19382 | 0.19456 | 0.18942 | 0.18986 | 0.15900 | 0.18290 | 0.18854 |
| 9:30 | 0.26982 | 0.22216 | 0.21037 | 0.16172 | 0.17914 | 0.25594 | 0.26176 | 0.17096 | 0.21970 | 0.19374 | 0.19451 | 0.18928 | 0.18986 | 0.15903 | 0.18289 | 0.18853 |
| 9:45 | 0.26831 | 0.22202 | 0.21056 | 0.16168 | 0.17909 | 0.25584 | 0.26175 | 0.17098 | 0.21953 | 0.19377 | 0.19442 | 0.18939 | 0.18984 | 0.15894 | 0.18288 | 0.18854 |
| 10:00 | 0.26643 | 0.22197 | 0.21066 | 0.16183 | 0.17905 | 0.25574 | 0.26174 | 0.17082 | 0.21950 | 0.19376 | 0.19436 | 0.18925 | 0.18968 | 0.15888 | 0.18291 | 0.18856 |
| 10:15 | 0.26449 | 0.22179 | 0.21080 | 0.16184 | 0.17891 | 0.25562 | 0.26160 | 0.17091 | 0.21933 | 0.19366 | 0.19435 | 0.18919 | 0.18967 | 0.15880 | 0.18297 | 0.18855 |
| 10:30 | 0.26268 | 0.22185 | 0.21098 | 0.16184 | 0.17884 | 0.25551 | 0.26154 | 0.17089 | 0.21937 | 0.19364 | 0.19435 | 0.18922 | 0.18960 | 0.15875 | 0.18289 | 0.18845 |
| 10:45 | 0.26063 | 0.22170 | 0.21106 | 0.16194 | 0.17874 | 0.25542 | 0.26153 | 0.17079 | 0.21918 | 0.19354 | 0.19434 | 0.18917 | 0.18947 | 0.15864 | 0.18278 | 0.18852 |
| 11:00 | 0.25885 | 0.22168 | 0.21110 | 0.16196 | 0.17858 | 0.25521 | 0.26153 | 0.17077 | 0.21897 | 0.19350 | 0.19429 | 0.18909 | 0.18947 | 0.15854 | 0.18277 | 0.18839 |
| 11:15 | 0.27952 | 0.22159 | 0.21127 | 0.16205 | 0.17856 | 0.25506 | 0.26153 | 0.17077 | 0.21882 | 0.19346 | 0.19425 | 0.18911 | 0.18935 | 0.15842 | 0.18278 | 0.18835 |
| 11:30 | 0.33268 | 0.22171 | 0.21148 | 0.16212 | 0.18265 | 0.25500 | 0.26165 | 0.17077 | 0.21882 | 0.19340 | 0.19420 | 0.18901 | 0.18940 | 0.15833 | 0.18279 | 0.18836 |
| 11:45 | 0.33919 | 0.22179 | 0.21173 | 0.16226 | 0.19436 | 0.25485 | 0.26173 | 0.17078 | 0.21871 | 0.19334 | 0.19417 | 0.18901 | 0.18925 | 0.15828 | 0.18277 | 0.18834 |
| 12:00 | 0.34025 | 0.22693 | 0.21188 | 0.16236 | 0.21133 | 0.25474 | 0.26174 | 0.17077 | 0.21869 | 0.19333 | 0.19420 | 0.18902 | 0.18921 | 0.15818 | 0.18272 | 0.18835 |
| 12:15 | 0.33988 | 0.24588 | 0.21228 | 0.16244 | 0.24381 | 0.25466 | 0.26172 | 0.17077 | 0.21913 | 0.19325 | 0.19415 | 0.18901 | 0.18913 | 0.15831 | 0.18266 | 0.18834 |

(continued)

**Table 3** (continued)

| Time | Water content at sensor 1 | Water content at sensor 2 | Water content at sensor 3 | Water content at sensor 4 | Water content at sensor 5 | Water content at sensor 6 | Water content at sensor 7 | Water content at sensor 8 | Water content at sensor 9 | Water content at sensor 10 | Water content at sensor 11 | Water content at sensor 12 | Water content at sensor 13 | Water content at sensor 14 | Water content at sensor 15 | Water content at sensor 16 |
|---|---|---|---|---|---|---|---|---|---|---|---|---|---|---|---|---|
| 12:30 | 0.33617 | 0.26697 | 0.21268 | 0.16252 | 0.25078 | 0.25463 | 0.26182 | 0.17078 | 0.22227 | 0.19318 | 0.19421 | 0.18901 | 0.18900 | 0.15814 | 0.18265 | 0.18837 |
| 12:45 | 0.32765 | 0.27315 | 0.21314 | 0.16258 | 0.24855 | 0.25459 | 0.26193 | 0.17082 | 0.22674 | 0.19310 | 0.19423 | 0.18885 | 0.18891 | 0.15814 | 0.18259 | 0.18834 |
| 13:00 | 0.32002 | 0.27325 | 0.21354 | 0.16269 | 0.24670 | 0.25447 | 0.26184 | 0.17095 | 0.22938 | 0.19307 | 0.19422 | 0.18894 | 0.18890 | 0.15804 | 0.18257 | 0.18836 |
| 13:15 | 0.31567 | 0.27250 | 0.21404 | 0.16277 | 0.24542 | 0.25456 | 0.26194 | 0.17085 | 0.23089 | 0.19298 | 0.19419 | 0.18898 | 0.18876 | 0.15795 | 0.18256 | 0.18838 |
| 13:30 | 0.31293 | 0.27169 | 0.21454 | 0.16291 | 0.24442 | 0.25458 | 0.26205 | 0.17085 | 0.23204 | 0.19292 | 0.19417 | 0.18880 | 0.18874 | 0.15793 | 0.18256 | 0.18837 |
| 13:45 | 0.31107 | 0.27093 | 0.21497 | 0.16302 | 0.24372 | 0.25446 | 0.26225 | 0.17099 | 0.23290 | 0.19291 | 0.19417 | 0.18879 | 0.18862 | 0.15787 | 0.18256 | 0.18854 |
| 14:00 | 0.30949 | 0.27024 | 0.21546 | 0.16300 | 0.24315 | 0.25450 | 0.26244 | 0.17095 | 0.23354 | 0.19282 | 0.19421 | 0.18879 | 0.18856 | 0.15789 | 0.18256 | 0.18838 |
| 14:15 | 0.30824 | 0.26969 | 0.21575 | 0.16315 | 0.24257 | 0.25460 | 0.26254 | 0.17101 | 0.23404 | 0.19278 | 0.19416 | 0.18881 | 0.18856 | 0.15787 | 0.18255 | 0.18835 |
| 14:30 | 0.30719 | 0.26919 | 0.21601 | 0.16327 | 0.24214 | 0.25469 | 0.26269 | 0.17104 | 0.23444 | 0.19278 | 0.19413 | 0.18879 | 0.18846 | 0.15786 | 0.18252 | 0.18835 |
| 14:45 | 0.30653 | 0.26870 | 0.21626 | 0.16346 | 0.24181 | 0.25482 | 0.26297 | 0.17115 | 0.23481 | 0.19276 | 0.19415 | 0.18882 | 0.18835 | 0.15786 | 0.18255 | 0.18836 |
| 15:00 | 0.30576 | 0.26827 | 0.21651 | 0.16357 | 0.24157 | 0.25495 | 0.26315 | 0.17119 | 0.23514 | 0.19261 | 0.19427 | 0.18878 | 0.18835 | 0.15787 | 0.18255 | 0.18835 |
| 15:15 | 0.30507 | 0.26785 | 0.21678 | 0.16351 | 0.24119 | 0.25501 | 0.26325 | 0.17123 | 0.23540 | 0.19267 | 0.19434 | 0.18879 | 0.18834 | 0.15787 | 0.18256 | 0.18834 |
| 15:30 | 0.30446 | 0.26740 | 0.21674 | 0.16363 | 0.24093 | 0.25512 | 0.26345 | 0.17131 | 0.23564 | 0.19276 | 0.19435 | 0.18879 | 0.18834 | 0.15786 | 0.18255 | 0.18834 |
| 15:45 | 0.30388 | 0.26703 | 0.21681 | 0.16376 | 0.24073 | 0.25525 | 0.26365 | 0.17139 | 0.23582 | 0.19266 | 0.19435 | 0.18879 | 0.18832 | 0.15786 | 0.18254 | 0.18835 |
| 16:00 | 0.30337 | 0.26660 | 0.21684 | 0.16419 | 0.24043 | 0.25539 | 0.26384 | 0.17147 | 0.23596 | 0.19258 | 0.19445 | 0.18886 | 0.18826 | 0.15786 | 0.18250 | 0.18835 |
| 16:15 | 0.30294 | 0.26635 | 0.21703 | 0.16447 | 0.24022 | 0.25550 | 0.26395 | 0.17149 | 0.23617 | 0.19265 | 0.19455 | 0.18889 | 0.18814 | 0.15786 | 0.18241 | 0.18834 |
| 16:30 | 0.30257 | 0.26601 | 0.21703 | 0.16437 | 0.24014 | 0.25568 | 0.26416 | 0.17158 | 0.23638 | 0.19268 | 0.19455 | 0.18881 | 0.18815 | 0.15787 | 0.18242 | 0.18835 |
| 16:45 | 0.30208 | 0.26568 | 0.21710 | 0.16453 | 0.23992 | 0.25584 | 0.26420 | 0.17166 | 0.23654 | 0.19262 | 0.19452 | 0.18879 | 0.18812 | 0.15787 | 0.18241 | 0.18834 |

(continued)

**Table 3** (continued)

| Time | Water content at sensor 1 | Water content at sensor 2 | Water content at sensor 3 | Water content at sensor 4 | Water content at sensor 5 | Water content at sensor 6 | Water content at sensor 7 | Water content at sensor 8 | Water content at sensor 9 | Water content at sensor 10 | Water content at sensor 11 | Water content at sensor 12 | Water content at sensor 13 | Water content at sensor 14 | Water content at sensor 15 | Water content at sensor 16 |
|---|---|---|---|---|---|---|---|---|---|---|---|---|---|---|---|---|
| 17:00 | 0.30170 | 0.26538 | 0.21707 | 0.16454 | 0.23974 | 0.25598 | 0.26422 | 0.17168 | 0.23666 | 0.19263 | 0.19457 | 0.18879 | 0.18812 | 0.15799 | 0.18241 | 0.18835 |
| 17:15 | 0.30125 | 0.26516 | 0.21704 | 0.16454 | 0.23963 | 0.25618 | 0.26441 | 0.17176 | 0.23672 | 0.19268 | 0.19458 | 0.18887 | 0.18812 | 0.15793 | 0.18243 | 0.18834 |
| 17:30 | 0.30097 | 0.26498 | 0.21701 | 0.16454 | 0.23956 | 0.25637 | 0.26444 | 0.17180 | 0.23683 | 0.19277 | 0.19457 | 0.18900 | 0.18812 | 0.15792 | 0.18246 | 0.18834 |
| 17:45 | 0.30067 | 0.26480 | 0.21683 | 0.16443 | 0.23955 | 0.25657 | 0.26463 | 0.17184 | 0.23703 | 0.19267 | 0.19468 | 0.18896 | 0.18812 | 0.15809 | 0.18250 | 0.18835 |
| 18:00 | 0.30040 | 0.26459 | 0.21676 | 0.16445 | 0.23953 | 0.25678 | 0.26464 | 0.17187 | 0.23717 | 0.19269 | 0.19479 | 0.18881 | 0.18812 | 0.15807 | 0.18254 | 0.18834 |
| 18:15 | 0.30033 | 0.26438 | 0.21677 | 0.16450 | 0.23949 | 0.25696 | 0.26464 | 0.17197 | 0.23737 | 0.19279 | 0.19479 | 0.18900 | 0.18817 | 0.15805 | 0.18256 | 0.18834 |
| 18:30 | 0.30012 | 0.26434 | 0.21659 | 0.16432 | 0.23946 | 0.25716 | 0.26467 | 0.17191 | 0.23749 | 0.19279 | 0.19487 | 0.18901 | 0.18832 | 0.15809 | 0.18256 | 0.18845 |
| 18:45 | 0.29991 | 0.26407 | 0.21644 | 0.16441 | 0.23946 | 0.25736 | 0.26485 | 0.17205 | 0.23755 | 0.19279 | 0.19501 | 0.18902 | 0.18834 | 0.15814 | 0.18256 | 0.18841 |
| 19:00 | 0.29972 | 0.26381 | 0.21625 | 0.16432 | 0.23948 | 0.25753 | 0.26483 | 0.17197 | 0.23775 | 0.19279 | 0.19501 | 0.18904 | 0.18834 | 0.15811 | 0.18256 | 0.18837 |
| 19:15 | 0.29968 | 0.26373 | 0.21616 | 0.16424 | 0.23949 | 0.25768 | 0.26483 | 0.17209 | 0.23796 | 0.19279 | 0.19501 | 0.18901 | 0.18834 | 0.15822 | 0.18256 | 0.18851 |
| 19:30 | 0.29953 | 0.26352 | 0.21610 | 0.16429 | 0.23940 | 0.25782 | 0.26486 | 0.17210 | 0.23818 | 0.19290 | 0.19513 | 0.18915 | 0.18834 | 0.15829 | 0.18256 | 0.18856 |
| 19:45 | 0.29933 | 0.26326 | 0.21591 | 0.16416 | 0.23928 | 0.25797 | 0.26483 | 0.17213 | 0.23835 | 0.19288 | 0.19512 | 0.18912 | 0.18847 | 0.15821 | 0.18256 | 0.18856 |
| 20:00 | 0.29912 | 0.26303 | 0.21570 | 0.16413 | 0.23928 | 0.25810 | 0.26485 | 0.17210 | 0.23839 | 0.19289 | 0.19506 | 0.18907 | 0.18856 | 0.15813 | 0.18261 | 0.18855 |
| 20:15 | 0.29898 | 0.26295 | 0.21549 | 0.16409 | 0.23928 | 0.25824 | 0.26486 | 0.17216 | 0.23850 | 0.19301 | 0.19510 | 0.18923 | 0.18856 | 0.15830 | 0.18274 | 0.18852 |
| 20:30 | 0.29894 | 0.26271 | 0.21543 | 0.16400 | 0.23917 | 0.25839 | 0.26486 | 0.17229 | 0.23873 | 0.19297 | 0.19522 | 0.18922 | 0.18856 | 0.15831 | 0.18278 | 0.18854 |
| 20:45 | 0.29874 | 0.26246 | 0.21531 | 0.16401 | 0.23906 | 0.25852 | 0.26475 | 0.17221 | 0.23885 | 0.19301 | 0.19524 | 0.18918 | 0.18856 | 0.15831 | 0.18278 | 0.18856 |
| 21:00 | 0.29853 | 0.26232 | 0.21512 | 0.16391 | 0.23906 | 0.25861 | 0.26464 | 0.17212 | 0.23904 | 0.19306 | 0.19524 | 0.18923 | 0.18866 | 0.15831 | 0.18278 | 0.18856 |
| 21:15 | 0.29836 | 0.26216 | 0.21494 | 0.16385 | 0.23905 | 0.25876 | 0.26464 | 0.17214 | 0.23911 | 0.19309 | 0.19524 | 0.18924 | 0.18869 | 0.15833 | 0.18278 | 0.18856 |

(continued)

**Table 3** (continued)

| Time | Water content at sensor 1 | Water content at sensor 2 | Water content at sensor 3 | Water content at sensor 4 | Water content at sensor 5 | Water content at sensor 6 | Water content at sensor 7 | Water content at sensor 8 | Water content at sensor 9 | Water content at sensor 10 | Water content at sensor 11 | Water content at sensor 12 | Water content at sensor 13 | Water content at sensor 14 | Water content at sensor 15 | Water content at sensor 16 |
|---|---|---|---|---|---|---|---|---|---|---|---|---|---|---|---|---|
| 21:30 | 0.29828 | 0.26190 | 0.21476 | 0.16382 | 0.23893 | 0.25883 | 0.26463 | 0.17223 | 0.23911 | 0.19304 | 0.19528 | 0.18943 | 0.18862 | 0.15831 | 0.18278 | 0.18857 |
| 21:45 | 0.29817 | 0.26166 | 0.21474 | 0.16372 | 0.23884 | 0.25894 | 0.26445 | 0.17230 | 0.23925 | 0.19315 | 0.19542 | 0.18927 | 0.18858 | 0.15831 | 0.18278 | 0.18857 |
| 22:00 | 0.29798 | 0.26159 | 0.21459 | 0.16364 | 0.23888 | 0.25904 | 0.26442 | 0.17233 | 0.23928 | 0.19318 | 0.19546 | 0.18926 | 0.18860 | 0.15832 | 0.18278 | 0.18857 |
| 22:15 | 0.29776 | 0.26134 | 0.21439 | 0.16360 | 0.23888 | 0.25913 | 0.26432 | 0.17233 | 0.23944 | 0.19323 | 0.19546 | 0.18944 | 0.18876 | 0.15847 | 0.18279 | 0.18873 |
| 22:30 | 0.29759 | 0.26105 | 0.21421 | 0.16355 | 0.23879 | 0.25921 | 0.26419 | 0.17233 | 0.23950 | 0.19324 | 0.19546 | 0.18945 | 0.18879 | 0.15850 | 0.18293 | 0.18875 |
| 22:45 | 0.29755 | 0.26093 | 0.21405 | 0.16346 | 0.23880 | 0.25927 | 0.26419 | 0.17233 | 0.23954 | 0.19332 | 0.19546 | 0.18941 | 0.18879 | 0.15835 | 0.18300 | 0.18866 |
| 23:00 | 0.29737 | 0.26077 | 0.21401 | 0.16339 | 0.23881 | 0.25937 | 0.26409 | 0.17233 | 0.23971 | 0.19326 | 0.19546 | 0.18934 | 0.18879 | 0.15832 | 0.18300 | 0.18857 |
| 23:15 | 0.29721 | 0.26053 | 0.21392 | 0.16334 | 0.23868 | 0.25947 | 0.26418 | 0.17233 | 0.23976 | 0.19325 | 0.19549 | 0.18945 | 0.18894 | 0.15836 | 0.18300 | 0.18856 |
| 23:30 | 0.29704 | 0.26029 | 0.21372 | 0.16325 | 0.23861 | 0.25954 | 0.26407 | 0.17233 | 0.23993 | 0.19341 | 0.19553 | 0.18945 | 0.18901 | 0.15852 | 0.18300 | 0.18867 |
| 23:45 | 0.29691 | 0.26022 | 0.21355 | 0.16316 | 0.23861 | 0.25959 | 0.26397 | 0.17229 | 0.23993 | 0.19341 | 0.19559 | 0.18945 | 0.18901 | 0.15853 | 0.18300 | 0.18878 |
| 0:00 | 0.29686 | 0.26004 | 0.21340 | 0.16304 | 0.23848 | 0.25971 | 0.26390 | 0.17215 | 0.23991 | 0.19332 | 0.19565 | 0.18945 | 0.18901 | 0.15852 | 0.18300 | 0.18879 |
| 0:15 | 0.29670 | 0.25981 | 0.21328 | 0.16293 | 0.23839 | 0.25974 | 0.26375 | 0.17211 | 0.23995 | 0.19338 | 0.19568 | 0.18947 | 0.18908 | 0.15843 | 0.18300 | 0.18876 |
| 0:30 | 0.29652 | 0.25958 | 0.21325 | 0.16279 | 0.23839 | 0.25983 | 0.26371 | 0.17212 | 0.24004 | 0.19346 | 0.19568 | 0.18961 | 0.18923 | 0.15845 | 0.18300 | 0.18870 |
| 0:45 | 0.29636 | 0.25954 | 0.21311 | 0.16270 | 0.23828 | 0.25985 | 0.26353 | 0.17220 | 0.24000 | 0.19346 | 0.19568 | 0.18964 | 0.18921 | 0.15853 | 0.18306 | 0.18878 |
| 1:00 | 0.29622 | 0.25935 | 0.21298 | 0.16261 | 0.23818 | 0.25991 | 0.26352 | 0.17227 | 0.23997 | 0.19346 | 0.19567 | 0.18954 | 0.18917 | 0.15853 | 0.18315 | 0.18879 |
| 1:15 | 0.29619 | 0.25914 | 0.21281 | 0.16252 | 0.23820 | 0.25997 | 0.26335 | 0.17231 | 0.23999 | 0.19349 | 0.19567 | 0.18946 | 0.18912 | 0.15853 | 0.18312 | 0.18879 |
| 1:30 | 0.29603 | 0.25893 | 0.21265 | 0.16248 | 0.23815 | 0.26004 | 0.26349 | 0.17219 | 0.24013 | 0.19348 | 0.19566 | 0.18945 | 0.18920 | 0.15854 | 0.18306 | 0.18879 |
| 1:45 | 0.29587 | 0.25884 | 0.21255 | 0.16245 | 0.23809 | 0.26006 | 0.26333 | 0.17211 | 0.24017 | 0.19346 | 0.19567 | 0.18945 | 0.18923 | 0.15861 | 0.18303 | 0.18879 |

(continued)

**Table 3** (continued)

| Time | Water content at sensor 1 | Water content at sensor 2 | Water content at sensor 3 | Water content at sensor 4 | Water content at sensor 5 | Water content at sensor 6 | Water content at sensor 7 | Water content at sensor 8 | Water content at sensor 9 | Water content at sensor 10 | Water content at sensor 11 | Water content at sensor 12 | Water content at sensor 13 | Water content at sensor 14 | Water content at sensor 15 | Water content at sensor 16 |
|---|---|---|---|---|---|---|---|---|---|---|---|---|---|---|---|---|
| 2:00 | 0.29570 | 0.25873 | 0.21254 | 0.16234 | 0.23815 | 0.26015 | 0.26330 | 0.17210 | 0.24017 | 0.19347 | 0.19568 | 0.18945 | 0.18923 | 0.15870 | 0.18302 | 0.18879 |
| 2:15 | 0.29555 | 0.25852 | 0.21243 | 0.16235 | 0.23811 | 0.26016 | 0.26313 | 0.17210 | 0.24017 | 0.19346 | 0.19567 | 0.18945 | 0.18923 | 0.15867 | 0.18301 | 0.18879 |
| 2:30 | 0.29547 | 0.25831 | 0.21231 | 0.16224 | 0.23796 | 0.26024 | 0.26308 | 0.17212 | 0.24025 | 0.19350 | 0.19567 | 0.18945 | 0.18935 | 0.15855 | 0.18300 | 0.18882 |
| 2:45 | 0.29538 | 0.25812 | 0.21214 | 0.16229 | 0.23794 | 0.26025 | 0.26293 | 0.17216 | 0.24039 | 0.19364 | 0.19567 | 0.18945 | 0.18945 | 0.15853 | 0.18300 | 0.18888 |
| 3:00 | 0.29523 | 0.25812 | 0.21200 | 0.16213 | 0.23789 | 0.26038 | 0.26286 | 0.17211 | 0.24039 | 0.19357 | 0.19567 | 0.18945 | 0.18945 | 0.15853 | 0.18300 | 0.18895 |
| 3:15 | 0.29506 | 0.25796 | 0.21191 | 0.16216 | 0.23773 | 0.26035 | 0.26272 | 0.17210 | 0.24039 | 0.19364 | 0.19567 | 0.18945 | 0.18945 | 0.15854 | 0.18305 | 0.18889 |
| 3:30 | 0.29489 | 0.25774 | 0.21173 | 0.16208 | 0.23772 | 0.26041 | 0.26269 | 0.17209 | 0.24039 | 0.19358 | 0.19568 | 0.18945 | 0.18949 | 0.15854 | 0.18317 | 0.18882 |
| 3:45 | 0.29474 | 0.25755 | 0.21178 | 0.16200 | 0.23762 | 0.26041 | 0.26272 | 0.17195 | 0.24039 | 0.19350 | 0.19568 | 0.18945 | 0.18966 | 0.15855 | 0.18322 | 0.18879 |
| 4:00 | 0.29472 | 0.25740 | 0.21169 | 0.16209 | 0.23750 | 0.26050 | 0.26263 | 0.17202 | 0.24039 | 0.19347 | 0.19567 | 0.18945 | 0.18968 | 0.15866 | 0.18322 | 0.18879 |
| 4:15 | 0.29457 | 0.25739 | 0.21157 | 0.16196 | 0.23752 | 0.26048 | 0.26253 | 0.17210 | 0.24039 | 0.19349 | 0.19568 | 0.18945 | 0.18968 | 0.15874 | 0.18322 | 0.18879 |
| 4:30 | 0.29440 | 0.25722 | 0.21147 | 0.16200 | 0.23743 | 0.26046 | 0.26241 | 0.17199 | 0.24040 | 0.19356 | 0.19568 | 0.18945 | 0.18968 | 0.15876 | 0.18322 | 0.18879 |
| 4:45 | 0.29422 | 0.25703 | 0.21129 | 0.16187 | 0.23741 | 0.26062 | 0.26234 | 0.17188 | 0.24041 | 0.19364 | 0.19567 | 0.18945 | 0.18980 | 0.15876 | 0.18322 | 0.18890 |
| 5:00 | 0.29404 | 0.25685 | 0.21122 | 0.16188 | 0.23746 | 0.26057 | 0.26219 | 0.17197 | 0.24045 | 0.19368 | 0.19567 | 0.18946 | 0.18979 | 0.15876 | 0.18322 | 0.18901 |
| 5:15 | 0.29405 | 0.25667 | 0.21103 | 0.16188 | 0.23736 | 0.26059 | 0.26216 | 0.17201 | 0.24046 | 0.19365 | 0.19567 | 0.18945 | 0.18971 | 0.15875 | 0.18322 | 0.18901 |
| 5:30 | 0.29393 | 0.25666 | 0.21101 | 0.16187 | 0.23728 | 0.26063 | 0.26197 | 0.17188 | 0.24049 | 0.19364 | 0.19563 | 0.18945 | 0.18968 | 0.15874 | 0.18322 | 0.18897 |
| 5:45 | 0.29377 | 0.25650 | 0.21101 | 0.16174 | 0.23727 | 0.26064 | 0.26201 | 0.17188 | 0.24040 | 0.19366 | 0.19559 | 0.18945 | 0.18975 | 0.15870 | 0.18322 | 0.18894 |
| 6:00 | 0.29357 | 0.25625 | 0.21086 | 0.16182 | 0.23709 | 0.26074 | 0.26195 | 0.17188 | 0.24039 | 0.19368 | 0.19551 | 0.18945 | 0.18990 | 0.15866 | 0.18322 | 0.18900 |
| 6:15 | 0.29340 | 0.25607 | 0.21079 | 0.16169 | 0.23706 | 0.26064 | 0.26197 | 0.17184 | 0.24039 | 0.19368 | 0.19548 | 0.18945 | 0.18990 | 0.15861 | 0.18322 | 0.18901 |

(continued)

**Table 3** (continued)

| Time | Water content at sensor 1 | Water content at sensor 2 | Water content at sensor 3 | Water content at sensor 4 | Water content at sensor 5 | Water content at sensor 6 | Water content at sensor 7 | Water content at sensor 8 | Water content at sensor 9 | Water content at sensor 10 | Water content at sensor 11 | Water content at sensor 12 | Water content at sensor 13 | Water content at sensor 14 | Water content at sensor 15 | Water content at sensor 16 |
|---|---|---|---|---|---|---|---|---|---|---|---|---|---|---|---|---|
| 6:30 | 0.29336 | 0.25593 | 0.21059 | 0.16172 | 0.23699 | 0.26065 | 0.26178 | 0.17186 | 0.24039 | 0.19368 | 0.19546 | 0.18945 | 0.18990 | 0.15857 | 0.18323 | 0.18901 |
| 6:45 | 0.29330 | 0.25593 | 0.21052 | 0.16165 | 0.23684 | 0.26066 | 0.26175 | 0.17188 | 0.24039 | 0.19368 | 0.19546 | 0.18946 | 0.18990 | 0.15858 | 0.18329 | 0.18901 |
| 7:00 | 0.29311 | 0.25575 | 0.21036 | 0.16152 | 0.23684 | 0.26066 | 0.26159 | 0.17172 | 0.24040 | 0.19368 | 0.19546 | 0.18948 | 0.19003 | 0.15859 | 0.18337 | 0.18901 |
| 7:15 | 0.29290 | 0.25558 | 0.21028 | 0.16163 | 0.23672 | 0.26064 | 0.26152 | 0.17174 | 0.24039 | 0.19366 | 0.19546 | 0.18960 | 0.19012 | 0.15861 | 0.18344 | 0.18899 |
| 7:30 | 0.29272 | 0.25547 | 0.21018 | 0.16164 | 0.23675 | 0.26072 | 0.26136 | 0.17176 | 0.24039 | 0.19365 | 0.19546 | 0.18954 | 0.19012 | 0.15858 | 0.18344 | 0.18896 |
| 7:45 | 0.29262 | 0.25529 | 0.21026 | 0.16145 | 0.23663 | 0.26076 | 0.26130 | 0.17166 | 0.24039 | 0.19362 | 0.19546 | 0.18945 | 0.19012 | 0.15854 | 0.18345 | 0.18901 |
| 8:00 | 0.29239 | 0.25519 | 0.21013 | 0.16144 | 0.23650 | 0.26067 | 0.26116 | 0.17171 | 0.24039 | 0.19364 | 0.19546 | 0.18945 | 0.19012 | 0.15863 | 0.18345 | 0.18898 |

## Irrigation Depth ($i_t$)

The irrigation depth ($I_t$) is the depth of water applied by a dripper for a duration of '$t$'. The following formula is used for arriving at the value of $I_t$.

$$I_t = \frac{q \times 1000 \times t}{A} \tag{35}$$

where $q$ is the discharge rate of dripper in l/h and $t$ is the time duration of dripper operation in hours and $A$ (=30 × 210 cm$^2$) is the area irrigated by each dripper in cm$^2$ (Fig. 10). Dripper discharge rate is 4.41 l/h. Irrigation is started at 11.00 and stopped at the 12.30 for a duration of 1.5 h. Hence, $I_t$ works out to 1.05 cm and 0.175 cm depth of water is applied for every 15 min.

## Average Water Content in Root Zone

Depth of root zone is 45 cm, and area of root zone is 4725 cm$^2$ (i.e. 45 cm × 105 cm). Areal weighted averaging technique is used for finding the spatial mean volumetric water content using the following expression for each time period:

$$\bar{\theta} = 0.048(\theta_1 + \theta_5 + \theta_9) + 0.095(\theta_2 + \theta_6 + \theta_{10}) + 0.119(\theta_3 + \theta_7 + \theta_{11})$$
$$+ 0.071(\theta_4 + \theta_8 + \theta_{12}) \tag{36}$$

where $\bar{\theta}$ is spatial mean volumetric water content and $\theta_i$ is the volumetric water content at sensor '$i$'. The coefficients in the preceding equations are obtained by dividing the cell area by the total root zone (i.e. 4725 cm$^2$). For instance, the area of cells for the sensors 1, 5 and 9 is 225 cm$^2$ (i.e. 225/4725 = 0.048).

## Results of the Case Study

Preceding equations are used to estimate ET, and drainage occurring for every 15 min interval is given in Table 4. It can be seen that total ET for one day is 0.65196 cm and the total drainage is 0.03734 cm and the proportion of drainage is 5.73% of ET.

**Table 4** Computations for numerical data in Table 3

| Time | Mean water content | $(\bar{\theta}_{t+1} - \bar{\theta}_t) \times z$ (cm for 15 min) | Sensor 9 $K(\theta)$ cm/day | Sensor 13 $K(\theta)$ cm/day | Drainage between 9 & 13 cm/day | Sensor 10 $K(\theta)$ cm/day | Sensor 14 $K(\theta)$ cm/day | Drainage between 10 & 14 cm/day | Sensor 11 $K(\theta)$ cm/day | Sensor 15 $K(\theta)$ cm/day | Drainage between 11 & 15 cm/day | Sensor 12 $K(\theta)$ cm/day | Sensor 16 $K(\theta)$ cm/day | Drainage between 12 & 16 cm/day | Mean drainage (cm for 15 min) | ET for 15 min interval (cm) |
|---|---|---|---|---|---|---|---|---|---|---|---|---|---|---|---|---|
| 8:00 | 0.21276 | − 0.0031 | 0.0038 | 0.00074 | 0.05982 | 0.00095 | 0.00009 | 0.03924 | 0.00098 | 0.00048 | 0.01192 | 0.00072 | 0.00069 | 0.00152 | 0.00025 | 0.00282 |
| 8:15 | 0.21269 | − 0.0001 | 0.0038 | 0.00074 | 0.0599 | 0.00094 | 0.00009 | 0.03901 | 0.00098 | 0.00048 | 0.01193 | 0.00073 | 0.00069 | 0.00155 | 0.00025 | − 0.00012 |
| 8:30 | 0.21269 | − 0.0032 | 0.00381 | 0.00074 | 0.06001 | 0.00095 | 0.00009 | 0.03949 | 0.00098 | 0.00048 | 0.01194 | 0.00072 | 0.00069 | 0.00139 | 0.00025 | 0.00297 |
| 8:45 | 0.21262 | − 0.0025 | 0.0038 | 0.00074 | 0.05983 | 0.00094 | 0.00009 | 0.03919 | 0.00098 | 0.00048 | 0.01196 | 0.00072 | 0.00069 | 0.00149 | 0.00025 | 0.00222 |
| 9:00 | 0.21256 | − 0.0042 | 0.00379 | 0.00074 | 0.05978 | 0.00094 | 0.00009 | 0.03896 | 0.00098 | 0.00048 | 0.01199 | 0.00072 | 0.00068 | 0.00145 | 0.00025 | 0.00399 |
| 9:15 | 0.21247 | − 0.007 | 0.00378 | 0.00074 | 0.05943 | 0.00094 | 0.00009 | 0.03904 | 0.00098 | 0.00048 | 0.01202 | 0.00072 | 0.00068 | 0.00151 | 0.00025 | 0.00679 |
| 9:30 | 0.21231 | − 0.004 | 0.00375 | 0.00074 | 0.05887 | 0.00093 | 0.00009 | 0.0388 | 0.00097 | 0.00048 | 0.01196 | 0.00072 | 0.00068 | 0.00139 | 0.00025 | 0.00372 |
| 9:45 | 0.21222 | − 0.0052 | 0.00372 | 0.00074 | 0.0583 | 0.00093 | 0.00009 | 0.03902 | 0.00097 | 0.00048 | 0.01186 | 0.00072 | 0.00068 | 0.00148 | 0.00025 | 0.00492 |
| 10:00 | 0.21211 | − 0.0065 | 0.00371 | 0.00073 | 0.05858 | 0.00093 | 0.00009 | 0.0391 | 0.00097 | 0.00048 | 0.01176 | 0.00071 | 0.00068 | 0.00133 | 0.00025 | 0.00623 |
| 10:15 | 0.21196 | − 0.0036 | 0.00368 | 0.00073 | 0.05798 | 0.00093 | 0.00009 | 0.03897 | 0.00097 | 0.00048 | 0.01169 | 0.00071 | 0.00068 | 0.00128 | 0.00025 | 0.00338 |
| 10:30 | 0.21188 | − 0.0063 | 0.00369 | 0.00073 | 0.05828 | 0.00093 | 0.00009 | 0.03902 | 0.00097 | 0.00048 | 0.01176 | 0.00071 | 0.00068 | 0.0014 | 0.00025 | 0.00608 |
| 10:45 | 0.21174 | − 0.0061 | 0.00365 | 0.00072 | 0.05788 | 0.00092 | 0.00009 | 0.03893 | 0.00097 | 0.00048 | 0.01185 | 0.00071 | 0.00068 | 0.00129 | 0.00025 | 0.00587 |
| 11:00 | 0.21161 | 0.04414 | 0.00361 | 0.00072 | 0.05712 | 0.00092 | 0.00009 | 0.039 | 0.00096 | 0.00048 | 0.0118 | 0.00071 | 0.00068 | 0.00134 | 0.00025 | 0.13061 |
| 11:15 | 0.21259 | 0.1251 | 0.00359 | 0.00072 | 0.05685 | 0.00092 | 0.00008 | 0.03909 | 0.00096 | 0.00048 | 0.01174 | 0.00071 | 0.00068 | 0.00139 | 0.00025 | 0.04965 |
| 11:30 | 0.21537 | 0.04059 | 0.00359 | 0.00072 | 0.05674 | 0.00091 | 0.00008 | 0.0391 | 0.00096 | 0.00048 | 0.01167 | 0.0007 | 0.00068 | 0.00129 | 0.00025 | 0.13416 |
| 11:45 | 0.21627 | 0.06171 | 0.00357 | 0.00071 | 0.05668 | 0.00091 | 0.00008 | 0.039 | 0.00096 | 0.00048 | 0.01165 | 0.0007 | 0.00068 | 0.0013 | 0.00025 | 0.11304 |
| 12:00 | 0.21764 | 0.15265 | 0.00357 | 0.00071 | 0.05671 | 0.00091 | 0.00008 | 0.03915 | 0.00096 | 0.00048 | 0.01173 | 0.0007 | 0.00068 | 0.0013 | 0.00025 | 0.02210 |
| 12:15 | 0.22103 | 0.10676 | 0.00364 | 0.00071 | 0.05847 | 0.00091 | 0.00008 | 0.03873 | 0.00095 | 0.00047 | 0.01172 | 0.0007 | 0.00068 | 0.0013 | 0.00025 | 0.06799 |
| 12:30 | 0.22341 | 0.01531 | 0.00425 | 0.0007 | 0.07109 | 0.0009 | 0.00008 | 0.03883 | 0.00096 | 0.00047 | 0.0118 | 0.0007 | 0.00068 | 0.00127 | 0.00027 | − 0.01558 |
| 12:45 | 0.22375 | − 0.0122 | 0.00527 | 0.0007 | 0.09247 | 0.0009 | 0.00008 | 0.03863 | 0.00096 | 0.00047 | 0.01187 | 0.0007 | 0.00068 | 0.00115 | 0.0003 | 0.01193 |
| 13:00 | 0.22347 | − 0.009 | 0.00597 | 0.0007 | 0.1073 | 0.0009 | 0.00008 | 0.03872 | 0.00096 | 0.00047 | 0.01188 | 0.0007 | 0.00068 | 0.00122 | 0.00032 | 0.00870 |
| 13:15 | 0.22327 | − 0.0062 | 0.0064 | 0.00069 | 0.11711 | 0.00089 | 0.00008 | 0.03863 | 0.00096 | 0.00047 | 0.01187 | 0.0007 | 0.00068 | 0.00124 | 0.00034 | 0.00583 |

(continued)

**Table 4** (continued)

| Time | Mean water content | $(\bar{\theta}_{i+1} - \bar{\theta}_i) \times z$ (cm for 15 min) | Sensor 9 $K(\theta)$ cm/day | Sensor 13 $K(\theta)$ cm/day | Drainage between 9 & 13 cm/day | Sensor 10 $K(\theta)$ cm/day | Sensor 14 $K(\theta)$ cm/day | Drainage between 10 & 14 cm/day | Sensor 11 $K(\theta)$ cm/day | Sensor 15 $K(\theta)$ cm/day | Drainage between 11 & 15 cm/day | Sensor 12 $K(\theta)$ cm/day | Sensor 16 $K(\theta)$ cm/day | Drainage between 12 & 16 cm/day | Mean drainage (cm for 15 min) | ET for 15 min interval (cm) |
|---|---|---|---|---|---|---|---|---|---|---|---|---|---|---|---|---|
| 13:30 | 0.22314 | −0.0034 | 0.00676 | 0.00069 | 0.12475 | 0.00089 | 0.00008 | 0.0385 | 0.00096 | 0.00047 | 0.01184 | 0.00069 | 0.00068 | 0.00108 | 0.00035 | 0.00310 |
| 13:45 | 0.22306 | −0.0027 | 0.00703 | 0.00069 | 0.13117 | 0.00089 | 0.00008 | 0.03858 | 0.00096 | 0.00047 | 0.01183 | 0.00069 | 0.00068 | 0.00092 | 0.00036 | 0.00235 |
| 14:00 | 0.223 | −0.0024 | 0.00724 | 0.00068 | 0.13601 | 0.00083 | 0.00008 | 0.03832 | 0.00096 | 0.00047 | 0.01188 | 0.00069 | 0.00068 | 0.00105 | 0.00036 | 0.00203 |
| 14:15 | 0.22295 | −0.0017 | 0.0074 | 0.00068 | 0.1397 | 0.00088 | 0.00008 | 0.03825 | 0.00095 | 0.00047 | 0.01183 | 0.00069 | 0.00068 | 0.0011 | 0.00037 | 0.00130 |
| 14:30 | 0.22291 | 0.00104 | 0.00754 | 0.00058 | 0.14311 | 0.00088 | 0.00008 | 0.03826 | 0.00095 | 0.00047 | 0.01182 | 0.00069 | 0.00068 | 0.00108 | 0.00037 | −0.00141 |
| 14:45 | 0.22293 | −0.0001 | 0.00767 | 0.00058 | 0.14646 | 0.00088 | 0.00008 | 0.03822 | 0.00095 | 0.00047 | 0.01181 | 0.0007 | 0.00068 | 0.0011 | 0.00038 | −0.00026 |
| 15:00 | 0.22293 | −0.0007 | 0.00778 | 0.00068 | 0.14904 | 0.00087 | 0.00008 | 0.03779 | 0.00096 | 0.00047 | 0.01197 | 0.00069 | 0.00068 | 0.00108 | 0.00038 | 0.00032 |
| 15:15 | 0.22291 | −0.0009 | 0.00788 | 0.00068 | 0.15115 | 0.00088 | 0.00008 | 0.03796 | 0.00097 | 0.00047 | 0.01205 | 0.00069 | 0.00068 | 0.00108 | 0.00039 | 0.00048 |
| 15:30 | 0.2229 | −0.0006 | 0.00796 | 0.00068 | 0.15308 | 0.00088 | 0.00008 | 0.03821 | 0.00097 | 0.00047 | 0.01206 | 0.00069 | 0.00068 | 0.00108 | 0.00039 | 0.00017 |
| 15:45 | 0.22288 | 0.00049 | 0.00803 | 0.00067 | 0.15457 | 0.00087 | 0.00008 | 0.03795 | 0.00097 | 0.00047 | 0.01207 | 0.00069 | 0.00068 | 0.00109 | 0.00039 | −0.00088 |
| 16:00 | 0.22289 | 0.0019 | 0.00808 | 0.00067 | 0.15604 | 0.00087 | 0.00008 | 0.03774 | 0.00097 | 0.00047 | 0.01224 | 0.0007 | 0.00068 | 0.00115 | 0.00039 | −0.00229 |
| 16:15 | 0.22294 | −0.0001 | 0.00816 | 0.00067 | 0.15836 | 0.00087 | 0.00008 | 0.0379 | 0.00098 | 0.00047 | 0.01244 | 0.0007 | 0.00068 | 0.00118 | 0.0004 | −0.00026 |
| 16:30 | 0.22293 | −0.0011 | 0.00823 | 0.00067 | 0.16006 | 0.00088 | 0.00008 | 0.03799 | 0.00098 | 0.00047 | 0.01244 | 0.00069 | 0.00068 | 0.0011 | 0.0004 | 0.00066 |
| 16:45 | 0.22291 | −0.0013 | 0.00829 | 0.00067 | 0.16156 | 0.00087 | 0.00008 | 0.03783 | 0.00098 | 0.00047 | 0.01241 | 0.00069 | 0.00068 | 0.00109 | 0.0004 | 0.00092 |
| 17:00 | 0.22288 | 0.00054 | 0.00834 | 0.00067 | 0.16254 | 0.00087 | 0.00008 | 0.03764 | 0.00098 | 0.00047 | 0.01247 | 0.00069 | 0.00068 | 0.00108 | 0.0004 | −0.00094 |
| 17:15 | 0.22289 | 0.00037 | 0.00836 | 0.00067 | 0.16312 | 0.00088 | 0.00008 | 0.03789 | 0.00098 | 0.00047 | 0.01247 | 0.0007 | 0.00068 | 0.00117 | 0.0004 | −0.00077 |
| 17:30 | 0.2229 | −0.0004 | 0.0084 | 0.00067 | 0.16405 | 0.00088 | 0.00008 | 0.03812 | 0.00098 | 0.00047 | 0.01243 | 0.0007 | 0.00068 | 0.00129 | 0.00041 | 0.00000 |
| 17:45 | 0.22289 | −0.0002 | 0.00848 | 0.00067 | 0.16569 | 0.00087 | 0.00008 | 0.03757 | 0.00098 | 0.00047 | 0.01253 | 0.0007 | 0.00068 | 0.00124 | 0.00041 | −0.00024 |
| 18:00 | 0.22289 | 0.00157 | 0.00853 | 0.00067 | 0.16688 | 0.00088 | 0.00008 | 0.03766 | 0.00099 | 0.00047 | 0.01264 | 0.00069 | 0.00068 | 0.00111 | 0.00041 | −0.00198 |
| 18:15 | 0.22292 | −0.0007 | 0.00861 | 0.00067 | 0.16836 | 0.00088 | 0.00008 | 0.03797 | 0.00099 | 0.00047 | 0.01263 | 0.0007 | 0.00068 | 0.00128 | 0.00041 | 0.00030 |
| 18:30 | 0.22291 | 0.00118 | 0.00865 | 0.00067 | 0.16856 | 0.00088 | 0.00008 | 0.03788 | 0.001 | 0.00047 | 0.01272 | 0.0007 | 0.00068 | 0.0012 | 0.00041 | −0.00159 |
| 18:45 | 0.22293 | −0.002 | 0.00868 | 0.00068 | 0.16903 | 0.00088 | 0.00008 | 0.03781 | 0.001 | 0.00047 | 0.01291 | 0.0007 | 0.00068 | 0.00124 | 0.00042 | 0.00155 |

(continued)

**Table 4** (continued)

| Time | Mean water content | $(\bar{\theta}_{i+1} - \bar{\theta}_i) \times z$ (cm for 15 min) | Sensor 9 $K(\theta)$ cm/day | Sensor 13 $K(\theta)$ cm/day | Drainage between 9 & 13 cm/day | Sensor 10 $K(\theta)$ cm/day | Sensor 14 $K(\theta)$ cm/day | Drainage between 10 & 14 cm/day | Sensor 11 $K(\theta)$ cm/day | Sensor 15 $K(\theta)$ cm/day | Drainage between 11 & 15 cm/day | Sensor 12 $K(\theta)$ cm/day | Sensor 16 $K(\theta)$ cm/day | Drainage between 12 & 16 cm/day | Mean drainage (cm for 15 min) | ET for 15 min interval (cm) |
|---|---|---|---|---|---|---|---|---|---|---|---|---|---|---|---|---|
| 19:00 | 0.22289 | 0.00024 | 0.00876 | 0.00068 | 0.17073 | 0.00088 | 0.00008 | 0.03786 | 0.001 | 0.00047 | 0.01291 | 0.0007 | 0.00068 | 0.0013 | 0.00042 | − 0.00066 |
| 19:15 | 0.22289 | 0.00119 | 0.00884 | 0.00068 | 0.17261 | 0.00088 | 0.00008 | 0.03768 | 0.001 | 0.00047 | 0.01292 | 0.0007 | 0.00068 | 0.00115 | 0.00042 | − 0.00161 |
| 19:30 | 0.22292 | − 0.0025 | 0.00893 | 0.00068 | 0.17453 | 0.00089 | 0.00008 | 0.03785 | 0.00101 | 0.00047 | 0.01307 | 0.00071 | 0.00068 | 0.00123 | 0.00042 | 0.00211 |
| 19:45 | 0.22286 | − 0.0024 | 0.00899 | 0.00068 | 0.17518 | 0.00089 | 0.00008 | 0.03792 | 0.00101 | 0.00047 | 0.01306 | 0.00071 | 0.00068 | 0.00121 | 0.00043 | 0.00195 |
| 20:00 | 0.22281 | 0.00032 | 0.00901 | 0.00068 | 0.17521 | 0.00089 | 0.00008 | 0.03809 | 0.00101 | 0.00047 | 0.01293 | 0.00071 | 0.00068 | 0.00117 | 0.00043 | − 0.00075 |
| 20:15 | 0.22282 | 0.00015 | 0.00906 | 0.00068 | 0.17615 | 0.00089 | 0.00008 | 0.0381 | 0.00101 | 0.00048 | 0.01285 | 0.00071 | 0.00068 | 0.00135 | 0.00043 | − 0.00058 |
| 20:30 | 0.22282 | − 0.0022 | 0.00915 | 0.00068 | 0.17828 | 0.00089 | 0.00008 | 0.03799 | 0.00102 | 0.00048 | 0.01298 | 0.00071 | 0.00068 | 0.00132 | 0.00043 | 0.00177 |
| 20:45 | 0.22277 | − 0.0022 | 0.0092 | 0.00068 | 0.17937 | 0.00089 | 0.00008 | 0.0381 | 0.00102 | 0.00048 | 0.013 | 0.00071 | 0.00068 | 0.00127 | 0.00043 | 0.00173 |
| 21:00 | 0.22273 | − 0.0013 | 0.00928 | 0.00069 | 0.18056 | 0.0009 | 0.00008 | 0.03823 | 0.00102 | 0.00048 | 0.013 | 0.00071 | 0.00068 | 0.00131 | 0.00043 | 0.00083 |
| 21:15 | 0.2227 | − 0.0013 | 0.00931 | 0.00069 | 0.18101 | 0.0009 | 0.00008 | 0.03828 | 0.00102 | 0.00048 | 0.013 | 0.00071 | 0.00068 | 0.00132 | 0.00043 | 0.00090 |
| 21:30 | 0.22267 | − 0.0012 | 0.00931 | 0.00069 | 0.18144 | 0.00089 | 0.00008 | 0.03818 | 0.00102 | 0.00048 | 0.01307 | 0.00072 | 0.00068 | 0.00149 | 0.00044 | 0.00073 |
| 21:45 | 0.22264 | − 0.0011 | 0.00937 | 0.00069 | 0.18298 | 0.0009 | 0.00008 | 0.03846 | 0.00103 | 0.00048 | 0.01324 | 0.00071 | 0.00068 | 0.00135 | 0.00044 | 0.00065 |
| 22:00 | 0.22262 | − 0.0017 | 0.00938 | 0.00069 | 0.18315 | 0.0009 | 0.00008 | 0.03852 | 0.00103 | 0.00048 | 0.0133 | 0.00071 | 0.00068 | 0.00133 | 0.00044 | 0.00123 |
| 22:15 | 0.22258 | − 0.0031 | 0.00945 | 0.00069 | 0.18369 | 0.0009 | 0.00009 | 0.0384 | 0.00103 | 0.00048 | 0.01329 | 0.00072 | 0.00069 | 0.00136 | 0.00044 | 0.00264 |
| 22:30 | 0.22251 | − 0.0012 | 0.00947 | 0.00069 | 0.1841 | 0.0009 | 0.00009 | 0.03837 | 0.00103 | 0.00048 | 0.01315 | 0.00072 | 0.00069 | 0.00135 | 0.00044 | 0.00075 |
| 22:45 | 0.22249 | − 0.0017 | 0.00949 | 0.00069 | 0.1845 | 0.00091 | 0.00008 | 0.03885 | 0.00103 | 0.00048 | 0.01309 | 0.00072 | 0.00069 | 0.00139 | 0.00044 | 0.00126 |
| 23:00 | 0.22245 | − 0.0008 | 0.00956 | 0.00069 | 0.18609 | 0.00091 | 0.00008 | 0.03874 | 0.00103 | 0.00048 | 0.01309 | 0.00072 | 0.00068 | 0.00141 | 0.00044 | 0.00038 |
| 23:15 | 0.22243 | − 0.0019 | 0.00958 | 0.0007 | 0.18558 | 0.00091 | 0.00008 | 0.03865 | 0.00103 | 0.00048 | 0.01313 | 0.00072 | 0.00068 | 0.00151 | 0.00044 | 0.00146 |
| 23:30 | 0.22239 | − 0.0019 | 0.00965 | 0.0007 | 0.18686 | 0.00091 | 0.00009 | 0.03878 | 0.00103 | 0.00048 | 0.01319 | 0.00072 | 0.00069 | 0.00142 | 0.00045 | 0.00141 |
| 23:45 | 0.22235 | − 0.0028 | 0.00965 | 0.0007 | 0.18682 | 0.00091 | 0.00009 | 0.03875 | 0.00104 | 0.00048 | 0.01327 | 0.00072 | 0.00069 | 0.00133 | 0.00045 | 0.00234 |
| 0:00 | 0.22228 | − 0.0028 | 0.00965 | 0.0007 | 0.18665 | 0.00091 | 0.00009 | 0.03853 | 0.00104 | 0.00048 | 0.01334 | 0.00072 | 0.00069 | 0.00132 | 0.00045 | 0.00232 |
| 0:15 | 0.22222 | − 0.0008 | 0.00966 | 0.00071 | 0.18657 | 0.00091 | 0.00008 | 0.03887 | 0.00104 | 0.00048 | 0.01338 | 0.00072 | 0.00069 | 0.00137 | 0.00045 | 0.00039 |

(continued)

**Table 4** (continued)

| Time | Mean water content | $(\bar{\theta}_{t+1} - \bar{\theta}_t) \times z$ (cm for 15 min) | Sensor 9 $K(\theta)$ cm/day | Sensor 13 $K(\theta)$ cm/day | Drainage between 9 & 13 cm/day | Sensor 10$K(\theta)$ cm/day | Sensor 14 $K(\theta)$ cm/day | Drainage between 10 & 14 cm/day | Sensor 11 $K(\theta)$ cm/day | Sensor 15 $K(\theta)$ cm/day | Drainage between 11 & 15 cm/day | Sensor 12 $K(\theta)$ cm/day | Sensor 16 $K(\theta)$ cm/day | Drainage between 12 & 16 cm/day | Mean drainage (cm for 15 min) | ET for 15 min interval (cm) |
|---|---|---|---|---|---|---|---|---|---|---|---|---|---|---|---|---|
| 0:30 | 0.2222 | − 0.0023 | 0.0097 | 0.00071 | 0.18652 | 0.00092 | 0.00009 | 0.03904 | 0.00104 | 0.00048 | 0.01339 | 0.00073 | 0.00069 | 0.00154 | 0.00045 | 0.00184 |
| 0:45 | 0.22215 | − 0.0024 | 0.00969 | 0.00071 | 0.18632 | 0.00092 | 0.00009 | 0.03889 | 0.00104 | 0.00049 | 0.01332 | 0.00073 | 0.00069 | 0.00151 | 0.00045 | 0.00193 |
| 1:00 | 0.2221 | − 0.0026 | 0.00967 | 0.00071 | 0.18617 | 0.00092 | 0.00009 | 0.03888 | 0.00104 | 0.00049 | 0.01323 | 0.00073 | 0.00069 | 0.0014 | 0.00045 | 0.00220 |
| 1:15 | 0.22204 | − 0.0016 | 0.00968 | 0.00071 | 0.18672 | 0.00092 | 0.00009 | 0.03898 | 0.00104 | 0.00049 | 0.01326 | 0.00072 | 0.00069 | 0.00133 | 0.00045 | 0.00114 |
| 1:30 | 0.22201 | − 0.0024 | 0.00974 | 0.00071 | 0.18757 | 0.00092 | 0.00009 | 0.03893 | 0.00104 | 0.00049 | 0.01329 | 0.00072 | 0.00069 | 0.00132 | 0.00045 | 0.00193 |
| 1:45 | 0.22195 | − 0.0009 | 0.00976 | 0.00071 | 0.18775 | 0.00092 | 0.00009 | 0.03875 | 0.00104 | 0.00049 | 0.01334 | 0.00072 | 0.00069 | 0.00132 | 0.00045 | 0.00047 |
| 2:00 | 0.22193 | − 0.0027 | 0.00976 | 0.00071 | 0.18775 | 0.00092 | 0.00009 | 0.03863 | 0.00104 | 0.00049 | 0.01336 | 0.00072 | 0.00069 | 0.00132 | 0.00045 | 0.00227 |
| 2:15 | 0.22187 | − 0.002 | 0.00976 | 0.00071 | 0.18775 | 0.00092 | 0.00009 | 0.03867 | 0.00104 | 0.00048 | 0.01336 | 0.00072 | 0.00069 | 0.00132 | 0.00045 | 0.00151 |
| 2:30 | 0.22183 | − 0.0015 | 0.00979 | 0.00072 | 0.18785 | 0.00092 | 0.00009 | 0.03898 | 0.00104 | 0.00048 | 0.01337 | 0.00072 | 0.0007 | 0.00129 | 0.00045 | 0.00108 |
| 2:45 | 0.22179 | − 0.0016 | 0.00985 | 0.00072 | 0.18849 | 0.00093 | 0.00009 | 0.03939 | 0.00104 | 0.00048 | 0.01336 | 0.00072 | 0.0007 | 0.00124 | 0.00045 | 0.00115 |
| 3:00 | 0.22176 | − 0.0027 | 0.00985 | 0.00072 | 0.18854 | 0.00093 | 0.00009 | 0.03944 | 0.00104 | 0.00048 | 0.01336 | 0.00072 | 0.0007 | 0.00117 | 0.00045 | 0.00229 |
| 3:15 | 0.2217 | − 0.0027 | 0.00985 | 0.00072 | 0.18854 | 0.00093 | 0.00009 | 0.03936 | 0.00104 | 0.00049 | 0.01333 | 0.00072 | 0.0007 | 0.00123 | 0.00045 | 0.00225 |
| 3:30 | 0.22164 | − 0.002 | 0.00985 | 0.00072 | 0.18831 | 0.00092 | 0.00009 | 0.03921 | 0.00104 | 0.00049 | 0.01322 | 0.00072 | 0.0007 | 0.0013 | 0.00045 | 0.00153 |
| 3:45 | 0.22159 | − 0.0012 | 0.00985 | 0.00073 | 0.1873 | 0.00092 | 0.00009 | 0.03897 | 0.00104 | 0.00049 | 0.01317 | 0.00072 | 0.00069 | 0.00132 | 0.00045 | 0.00071 |
| 4:00 | 0.22157 | − 0.0017 | 0.00985 | 0.00073 | 0.1872 | 0.00092 | 0.00009 | 0.0387 | 0.00104 | 0.00049 | 0.01316 | 0.00072 | 0.00069 | 0.00132 | 0.00045 | 0.00121 |
| 4:15 | 0.22153 | − 0.0024 | 0.00985 | 0.00073 | 0.1872 | 0.00092 | 0.00009 | 0.03863 | 0.00104 | 0.00049 | 0.01317 | 0.00072 | 0.00069 | 0.00132 | 0.00045 | 0.00197 |
| 4:30 | 0.22148 | − 0.0024 | 0.00986 | 0.00073 | 0.18726 | 0.00092 | 0.00009 | 0.03878 | 0.00104 | 0.00049 | 0.01317 | 0.00072 | 0.00069 | 0.00132 | 0.00045 | 0.00193 |
| 4:45 | 0.22142 | − 0.0019 | 0.00986 | 0.00074 | 0.18663 | 0.00093 | 0.00009 | 0.03898 | 0.00104 | 0.00049 | 0.01316 | 0.00072 | 0.0007 | 0.00122 | 0.00045 | 0.00144 |
| 5:00 | 0.22138 | − 0.002 | 0.00987 | 0.00074 | 0.1869 | 0.00093 | 0.00009 | 0.03909 | 0.00104 | 0.00049 | 0.01316 | 0.00072 | 0.0007 | 0.00113 | 0.00045 | 0.00158 |
| 5:15 | 0.22134 | − 0.0021 | 0.00988 | 0.00073 | 0.18768 | 0.00093 | 0.00009 | 0.03902 | 0.00104 | 0.00049 | 0.01315 | 0.00072 | 0.0007 | 0.00113 | 0.00045 | 0.00161 |
| 5:30 | 0.22129 | − 0.0015 | 0.0099 | 0.00073 | 0.18816 | 0.00093 | 0.00009 | 0.03901 | 0.00104 | 0.00049 | 0.0131 | 0.00072 | 0.0007 | 0.00116 | 0.00045 | 0.00108 |
| 5:45 | 0.22126 | − 0.0027 | 0.00986 | 0.00074 | 0.18686 | 0.00093 | 0.00009 | 0.03914 | 0.00104 | 0.00049 | 0.01305 | 0.00072 | 0.0007 | 0.00119 | 0.00045 | 0.00224 |

(continued)

**Table 4** (continued)

| Time | Mean water content | $(\bar{\theta}_{t+1} - \bar{\theta}_t) \times z$ (cm for 15 min) | Sensor 9 $K(\theta)$ cm/day | Sensor 13 $K(\theta)$ cm/day | Drainage between 9 & 13 cm/day | Sensor 10 $K(\theta)$ cm/day | Sensor 14 $K(\theta)$ cm/day | Drainage between 10 & 14 cm/day | Sensor 11 $K(\theta)$ cm/day | Sensor 15 $K(\theta)$ cm/day | Drainage between 11 & 15 cm/day | Sensor 12 $K(\theta)$ cm/day | Sensor 16 $K(\theta)$ cm/day | Drainage between 12 & 16 cm/day | Mean drainage (cm for 15 min) | ET for 15 min interval (cm) |
|---|---|---|---|---|---|---|---|---|---|---|---|---|---|---|---|---|
| 6:00 | 0.2212 | − 0.0026 | 0.00985 | 0.00074 | 0.1859 | 0.00093 | 0.00009 | 0.03926 | 0.00103 | 0.00049 | 0.01295 | 0.00072 | 0.0007 | 0.00113 | 0.00044 | 0.00217 |
| 6:15 | 0.22114 | − 0.0029 | 0.00985 | 0.00074 | 0.18587 | 0.00093 | 0.00009 | 0.03935 | 0.00103 | 0.00049 | 0.01291 | 0.00072 | 0.0007 | 0.00112 | 0.00044 | 0.00243 |
| 6:30 | 0.22108 | − 0.0011 | 0.00985 | 0.00074 | 0.18587 | 0.00093 | 0.00009 | 0.03941 | 0.00103 | 0.00049 | 0.01288 | 0.00072 | 0.0007 | 0.00113 | 0.00044 | 0.00067 |
| 6:45 | 0.22105 | − 0.0037 | 0.00985 | 0.00074 | 0.18585 | 0.00093 | 0.00009 | 0.0394 | 0.00103 | 0.00049 | 0.01282 | 0.00072 | 0.0007 | 0.00113 | 0.00044 | 0.00326 |
| 7:00 | 0.22097 | − 0.0016 | 0.00986 | 0.00075 | 0.1852 | 0.00093 | 0.00009 | 0.03938 | 0.00103 | 0.0005 | 0.01274 | 0.00072 | 0.0007 | 0.00115 | 0.00044 | 0.00113 |
| 7:15 | 0.22093 | − 0.002 | 0.00985 | 0.00075 | 0.18456 | 0.00093 | 0.00009 | 0.03929 | 0.00103 | 0.0005 | 0.01268 | 0.00073 | 0.0007 | 0.00128 | 0.00044 | 0.00161 |
| 7:30 | 0.22089 | − 0.0023 | 0.00985 | 0.00075 | 0.18456 | 0.00093 | 0.00009 | 0.03931 | 0.00103 | 0.0005 | 0.01268 | 0.00073 | 0.0007 | 0.00126 | 0.00044 | 0.00186 |
| 7:45 | 0.22084 | − 0.0028 | 0.00985 | 0.00075 | 0.18456 | 0.00093 | 0.00009 | 0.03932 | 0.00103 | 0.0005 | 0.01267 | 0.00072 | 0.0007 | 0.00113 | 0.00044 | 0.00239 |
| 8:00 **Sum** | | | | | | | | | | | | | | | 0.03734 | 0.65196 |

## Multiple Cell–Multiple Step Model

In this model, root zone as well as zone surrounding the root zone is divided into number of cells and soil moisture sensing is done in each cell at frequent intervals. The change in water content occurring in each cell is due to ET as well as movement to and fro due to potential difference between the cells. The flow in root zone in vertical two dimensions is represented as below:

$$\frac{\partial}{\partial X}\left(K(h)\frac{\partial h}{\partial X}\right) + \frac{\partial}{\partial Z}\left(K(h)(\frac{\partial h}{\partial Z} - 1)\right) - \text{ETU} + \text{IRU} = \frac{\partial \theta}{\partial t} \qquad (37)$$

where ETU is evapotranspiration per unit volume of soil per unit time ($L^3 L^{-3} T^{-1}$) and IRU is irrigation per unit volume of soil per unit time. This equation is solved for each cell using finite difference methods to estimate ETU for each cell. Conventionally Evapotranspiration is expressed in depth units considering the total area (Fig. 11).

For application of finite difference method, for solving Eq. 37, spatial discretization and temporal discretization must be adopted. Notation $(i, j)$ is used for spatial discretization where '$i$' is used to denote column and '$j$' is used to denote row. $t$ and $t + 1$ are time steps. $x_{i,j}$ is distance of node $(i, j)$ in $X$-direction from origin. $z_{i,j}$ is distance of node $(i, j)$ in $Z$-direction from origin. $\theta_{i,j}^t$ is the water content of node $(i, j)$ at the start of time step '$t$'. $\text{ETU}_{i,j}^t$ is evapotranspiration per unit volume of soil per unit time at node $(i, j)$ between the start of time periods $t$ and $t + 1$. $\text{IRU}_{i,j}^t$ is irrigation per unit volume of soil per unit time happening at node $(i, j)$ [5] (Angaleeswari & Ravikumar, 2019).

**Fig. 11** Cell representation for finite difference form

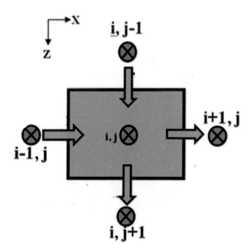

$$
\mathrm{ETU}^t_{i,j} = \mathrm{IRU}^t_{i,j} + \left\{ \left( K \left[ \frac{h^t_{i+1,j} + h^{t+1}_{i+1,j} + h^t_{i,j} + h^{t+1}_{i,j}}{4} \right] \left( \frac{\bar{h}^{t,t+1}_{i+1,j} - \bar{h}^{t,t+1}_{i,j}}{x_{i+1,j} - x_{i,j}} \right) \right. \right.
$$

$$
\left. \left. - K \left[ \frac{h^t_{i,j} + h^{t+1}_{i,j} + h^t_{i-1,j} + h^{t+1}_{i-1,j}}{4} \right] \left( \frac{\bar{h}^{t,t+1}_{i,j} - \bar{h}^{t,t+1}_{i-1,j}}{x_{i,j} - x_{i-1,j}} \right) \right) \right.
$$

$$
\left. \frac{1}{\frac{x_{i,j} + x_{i+1,j}}{2} - \frac{x_{i-1,j} + x_{i,j}}{2}} \right\}
$$

$$
+ \left\{ \left( K \left[ \frac{h^t_{i,j+1} + h^{t+1}_{i,j+1} + h^t_{i,j} + h^{t+1}_{i,j}}{4} \right] \right. \right.
$$

$$
\left( \frac{\bar{h}^{t,t+1}_{i,j+1} - \bar{h}^{t,t+1}_{i,j} - \left( z_{i,j+1} - z_{i,j} \right)}{z_{i,j+1} - z_{i,j}} \right)
$$

$$
- K \left[ \frac{h^t_{i,j} + h^{t+1}_{i,j} + h^t_{i,j-1} + h^{t+1}_{i,j-1}}{4} \right]
$$

$$
\left. \left. \left( \frac{\bar{h}^{t,t+1}_{i,j} - \bar{h}^{t,t+1}_{i,j-1} - \left( z_{i,j} - z_{i,j-1} \right)}{z_{i,j} - z_{i,j-1}} \right) \right) \right.
$$

$$
\left. \frac{1}{\frac{z_{i,j} + z_{i,j+1}}{2} - \frac{z_{i,j-1} + z_{i,j}}{2}} \right\} - \left\{ \frac{\theta^{t+1}_{i,j} - \theta^t_{i,j}}{\Delta t} \right\} \tag{38}
$$

where

$$
\bar{h}^{t,t+1}_{i,j} = \frac{h^t_{i,j} + h^{t+1}_{i,j}}{2}
$$

Note that $K[\dots]$ is used to denote $K$ as a function of the terms inside the brackets in Eq. 38.

When a cell for which ET is estimated, if it is surrounded with cells all around Eq. 38 can be used as such. Equation 38 must be suitably modified to take into account the boundary conditions.

For the left top most cell (sensor 1), the equation becomes as follows:

$$
\mathrm{ETU}^t_{i,j} = \mathrm{IRU}^t_{i,j} + \left\{ \left( K \left[ \frac{h^t_{i+1,j} + h^{t+1}_{i+1,j} + h^t_{i,j} + h^{t+1}_{i,j}}{4} \right] \left( \frac{\bar{h}^{t,t+1}_{i+1,j} - \bar{h}^{t,t+1}_{i,j}}{(x_{i+1,j} - x_{i,j})^2} \right) \right) \right\}
$$

$$
+ \left\{ \left( K \left[ \frac{h^t_{i,j+1} + h^{t+1}_{i,j+1} + h^t_{i,j} + h^{t+1}_{i,j}}{4} \right] \right. \right.
$$

$$
\left. \left. \left( \frac{\bar{h}^{t,t+1}_{i,j+1} - \bar{h}^{t,t+1}_{i,j} - \left( z_{i,j+1} - z_{i,j} \right)}{\left( z_{i,j+1} - z_{i,j} \right)^2} \right) \right) \right\}
$$

$$-\left\{\frac{\theta_{i,j}^{t+1} - \theta_{i,j}^{t}}{\Delta t}\right\} \tag{39}$$

Note that for the term $\frac{x_{i,j}+x_{i+1,j}}{2} - \frac{x_{i-1,j}+x_{i,j}}{2}$ has been made equal to $(x_{i+1,j} - x_{i,j})$ and also the term $\frac{z_{i,j}+z_{i,j+1}}{2} - \frac{z_{i,j-1}+z_{i,j}}{2}$ has been made equal to $(z_{i,j+1} - z_{i,j})$ in the preceding equation. This is because for every node for which ET is calculated at the boundaries, if we assume the existence of fictitious nodes with equal potential of the node on the no-flow boundary, flow would not occur from the fictitious nodes (Fig. 13).

For other surface cells except corners (sensors 2 and 3), the equation becomes as follows:

$$\text{ETU}_{i,j}^{t} = \text{IRU}_{i,j}^{t} + \left\{ \left( K\left[\frac{h_{i+1,j}^{t} + h_{i+1,j}^{t+1} + h_{i,j}^{t} + h_{i,j}^{t+1}}{4}\right] \left(\frac{\bar{h}_{i+1,j}^{t,t+1} - \bar{h}_{i,j}^{t,t+1}}{x_{i+1,j} - x_{i,j}}\right) \right.\right.$$
$$\left.\left. - K\left[\frac{h_{i,j}^{t} + h_{i,j}^{t+1} + h_{i-1,j}^{t} + h_{i-1,j}^{t+1}}{4}\right] \left(\frac{\bar{h}_{i,j}^{t,t+1} - \bar{h}_{i-1,j}^{t,t+1}}{x_{i,j} - x_{i-1,j}}\right)\right) \right.$$
$$\left. \frac{1}{\frac{x_{i,j}+x_{i+1,j}}{2} - \frac{x_{i-1,j}+x_{i,j}}{2}} \right\}$$
$$+ \left\{ \left( K\left[\frac{h_{i,j+1}^{t} + h_{i,j+1}^{t+1} + h_{i,j}^{t} + h_{i,j}^{t+1}}{4}\right] \right.\right.$$
$$\left.\left. \left(\frac{\bar{h}_{i,j+1}^{t,t+1} - \bar{h}_{i,j}^{t,t+1} - (z_{i,j+1} - z_{i,j})}{(z_{i,j+1} - z_{i,j})^2}\right)\right) \right\}$$
$$+ -\left\{\frac{\theta_{i,j}^{t+1} - \theta_{i,j}^{t}}{\Delta t}\right\} \tag{40}$$

For the right most corner surface cell (sensor 4), the equation becomes as follows:

$$\text{ETU}_{i,j}^{t} = \text{IRU}_{i,j}^{t} + \left\{ \left( - K\left[\frac{h_{i,j}^{t} + h_{i,j}^{t+1} + h_{i-1,j}^{t} + h_{i-1,j}^{t+1}}{4}\right] \left(\frac{\bar{h}_{i,j}^{t,t+1} - \bar{h}_{i-1,j}^{t,t+1}}{(x_{i,j} - x_{i-1,j})^2}\right)\right) \right\}$$
$$+ \left\{ \left( K\left[\frac{h_{i,j+1}^{t} + h_{i,j+1}^{t+1} + h_{i,j}^{t} + h_{i,j}^{t+1}}{4}\right] \right.\right.$$
$$\left.\left. \left(\frac{\bar{h}_{i,j+1}^{t,t+1} - \bar{h}_{i,j}^{t,t+1} - (z_{i,j+1} - z_{i,j})}{(z_{i,j+1} - z_{i,j})^2}\right)\right) \right\} - \left\{\frac{\theta_{i,j}^{t+1} - \theta_{i,j}^{t}}{\Delta t}\right\} \tag{41}$$

For the other boundary cells 5, 9, 8 and 12, the same principles can be adopted to write the correct finite difference form.

For subsurface drip irrigation, the cells may be so selected to accommodate the dripper in any one cell. For surface drip irrigation, if the wetted surface falls within the area of one surface cell, it is easy to attach the irrigation to that cell. If the area of wetted surface occupies more than one surface cell, the irrigation from the dripper must be accordingly shared to each surface cell. The wetted surface itself is dynamic and varies with time. So when we want to account all these conditions, the computations become still more complicated. But computations for all these situations are very well accounted in the software packages like Hydrus.

By using the preceding equations, during non-rainy periods and during the time when irrigation is not done, we can get a good estimate of ET. Most of the time, significant amount of error comes in the result due to erroneous representation of unsaturated hydraulic conductivity parameters.

The following equation is used for finding normalized ET distribution ($ETD_{i,j}^t$):

$$ETD_{i,j}^t = \frac{ETU_{i,j}^t}{\sum_{i=1}^n \sum_{j=1}^n ETU_{i,j}^t} \tag{42}$$

For root zone modelling, we need to express the root water uptake as a function of space. The root water uptake function used in Hydrus for two-dimensional flow is as follows [22]:

$$b(x,z) = \left(1 - \frac{x}{x_m}\right)\left(1 - \frac{z}{z_m}\right)e^{-\left\{\left(\frac{p_x}{x_m}\right)|x^*-x| + \left(\frac{p_z}{z_m}\right)|z^*-z|\right\}} \tag{43}$$

where

$x_m$                                      is the maximum distance to which root zone exists in X-direction;

$z_m$                                       is the maximum depth of root zone in Z-direction (Fig. 12);

$x^*, z^*, p_x$ and $p_z$       are fitting shape parameters.

Depth of evapotranspiration considering only rooting area

$$ET = \frac{\sum_{i=1}^n \sum_{j=1}^n ETU_{i,j}^t}{x_m \times 1} \tag{44}$$

But conventionally, the depth of transpiration is expressed considering total area. In that case, the depth of transpiration 'ET' is as follows (Fig. 12).

$$ET = \frac{\sum_{i=1}^n \sum_{j=1}^n ETU_{i,j}^t}{\frac{x_p}{2} \times 1} \tag{45}$$

**Fig. 12** Rooting zone and crop row spacing

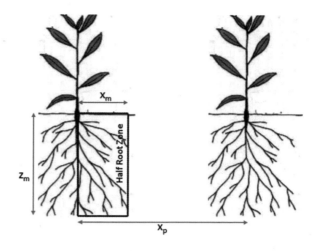

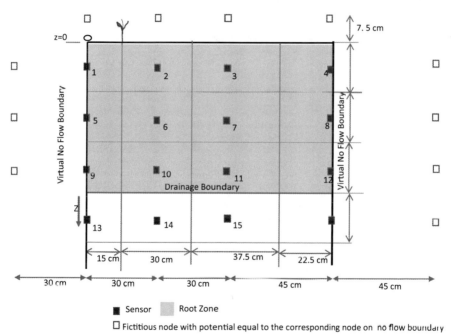

**Fig. 13** Fictitious nodes for boundary nodes for accounting boundary conditions

where $x_p$ is spacing between plant rows.

## Problem

For the case study discussed in the single cell–multiple step model section, find out the ET for the surface cells alone for the time between 13.30 and 13.45. During this time period, no irrigation is done.

**Solution**

For calculating ET for surface cells, the water content data for the sensors 1 to 8 is needed, and they are reproduced below:

| Time | Water contents at | | | | | | | |
|------|-----------|-----------|-----------|-----------|-----------|-----------|-----------|-----------|
|      | Sensor 1 | Sensor 2 | Sensor 3 | Sensor 4 | Sensor 5 | Sensor 6 | Sensor 7 | Sensor 8 |
| 13:30 | 0.31293 | 0.27169 | 0.21454 | 0.16291 | 0.24442 | 0.25458 | 0.26205 | 0.17085 |
| 13:45 | 0.31107 | 0.27093 | 0.21497 | 0.16302 | 0.24372 | 0.25446 | 0.26225 | 0.17099 |

**Step 1**

Convert all the water contents to pressure heads by van Genuchten–Mualem equation.

$$h = -\frac{1}{\alpha}\left[\left(\frac{\theta_s - \theta_r}{\theta - \theta_r}\right)^{\frac{1}{m}} - 1\right]^{\frac{1}{n}}$$

$\theta_r = 0.063$, $\theta_s = 0.384$, $\alpha = 0.021$ cm$^{-1}$, n = 1.33 and K$_s$ = 13.2 cm/d.

| Time | Pressure heads at | | | | | | | |
|------|-----------|-----------|-----------|----------|-----------|-----------|-----------|----------|
|      | Sensor 1 | Sensor 2 | Sensor 3 | Sensor 4 | Sensor 5 | Sensor 6 | Sensor 7 | Sensor 8 |
| 13:30 | − 94.34 | − 179.13 | − 440.61 | − 1576.26 | − 274.05 | − 191.41 | − 171.89 | − 1240.32 |
| 13:45 | − 94.63 | − 179.65 | − 442.28 | − 1579.13 | − 274.10 | − 190.90 | − 171.89 | − 1239.63 |

For cell 1, apply Eq. 39; for cells 2 and 3 apply Eq. 40; and for cell 4, apply Eq. 41. The results are as follows:

| Cell number | 1 | 2 | 3 | 4 |
|-------------|------|------|------|------|
| ETU during time 13:30 and 13:45 (cm$^3$/cm$^3$ of soil/ day) | 0.0020 | 0.0171 | 0.0028 | 0.0056 |
| Evapotranspiration during time 13:30 and 13:45 for the cell depth (cm/day) = ETU × Depth of cell (i.e. 15 cm) | 0.0300 | 0.2565 | 0.0420 | 0.0840 |

**Problem**

In the following table, the evapotranspiration per unit volume of soil per day is provided for the case study in Fig. 13. Find out the optimal Vrugt's water uptake parameters.

| Sensor | 1 | 2 | 3 | 4 | 5 | 6 | 7 | 8 | 9 | 10 | 11 | 12 |
|---|---|---|---|---|---|---|---|---|---|---|---|---|
| ETU (cm³/(cm³-d) | 0.3 | 0.2 | 0.1 | 0.05 | 0.2 | 0.1 | 0.05 | 0.03 | 0.02 | 0.01 | 0.002 | 0.001 |

## Solution

First convert all ETU values into normalized evapotranspiration distribution values using the following equation:

$$ETD_{i,j}^t = \frac{ETU_{i,j}^t}{\sum_{i=1}^{n} \sum_{j=1}^{n} ETU_{i,j}^t}$$

| Sensor | 1 | 2 | 3 | 4 | 5 | 6 | 7 | 8 | 9 | 10 | 11 | 12 |
|---|---|---|---|---|---|---|---|---|---|---|---|---|
| ETD (cm³/(cm³-d) | 0.282 | 0.188 | 0.094 | 0.047 | 0.188 | 0.094 | 0.047 | 0.028 | 0.019 | 0.009 | 0.002 | 0.001 |

The fitted parameters using MS-Excel Solver are as follows:

| Px | Pz | x* | z* |
|---|---|---|---|
| 0.69 | 6.11 | 4.33 | 15.07 |

The following table provides the data for comparison of estimated $b(x, z)$ using the fitted parameters and the observed ETD $(x, z)$.

| Distance in X-direction (cm) | Distance in Y-direction (cm) | Estimated $b(x, z)$ | Observed ETD $(x, z)$ | Squared deviation |
|---|---|---|---|---|
| (1) | (2) | (3) | (4) | $((3)-(4))^2$ |
| 0 | 7.5 | 0.2895 | 0.2822 | 5.3607E−05 |
| 30 | 7.5 | 0.1797 | 0.1881 | 7.1443E−05 |
| 60 | 7.5 | 0.0885 | 0.0941 | 3.1227E−05 |
| 105 | 7.5 | 0.0000 | 0.0470 | 2.2125E−03 |
| 0 | 22.5 | 0.1773 | 0.1881 | 1.1825E−04 |
| 30 | 22.5 | 0.1100 | 0.0941 | 2.5424E−04 |
| 60 | 22.5 | 0.0542 | 0.0470 | 5.0958E−05 |
| 105 | 22.5 | 0.0000 | 0.0282 | 7.9648E−04 |
| 0 | 37.5 | 0.0077 | 0.0188 | 1.2339E−04 |
| 30 | 37.5 | 0.0048 | 0.0094 | 2.1387E−05 |
| 60 | 37.5 | 0.0024 | 0.0019 | 2.2433E−07 |
| 105 | 37.5 | 0.0000 | 0.0009 | 8.8498E−07 |
| | | | Sum | 3.7345E−03 |

## Derivation of Reactive Transport Solute Transport Equation

Any chemical or fertilizer applied to soil undergoes processes such as adsorption to soil particles, chemical reactions, convective transport and diffusion. Accounting all these processes in a soil medium accurately as of now is impossible because of the complexities in flow path as well as the complexities in different processes. However, the research in reactive transport modelling in soils continues to grow and useful real-time applications are happening nowadays.

## *Diffusion*

In a multicomponent environment in soil water, due to repeated collisions, solutes with higher concentration tends to move to lower concentration. The diffusive flux is conventionally expressed by Fick's law and is as follows:

$$F_d = -\theta D_0 \beta \frac{\partial c}{\partial x} \tag{46}$$

where $F_d$ is called diffusive flux density ($ML^{-2} T^{-1}$); '$c$' is concentration of solute ($ML^{-3}$); $\frac{\partial c}{\partial x}$ is concentration gradient in $X$-direction; $D_0$ is called as diffusion coefficient for free water. Its value for many chemicals is very low and it is around 1 cm$^2$/day. In soils, the diffusive flux gets reduced due to the tortuous flow in the soil. $\beta$ is reduction factor to account for the tortuous flow in soils. Tortuosity is an intrinsic property of a soil defined as the ratio of actual flow path length to the straight distance between the ends. Millington and Quirk [19] has obtained a relationship for $\beta$ as a function of water content ($\theta$) and saturated water content ($\theta_s$) which is as follows:

$$\beta(\theta) = \frac{\theta^{7/3}}{(\theta_s)^2} \tag{47}$$

## *Convection*

Recall that the Darcian velocity does not account for the porosity of the soil and average pore velocity in soils is Darcian velocity divided by the porous cross-sectional area in which flow occurs. In case of saturated flow, flow occurs through total porous area. In case of unsaturated flow, flow occurs only through the water filled cross-sectional area in pores. Hence, the average pore velocity of flow in soils is the Darcian velocity divided by the water content. When flow occurs, solute is carried along with the flow. This process is called as convective transport. Term advective transport is also synonymously used.

$$F_c = v\theta c \tag{48}$$

where $F_c$ is called convective flux density ($ML^{-2} T^{-1}$); '$v$' is average pore velocity in the direction of flow.

Flow occurring through larger pores are faster than smaller pores due to friction effect of soil. Some path lengths are smaller and some path lengths are larger (Fig. 14). In reality convective effect of carrying the solute by the water does not happen with average pore velocity. Therefore convective flux accounted by Eq. 48 is not a complete representation.

This fact can be easily understood in the leaching column studies (Fig. 18). In a packed soil column of length '$L$', let us assume that water is flowing with a steady

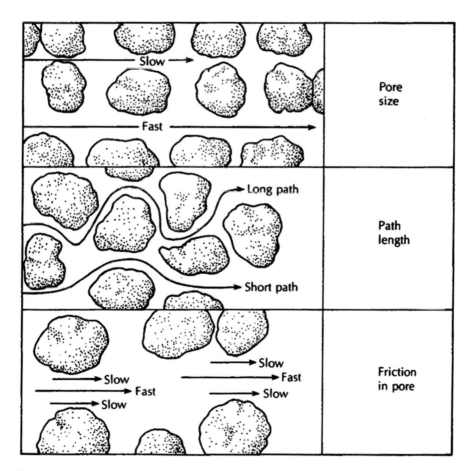

**Fig. 14** Sources for hydrodynamic dispersion

state. Let us assume a solute of concentration ($c_0$) is introduced at time $t_0$. If water through the porous soil column flows with average pore velocity ($v$), we must be able to collect the effluent with concentration ($c_0$) after the time of $v/L$. Before $v/L$ time, we must not get any solute in the leachate. But in reality, we would get leachate with concentration less than $c_0$ earlier than $v/L$ time. This is obviously because through some pores, water passes faster and the solute concentration gets reduced along the way due to mixing of flow from many other pores. Therefore a new terminology, namely hydrodynamic, dispersion was introduced.

## *Hydrodynamic Dispersion*

Hydrodynamic dispersion accounts for variations in pore velocities and consequent mixing apart from the convective flux. It must be understood that accurate representation of this process is almost impossible and hence only approximate representations of this processes are implemented. Many laboratory and field experiments are were done by scientists to understand this process.

In one-dimensional vertical flow, if a solute plume is introduced as in Fig. 15, if only convection as we defined in our convective flux occurs, the plume after time '$t$' would retain the shape and concentration. But in reality, the plume would get enlarged and assume an ellipsoid shape with reduction in concentration. It was found from experiments that the change in shape, size and concentration depended on soil characteristics and pore velocity. The dispersion was seen in the general direction of flow as well as transverse to the general direction of flow. Hence dispersive flux was also defined to occur similar to Fick's law.

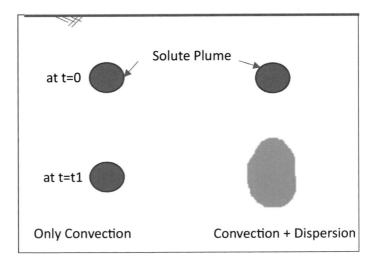

**Fig. 15** Convection and dispersion in a steady one-dimensional vertical flow

Dispersive flux ($F_h^L$) in the direction of flow is defined with the following equation:

$$F_h^L = -\theta D_L \frac{\partial c}{\partial l} \tag{49}$$

where $D_L$ is longitudinal hydrodynamic dispersion coefficient and it is a product of longitudinal ($\varepsilon_L$) dispersivity and average pore velocity in the direction of flow ($v_L$). $\varepsilon_L$ is a property of soil with dimension of length.

Dispersive flux ($F_h^T$) in the transverse direction of flow is defined with the following equation:

$$F_h^T = -\theta D_T \frac{\partial c}{\partial T} \tag{50}$$

where $D_T$ is transverse hydrodynamic dispersion coefficient, and it is product of longitudinal ($\varepsilon_T$) dispersivity and average pore velocity in the direction of flow ($v_L$).

One very big difficulty in developing methodologies for handling dispersive flux arises due to a fact that the when continuity equations are derived we need to resolve the dispersive flux for the Cartesian coordinate directions. The concept of dispersion hinges on the dispersion in the direction of flow and in the direction transverse to flow. When we want to estimate the dispersive flux for any coordinate directions, the dispersive flux along the general direction of flow at any location must be resolved.

Bear [6, 7] resolved this issue and provided the following equations:

If v is velocity of flow and has any arbitrary direction with components $v_x$, $v_y$, $v_z$ in x-, y- *and* z-directions, then the dispersive fluxes in X-, Y- and Z-directions are as given below (Fig. 16).

**Fig. 16** Resolving
dispersive flux to coordinate
directions

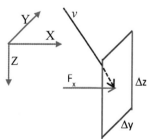

Plane is parallel to YZ plane
$F_x$ is perpendicular to YZ plane
$v$ has any arbitrary direction
with components $v_x, v_y, v_z$ in X, Y, Z
directions

$$F_x = -\left(D_{xx}\frac{\partial c}{\partial x} + D_{xy}\frac{\partial c}{\partial y} + D_{xz}\frac{\partial c}{\partial z}\right) \tag{51}$$

$$F_y = -\left(D_{yy}\frac{\partial c}{\partial x} + D_{yz}\frac{\partial c}{\partial y} + D_{yx}\frac{\partial c}{\partial z}\right) \tag{52}$$

$$F_z = -\left(D_{zz}\frac{\partial c}{\partial x} + D_{zy}\frac{\partial c}{\partial y} + D_{zx}\frac{\partial c}{\partial z}\right) \tag{53}$$

where the $D_{i,j}$ is called as dispersivity tensor and found out using the following relations:

$$D_{xx} = \varepsilon_L \frac{v_x^2}{|v|} + \varepsilon_T \frac{v_y^2}{|v|} + \varepsilon_T \frac{v_z^2}{|v|} \tag{54}$$

$$D_{yy} = \varepsilon_L \frac{v_y^2}{|v|} + \varepsilon_T \frac{v_x^2}{|v|} + \varepsilon_T \frac{v_z^2}{|v|} \tag{55}$$

$$D_{zz} = \varepsilon_L \frac{v_z^2}{|v|} + \varepsilon_T \frac{v_x^2}{|v|} + \varepsilon_T \frac{v_y^2}{|v|} \tag{56}$$

$$D_{xy} = D_{yx} = (\varepsilon_L - \varepsilon_T)\frac{v_x v_y}{|v|} \tag{57}$$

$$D_{xz} = D_{zx} = (\varepsilon_L - \varepsilon_T)\frac{v_x v_z}{|v|} \tag{58}$$

$$D_{yz} = D_{yz} = (\varepsilon_L - \varepsilon_T)\frac{v_y v_z}{|v|} \tag{59}$$

## *Adsorption*

Certain solutes have the tendency to adsorb to soil matrix, and a simple linear equation as mentioned below is used often:

$$s = k_d c \tag{60}$$

where '$s$' is adsorbed solute in soil per unit mass of soil when '$c$' is the concentration of solute in soil water. $k_d$ is called as distribution coefficient ($M^{-1} L^3$).

## *Chemical Reaction*

First-order reactions are often considered. As per the first-order reaction, the rate of change of concentration is proportional to the concentration of solute. Equation for the first-order reaction is as follows:

$$\frac{\partial c}{\partial t} = -\mu c \tag{61}$$

where $\mu$ is first-order reaction constant for any solute. A solute entering into a reaction gets converted into another solute. The mass of one solute lost in a chemical reaction is to be accounted as another solute, and the produced mass must be accounted in a chain of reactions as the reaction of urea in soil.

## *Urea Reactions in Soil*

Urea is the primary nitrogen fertilizer used by farmers for almost all crops and is also highly reactive fertilizer. For optimal management of urea application to crops, reactive transport modelling in root zone is very much useful.

When urea dissolves in water, it gets converted into ammonium by hydrolysis process with a reaction rate constant of $(\mu_a)$ (Fig. 17). Part of this ammonium gets volatilized into ammonia and lost into the atmosphere with reaction rate of $(\mu_v)$. Ammonium has a property of getting adsorbed to the soil matrix. The ammonium then converts into nitrate by nitrification process with reaction rate $(\mu_n)$. The nitrate may also get denitrified into nitrogen gas. In drip and sprinkler irrigation, mostly denitrification does not occur significantly, and hence, denitrification can be neglected.

**Fig. 17**  Urea transformation in soils

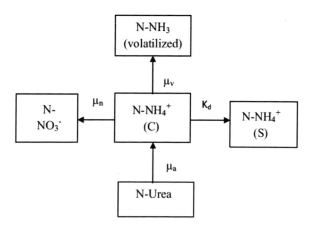

**Fig. 18** Typical soil column
leaching experimental set-up
[8]

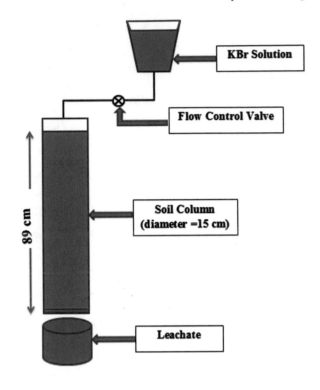

## *Continuity Equations for Reactive Solute Transport*

We will consider the plant uptake of nutrients passively with root water uptake
(transpiration) by plants with the same concentration of the nutrient in soil water.
The root water uptake (TU) which is defined as the volumetric water uptake per unit
volume of soil. Let '$\rho$' be the bulk density of the soil. Let $c^u$, $c^a$ and $c^n$ be soil solute
concentrations of urea, ammonium and nitrate, respectively. Let '$s^a$' be the adsorbed
ammonium per unit mass of soil.

By the rule of conservation of mass, we can write the following equations for the
reactive transport for urea, ammonium and nitrate separately as follows:

**For Urea**:

$$
\begin{aligned}
\frac{\partial(\theta c^u)}{\partial t} &= \frac{\partial}{\partial x}\left(\theta\left(D_{xx}\frac{\partial c^u}{\partial x} + D_{xy}\frac{\partial c^u}{\partial y} + D_{xz}\frac{\partial c^u}{\partial z}\right)\right) \\
&+ \frac{\partial}{\partial y}\left(\theta\left(D_{yy}\frac{\partial c^u}{\partial y} + D_{xy}\frac{\partial c^u}{\partial x} + D_{yz}\frac{\partial c^u}{\partial z}\right)\right) \\
&+ \frac{\partial}{\partial z}\left(\theta\left(D_{zz}\frac{\partial c^u}{\partial z} + D_{xz}\frac{\partial c^u}{\partial x} + D_{zy}\frac{\partial c^u}{\partial y}\right)\right)
\end{aligned}
$$

$$-\frac{\partial}{\partial x}\left(\theta v_x c^u\right) - \frac{\partial}{\partial y}\left(\theta v_y c^u\right) - \frac{\partial}{\partial z}\left(\theta v_z c^u\right)$$

$$-\mu_a \theta c^u - (TU)\, c^u \tag{62}$$

**For Ammonium:**

$$\frac{\partial(\theta c^a)}{\partial t} + \rho\frac{\partial s^a}{\partial t} = \frac{\partial}{\partial x}\left(\theta\left(D_{xx}\frac{\partial c^a}{\partial x} + D_{xy}\frac{\partial c^a}{\partial y} + D_{xz}\frac{\partial c^a}{\partial z}\right)\right)$$

$$+ \frac{\partial}{\partial y}\left(\theta\left(D_{yy}\frac{\partial c^a}{\partial y} + D_{xy}\frac{\partial c^a}{\partial x} + D_{yz}\frac{\partial c^a}{\partial z}\right)\right)$$

$$+ \frac{\partial}{\partial z}\left(\theta\left(D_{zz}\frac{\partial c^a}{\partial z} + D_{xz}\frac{\partial c^a}{\partial x} + D_{zy}\frac{\partial c^a}{\partial y}\right)\right) - \frac{\partial}{\partial x}\left(\theta v_x c^a\right)$$

$$- \frac{\partial}{\partial y}\left(\theta v_y c^a\right) - \frac{\partial}{\partial z}\left(\theta v_z c^a\right) - \mu_v \theta c^a - \mu_n \theta c^a$$

$$- (TU)c^a + \mu_a \theta c^u \tag{63}$$

**For Nitrate:**

$$\frac{\partial(\theta c^n)}{\partial t} = \frac{\partial}{\partial x}\left(\theta\left(D_{xx}\frac{\partial c^n}{\partial x} + D_{xy}\frac{\partial c^n}{\partial y} + D_{xz}\frac{\partial c^n}{\partial z}\right)\right)$$

$$+ \frac{\partial}{\partial y}\left(\theta\left(D_{yy}\frac{\partial c^n}{\partial y} + D_{xy}\frac{\partial c^n}{\partial x} + D_{yz}\frac{\partial c^n}{\partial z}\right)\right)$$

$$+ \frac{\partial}{\partial z}\left(\theta\left(D_{zz}\frac{\partial c^n}{\partial z} + D_{xz}\frac{\partial c^n}{\partial x} + D_{zy}\frac{\partial c^n}{\partial y}\right)\right)$$

$$- \frac{\partial}{\partial x}\left(\theta v_x c^n\right) - \frac{\partial}{\partial y}\left(\theta v_y c^n\right) - \frac{\partial}{\partial z}\left(\theta v_z c^n\right)$$

$$- (TU)c^n + \mu_n \theta c^a \tag{64}$$

With the dispersive coefficients, the diffusion coefficient $(D_0)$ have to be added. But practically, the effect of diffusivity becomes negligible as velocity gets increased. Hence the diffusivity term is neglected. It must be noted that Eqs. (62) to (64) are not independent and they are coupled equations.

## Boundary Conditions

There are three types of boundary conditions with which the convection–dispersion equation is usually solved. They are Dirichlet, Neumann and Cauchy boundary conditions.

If at boundaries, concentrations are known for different time periods, those boundary conditions are called as Dirichlet boundary condition (Type 1). It is mathematically expressed as follows:

$$C(x_b, y_b, z_b, t) = C_b \qquad (65)$$

At the boundary node $(x_b, y_b, z_b)$ at time '$t$', the concentration to be specified is denoted as $C_b$.

If at boundaries, concentration gradients is zero, for different time periods, that boundary conditions is called as Neumann boundary condition (Type 2). It is mathematically expressed as follows:

$$\text{For instance in } X \text{ - direction, } \frac{\partial c}{\partial x} = 0 \qquad (66)$$

If at boundaries, chemical flux across boundaries are known for different time periods that boundary conditions is known as Cauchy boundary condition (Type 3). It is mathematically expressed as follows:

$$F(x_b, y_b, z_b, t) = F_b \qquad (67)$$

At the boundary node $(x_b, y_b, z_b)$ at time '$t$', the flux to be specified is denoted as $F_b$.


## Estimation of Dispersivity

Solution of convection–dispersion equation is done normally by numerical methods like finite difference and finite element methods. For very simple conditions, analytical solutions exist. One analytical solution is very useful for determining dispersivity in column leaching experiments. A typical experimental set-up for conducting a leaching experiment is shown in Fig. 18

For one-dimensional steady flow with only advection and convection without any reaction, the following is the equation:

$$\frac{\partial c}{\partial t} = D\frac{\partial^2 c}{\partial z^2} - \frac{\partial}{\partial z}(v_z c) \qquad (68)$$

Before any solute is applied to the column of soil, steady flow of water in unsaturated or saturated condition will be maintained. Then, a steady concentration of a solute will be applied from the upper soil boundary. Therefore, the initial condition becomes zero concentration of solute in the flow.

The top boundary condition in the soil column is constant solute flux (Cauchy), and the bottom boundary condition can be approximated as Neumann boundary condition with the rate of change of concentration to be zero.

For this initial and boundary conditions, approximate analytical solution of Eq. 68 is as follows:

$$\frac{C(z,t)}{C_0} = \left(0.5 erfc\left(\frac{z - v_z t}{\sqrt{4Dt}}\right)\right) \tag{69}$$

where $c_0$ is input concentration. erfc is known as complementary error function. If MS-Excel is used to solve this equation, as MS-Excel does not have the ability to evaluate erfc function for negative arguments, the following relationship is to be used.

$$erfc(-x) = 1 + erf(|-x|) \tag{70}$$

Column leaching experiment is done to find out the leachate concentration at different times until the leachate received becomes equal to $c_0$. Dispersion coefficient can be found out as all other variables in Eq. 69 are measurable in this experiment. From the dispersion coefficient, the dispersivity value can be found out as the velocity of flow also is known.

The tracer used to estimate the dispersivity must be conservative and must be non-reactive with the soil. One such tracer is potassium bromide. When filling the column, it is preferable to use undisturbed core sample. In case if disturbed soil is used, the soil must be packed in the column so that the field bulk density is maintained in the packed soil.

An experimental result of conducting a column leaching study with potassium bromide is presented here. The soil is silty clay loam, the depth of soil column is 27.5 cm, and pore velocity of flow is 120 cm/day. In Columns 1 and 2 of Table 5, the time versus experimental relative concentration of leachate obtained data is shown. Column 3 of Table 5 shows the result of calculation using Eq. 69 for a trial value of $D = 200$ cm²/day. In Fig. 19, the fitted and observed relative concentration plot is shown (known as breakthrough curve). Trial value selected for fitting must be so selected that fitted curve and observed curve closely match together. For natural soil columns, it is very difficult to a very good match. For this data for the value of $D = 200$ cm²/day, the match is satisfactory.

Then, from the following relationship, longitudinal dispersivity can be found out.

$$\varepsilon_L = \frac{D}{v} = \frac{200}{120} = 1.67 \text{ cm} \tag{71}$$

## Estimation of Urea Reaction Rate Constants

For rhizosphere modelling, estimation of parameters like first-order reaction rate constants such as ammonification rate from urea ($\mu_a$), volatilization rate of ammonia from ammonium ($\mu_v$), nitrification rate from ammonium ($\mu_n$) and ammonium distribution coefficient with soil ($k_d$) are to be estimated for improving the accuracy of modelling.

**Table 5** Experimental and fitted data for a soil column leaching experiment

| Time (days) | Relative concentration ($C/C_0$) | Fitted relative concentration ($C/C_0$) for $D = 200$ cm$^2$/day |
|---|---|---|
| 0.09 | 0.061 | 0.003 |
| 0.10 | 0.073 | 0.007 |
| 0.10 | 0.085 | 0.010 |
| 0.11 | 0.098 | 0.013 |
| 0.11 | 0.122 | 0.017 |
| 0.11 | 0.134 | 0.020 |
| 0.13 | 0.171 | 0.046 |
| 0.14 | 0.244 | 0.073 |
| 0.14 | 0.256 | 0.082 |
| 0.15 | 0.268 | 0.126 |
| 0.16 | 0.317 | 0.170 |
| 0.18 | 0.354 | 0.229 |
| 0.19 | 0.378 | 0.282 |
| 0.20 | 0.451 | 0.337 |
| 0.21 | 0.488 | 0.392 |
| 0.23 | 0.585 | 0.500 |
| 0.25 | 0.683 | 0.599 |
| 0.27 | 0.756 | 0.685 |
| 0.29 | 0.866 | 0.756 |
| 0.31 | 0.939 | 0.814 |
| 0.33 | 0.963 | 0.860 |
| 0.35 | 0.988 | 0.896 |
| 0.36 | 0.994 | 0.911 |
| 0.38 | 0.995 | 0.923 |
| 0.52 | 1.000 | 0.992 |

## Experimental Set-up for Estimation of Urea Reaction Rate Constants

A simple and easy method recommended by [9] is presented here. In this method, 500 ml of glass bottles filled with soil and suitable dose of urea with the soil are used. Before application of urea, ensure that the residual ammonium in the soil is insignificant. The soil must be packed in the bottle so that the field bulk density and the field soil moisture are maintained within the bottles. The urea-applied bottles are buried in soil to maintain the microclimatic conditions in the field (Figs. 20 and 21). The ammonia gas evolved from the urea spiked bottles are collected by passing air from compressor, and the ammonia-collected air passes through 2% boric acid. In

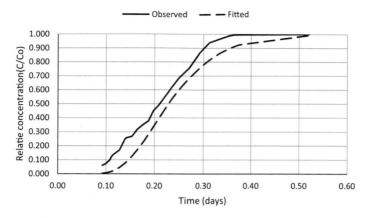

Fig. 19 Relative concentration of leachate after different time periods in a column leaching test

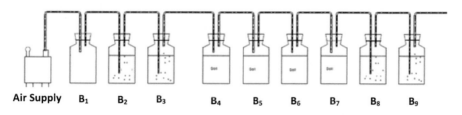

B₁ - Safety trap

B₂ - Distilled water

B₄, B₅, B₆ & B₇ –Urea spiked with soil

B₃, B₈ & B₉- Boric acid

Fig. 20 Line sketch of experimental set-up for estimation of reaction rate constants

Fig. 21 Experimental set-up for estimation of reaction rate constants installed in the field [10]

order to ensure that all the ammonia is entrapped in the boric acid, there is another boric acid filled bottle ($B_9$) which is periodically checked. Air is passed from a compressor through a distilled water bottle ($B_2$) and also boric acid bottle ($B_3$) in order to remove residual ammonium in the air. Emission of volatilized ammonia is determined by periodic titration of the boric acid solution collected from $B_8$ and $B_9$ bottles against 0.1 N $H_2SO_4$.

Figure 22 shows a result of a typical experiment conducted in India in a field during May and June months. The soil is sandy loam with a bulk density of 1.58 gm cm$^{-3}$, and volumetric moisture content was maintained at field capacity 0.29 cm$^3$ cm$^{-3}$. The initial concentration of urea-N in the soil was taken as 20.9 mg cm$^{-3}$. From the observations of cumulative ammonia evolved, the urea reaction rate values can be estimated.

If we do not consider the convection and dispersion and also plant uptake in Eqs. 62 to 64, we get the urea reaction rate equations which are as follows:

**For Urea**:

$$\frac{\partial(\theta c^u)}{\partial t} = -\mu_a \theta c^u \tag{72}$$

**For Ammonium**:

$$\frac{\partial(\theta c^a)}{\partial t} + \rho \frac{\partial s^a}{\partial t} = -\mu_v \theta c^a - \mu_n \theta c^a + \mu_a \theta c^u \tag{73}$$

**For Nitrate**:

$$\frac{\partial(\theta c^n)}{\partial t} = \mu_n \theta c^a \tag{74}$$

**Fig. 22** Cumulative ammonia volatilized from an experiment

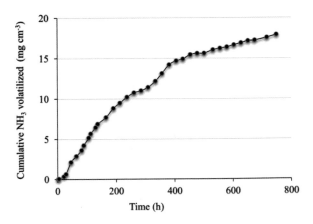

Where ($\theta$) is volumetric soil moisture content. Integral equation for Eq. 72 if the soil moisture content is maintained constant is as follows:

$$\theta c_t^u = \theta c_0^u e^{-\mu_a t} \tag{75}$$

where $c_t^u$ is concentration of urea in soil solution at time '$t$' and $c_0^u$ is initial concentration of urea in soil solution at time zero.

Equation 73 cannot be integrated, and hence finite difference method is used for solving this equation. This equation is amenable for explicit finite difference method with time steps of the order of minutes and hours. The explicit finite difference form for Eq. 73 is as follows:

$$\left(\theta c^a\right)_{t+\Delta t} = \left(\theta c^a\right)x + \left(\frac{\Delta t}{1 + \frac{\rho K_d}{\theta}}\right)\left[\mu_a \theta c_0^u c^{-\mu_a t} - (\mu_v + \mu_n)\left(\theta c^a\right)_t\right] \tag{76}$$

Estimated cumulative ammonium volatilized per unit volume of soil till time '$t$' $(AV)_t$ is as follows:

$$(AV)_t = \sum_0^t \mu_v\left(\theta c^a\right)\Delta t \tag{77}$$

Adsorbed ammonium in solid phase per unit volume of soil

$$(AA)_t = \rho K_d c_t^a \tag{78}$$

Finite difference form for Eq. 74 is as follows:

$$\left(\theta c^n\right)_{t+\Delta t} = \left(\theta c^n\right)_t + \mu_n\left(\theta c^a\right)_t \Delta t \tag{79}$$

Equations 75 to 79 have to be solved sequentially with a suitable set of reaction rate constants so that volatilized ammonia obtained from Eq. 77 matches well with the observed cumulative ammonium from the field experiment. Searching of optimal set of parameters can be easily done with an add-on tool in MS-Excel, namely 'Solver'. In this tool, we can set an objective function to minimize sum of squared deviations between the estimated cumulative ammonia evolved and the observed cumulative ammonia evolved for every time step. When the maximum and minimum values of urea reaction rate constants are specified within the software, the software does the search of optimal set of reaction rate constants for which the sum of squared deviations is minimum. There is no guarantee that global optimum solution is achieved

**Table 6** Search range of urea reaction rate constants

| Ammonification rate from urea ($\mu_a$) [h$^{-1}$] | Volatilization rate of ammonia from ammonium ($\mu_v$) [h$^{-1}$] | Nitrification rate from ammonium ($\mu_n$) [h$^{-1}$] | Ammonium distribution coefficient with soil ($k_d$) [cm$^3$/g] |
|---|---|---|---|
| 0.01–1.0 | 0.0001–0.05 | 0.0001–0.1 | 0.005–0.5 |

**Table 7** Fitted urea reaction rate constants

| Ammonification rate from urea ($\mu_a$) [h$^{-1}$] | Volatilization rate of ammonia from ammonium ($\mu_v$) [h$^{-1}$] | Nitrification rate from ammonium ($\mu_n$)[h$^{-1}$] | Ammonium distribution coefficient with soil (k$_d$) [cm$^3$/g] |
|---|---|---|---|
| 0.014 | 0.311 | 0.028 | 0.05 |

while solving these equations. But many advanced search engines with in the software provides most likely the global optimum solutions.

### Problem

Urea was added to soil at a dose of 14.076 mg N per one cubic centimetre of soil. Bulk density of soil is 1.53 g/cm$^3$. The expected range of reaction rate constants is as given in Table 6. Ammonia evolved from the soil per cubic centimetre of soil was estimated and shown in Table 7. The water content in the soil was maintained at 0.3. Estimate the reaction rate constants of urea.

This problem can be solved in MS-Excel with an add-on tool, namely solver by minimizing sum of squared deviation of estimated ammonia volatilized from observed ammonia volatilized. The fitted reaction rate constants are shown in Table 8. For the fitted urea reaction rate parameters, the computation is done to see how the urea-N content in soil, ammonium-N, ammonia-N volatilized, adsorbed ammonium and nitrate-N vary at different time steps at which observations of ammonia volatilized was estimated (Table 8; Fig. 23).

# Parameter Set Needed for Rhizosphere Modelling

Hydrus is a general-purpose software to simulate flow and solute transport in variably saturated flow domains. This software has a very good capability to do rhizosphere modelling. The quality of the simulation would get improved if the parameters such as unsaturated hydraulic conductivity data, dispersivity of soil data, root water uptake parameters data as well as urea reaction rate data are estimated in field conditions. For simulation of phosphorus and potash assumption of non-reactive transport is often made. In the preceding sections, simple procedures for collecting these data from the field have been discussed.

When rhizosphere modelling is done, it is possible to study the nutrient balance and also the plant uptake of nutrients. Hence, it is possible to develop fertigation schedules

**Table 8** Calculation of squared deviation of estimated ammonia volatilized from observed ammonia volatilized for the fitted urea reaction rate parameters

| No (1) | Time (h) (2) | Exp. ammonia volatzd (mg N/cm³) (3) | Urea-N (mg N/cm³) (Eq. 75) (4) | Estimated ammonium-N in soil solution (mg N/cm³) (Eq. 76) (5) | Cum amm.-N volatilized (mg N/cm³) (Eq. 77) (6) | Ammonium adsorbed-N (mg N/cm³) (Eq. 78) (7) | Nitrate-N (mg N/cm³) (Eq. 79) (8) | Squared deviation (mg N/cm³) [(3)-(5)]² (9) |
|---|---|---|---|---|---|---|---|---|
| 1 | 1 | 0.005 | 13.877 | 0.158 | 0.000 | 0.000 | 0.000 | 0.0000 |
| 2 | 2 | 0.011 | 13.681 | 0.311 | 0.001 | 0.012 | 0.001 | 0.0001 |
| 3 | 4 | 0.031 | 13.296 | 0.604 | 0.009 | 0.035 | 0.008 | 0.0005 |
| 4 | 24 | 0.123 | 10.000 | 2.724 | 0.330 | 0.203 | 0.298 | 0.0429 |
| 5 | 42 | 0.218 | 7.738 | 3.689 | 0.874 | 0.279 | 0.789 | 0.4300 |
| 6 | 65 | 1.268 | 5.576 | 4.114 | 1.721 | 0.314 | 1.555 | 0.2055 |
| 7 | 89 | 2.348 | 3.962 | 4.002 | 2.638 | 0.307 | 2.382 | 0.0839 |
| 8 | 113 | 3.468 | 2.814 | 3.610 | 3.495 | 0.278 | 3.157 | 0.0007 |
| 9 | 140 | 4.618 | 1.916 | 3.044 | 4.336 | 0.235 | 3.917 | 0.0793 |
| 10 | 164 | 5.598 | 1.361 | 2.534 | 4.963 | 0.195 | 4.482 | 0.4037 |
| 11 | 192 | 5.838 | 0.913 | 1.991 | 5.554 | 0.154 | 5.017 | 0.0805 |
| 12 | 216 | 6.058 | 0.649 | 1.591 | 5.956 | 0.123 | 5.379 | 0.0105 |
| 13 | 240 | 6.066 | 0.461 | 1.256 | 6.275 | 0.097 | 5.668 | 0.0436 |
| 14 | 264 | 6.070 | 0.327 | 0.981 | 6.525 | 0.076 | 5.894 | 0.2073 |
| Sum |  |  |  |  |  |  |  | 1.5885 |

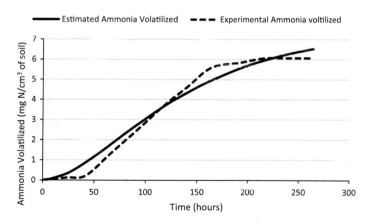

**Fig. 23** Estimated and experimental ammonia volatilized for the numerical problem

by rhizosphere modelling. If we do rhizosphere modelling, fieldwork needed for verification of fertigation schedules gets significantly reduced.

## Development of Fertigation Schedule by Modelling with Hydrus

Good fertigation scheduling is application of fertilizer at any point in time so that before the next fertigation, the plant has been able to assimilate a sufficient quantity of the fertilizer. Figure 24 shows a typical cumulative $N$ assimilation curve with respect to days after planting. For instance, let us assume that during the time period between $t_1$ and $t_2$, a plant assimilates 20 kg/ha of nitrogen. Let us also assume that during this time period, one expects 4 kg/ha of volatilization loss and 1 kg/ha of deep drainage loss. If, in this situation, 25 kg/ha of nitrogen is applied at time $t_1$, 20 kg/ha of nitrogen will be available for the plant in the root zone. This method of applying fertilizers is called a "growth curve nutrition approach". One disadvantage of this approach is that it does not take into account the carry-over of a nutrient from previous periods. Additionally, for the situation discussed above, the spatial and temporal distribution of urea, ammonium and nitrate may be such that less than 20 kg/ha of nitrogen only can be taken up by the roots. Therefore, a certain quantity of excess nitrogen must be applied in order to provide sufficient amount of nitrogen to meet assimilation requirements. How much excess application will be optimal can be estimated by rhizosphere modelling.

In Hydrus, while solving the Richard's equation considering fertigation in root zone, evaporation and root water uptake are separately considered. Evaporation from soil occurs from soil surface as well as from beneath the surface. It is usual practice

**Fig. 24** Typical N assimilation curve

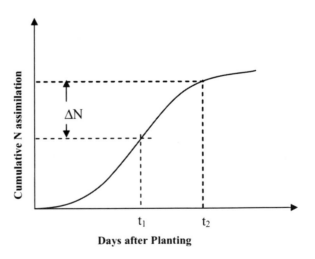

to assume evaporation as to occur only from the surface. The Richard's equation incorporating the evaporation and root water uptake separately is as follows:

$$\frac{\partial}{\partial x}\left(K(h)\frac{\partial h}{\partial x}\right) + \frac{\partial}{\partial y}\left(K(h)\frac{\partial h}{\partial y}\right) + \frac{\partial}{\partial Z}\left(K(h)(\frac{\partial h}{\partial Z} - 1)\right) - EU - TU = \frac{\partial \theta}{\partial t}$$
(80)

where EU and TU are evaporation and transpiration from unit volume of soil, respectively. When we separate evaporation and transpiration separately, finer discretization of nodes must be done and surface nodes may be assumed to have evaporation alone and the nodes beneath the surface nodes may be assumed to have transpiration alone. Computational procedures for finding evaporation and transpiration using meteorological data separately are discussed in the basic section. Hydrus needs evaporation transpiration data in depth units as input data for solving Eq. 80.

Hydrus-2D and Hydrus-3D are capable of simulating all types of root zones those are commonly encountered. A sample root zone modelled in Hydrus is shown in Fig. 25. By solving the Richard's equation we can get water content, pressure head distributions and water flux across any flow section. A sample water content distribution simulated with Hydrus is shown in Fig. 26. These kinds of results are very useful in deriving best water management practices.

Similarly, by solving convection–dispersion equation with urea reactions in Hydrus, we can get spatial and temporal distribution of different forms of solute such as urea, ammonium and nitrate (Fig. 27 and Fig. 28). From the results of Hydrus, it is also possible to compute the plant uptake of all forms of solute (Fig. 29). From these

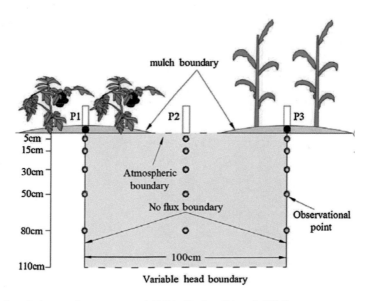

**Fig. 25** Sample image of root zone modelled in Hydrus (Li et al. 2015)

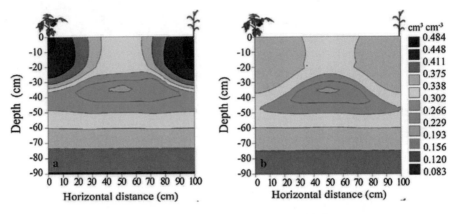

**Fig. 26** Sample water content distribution from simulation using Hydrus

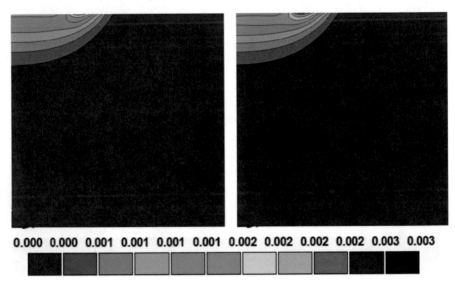

**Fig. 27** Sample spatial distribution of ammonium concentrations in mmol/cm$^3$ in the soil solution on two different days

results, by trial and error, one will be able to devise a suitable fertigation schedule which matches with plants need (Table 9 and Fig. 30). Literature is abundant with these kinds of real-time applications [11].

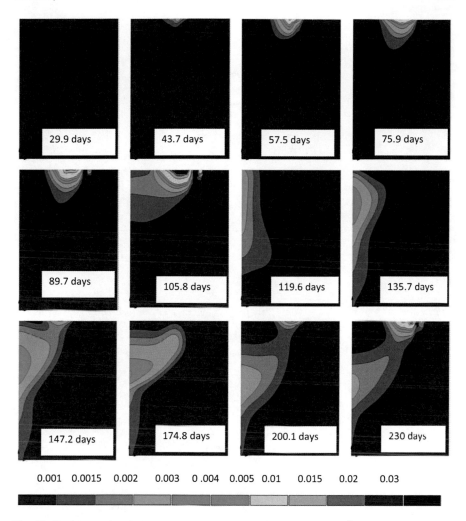

**Fig. 28** Sample spatial distribution of nitrate concentrations in mmol/cm³ in the soil solution on different days

## Summary

Richard's equation and convection–dispersion equation with solute transport are essential equations for rhizosphere modelling. We derive both the equations in this chapter. Evapotranspiration estimation using soil water content data by application of water balance models is demonstrated with numerical examples. Hydrus is one of the important software packages used for rhizosphere modelling. Presently, computing rhizosphere modelling parameters is an essential task in application of rhizosphere models for field situations. With regard to application of Hydrus at least the following

**Fig. 29** Sample result from
Hydrus of root uptake of
nitrogen in different forms

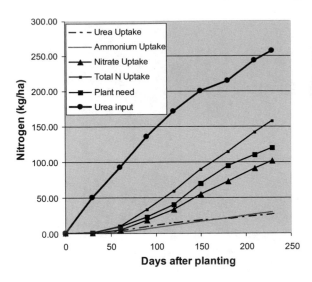

**Table 9** Sample optimal
fertigation schedule for
sugarcane [12]

| Fortnightly periods | Fertigation day | Fertilizer to be applied urea (kg/ha) |
|---|---|---|
| 1 | 1 | 2 |
| 2 | 15 | 15 |
| 3 | 28 | 15 |
| 4 | 42 | 22 |
| 5 | 57 | 22 |
| 6 | 74 | 25 |
| 7 | 88 | 70 |
| 8 | 102 | 70 |
| 9 | 116 | 70 |
| | 135 | 70 |
| 11 | 146 | 19 |

parameters are essential to run the software package. They are as follows: unsaturated
hydraulic conductivity parameters, root water uptake parameters, dispersivity of soil
and reaction rate constants of urea. Simple and easy field methods for computing
these data are demonstrated with real-time data. Sample results of Hydrus with regard
to development of best management practices for water and fertilizers are shown.

**Fig. 30** Sample result of matching the uptake of nitrogen to plant's need of nitrogen

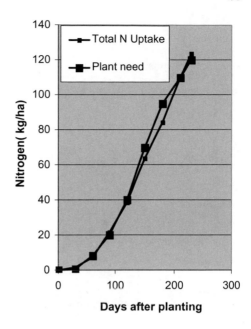

# Appendix I

The detailed procedure for optimization of unsaturated hydraulic parameters through inverse modelling for a double-ring infiltrometer test data using Hydrus-1D is described here.

The soil is sandy loam with sand content of 72.5%, silt content of 16.5% and clay content of 11%. The initial suction pressure head at the top soil and at 75 cm below the surface is − 1250 cm and − 300 cm, respectively. Data from a double-ring infiltration experiment is provided in Tables 10 and 11.

To begin a new project, the project data manager icon in Hydrus main window is opened and the new project is given with a name and a brief description. The project manager window (Fig. 31) consists of some inbuilt projects already existing in Hydrus. The window allows the user to open, rename, copy or delete an existing project and also to create a new project.

Here, the name and brief description given to the new project are 'inverse modelling' and 'inverse modelling using infiltration data' respectively.

### Preprocessing

Preprocessing tab has option for entering necessary data (Fig. 32). Then after selecting 'main processes', opt for 'Water flow' and 'Inverse solution' (Fig. 33).

The preprocessing data for inverse solution is shown in Fig. 34. The maximum number of iterations for the inverse solution is to be specified here. If one selects zero, then only the direct simulation is carried out. The maximum number of iterations selected for this problem is '20'. The number of data points in objective function is

**Table 10** Cumulative infiltration depths observations from double-ring infiltrometer test

| No. | Time (min) | Cumulative infiltration depth (cm) |
|-----|-----------|-----------------------------------|
| 1   | 0         | 0.0                               |
| 2   | 5         | 1.1                               |
| 3   | 10        | 0.9                               |
| 4   | 20        | 0.7                               |
| 5   | 30        | 0.6                               |
| 6   | 40        | 0.8                               |
| 7   | 50        | 0.7                               |
| 8   | 65        | 0.7                               |
| 9   | 80        | 0.6                               |
| 10  | 110       | 1.2                               |
| 11  | 170       | 2.3                               |
| 12  | 230       | 2.2                               |
| 13  | 290       | 2.0                               |

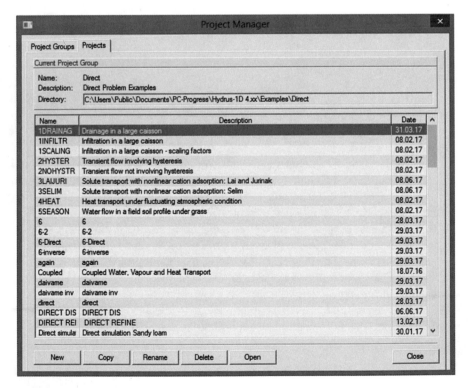

**Fig. 31** Project manager window of the Hydrus-1D module

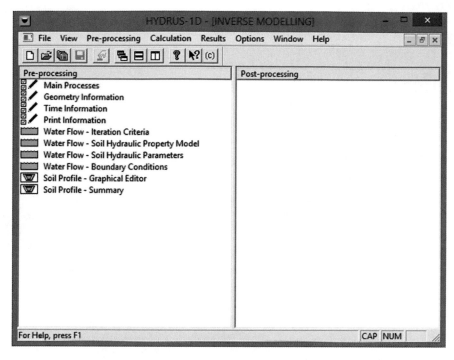

**Fig. 32** Hydrus window showing the preprocessing menu

also specified here. In this study, cumulative infiltration data for 13 time intervals is taken as the input data. So, the number of data points is '13'.

The geometry information window (Fig. 35) selects the length unit, specifies the depth and inclination of the soil profile to be analysed and determines the number of materials to be used. The 'decline from vertical axis' allows the user to choose between a vertical, horizontal or inclined soil profile. The inclination is specified in terms of the cosine of the angle between the vertical axis and the axis of the soil profile. Its value is equal to one for vertical soil columns and zero for horizontal soil columns. In this problem, length units are chosen as centimetres and a soil profile with a single material having 75 cm depth is considered as the domain of study. Since a vertical soil profile is considered, the value of 'decline from vertical axis' is specified as 1.

The time information window data used is in Fig. 36. The number of time variable boundary conditions corresponds to number of data in Table 11.

The print information window (Figs. 37 and 38) allows the user to enter the variables governing the output from Hydrus. The T-level information checkbox is used to decide whether certain information concerning the mean pressure head, mean water, cumulative water fluxes and time and iteration information are to be printed after each time step or only at preselected times. One can also specify the number of time steps after which the output is to be printed on the screen.

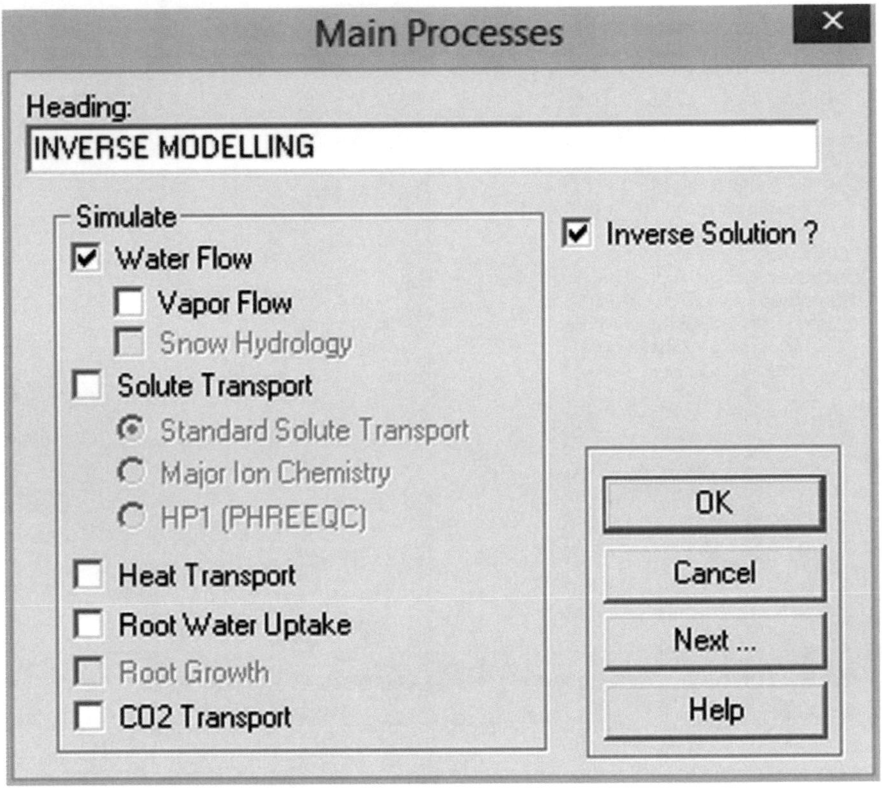

**Fig. 33** Main processes window of Hydrus-1D

The iteration criteria window (Fig. 39) specifies the iteration criteria for the solution precision and parameters for the time step control. An iterative process must be used to obtain solutions at each new time step because of the nonlinear nature of the Richards' equation. The iterative process continues until a satisfactory degree of convergence is obtained, i.e. until at all nodes in the saturated (unsaturated) region the absolute change in pressure head (water content) between two successive iterations becomes less than some small value determined by the imposed absolute pressure head (or water content) tolerance. The water content tolerance represents the maximum desired absolute change in the value of the water content between two successive iterations during a particular time step. Its recommended value is 0.001. The pressure head tolerance represents the maximum desired absolute change in the value of the pressure head between two successive iterations during a particular time step. Its recommended value is 1 cm.

The lower optimal iteration range represents the minimum number of iterations necessary to reach convergence for water flow, below which the time step have to be increased by multiplying it with a lower time step multiplication factor. The recommended and default value for the lower optimal iteration range is 3, and the lower

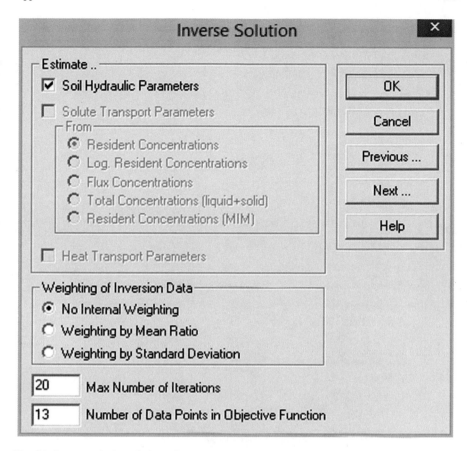

**Fig. 34** Inverse solution window of Hydrus

**Fig. 35** Geometry information window

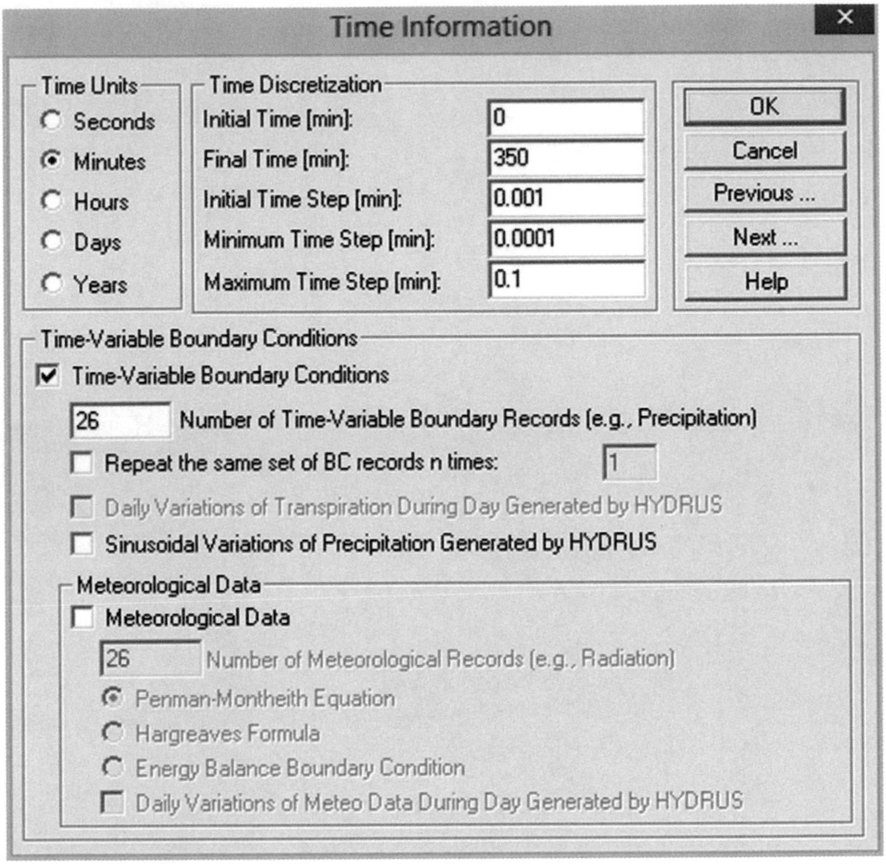

**Fig. 36** Time information window

time step multiplication factor is 1.3. The upper optimal iteration range represents the maximum number of iterations necessary to reach convergence for water flow, above which the time step has to be decreased by multiplying it with an upper time step multiplication factor. The recommended and default value for the upper optimal iteration range is 7, and upper time step multiplication factor is 0.7.

At the beginning of a numerical simulation, Hydrus generates for each soil type in the flow domain a table of water contents, hydraulic conductivities and specific water capacities from the specified set of hydraulic parameters. Values of the hydraulic properties are then computed during the iterative solution process using linear interpolation between entries in the table. So, for facilitating the interpolation, a prescribed interval for the pressure head value is to be given. The lower and upper limits of the tension interval represent the absolute value of the lower and upper limits of the pressure head interval for which a table of hydraulic properties will be generated internally for each material. The lower limit of the tension interval is usually selected

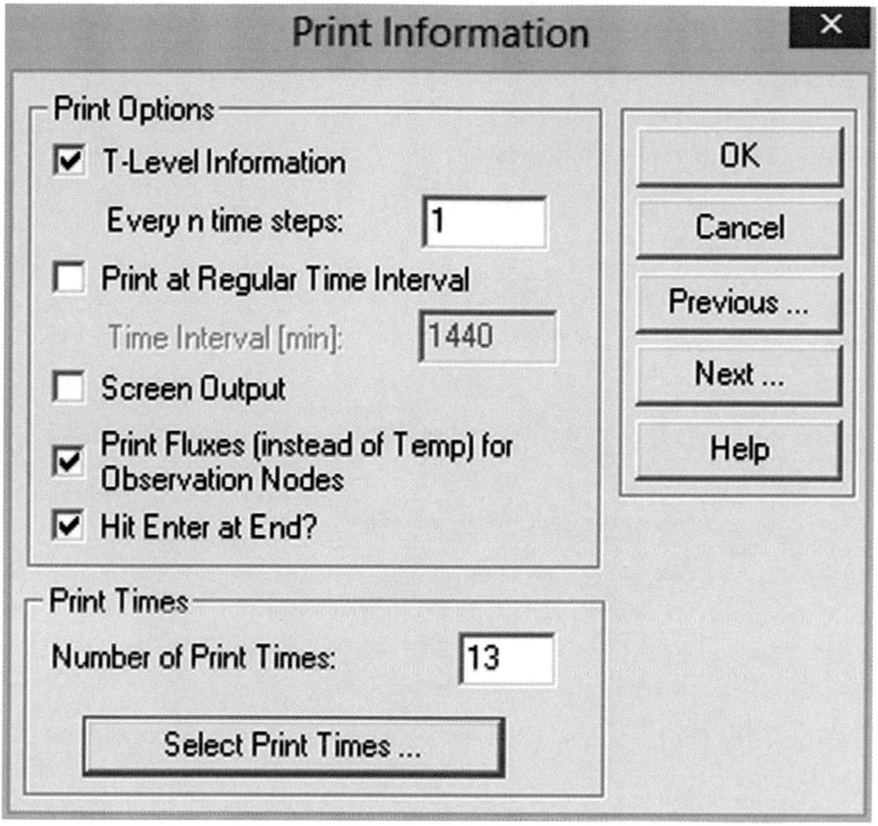

**Fig. 37** Print information window

to be a very small number. The upper limit can be modified so that it encompasses most pressure heads encountered during the simulation.

In the soil hydraulic model window (Fig. 40), van Genuchten–Mualem model and 'No hysteresis' are selected.

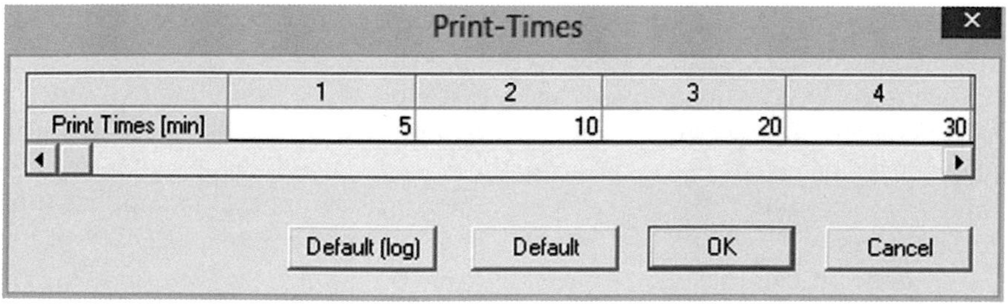

**Fig. 38** Print times window

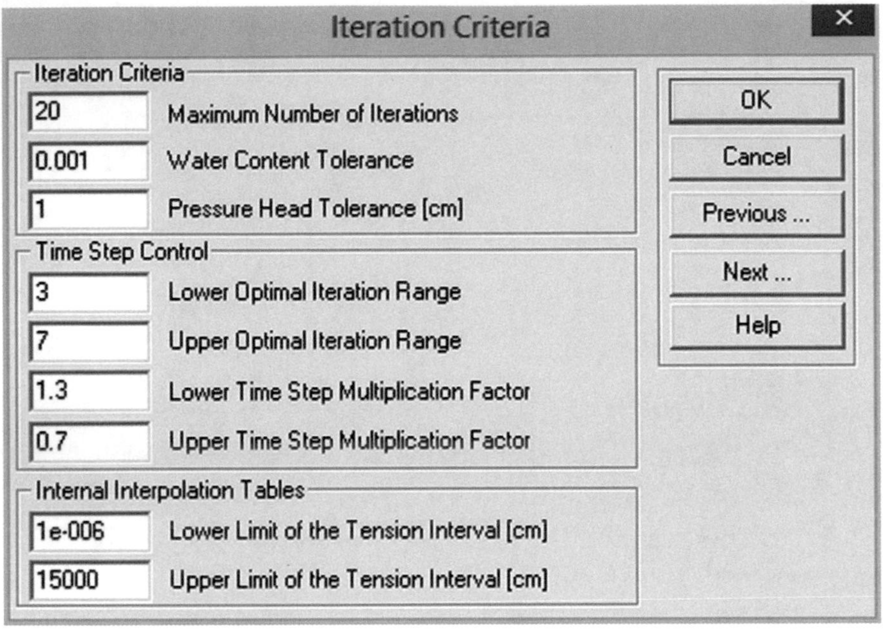

| Iteration Criteria | | |
|---|---|---|
| Iteration Criteria | | |
| 20 | Maximum Number of Iterations | OK |
| 0.001 | Water Content Tolerance | Cancel |
| 1 | Pressure Head Tolerance [cm] | Previous ... |
| Time Step Control | | Next ... |
| 3 | Lower Optimal Iteration Range | |
| 7 | Upper Optimal Iteration Range | Help |
| 1.3 | Lower Time Step Multiplication Factor | |
| 0.7 | Upper Time Step Multiplication Factor | |
| Internal Interpolation Tables | | |
| 1e-006 | Lower Limit of the Tension Interval [cm] | |
| 15000 | Upper Limit of the Tension Interval [cm] | |

**Fig. 39** Iteration criteria window

The user has to provide initial estimates of the soil hydraulic parameters. The Rosetta Lite software can also be accessed from within Hydrus and the soil constituent data to get initial estimate of hydraulic parameters of the soil. The initial parameters used here are as below (Fig. 41).

$\theta_r = 0.045$, $\theta_s = 0.3684$, $\alpha = 0.0356$ cm$^{-1}$, $l = 0.5$, $n = 1.4884$ and $K_s = 1.4884$ cm min$^{-1}$

In water flow boundary conditions window (Fig. 42), since the ponding depth varies with time, a variable pressure head upper boundary condition is given and the lower boundary condition is free drainage. The initial conditions are specified in terms of pressure head.

The time variable boundary condition window (Fig. 43) allows the user to enter the time dependent values of boundary conditions, and the data of Table 11 is entered.

In the 'Data for Inverse Solution Window', the data of cumulative depth of water infiltrated in Table 11 is given as shown in Fig. 44.

The profile information (Fig. 45) module discretizes soil profile into a series of finite elements. If no previous nodal distribution exists, the programme generates automatically an equidistant nodal distribution with a default number of nodes. The number and location of the nodes can then be edited by the user to optimize finite element lengths.

The initial pressure head at the top and bottom are given as per the data as − 1250 cm and − 300 cm, respectively.

After running the Hydrus-1D, the optimized parameters obtained is as follows:

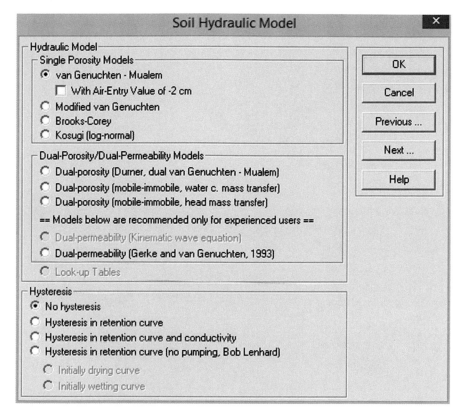

**Fig. 40** Soil hydraulic model window

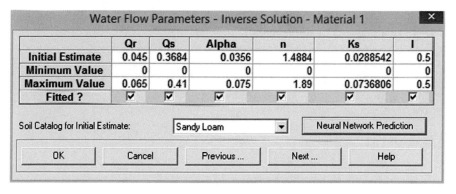

**Fig. 41** Water flow parameters

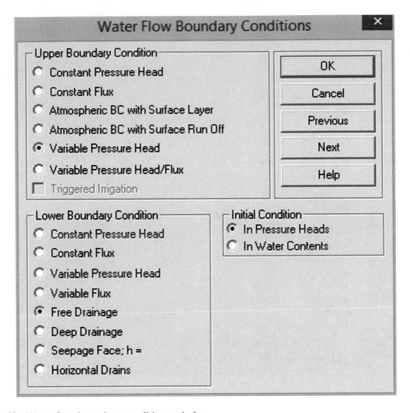

**Fig. 42** Water flow boundary conditions window

$\theta_r = 0.0445$, $\theta_s = 0.3719$, $\alpha = 0.0251$ cm$^{-1}$, $l = 0.0003$, $n = 1.5181$ and $K_s = 1.5181$ cm min$^{-1}$.

Figure 46 shows the comparison between the simulated cumulative infiltration depth using optimized hydraulic parameters and observed cumulative infiltration depth.

# Appendix II

## *Introduction to Finite Difference Methods*

The basis of finite difference method (FDM) is expression of partial differential equations like Richard's equation and convection–dispersion equation into finite difference equation by using truncated form of Taylor series expansion.

Taylor series expansion for pressure head ($h$) can be expressed as follows:

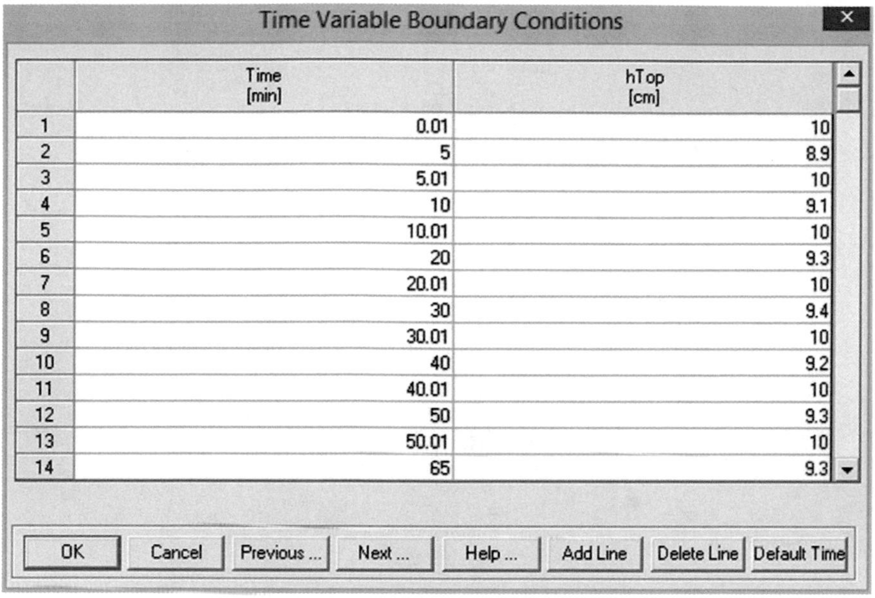

**Fig. 43** Time variable boundary condition window

$$h(x + \Delta x) = h(x) + \Delta x \left(\frac{\partial h}{\partial x}\right)_x + \frac{(\Delta x)^2}{2!} \frac{\partial^2 h}{\partial x^2} + \dots \qquad (80)$$

If we neglect second-order terms in the preceding equation and rearrange, we get as follows:

$$\left(\frac{\partial h}{\partial x}\right)_x = \frac{h(x + \Delta x) - h(x)}{\Delta x} \qquad (81)$$

If we express the $\left(\frac{\partial h}{\partial x}\right)$ of Richard's equation or any other first-order spatial derivative like this, it is said to have been expressed in forward difference form.

If we substitute $(-\Delta x)$ in the place of $(\Delta x)$ in Eq. 1, we get as follows:

$$h(x - \Delta x) = h(x) - \Delta x \left(\frac{\partial h}{\partial x}\right)_x + \frac{(\Delta x)^2}{2!} \frac{\partial^2 h}{\partial x^2} + \dots \qquad (82)$$

If we neglect second-order terms in the preceding equation and rearrange, we get as follows:

$$\left(\frac{\partial h}{\partial x}\right)_x = \frac{h(x) - h(x - \Delta x)}{\Delta x} \qquad (83)$$

Instead of using Eq. 81, if we use Eq. 83, it is called as backward difference form.

**Table 11** Depth of standing water in inner ring

| S. no. | Time (minutes) | Depth of standing water in inner ring (cm) |
|---|---|---|
| 1 | 0.01 | 10 |
| 2 | 5.0 | 8.9 |
| 3 | 5.01 | 10 |
| 4 | 10 | 9.1 |
| 5 | 10.01 | 10 |
| 6 | 20 | 9.3 |
| 7 | 20.01 | 10 |
| 8 | 30 | 9.4 |
| 9 | 30.01 | 10 |
| 10 | 40 | 9.2 |
| 11 | 40.01 | 10 |
| 12 | 50 | 9.3 |
| 13 | 50.01 | 10 |
| 14 | 65 | 9.3 |
| 15 | 65.01 | 10 |
| 16 | 80 | 9.4 |
| 17 | 80.01 | 10 |
| 18 | 110 | 8.8 |
| 19 | 110.01 | 10 |
| 20 | 170 | 7.7 |
| 21 | 170.01 | 10 |
| 22 | 230 | 7.8 |
| 23 | 230.01 | 10 |
| 24 | 290 | 8 |
| 25 | 290.01 | 10 |
| 26 | 350 | 8 |

If we add Eqs. 81 and 83, we get as follows:

$$\left(\frac{\partial h}{\partial x}\right)_x = \frac{h(x + \Delta x) - h(x - \Delta x)}{2\Delta x} \tag{84}$$

The preceding form of equation is called as central difference.

By following the same principles followed so far, we can write the following equation for second-order derivatives:

$$\left(\frac{\partial^2 h}{\partial x^2}\right)_x = \frac{h(x + \Delta x) - 2h(x) + h(x - \Delta x)}{(\Delta x)^2} \tag{85}$$

## Data for Inverse Solution

| | X | Y | Type | Position | Weight |
|---|---|---|---|---|---|
| 1 | 5 | -1.1 | 0 | 1 | 1 |
| 2 | 10 | -2 | 0 | 1 | 1 |
| 3 | 20 | -2.7 | 0 | 1 | 1 |
| 4 | 30 | -3.3 | 0 | 1 | 1 |
| 5 | 40 | -4.1 | 0 | 1 | 1 |
| 6 | 50 | -4.8 | 0 | 1 | 1 |
| 7 | 65 | -5.5 | 0 | 1 | 1 |
| 8 | 80 | -6.1 | 0 | 1 | 1 |
| 9 | 110 | -7.3 | 0 | 1 | 1 |
| 10 | 170 | -9.6 | 0 | 1 | 1 |
| 11 | 230 | -11.8 | 0 | 1 | 1 |
| 12 | 290 | -13.8 | 0 | 1 | 1 |
| 13 | 350 | -15.8 | 0 | 1 | 1 |

OK
Cancel
Previous ...
Next ...
Add Line
Delete Line
Help ...

☐ Show list boxes (not recommended for large data files)

**Fig. 44** Data for inverse solution window

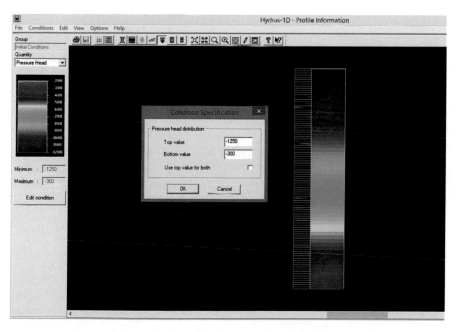

**Fig. 45** Profile information window

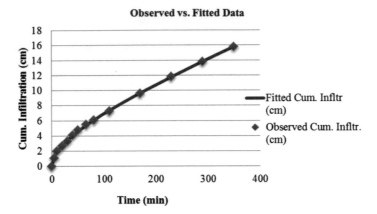

**Fig. 46** Goodness of parameter estimation

As far as second-order derivative are concerned, Eq. 85 is the only one form, whereas for the first-order derivative, there are three forms.

Let us express the Richard's equation in one-dimensional form as follows:

$$\frac{\partial}{\partial X}\left(K(h)\frac{\partial h}{\partial X}\right) = C_w \frac{\partial h}{\partial t} \tag{86}$$

Let us assume K (h) and $C_w$ to be constants for simplicity. Then the preceding equation can be expressed as follows:

$$\frac{\partial^2 h}{\partial x^2} = k\frac{\partial h}{\partial t} \tag{87}$$

When we want to find out the pressure head distribution in a one-dimensional flow field, the flow domain is discretized a shown in Fig. 47. There are two methods by which equations with time derivative are represented in FDM. One is called as explicit method, and another is called as implicit method. Equation 87 is written using explicit method as follows:

$$\frac{h^t(x + \Delta x) - 2h^t(x) + h^t(x - \Delta x)}{(\Delta x)^2} = k\frac{h^{t+1}(x) - h^t(x)}{\Delta t} \tag{88}$$

In Eq. 88, it must be noted that the time period for which the spatial derivatives are written is for the time period '*t*'. Therefore, when Eq. 88 is solved with necessary initial conditions, the pressure head distribution in flow domain for all the nodes at time period '*t*' is known and only unknown is pressure head for time period '*t* + 1'. Hence Eq. 88 can be written explicitly as follows:

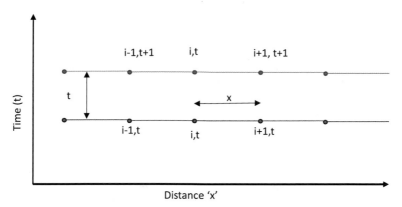

**Fig. 47** One-dimensional finite difference grid

$$h^{t+1}(x) = h^t(x) + \left( \frac{h^t(x + \Delta x) - 2h^t(x) + h^t(x - \Delta x)}{(\Delta x)^2} \right) \Delta t / k \qquad (89)$$

Equation 89 is solved progressively for all the nodes in the flow domain, and pressure head distributions are found out.

Equation 87 is written using implicit method as follows:

$$\frac{h^{t+1}(x + \Delta x) - 2h^{t+1}(x) + h^{t+1}(x - \Delta x)}{(\Delta x)^2} = k \frac{h^{t+1}(x) - h^t(x)}{\Delta t} \qquad (90)$$

In Eq. 90, it must be noted that the time period for which the spatial derivatives are written is for the time period '$t + 1$'. So in Eq. 90, for every time period only $h^t(x)$ is known and $h^{t+1}(x - \Delta x)$, $h^{t+1}(x)$ and $h^{t+1}(x + \Delta x)$ are unknowns.

When we write Eq. 90 for all the nodes considering initial and boundary conditions for every time step, the number of linearly independent simultaneous equations obtained will be equal to the number of unknowns. These simultaneous equations can be solved using matrix methods. These matrices are diagonally dominant matrices, and hence they can also be solved by iterative methods.

## Truncation Errors and Numerical Dispersion

Since in FDM, truncation of Taylor series is done, these methods are bound to have errors. These errors are called truncation errors, and they can be minimized by selecting finer grids to the extent possible. When we solve the convection–dispersion equation, due to truncation of Taylor series, the error caused is termed as numerical dispersion. From the following derivation, we would be able to understand why this error is called as numerical dispersion.

The following is the two-dimensional form of convection–dispersion equation:

$$\frac{\partial c}{\partial t} = D\left(\frac{\partial^2 c}{\partial x^2} + \frac{\partial^2 c}{\partial y^2}\right) - v_x \frac{\partial c}{\partial x} - v_y \frac{\partial c}{\partial y} \tag{91}$$

Let us assume dispersive flux is zero and velocity of flow is also a constant. Then the convection–dispersion equation becomes a pure convection equation which is as follows:

$$\frac{\partial c}{\partial t} = -v_x \frac{\partial c}{\partial x} \tag{92}$$

When we solve Eq. 92 using FDM by using the implicit method for time derivative and backward difference for the spatial derivative, we get as follows:

$$\frac{c_x^t - c_x^{t-\Delta t}}{\Delta t} = -v_x \frac{c_x^t - c_{x-\Delta x}^t}{\Delta x} \tag{93}$$

Equation 93 is obtained by truncating Taylor series expansion. When we solve Eq. 93, what partial differential equation is exactly solved is little different. It may sound very odd at this place. But continuing the derivation would dispel all our doubts.

Taylor series expansion for time derivative and neglecting third-order terms is as follows:

$$c^{t-\Delta t} = c^t - \Delta t \frac{\partial c}{\partial t} + \frac{\Delta t^2}{2!} \frac{\partial^2 c}{\partial t^2} \tag{94}$$

$$\frac{c^t - c^{t-\Delta t}}{\Delta t} = \frac{\partial c}{\partial t} - \frac{\Delta t}{2} \frac{\partial^2 c}{\partial t^2} \tag{95}$$

Similarly, we can write using Taylor series expansion for writing the spatial derivative as follows:

$$\frac{c_x - c_{x-\Delta x}}{\Delta x} = \frac{\partial c}{\partial x} - \frac{\Delta x}{2} \frac{\partial^2 c}{\partial x^2} \tag{96}$$

Put Eqs. 95 and 96 in Eq. 93 and we get,

$$\frac{\partial c}{\partial t} - \frac{\Delta t}{2} \frac{\partial^2 c}{\partial t^2} = -v_x \left(\frac{\partial c}{\partial x} - \frac{\Delta x}{2} \frac{\partial^2 c}{\partial x^2}\right) \tag{97}$$

We can write following equations by partially differentiating Eq. 92 with respect to time once and with respect to $x$ once:

$$\frac{\partial^2 c}{\partial t^2} = \frac{\partial}{\partial t}\left(-v_x\left(\frac{\partial c}{\partial x}\right)\right) \tag{98}$$

$$\frac{\partial^2 c}{\partial x \partial t} = -v_x \frac{\partial^2 c}{\partial x^2} \tag{99}$$

Put Eq. 99 in Eq. 98

$$\frac{\partial^2 c}{\partial t^2} = v_x^2 \left( \frac{\partial^2 c}{\partial x^2} \right) \tag{100}$$

Putting Eq. 100 in Eq. 97, we get as follows:

$$\frac{\partial c}{\partial t} - v_x^2 \frac{\Delta t}{2} \frac{\partial^2 c}{\partial x^2} = -v_x \left( \frac{\partial c}{\partial x} - \frac{\Delta x}{2} \frac{\partial^2 c}{\partial x^2} \right) \tag{101}$$

Rearranging the preceding equation we get,

$$\frac{\partial c}{\partial t} = -v_x \left( \frac{\partial c}{\partial x} - \frac{\Delta x}{2} \frac{\partial^2 c}{\partial x^2} \right) + v_x^2 \frac{\Delta t}{2} \frac{\partial^2 c}{\partial x^2} \tag{102}$$

Rearranging the preceding equation, we get,

$$\frac{\partial \boldsymbol{c}}{\partial t} = -\boldsymbol{v_x} \left( \frac{\partial \boldsymbol{c}}{\partial x} \right) + \left( v_x^2 \frac{\Delta t}{2} + \frac{\Delta x v_x}{2} \right) \frac{\partial^2 c}{\partial x^2} \tag{103}$$

We intended to use FDM for solving the partial differential Eq. 92 (shown in bold in Eq. 103). But while doing so, we are solving the partial differential Eq. 103 which has an additional dispersive flux terms with artificially induced dispersion coefficient of $\left( v_x^2 \frac{\Delta t}{2} + \frac{\Delta x v_x}{2} \right)$.

Hence numerical dispersion is defined as the dispersion caused due to application of FDM to the convective terms. Obviously, in order to keep numerical dispersion at a satisfactory level, the spatial discretization and temporal discretization level must be as low as possible. This fact can also be verified by doing a numerical experiment by solving a pure convective equation for modelling a sharp concentration front. We will be able to see from the results that after some time period from the initial time, the sharp concentration front would have got diffused.

Oscillation of results is another term used in modelling. When the concentration results go above the maximum expected level and go below the minimum expected level, the solution is said to be oscillating. Oscillation can be totally eliminated by adopting upstream weighting methods. Upstream weighting methods represent the convective term for each node based on the direction of flow at each node from the surrounding nodes. For instance in Fig. 48, four possible flow patterns that may occur at a node are shown. While writing the FD terms for $\left( v \frac{\partial c}{\partial x} \right)$, direction of velocity of flow is considered and the convective flow is computed based on the node from which flow occurs. Hence, this method is called as upstream weighting method. Upstream weighting methods severely suffer from numerical dispersion.

**Fig. 48** Different flow
patterns at a node for
one-dimensional flow

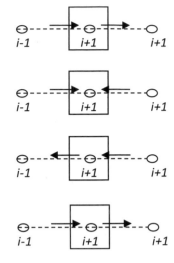

A non-dimensional called as Peclet number (Pe) is used to characterize spatial discretization. It is defined as follows:

$$Pe = \frac{v_x \Delta x}{D} \tag{24}$$

If Peclet number is maintained below 5, the oscillations can be kept at a satisfactory level for the other FDM methods.

Stability of numerical solution is another term commonly used. Stability is an issue for explicit methods. When the magnitudes of spatial and temporal discretization are above certain critical level, the solution from explicit method becomes unstable. Instability may be identified by observing the solution, and the solution will have very big positive and negative concentration values. The reason for instability in explicit method is due to the error caused due to truncation, and the error gets propagated unbounded. Implicit methods do not suffer from instability.

Courant number (Cr) is used to characterize the spatial and temporal discretization and is as follows:

$$Cr = \frac{v_x \Delta t}{\Delta x} \tag{104}$$

A Courant value of less than one is one necessary condition for avoiding instability in explicit methods. Another condition called as Neumann condition must also be satisfied for getting a stable solution which is as follows:

$$\frac{D \Delta t}{(\Delta x)^2} < 0.5 \tag{105}$$

Studying the subject of numerical modelling rigorously is always useful in modelling. However, the material presented here will be very useful for the readers who do not intend to write any computer code for modelling and use the standard software.

# References and Further Reading

1. Schaap, M. G., Leij, F. J., & van Genuchten, M. T. (2001). Rosetta: A computer program for estimating soil hydraulic parameters with hierarchical pedotransfer functions. *Journal of hydrology, 251*, 163–176.
2. Li, S.-L., & Liang, W.-L. (2019). Spatial–temporal soil water dynamics beneath a tree monitored by tensiometer-time domain reflectometry probes. *Water, 11*, 1662. https://doi.org/10.3390/w11081662
3. Campbell Gaylon, S. (1974). Unsaturated conductivity from moisture retention data. *Soil Science, 117*(6), 311–314. https://doi.org/10.1097/00010694-197406000-00001CIT
4. Ravikumar, V., Angaleeswari, M., & Vallalkannan, S. (2021). *Design and evaluation of drip irrigation system for sugarcane in India* (pp. 1–12). Sugar Tech, Springer. https://doi.org/10.1007/s12355-021-00983-7
5. Angaleeswari, M., Ravikumar, V., & Kannan, S. V. (2021). Evapotranspiration estimation by inverse soil water flow modelling. *Irrigation Science, 1–17*, 633–649. https://doi.org/10.1007/s00271-021-00734-2
6. Bear, J. (1972). *Dynamics of fluids in porous media* (764 pp). American Elsevier.
7. Bear, J. (1979). Hydraulics of groundwater. McGraw-Hill series in water resources and environmental engineering. McGraw-Hill.
8. Namitha, M. R., & Ravikumar, V. (2018). Determination of dispersivity in a laboratory soil column. *International Journal of Trend in Scientific Research and Development, 2*(4), 2731–2735. https://www.ijtsrd.com/papers/ijtsrd14580.pdf
9. Bolado, R., Alonso, G., & Alvarez, B. (2005). Characterization of nitrogen transformations, sorption and volatilization process in urea fertilized soils. *Vadose Zone Journal, 4*, 329–336.
10. Ravikumar, V. (2011). Modelling urea transport for nitrogen scheduling in drip irrigation. Scheme Completion report submitted to Department of Science and Technology, Government of India. (Unpublished).
11. Sharmiladevi, R., & Ravikumar, V. (2021). Simulation of nitrogen fertigation schedule for drip irrigated paddy. *Agricultural Water Management, 252*(3), 106841. https://doi.org/10.1016/j.agwat.2021.106841
12. Ravikumar, V., Vijayakumar, G., Šimůnek, J., Chellamuthu, S., Santhi, R., & Appavu, K. (2011). Evaluation of fertigation scheduling for sugarcane using a vadose zone flow and transport model. *Agricultural Water Management, 98*(9), 1431–1440. https://doi.org/10.1016/j.agwat.2011.04.012
13. Li, X., Shi, H., Šimůnek, J., Gong, X., & Peng, Z. (2015). Modeling soil water dynamics in a drip-irrigated intercropping field under plastic mulch. *Irrigation Science, 33*, 289–302. https://doi.org/10.1007/s00271-015-0466-4
14. Radcliffe David, E., & Simunek, J. (2010).*Soil physics with HYDRUS: Modeling and applications*. CRC Press.
15. Angaleeswari, M., and Ravikumar, V. (2019). Estimating evapotranspiration parameters by inverse modelling and non-linear optimization. *Agricultural Water Management, 223*(20), 105681. https://doi.org/10.1016/j.agwat.2019.06.016
16. Simunek, J., Sejna, M., & van Genuchten, M. T. (2006). The HYDRUS software package for simulating two- and three-dimensional movement of water, heat, and multiple solutes in variably-saturated media. In: *User Manual, Version 1.0. PC Progress*, Prague, Czech Republic.

17. Simunek, J., van Genuchten, M., & Sejna, T. M. (2008). Development and applications of the HYDRUS and STANMOD software packages, and related codes, special issue "Vadose Zone Modeling". *Vadose Zone Journal, 7*(2), 587–600. https://doi.org/10.2136/VZJ2007.0077

18. Wegehenkel, M., Luzi, K., Sowa, D., Barkusky, D., & Mirschel, W. (2019). Simulation of long-term soil hydrological conditions at three agricultural experimental field plots compared with measurements. *Water, 11*, 989. https://doi.org/10.3390/w11050989

19. Millington, R., J. & Quirk, J., P. (1961). Permeability of porous solids. *Transactions of The Faraday Society* 57, 1200–1207.

20. Rawls, W. J., Ahuja, L. R., Brakensiek, D. L., & Shirmohammadi, A. (1992). *Infiltration and soil water movement.* pp. 5.1–5.5. In D.R. Maidment (ed.) Handbook of hydrology. McGraw-Hill, Inc., New York.

21. van Genuchten, M.Th., Leij, F. J., & Yates, S. R. (1991). The RETC code for quantifying the hydraulic functions of unsaturated soils. Rep. EPA/600/2–91/065. Kerr, R. S. Environmental Research Laboratory, USEPA, Ada, OK.

22. Vrugt, J. A., Hopmans, J. W., & Simunek, J. (2001). Calibration of a two dimensional root water uptake model for a sprinkler-irrigated almond tree, Soil Sci. Soc. Am. J. 65(4), 1027–1037. https://doi.org/10.2136/sssaj2001.6541027x

# Water Budgeting and Economic Evaluation

## Introduction

Design and installation of an irrigation system must precede with water budgeting and economic analysis. Water budgeting is a process of ascertaining the gap between water availability in a farm and the demand due to the proposed crop plan. The irrigation requirement on any day depends not only on the crop area and crop type, but also on rainfall in the farm. Uncertainty in rainfall is inherent, and analytical tools to estimate the uncertainty and use of uncertainty estimate in decision-making are presented in this chapter.

Investing in installation of drip and sprinkler irrigation systems entails economic analysis to be done because the capital cost and also the operating cost involved are very high. Like any other investment decision, here also fund-flow occurs over a period of time and so rigorous economic analysis tools are needed to be known for making an informed investment plan. Practising irrigation engineers are also involved in providing service to farmers for getting loans from banks. The projects which are submitted to banks must have economic analysis. Therefore rudimentary tools in estimating the economic viability are presented in this chapter.

## Water Budgeting

The first step for water budgeting is to collect the information regarding the availability of water during different periods in a year. For instance, if a farmer has a well in his farm, yield of the well during each day/week/month/season of a year should be known. Also historical rainfall data on the land should be collected, and then rainfall analysis must be done using the data.

The next step is to calculate the irrigation requirements for the planned cropping pattern. The final step is to check whether the irrigation requirements and the water availability match. If there is a deficit, how much of deficit occurs, how much is the risk for the occurrence of the deficit must be calculated. If the deficit is above permitted threshold level, cropping pattern must be adjusted to bring down the crop demand with in permissible level. A water budgeting exercise for a real data is discussed and presented below.

## Rainfall Analysis

Table 1 shows the monthly rainfall occurred in a rain gauge station near a farm for 36 years. Normally the tendency is to use average rainfall in calculation of the irrigation requirement. But it is not advisable to use average rainfall. Normally average rainfall will be higher values due to the influence of very high rainfall values during good rainfall years.

It is always better to use median values. The median value of rainfall can be found out by arranging monthly rainfall in descending order as shown in Table 1. The median values are the 19th row in the Table. The median rainfall has 50% chance for occurrence and 50% chance for not occurring. Median rainfall for any month is the central value after arranging in descending order. When the number of years is even, it is not possible to get central value. In that case, first find out 50% of 36. That is equal to 18. Take average of 18th and 19th to get median. If a farmer wants to take less risk, he can also use 70% or 90% dependable rainfall. The 70% dependable rainfall is found out by multiplying 0.7 with total number of years data, i.e. 0.7 × 36 = 25.2. Therefore, take average of 25th and 26th row of rainfall for 70%. The physical significance of 70% dependable rainfall value is that 70% of years, the rainfall obtained will be more than this value. Using the same procedure 90% dependable rainfall can be calculated. The 100% dependable rainfall is the minimum rainfall expected which is the last and also the least value in Table 2.

## Water Balancing

Let us assume that a farmer owns an area of 2.5 ha, and the land is in the vicinity of the rain gauge data we analysed in the preceding section. He wants to grow crops such as beetroot, potato and onion. He wants to know, in each season if he grows certain area of each crop, how the water availability and the crop demand will match. In Table 3, one set of the planned crops to be planted and their extents of area, and the corresponding seasons and also crop factors are presented. In Table 4, monthly pan evaporation values, pan coefficients and yield from the well for each month are provided. In Table.5, the evapotranspiration, irrigation requirement, net irrigation requirement are calculated, and also gross irrigation requirements are provided for the average rainfall values. Already, it has been stated that these kinds of calculations using average rainfall are not useful. But for the sake of discussion, it has been presented here.

**Table 1** Monthly rainfall

| Year | January | February | March | April | May | June | July | August | September | October | November | December |
|------|---------|----------|-------|-------|-----|------|------|--------|-----------|---------|----------|----------|
| 1981 | 10 | 0 | 9 | 87 | 43 | 39 | 20 | 26 | 1385 | 244 | 59 | 81 |
| 1982 | 1 | 0 | 0 | 30 | 119 | 37 | 56 * | 18 | 135 | 219 | 207 | 228 |
| 1983 | 0 | 0 | 0 | 37 | 97 | 76 | 69 | 19 | 38 | 230 | 56 | 135 |
| 1984 | 0 | 0 | 0 | 38 | 6 | 3 | 41 | 24 | 90 | 28 | 46 | 4 |
| 1985 | 0 | 1 | 85 | 6 | 135 | 86 | 69 | 60 | 109 | 53 | 81 | 4 |
| 1986 | 0 | 0 | 0 | 71 | 39 | 23 | 6 | 53 | 74 | 70 | 185 | 5 |
| 1987 | 0 | 12 | 1 | 93 | 209 | 17 | 35 | 8 | 101 | 417 | 196 | 4 |
| 1988 | 0 | 20 | 5 | 11 | 71 | 30 | 54 | 20 | 25 | 167 | 174 | 51 |
| 1989 | 0 | 122 | 18 | 44 | 79 | 94 | 40 | 24 | 101 | 164 | 664 | 5 |
| 1990 | 0 | 0 | 25 | 84 | 100 | 33 | 26 | 16 | 68 | 218 | 177 | 7 |
| 1991 | 0 | 0 | 21 | 27 | 106 | 74 | 25 | 46 | 119 | 119 | 235 | 59 |
| 1992 | 0 | 0 | 16 | 60 | 54 | 34 | 37 | 51 | 36 | 107 | 237 | 8 |
| 1993 | 0 | 0 | 0 | 2 | 81 | 42 | 31 | 28 | 56 | 183 | 105 | 136 |
| 1994 | 10 | 54 | 122 | 53 | 53 | 41 | 44 | 9 | 83 | 290 | 86 | 41 |
| 1995 | 70 | 0 | 0 | 44 | 25 | 48 | 20 | 24 | 99 | 95 | 120 | 8 |
| 1996 | 38 | 5 | 2 | 6 | 43 | 35 | 24 | 14 | 124 | 125 | 81 | 10 |
| 1997 | 0 | 12 | 1 | 24 | 43 | 24 | 8 | 15 | 44 | 237 | 127 | 145 |
| 1998 | 0 | 0 | 2 | 144 | 23 | 43 | 82 | 110 | 22 | 53 | 44 | 0 |
| 1999 | 0 | 0 | 46 | 26 | 69 | 10 | 59 | 16 | 57 | 213 | 68 | 18 |
| 2000 | 0 | 0 | 0 | 0 | 0 | 0 | 0 | 0 | 0 | 0 | 0 | 0 |
| 2001 | 27 | 0 | 8 | 92 | 19 | 108 | 67 | 24 | 30 | 82 | 26 | 1 |

(continued)

**Table 1** (continued)

| Year | January | February | March | April | May | June | July | August | September | October | November | December |
|------|---------|----------|-------|-------|-----|------|------|--------|-----------|---------|----------|----------|
| 2002 | 0 | 0 | 0 | 13 | 35 | 50 | 75 | 16 | 155 | 80 | 307 | 8 |
| 2003 | 0 | 18 | 40 | 10 | 56 | 22 | 31 | 29 | 24 | 155 | 248 | 29 |
| 2004 | 73 | 29 | 6 | 138 | 31 | 28 | 126 | 1 | 49 | 261 | 172 | 2 |
| 2005 | 4 | 0 | 0 | 16 | 126 | 8 | 37 | 49 | 28 | 100 | 110 | 0 |
| 2006 | 0 | 0 | 0 | 168 | 7 | 56 | 48 | 36 | 53 | 198 | 61 | 117 |
| 2007 | 0 | 0 | 11 | 30 | 53 | 17 | 122 | 25 | 17 | 259 | 299 | 37 |
| 2008 | 0 | 0 | 0 | 96 | 53 | 58 | 52 | 25 | 171 | 30 | 304 | 162 |
| 2009 | 0 | 3 | 0 | 29 | 33 | 24 | 26 | 22 | 28 | 306 | 104 | 23 |
| 2010 | 2 | 37 | 0 | 20 | 14 | 27 | 16 | 164 | 210 | 37 | 75 | 23 |
| 2011 | 0 | 0 | 0 | 96 | 7 | 53 | 20 | 23 | 94 | 287 | 137 | 17 |
| 2012 | 0 | 0 | 69 | 3 | 29 | 7 | 7 | 69 | 20 | 228 | 100 | 2 |
| 2013 | 0 | 25 | 119 | 63 | 20 | 34 | 44 | 25 | 0 | 210 | 99 | 6 |
| 2014 | 10 | 0 | 1 | 115 | 168 | 52 | 46 | 27 | 97 | 289 | 156 | 0 |
| 2015 | 10 | 1 | 46 | 80 | 102 | 17 | 40 | 69 | 21 | 333 | 196 | 60 |
| 2016 | 28 | 0 | 151 | 29 | 84 | 48 | 9 | 6 | 69 | 205 | 294 | 1 |

**Table 2** Rainfall in descending order

| Rank | January | February | March | April | May | June | July | August | September | October | November | December |
|---|---|---|---|---|---|---|---|---|---|---|---|---|
| 1 | 73 | 122 | 151 | 168 | 209 | 108 | 126 | 164 | 1385 | 417 | 664 | 228 |
| 2 | 70 | 54 | 122 | 144 | 168 | 94 | 122 | 110 | 210 | 333 | 307 | 162 |
| 3 | 38 | 37 | 119 | 138 | 135 | 86 | 82 | 84 | 171 | 306 | 304 | 145 |
| 4 | 28 | 29 | 85 | 115 | 126 | 76 | 82 | 69 | 155 | 290 | 299 | 136 |
| 5 | 27 | 25 | 69 | 96 | 119 | 74 | 75 | 69 | 135 | 289 | 294 | 135 |
| 6 | 10 | 22 | 46 | 96 | 106 | 61 | 69 | 60 | 124 | 287 | 248 | 117 |
| 7 | 10 | 20 | 46 | 93 | 102 | 58 | 69 | 53 | 119 | 279 | 237 | 115 |
| 8 | 10 | 18 | 40 | 92 | 100 | 56 | 67 | 51 | 109 | 261 | 235 | 81 |
| 9 | 10 | 12 | 25 | 87 | 97 | 53 | 59 | 49 | 101 | 259 | 207 | 60 |
| 10 | 10 | 12 | 21 | 84 | 84 | 52 | 56 | 46 | 101 | 244 | 196 | 59 |
| 11 | 4 | 5 | 18 | 80 | 81 | 50 | 54 | 36 | 99 | 237 | 196 | 51 |
| 12 | 2 | 3 | 16 | 71 | 81 | 48 | 52 | 29 | 97 | 230 | 185 | 41 |
| 13 | 1 | 1 | 11 | 63 | 79 | 48 | 48 | 28 | 94 | 228 | 177 | 37 |
| 14 | 0 | 1 | 9 | 60 | 71 | 43 | 46 | 27 | 90 | 219 | 174 | 29 |
| 15 | 0 | 0 | 8 | 58 | 69 | 42 | 44 | 26 | 83 | 218 | 172 | 23 |
| 16 | 0 | 0 | 6 | 53 | 56 | 41 | 44 | 25 | 74 | 213 | 156 | 23 |
| 17 | 0 | 0 | 5 | 44 | 54 | 39 | 41 | 25 | 69 | 210 | 137 | 18 |
| 18 | 0 | 0 | 2 | 44 | 53 | 37 | 40 | 25 | 68 | 205 | 127 | 17 |
| 19 | 0 | 0 | 2 | 38 | 53 | 35 | 40 | 24 | 57 | 198 | 120 | 10 |
| 20 | 0 | 0 | 1 | 37 | 53 | 34 | 37 | 24 | 56 | 183 | 110 | 8 |
| 21 | 0 | 0 | 1 | 30 | 43 | 34 | 37 | 24 | 53 | 167 | 105 | 8 |

(continued)

**Table 2** (continued)

| Rank | January | February | March | April | May | June | July | August | September | October | November | December |
|---|---|---|---|---|---|---|---|---|---|---|---|---|
| 22 | 0 | 0 | 1 | 30 | 43 | 33 | 35 | 24 | 49 | 164 | 104 | 8 |
| 23 | 0 | 0 | 0 | 29 | 43 | 30 | 31 | 23 | 44 | 155 | 100 | 7 |
| 24 | 0 | 0 | 0 | 29 | 39 | 28 | 31 | 22 | 38 | 125 | 99 | 6 |
| 25 | 0 | 0 | 0 | 27 | 35 | 27 | 26 | 20 | 36 | 119 | 86 | 5 |
| 26 | 0 | 0 | 0 | 26 | 33 | 24 | 26 | 19 | 30 | 107 | 81 | 5 |
| 27 | 0 | 0 | 0 | 24 | 31 | 24 | 25 | 18 | 28 | 100 | 81 | 4 |
| 28 | 0 | 0 | 0 | 20 | 29 | 23 | 24 | 16 | 28 | 95 | 75 | 4 |
| 29 | 0 | 0 | 0 | 16 | 25 | 22 | 20 | 16 | 25 | 82 | 68 | 4 |
| 30 | 0 | 0 | 0 | 13 | 23 | 17 | 20 | 16 | 24 | 80 | 61 | 2 |
| 31 | 0 | 0 | 0 | 11 | 20 | 17 | 20 | 15 | 22 | 70 | 59 | 2 |
| 32 | 0 | 0 | 0 | 10 | 19 | 17 | 16 | 14 | 21 | 53 | 57 | 1 |
| 33 | 0 | 0 | 0 | 6 | 14 | 10 | 9 | 9 | 20 | 53 | 56 | 1 |
| 34 | 0 | 0 | 0 | 6 | 7 | 8 | 8 | 8 | 17 | 37 | 46 | 0 |
| 35 | 0 | 0 | 0 | 3 | 7 | 7 | 7 | 6 | 14 | 30 | 44 | 0 |
| 36 | 0 | 0 | 0 | 2 | 6 | 3 | 6 | 1 | 0 | 28 | 26 | 0 |
| Average RF | 8 | 10 | 22 | 52 | 62 | 39 | 43 | 34 | 104 | 178 | 154 | 42 |
| 30% dependent RF | 7 | 8 | 20 | 82 | 83 | 51 | 55 | 41 | 100 | 241 | 196 | 55 |
| Median 50% dependent RF | 0 | 0 | 2 | 41 | 53 | 36 | 40 | 24 | 63 | 201 | 123 | 14 |
| 70% dependent RF | 0 | 0 | 0 | 26 | 34 | 26 | 26 | 20 | 33 | 113 | 84 | 5 |
| 90% dependent RF | 0 | 0 | 0 | 8 | 17 | 13 | 12 | 11 | 21 | 53 | 56 | 1 |
| 100% dependent RF | 0 | 0 | 0 | 2 | 6 | 3 | 6 | 1 | 0 | 28 | 26 | 0 |

**Table 3** Crops grown and $Kc$ values

| Crop | Irrigation | Application efficiency (%) | Area (ha) | Season | $K_c$ |
|------|-----------|---------------------------|-----------|--------|-------|
| Beetroot | Conventional | 50 | 1.5 | July<br>Aug.<br>Sep.<br>Oct. | 0.5<br>1.05<br>1.05<br>0.9 |
| Potato | Drip | 90 | 1.3 | Nov.<br>Dec.<br>Jan.<br>Feb. | 0.5<br>1.15<br>1.15<br>0.75 |
| Onion season I | Sprinkler | 80 | 1.0 | Mar.<br>Apr.<br>May<br>June<br>July | 0.7<br>1.05<br>1.05<br>1.05<br>0.95 |
| Onion season II | Sprinkler | 80 | 1.1 | Oct.<br>Nov.<br>Dec.<br>Jan.<br>Feb. | 0.7<br>1.05<br>1.05<br>1.05<br>0.95 |

**Table 4** Monthly pan evaporation values and well yield

| Month | Pan coefficient ($Kp$) | Pan evaporation ($Ep$) mm | Well yield ($l/s$) |
|-------|------------------------|---------------------------|--------------------|
| January | 0.6 | 126 | 3.0 |
| February | 0.6 | 159 | 2.9 |
| March | 0.7 | 186 | 2.5 |
| April | 0.7 | 189 | 2.2 |
| May | 0.8 | 201 | 2.0 |
| June | 0.8 | 189 | 2.2 |
| July | 0.8 | 144 | 2.3 |
| August | 0.8 | 141 | 2.4 |
| September | 0.7 | 156 | 2.6 |
| October | 0.7 | 120 | 2.7 |
| November | 0.6 | 111 | 2.9 |
| December | 0.6 | 117 | 3.0 |

In Table 5, for calculating the water availability from the well, the number of hours of power available is taken as 6 h per day. The water available for each month is a product of the well yield and hours of power available per day and number of days in the corresponding month. Table 6 provides the water deficit for each month for different dependable rainfalls. It is better to adjust the cropping pattern for meeting the demand for the median rainfall. This kind of analysis provides information regarding the extent and frequency of deficit during situations of failure. How much will be

**Table 5** Water balance for average rainfall

| Month | ETc (mm) = $K_p \times K_c \times E_p$ | | | Rainfall (RF) (mm) | Irrigation requirement (IR) = $Et_c -$ RF (mm) | | | Irrigation requirement IR (litres) = IR (mm) × Area × 10,000 | | | Net irrigation requirement NIR (litres) = IR (litres)/application eff | | | Gross irrigation requirement (L) | Water available from well (L) | Deficit (%) |
|---|---|---|---|---|---|---|---|---|---|---|---|---|---|---|---|---|
| | Beetroot | Potato | Onion | | Beetroot | Potato | Onion | Beetroot | Potato | Onion | Beetroot | Potato | Onion | 12 + 13 + 14 | | |
| (1) | (2) | (3) | (4) | (5) | (6) | (7) | (8) | (9) | (10) | (11) | (12) | (13) | (14) | (15) | (16) | (17) |
| 1 | 0 | 87 | 79 | 8 | 0 | 79 | 71 | 0 | 1,026,220 | 785,180 | 0 | 1,140,244 | 981,475 | 212,719 | 2,008,800 | − 6 |
| 2 | 0 | 72 | 91 | 10 | 0 | 62 | 81 | 0 | 800,150 | 886,930 | 0 | 889,056 | 1,108,663 | 1,997,718 | 1,753,920 | − 14 |
| 3 | 0 | 0 | 91 | 22 | 0 | 0 | 69 | 0 | 0 | 691,400 | 0 | 0 | 864,250 | 864,250 | 1,674,000 | 0 |
| 4 | 0 | 0 | 139 | 54 | 0 | 0 | 85 | 0 | 0 | 849,150 | 0 | 0 | 1,061,438 | 1,061,438 | 1,425,600 | 0 |
| 5 | 0 | 0 | 169 | 64 | 0 | 0 | 105 | 0 | 0 | 1,048,400 | 0 | 0 | 1,310,500 | 1,310,500 | 1,339,200 | 0 |
| 6 | 0 | 0 | 159 | 40 | 0 | 0 | 119 | 0 | 0 | 1,187,600 | 0 | 0 | 1,484,500 | 1,484,500 | 1,425,600 | − 4 |
| 7 | 58 | 0 | 109 | 44 | 14 | 0 | 65 | 204,000 | 0 | 654,400 | 408,000 | 0 | 818,000 | 1,226,000 | 1,540,080 | 0 |
| 8 | 118 | 0 | 0 | 35 | 83 | 0 | 0 | 1,251,600 | 0 | 0 | 2,503,200 | 0 | 0 | 2,503,200 | 1,607,040 | − 56 |
| 9 | 115 | 0 | 0 | 107 | 8 | 0 | 0 | 114,900 | 0 | 0 | 229,800 | 0 | 0 | 229,800 | 1,684,800 | 0 |
| 10 | 76 | 0 | 59 | 182 | 0 | 0 | 0 | 0 | 0 | 0 | 0 | 0 | 0 | 0 | 1,807,920 | 0 |
| 11 | 60 | 33 | 70 | 158 | 0 | 0 | 0 | 0 | 0 | 0 | 0 | 0 | 0 | 0 | 1,879,200 | 0 |
| 12 | 0 | 81 | 74 | 43 | 0 | 38 | 31 | 0 | 490,490 | 337,810 | 0 | 544,989 | 422,263 | 967,251 | 2,008,800 | 0 |

**Table 6** Deficit of irrigation for different dependable rainfalls

| Months | Deficit in per cent for | | | | | |
|--------|--------------------|------------------------------|----------------------------------------|------------------------------|------------------------------|-------------------------------|
| | Average rainfall | 30% dependent rainfall | 50% dependent (median) rainfall | 70% dependent rainfall | 90% dependent rainfall | 100% dependent rainfall |
| January | − 6 | − 13 | − 17 | − 17 | − 11 | − 17 |
| February | − 4 | − 24 | − 30 | − 30 | − 19 | − 30 |
| March | 0 | 0 | 0 | 0 | 0 | 0 |
| April | 0 | 0 | 0 | 0 | − 21 | − 22 |
| May | 0 | 0 | − 8 | 0 | − 36 | − 58 |
| June | − 4 | 0 | − 9 | − 18 | − 29 | − 39 |
| July | 0 | 0 | 0 | − 31 | 0 | − 101 |
| August | − 56 | − 60 | − 76 | − 87 | 0 | − 121 |
| September | 0 | 0 | 0 | − 45 | − 40 | − 104 |
| October | 0 | 0 | 0 | 0 | 0 | − 70 |
| November | 0 | 0 | 0 | 0 | 0 | 0 |
| December | 0 | 0 | 0 | − 2 | − 3 | − 9 |

the extent of deficit and how many times, the water available may not meet the total requirement can be found out from this analysis. If a farmer grows very high-value crops and wants to take less risk, he can use 70% or 90% dependent rainfall and for the selected rainfall, he may adjust the cropped areas, to meet the demand closely with the water available.

# Basics of Economic Analysis

Money functions as a medium of exchange, a store of value, a unit of account, and also a standard of deferred payment. In a growing economy, the value of money to day would definitely tend to get reduced compared to the future. The process of losing the value of money as time passes is called as inflation. The rate of loss of value for money is called as inflation rate.

We can also purchase goods and services today and agree for deferred payment by taking into account the expected inflation rate and also the risks in the dynamics of local and global economy. When we get certain goods and services today without paying today means that we have used some body's money to day and we promise to pay the money in future. Hence, the rate of interest we need to pay back equivalent to the goods and services purchased on today is usually higher than the inflation rate. Higher the risk involved in payment, higher will be the interest rate. Interest rate is levied on the borrower of money by the lender.

Another term mostly used is discount rate. Interest rate and discount rate are not one and the same, but there is a difference. Let us assume I have Rs. 100 million/– today in hand and if I invest in banks, I get 6% interest per year with less risk. If I invest in real-estate, I get 20% return with more risk. If I invest in stock market, I get 30% return with more risk. Usually the discount rate is the opportunity cost of the money in hand with the lowest risk and lowest effort. In this case, the discount rate is 6%. But let us assume that the investor wants to do an investment analysis of investing in stock market but at the same time he believes that if his investment in real-estate poses zero risk and he is assured of getting 20%, then he can use a discount rate of 20%. But conventionally for doing economic analysis, the interest rate paid by banks to the investor is taken as discount rate.

Discount rate is very much useful in finding the net present value (NPV) of investments and expenditures when investments, expenditures and incomes happen at different time periods.

Expenditure on purchase of machinery and equipment are called as capital cost and the recurring expenditure like cost of power, maintenance costs, etc. are classified under operating costs.

Machinery and pipes last for a specified number of years called as life period of the project. After life period of the project is completed, the machinery and pipes would have to be disposed properly. Many a times the machinery and equipment after the end of useful life can be sold for value and that value is called as Salvage value.

If $P_n$ is the income generated from a project after $n$ years of establishment, then the NPV of the income is as follows (Fig. 1) (Peter Walner and Muluneh Yitayew. 2016) [1]:

$$NPV = \frac{P_n}{(1+i)^n} \tag{1}$$

where '$i$' is discount rate.

If $P_t$ is the income generated from a project after t years and the income accrues in every year, then the NPV of the income is as follows (Fig. 2).

**Fig. 1** Income generated on any year converted to net present value (NPV)

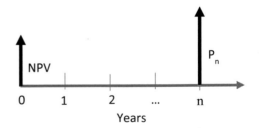

**Fig.2** Incomes generated on many years converted to net present value (NPV)

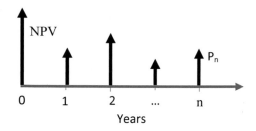

$$NPV = \sum_{t=1}^{n} \frac{P_t}{(1+i)^t} \tag{2}$$

On the same principle, the capital investments and the operating cost can be converted into NPV of cost (https://en.wikipedia.org/wiki/Net_present_value). The NPV of the project is calculated as follows:

$$NPV \text{ of the project} = NPV \text{ Benefits}$$

Benefit–cost (BC) ratio is the ratio of NPV of benefits to NPV of costs. A BC ratio greater than one is a viable project.

Another term internal rate of return (IRR) is also important in the economic analysis, which is defined as the discount rate at which the NPV is zero (https://en.wikipedia.org/wiki/Internal_rate_of_return). It is the maximum rate of return that a project could pay for the resources used. The selection criterion based on IRR is that it should be greater than the opportunity cost of the capital.

## Problem

A farmer owns an agricultural field of 1 ha area which is irrigated at present with canal supply. At present he does irrigation using manual labourers by adopting ridges and furrows method and wants to install sprinkler irrigation. He grows groundnut crop for 2 seasons per year and expects a total yield of 1800 kg/year. The selling price of groundnut is Rs.50/kg at present. The selling price has been found to increase at a rate of 3% per year. The other cost of cultivation is Rs.10,000/year, and this cost is expected to increase 3% per year.

The capital investment for installation of sprinkler system is Rs.100,000. The cost of electricity for pressurizing water has been worked out to be Rs. 15,000/year. The cost of electricity is expected to increase at a rate of 3% per year. The annual maintenance cost of sprinkler irrigation is Rs.2000/year, and it is expected to increase at a rate of 5%. The life period of the project is 20 years. Take the salvage value of the project components after 20 years is 10% of the present worth of capital investment.

Take a discount rate of 8%. Calculate NPV and BC ratio and IRR and evaluate the economic worthiness of installing the sprinkler irrigation system.

## Solution

The yearly benefit for the year 0 ($B_0$) is as follows:

$$= \text{Selling price of groundnut (Rs./kg)} \times \text{yield/year}$$

$$= \text{Rs.50/kg} \times 1800\text{kg/ year} = \text{Rs.90,000}$$

$$\text{Salvage value after 20 years} = \text{Rs.100,000} \times 0.1 = \text{Rs.10,000}$$

NPV of Salvage value $= \frac{10,000}{(1+0.08)^{20}} = \text{Rs.2145}$

NPV of installation of sprinkler irrigation system

$= \text{NPV of Benefits} - \text{NPV of Operating Cost} - \text{NPV of Salvage value}$

$- \text{Capital cost} = \text{Rs. } 1,135,579 - \text{Rs. } 345,590 - \text{Rs. } 2145$

$- \text{Rs. } 100,000 = \text{Rs. } 691,061$

$$\text{BC ratio } = 3.32$$

See Table 7.

**Calculation of IRR**

While calculating IRR, select a discount rate by trial and error so that the NPV is near zero. Select two discount rates, so that for one discount rate NPV is positive and for another discount rate NPV is negative. Then use the following interpolation formula to get the discount rate at which NPV is zero.

$$\text{IRR} = \text{LDR} + (\text{HDR} - \text{LDR})\frac{\text{NPV at LDR}}{\text{Sum of NPV at HDR and LDR(signs ignored)}}$$

LDR    Lower Discount Rate

HDR    Higher Discount Rate

For the preceding problem for the lower discount rate of 65%, NPV is Rs.4476 and for the higher discount rate of 70%, NPV is Rs.3309 (negative). Hence by applying the preceding interpolation formula, we get the IRR to be equal to 67.87%.

From the calculations of NPV, BC ratio as well as IRR, it is obvious that the conversion of irrigation system is economically viable.

## Summary

For doing water budget analysis, estimation of water availability from all the sources and estimation of crop demand are necessary. Estimating the uncertainty of rainfall is useful in arriving at irrigation requirements. Routine use of average rainfall in estimation of crop demand may yield lower crop demand. Median rainfall is a better estimate in arriving at crop demand than using average rainfall. Rainfall with higher

**Table 7** NPV computation

| Year (t) | Cost of power ($P_t$) (Rs.) (2) | Cost of maint ($M_t$) (Rs.) (3) | Cost of labour ($L_t$) (Rs.) (4) | Oper. cost ($OC_t$) (Rs.) (4) = (2) + (3) + (4) | Benefit ($B_t$) (Rs.) (5) | NPV of oper. cost (Rs.) (6) | NPV of benefits (Rs.) (7) |
|---|---|---|---|---|---|---|---|
| | $P_0(1.03)^n$ | $M_0(1.05)^n$ | $L_0(1.05)^n$ | | $B_0(1.03)^n$ | $\frac{OC_t}{(1+0.08)^t}$ | $\frac{B_t}{(1+0.08)^t}$ |
| 0 | $P_0 =$ 15,000 | $M_0 =$ 2000 | $L_0 =$ 20,000 | | $B_0 =$ 90,000 | – | – |
| 1 | 15,450 | 2100 | 10,300 | 27,850 | 92,700 | 25,787 | 85,833 |
| 2 | 15,914 | 2205 | 10,609 | 28,728 | 95,481 | 24,629 | 81,860 |
| 3 | 16,391 | 2315 | 10,927 | 29,633 | 98,345 | 23,524 | 78,070 |
| 4 | 16,883 | 2431 | 11,255 | 30,569 | 101,296 | 22,469 | 74,455 |
| 5 | 17,389 | 2553 | 11,593 | 31,534 | 104,335 | 21,462 | 71,008 |
| 6 | 17,911 | 2680 | 11,941 | 32,531 | 107,465 | 20,500 | 67,721 |
| 7 | 18,448 | 2814 | 12,299 | 33,561 | 110,689 | 19,583 | 64,586 |
| 8 | 19,002 | 2955 | 12,668 | 34,624 | 114,009 | 18,706 | 61,596 |
| 9 | 19,572 | 3103 | 13,048 | 35,722 | 117,430 | 17,870 | 58,744 |
| 10 | 20,159 | 3258 | 13,439 | 36,856 | 120,952 | 17,071 | 56,024 |
| 11 | 20,764 | 3421 | 13,842 | 38,027 | 124,581 | 16,309 | 53,431 |
| 12 | 21,386 | 3592 | 14,258 | 39,236 | 128,318 | 15,581 | 50,957 |
| 13 | 22,028 | 3771 | 14,685 | 40,485 | 132,168 | 14,886 | 48,598 |
| 14 | 22,689 | 3960 | 15,126 | 41,775 | 136,133 | 14,223 | 46,348 |
| 15 | 23,370 | 4158 | 15,580 | 43,107 | 140,217 | 13,589 | 44,202 |
| 16 | 24,071 | 4366 | 16,047 | 44,483 | 144,424 | 12,984 | 42,156 |
| 17 | 24,793 | 4584 | 16,528 | 45,905 | 148,756 | 12,407 | 40,204 |
| 18 | 25,536 | 4813 | 17,024 | 47,374 | 153,219 | 11,855 | 38,343 |
| 19 | 26,303 | 5054 | 17,535 | 48,892 | 157,816 | 11,329 | 36,568 |
| 20 | 27,092 | 5307 | 18,061 | 50,459 | 162,550 | 10,826 | 34,875 |
| **Sum** | | | | | | **345,590** | **1,135,579** |

dependability may be used when we want to arrive at conservative crop plans where in uncertainty in meeting the crop demand is to be reduced.

Cash flow analysis is useful in determining the economic viability of investments in irrigation systems installation. Determining net present value (NPV), benefit–cost (BC) ratio and internal rate of return (IRR) was demonstrated for a numerical example.

## Assessment

### Part A
### Say true or false for the following questions:

1. In a rainfall data set, the median rainfall value has 50% chance of occurrence (True)
2. Usually mean value of rainfall is greater than the median rainfall (True)
3. Using median rainfall for 50% chance of occurrence is correct rather than using median rainfall. (True)
4. Ninety per cent dependable rainfall value will be less than the fifty per cent dependable value (True)
5. The process of losing the value of money as time passes is called as inflation. (True)
6. Higher the risk involved in payment, higher will be the interest rate. (True)
7. Interest rate is levied on the borrower of money by the lender. (True)
8. Conventionally for doing economic analysis, the interest rate paid by banks to the investor is taken as discount rate. (True)
9. For calculation of internal rate of return (IRR), discount rate is needed to be known (False).

### Part B
### Short/Brief Answers:

1. What do you understand from the term capital cost?
2. What do you understand from the term operating cost?
3. What do you understand from the term salvage value?
4. Define net present value of benefits accrued from a project?
5. Define internal rate of return?
6. Define benefit–cost ratio?
7. Why is interest rate normally higher than the inflation rate?
8. Distinguish between interest rate and discount rate and discuss with a real life example?
9. Discuss about the ways by which evaluation of viability of a project is done using NPV, IRR and BC ratio?

## References and Further Reading

1. Walner, P., & Yitayew, M. (2016). Irrigation and drainage engineering. Springer. https://doi.org/10.1007/978-3-319-05699-9. https://en.wikipedia.org/wiki/Net_present_value https://en.wikipedia.org/wiki/Internal_rate_of_return

# Basic Section

# Pressure, Energy and Power in Water

## Introduction

In this chapter you will learn about pressure, energy and power and their applications in sprinkler and drip irrigation system. Sprinkler and drip irrigation system are made of pipeline network to deliver precisely the water required by crops. To design sprinkler and drip irrigation systems, basic understanding of pressure and energy is very much essential. Pascal's law is useful in understanding about pressure inside the water. Potential energy and kinetic energy are the two forms of energy often encountered in solid mechanics. Along with these two energy terms, concept of pressure energy is also very much used in sprinkler and drip irrigation.

**Learning Outcomes**

- Differentiate different forms of energy
- Describe the role of datum, kinetic and pressure energy in sprinkler and drip irrigation
- Give examples of application of various forms of energy in sprinkler and drip irrigation
- Describe and demonstrate pressure measurement
- Solve problems involving application of pressure, energy and power.

## *Units and Terminologies Used*

- In system international (SI), unit for mass is kilogram and is denoted with 'kg'.
- Length is measured in metre and is denoted with 'm'.
- Time is measured in second and is denoted with 's'.
- Pressure is force per unit area.
- The unit for pressure is $(kg\text{-}m/s^2)/m^2$ or $N/m^2$. One $N/m^2$ is also called as one pascal.

© The Author(s), under exclusive license to Springer Nature Singapore Pte Ltd. 2023
V. Ravikumar, *Sprinkler and Drip Irrigation*,
https://doi.org/10.1007/978-981-19-2775-1_18

- Work is the product of force and distance moved.
- One Joule of work done is when one newton force acts on a body in the direction of movement of the body for one metre distance.
- Energy is the capacity to do work. Energy and work done have the same units.
- Approximately one atmospheric pressure is approximately equal to 10 m of water column.

## Laws of Newton

Newton's laws of motion are three physical laws that laid the foundation for classical mechanics. They describe the relationship between a body and the forces acting upon it, and its motion in response to those forces. The first law defines the force qualitatively, the second law offers a quantitative measure of the force, and the third asserts that a single isolated force doesn't exist. These three laws are summarized as follows:

**First law**

In an inertial frame of reference, an object either remains at rest or continues to move at a constant velocity, unless acted upon by a force.

**Second law**

In an inertial frame of reference, the vector sum of the forces **F** on an object is equal to the mass 'm' of that object multiplied by the acceleration 'a' of the object: $\mathbf{F} = \mathbf{ma}$.

**Third law**

When one body exerts a force on a second body, the second body simultaneously exerts a force equal in magnitude and opposite in direction on the first body.

Figure 1 is used to illustrate Newton's laws with a real-life example.

## Pressure

Pressure of fluid mass in a container is due to the impinging force of the fluid molecules on the wall (Fig. 2). When a particle hits on a wall in any arbitrary direction as in Fig. 3, in an elastic collision, the direction of particle changes with angles of incidence and reflection being same. The change in velocity before and after the hit is only in the direction perpendicular to the wall not in the direction parallel to the wall ($V_x$ changes to $-V_x$ and $V_y$ remains $V_y$). Therefore it is proved that the wall exerts force on the particle only in the direction perpendicular to the wall. Therefore the reaction of the particle must be in the opposite direction perpendicular to the wall. Hence it is proved that the direction of pressure must be perpendicular to the surface of the wall. The first fallacy anyone falls into is misunderstanding this fact. Most of the time, people think that the wall's reaction at the time of hit is perpendicular to the incident angle.

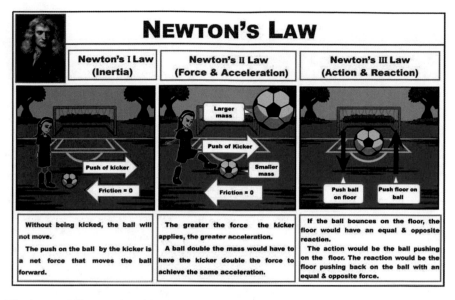

Fig. 1 A real-life example to illustrate Newton's laws

Fig. 2 Pressure inside a fluid mass. By Becarlson Own work, see http://www.becarlson.com/, CC BY-SA 3.0, https://commons.wikimedia.org/w/index.php?curid=14728470

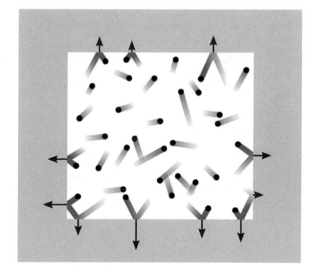

For a small spherical water mass at any depth, surrounding water will exert pressure on the water mass from all the directions equally and also perpendicular to the surface of the water mass (Fig. 4). This pressure is equal in all directions. Because of that, the water mass would be static at that place only if pressure acting on all directions is equal. If we consider the small spherical water mass to be infinitesimally

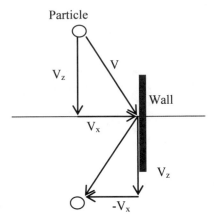

**Fig. 3** Velocity change of a particle colliding on a surface

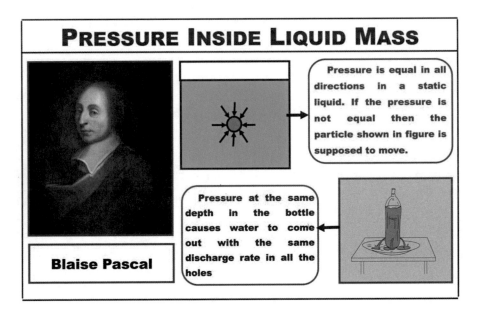

**Fig. 4** Pressure equal in all directions inside a fluid mass

small, then the mass becomes a point. Hence the pressure at any point is equal in all directions.

The pressure acting at the bottom of the tank is equal to the weight of water divided by the cross-sectional area of tank (Fig. 5). If the cross-sectional area of the tank is 1 m², depth of water standing in the tank is say 10 m, then the volume of the water is 10 m³.

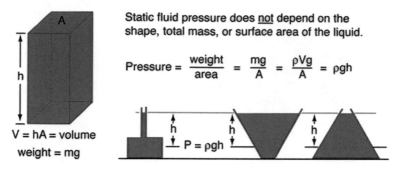

**Fig. 5** Pressure of water at any point

We know that mass density of water ($\rho$) is 1000 kg m$^{-3}$ and so the total mass of water standing above the bottom is 10,000 kg (mass = volume × mass density). Weight ($W$) is a force, and weight is due to the acceleration due to the gravity of earth on any mass m ($W = $ mg).

One Newton force is equal to the force required to move one kg of mass with one m/s$^2$ of acceleration. If '$x$' kg mass is given one m/s$^2$ of acceleration, then the force acting on the body is '$x$' Newton. If one kg mass is given '$y$' m/s$^2$ of acceleration, the force acting on the body is '$y$' Newton. If '$x$' kg of mass is given, '$y$' m/s$^2$ of acceleration, '$xy$' Newton of force is acting on the body. Therefore, force (F) is the product of mass (m) and acceleration (a) and the equivalent unit for Newton is kg-m/s$^2$.

Weight density ($\gamma$) is the product of mass density ($\rho = 1000$ kg m$^{-3}$) and acceleration due to the gravity of earth ($g = 9.81$ m/s$^2$), and hence weight density of water is 9810 N/m$^3$ ($\gamma = \rho g$).

Therefore, weight of the water $= \gamma hA = 9810 \times 10 \times 1 = 98{,}100$ N.

Pressure acting on the bottom of the tank = weight of the water/cross-sectional area of the tank

$$98,100\,\text{N}/1\text{m}^2 = 98,100\,\text{N/m}^2 = \rho gh$$

So, from the preceding discussion we understand that pressure at any point within a mass of water is the product of weight density of water ($\rho$g) and height of water standing above the point h. It should also be understood that at any point below the surface of the water the pressure is due to the column of water standing above the point.

Usually students misunderstand the preceding statement when they come across a situation as shown in Fig. 6. When we measure the distance straight above point $A$, we come across the boundary of the container at point $B$. In this case pressure at the point $A$ should be measured as sum of the pressure until point $B$ from $A$ and the pressure at point $B$. It is because; the pressure at $B$ is influenced by the height of standing water above $B$. If a pipe section as shown in Fig. 6 is inserted at point $B$, water will stand in the pipe until water surface. Therefore, pressure at point $A$ is sum

Fig. 6 Pressure in a conical container

Point B
Point A
Point B
Point A

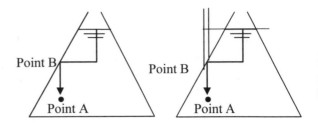

of water column from point *A* to *B* and the column of water that would stand at point *B*, if a pipe is placed at point *B*.

Pascal law states that, in a fluid at rest in a closed container, a pressure change in one part is transmitted without loss to every portion of the fluid and to the walls of the container. Pascal demonstrated this law by constructing a 10 m high rising pipe over a barrel of water (Fig. 7). When water was poured into the vertical tube, Pascal

Fig. 7 Pascal's barrel (Wikipedia)

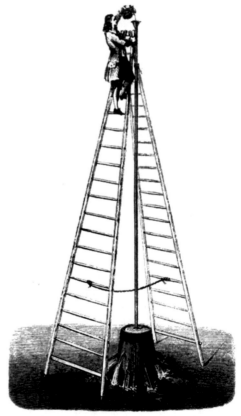

Fig. 45.—Hydrostatic paradox. Pascal's experiment.

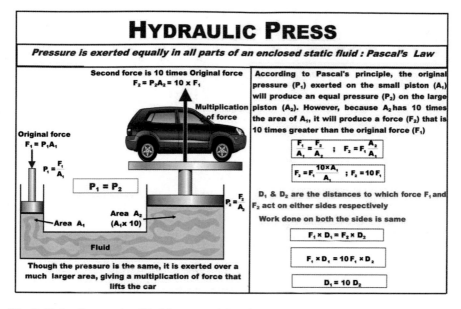

**Fig. 8** Hydraulic press (modified from www.britannica.com)

found that the increase in hydrostatic pressure caused the barrel to burst. Hydraulic press is one of the useful inventions based on Pascal's law (Fig. 8).

## *Pressure Measurement*

Pressure at any location can be found out in any pipe flowing with water or water in static condition by inserting a small piece of pipe (called as piezometer) inserted as shown in Fig. 9.

$$\text{Pressure at } A(P_A) = \gamma h + P_{\text{atm}} \tag{1}$$

**Fig. 9** Piezometer

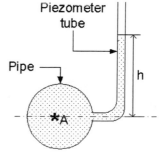

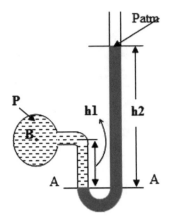

**Fig. 10** U-tube manometer

Limitation with the piezometer is that when pressure is high we need longer pipes. So in order to circumvent this issue U-tube manometer was invented (Fig. 10). U-tube manometer contains a gauge liquid which is different from flowing liquid. For U-tube manometer, section A-A is used for writing a pressure balance equation which is as follows:

$$P_B + \gamma_f h_1 = P_{\text{atm}} + \gamma_g h_2 \tag{2}$$

where $\gamma_f$ and $\gamma_g$ are weight densities of flowing fluid and gauge liquid. Mostly mercury is used as gauge liquid with a magnitude of 136,000 N-m$^{-3}$ (approximate).

The reference pressure for measurement is taken as atmospheric pressure. Such pressure measurement method is called as gauge pressure system. In gauge pressure system, the atmospheric pressure is assumed as zero. Another pressure measurement system called as absolute pressure system in which the absolute vacuum is taken as zero. Pressures below atmospheric pressures are termed as negative (Fig. 11).

It is noteworthy to see how the pressure of atmosphere is measured just by using a simple bowl and a transparent tube with mercury (Fig. 12).

In gauge pressure system, the preceding equations are written as follows:

$$\text{Pressure at } A = \gamma h \tag{3}$$

$$P_B + \gamma_f h_1 = \gamma_g h_2 \tag{4}$$

In real-life situations, mostly pressure measurement is done with pressure Bourdon-type pressure and vacuum gauges (Figs. 13 and 14).

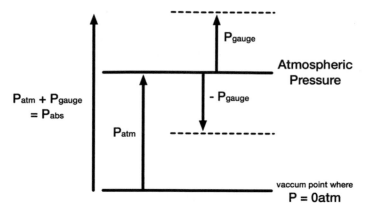

Fig. 11  Gauge pressure system and absolute pressure system

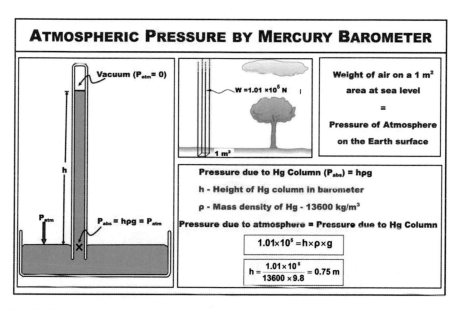

Fig. 12  Measurement of atmospheric pressure

## Energy of Water

Water is considered to possess three forms of energy. They are gravitational (also called as potential or datum) energy, kinetic energy and pressure energy.

**Fig. 13** Pressure gauge

**Fig. 14** Vacuum gauge

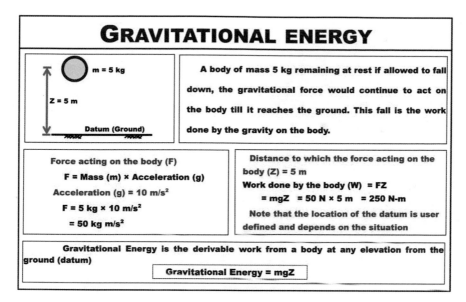

**Fig. 15** Gravitational energy

## *Potential or Gravitational or Datum Energy*

If we consider m mass falling freely from height '*z*' to a datum, it will get accelerated till it reaches the datum.

Force acting on the mass = mass (m) × acceleration due to the gravity (g)

Work done on the mass till it reaches the datum = Force (mg) × Distance (z)

If the mass is brought to the datum, the work that can be derived from the mass is mgz. The concept of gravitational energy is illustrated with a numerical example in Fig. 15.

It is conventional to express all forms of energy possessed by water in terms of energy per unit weight of water. Energy per unit weight of water has a unit of 'metres' because it is the ratio of N-m to N. The work that can be derived from a unit weight of water at a height from datum is z (= mgz/mg) and is called as datum energy head or simply datum head. It is also called as gravitational head.

## *Kinetic Energy*

Velocity of a body is the distance travelled by the body per unit of time, and the unit used is metre (m)/second(s). Acceleration is the change in velocity in unit time, and the unit used is $m/s^2$. The expression for kinetic energy (KE) added to a stationary body is derived in Fig. 16. Conversely if a body of mass m is moving with a velocity

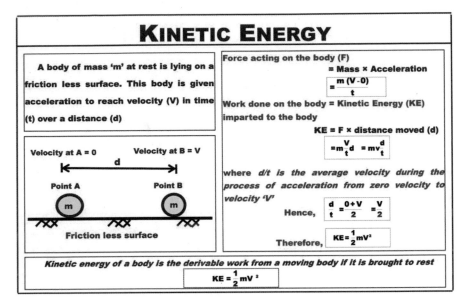

**Fig. 16** Kinetic energy

v, then if the body is brought to rest, $(\frac{1}{2})mv^2$ of work can be extracted from it. This $(\frac{1}{2})mv^2$ is called as kinetic energy possessed by the body.

If the body is water mass, the kinetic energy possessed by a unit weight of water is obtained by dividing the term $(\frac{1}{2})mv^2$ by mg and the result is $v^2/2\,g$. The kinetic energy possessed by unit weight of water is called as kinetic head expressed in m. The magnitude of velocity head is very less compared to pressure head for flow of water in pipe in microirrigation. Figure 17 illustrates how in a falling solid body the gravitational energy and kinetic energy keep changing and at the same time the total energy remains same. Understanding this concept with solid mechanics is fundamental in understanding Bernoulli's law in succeeding sections.

## *Pressure Energy*

The effect of pressure energy for water was seen when we took a container with water and holes in the container. The pressure increases with increase in depth and so the velocity of water issued from the container is the maximum for the bottom-most hole. For pressurizing water energy is needed. The expression for pressure energy and pressure head is derived in Fig. 18.

# ENERGY TRANSFER IN A FALLING SOLID

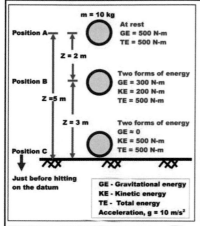

m = 10 kg

Position A

Z = 2 m

At rest
GE = 500 N-m
TE = 500 N-m

Position B

Z = 5 m

Two forms of energy
GE = 300 N-m
KE = 200 N-m
TE = 500 N-m

Z = 3 m

Position C

Two forms of energy
GE ≈ 0
KE = 500 N-m
TE = 500 N-m

Just before hitting
on the datum

GE - Gravitational energy
KE - Kinetic energy
TE - Total energy
Acceleration, g = 10 m/s²

A body of mass (m) = 10 kg is at rest at an elevation of 5 m from ground

Gravitational energy possessed by the body when it is at 5 m above the datum      = mg Z

= 10 × 10 × 5

= 500 N-m

If the body is allowed to fall, when the body is at 3 m above the datum, the body possesses 2 forms of energy. One is gravitational energy due to its 3 m elevation from the datum. Two is kinetic energy due to its velocity. This kinetic energy is derived from the loss of gravitational energy due to fall of 2 m from position A to B.

Gravitational energy at B   = mg Z   = 10 × 10 × 3

= 300 N-m

Kinetic energy at B   = mgZ = 10 × 10 × 2

= 200 N-m

*Total energy of a falling body remains constant but the individual components of energy such as gravitational energy and kinetic energy change along the path of falling body*

**Fig. 17**  Energy transfer in a falling solid

# PRESSURE ENERGY

For pressurizing water energy is needed. Assume that a small cylindrical segment of water of cross sectional area (A) and length (L) is forced through a orifice at a depth (h) in the tank with water. Weight density of water (Υ) is approximately 10000 N/m³

Force needed to push the water
inside through the orifice

$$= γhA$$

(Pressure × area of cross
section of orifice)

Work done to push the water
(inside through the orifice)

$$= γhAL$$

(Force × length of the water
segment (L))

Weight of water pushed

$$= γAL$$

Work done to push the water per
unit weight of water handled

$$= \frac{γhAL}{γAL} = h$$

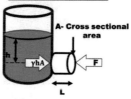

A- Cross sectional
area

Pressure energy possessed by unit weight of water at any depth of "h" from the free surface is the depth of the water from the free surface. This pressure energy per unit weight of water is termed as pressure head possessed by the water.

**Fig. 18**  Pressure energy

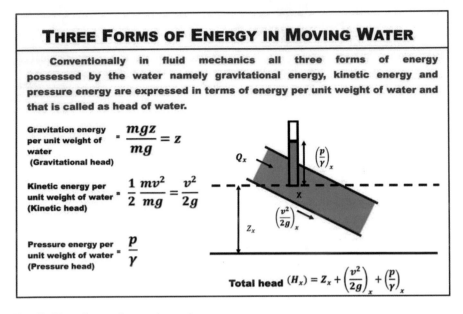

**THREE FORMS OF ENERGY IN MOVING WATER**

Conventionally in fluid mechanics all three forms of energy possessed by the water namely gravitational energy, kinetic energy and pressure energy are expressed in terms of energy per unit weight of water and that is called as head of water.

Gravitation energy
per unit weight of       $= \dfrac{mgz}{mg} = z$
water
(Gravitational head)

Kinetic energy per        $= \dfrac{1}{2}\dfrac{mv^2}{mg} = \dfrac{v^2}{2g}$
unit weight of water
(Kinetic head)

Pressure energy per       $= \dfrac{p}{\gamma}$
unit weight of water
(Pressure head)

Total head $(H_x) = Z_x + \left(\dfrac{v^2}{2g}\right)_x + \left(\dfrac{p}{\gamma}\right)_x$

**Fig. 19**  Three forms of energy in moving water

## Total Energy of Water

The total energy possessed by water per unit weight of water $(H_x)$ (called as total head) is the sum of datum head $(z_x)$, velocity head $(v^2/2\,g)$ and pressure head $(h_x = P_x/\gamma)$

$$H_x = z_x + \left(\frac{v^2}{2g}\right)_x + \left(\frac{P}{\gamma}\right)_x$$

Figure 19 illustrates all three forms of energy in moving water in a pipe.

## Power of Water

Power is rate of transfer of energy per unit time. One Watt of power is one Joule of energy flow per second. An expression for power of water is derived in Fig. 20

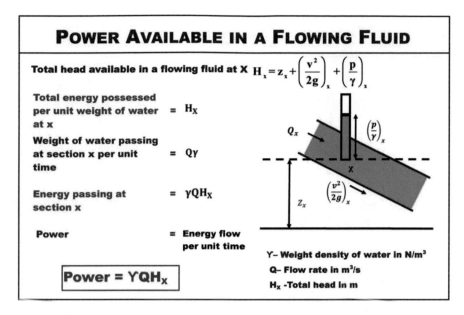

**Fig. 20** Power available in a flowing fluid

## Summary

Pressure at any point within static water mass is due to the height of water standing above the point. Pressure at any point in a flowing pipe is the height of water that would stand if a standing pipe is inserted at the point.

Water flowing in a pipe possesses three forms of energy namely datum, kinetic and pressure energies. It is conventional in fluid mechanics to express these energies in terms of energy possessed per unit weight of water called as head. Total head of flowing water is sum of datum head, kinetic head and pressure head. Power of water flow is rate of flow of energy across a pipe section.

## *Solved Exercises*

1. A circular tank is filled with water to a depth of 10 m. The radius of the tank is 2 m. Find out the pressure of water at a depth of 2 m and also find out the force acting on the bottom of the water tank. Weight density of water $= 9810$ N/m$^3$.

**Solution:**

Pressure at a depth of 2 m $= \gamma h = 9810 \times 2 = 19620$ N/m$^2$

(continued)

(continued)

| Pressure at a depth of 10 m $= \gamma h = 9810 \times 10 = 98100$ N/m$^2$ |
| Circular area of tank $= \pi r^2 = 3.14 \times 2 \times 2 = 12.56$ m$^2$ |
| Force acting on the bottom of tank = Pressure $\times$ Area of tank |
| $= 98{,}100 \times 12.56$ |
| $= 1{,}232{,}136$ N |

2. A pressure gauge is fitted at a pipe section carrying water. The pressure gauge shows a pressure of 2 kg(f)/cm$^2$. The discharge rate at the section is 0.01m$^3$/s and pipe diameter is 100 mm. The pipe section is at a height of 2 m from an arbitrary datum. Find out the datum head, pressure head, kinetic head and power of water flow at the section?

| **Solution**: |
| Datum head $= 2$ m |
| Diameter of pipe $= 100$ mm $= 0.1$ m |
| Radius of pipe $= 0.1/2 = 0.05$ m |
| Area of water flow $= \pi r^2 = 3.14 \times 0.05 \times 0.05 = 0.00785$ m$^2$ |
| Velocity of water = Discharge rate/area $= 0.01/0.00785 = 1.274$ m/s |
| Kinetic head $= v^2/2g = 0.0827$ m |
| Pressure $= 2$ kg(f)/cm$^2$ $= 2 * 10 = 20$ m (Because 1 kg(f)/cm$^2$ $= 10$ m of water) |
| Pressure head $= 20$ m |
| Total head = Datum head + kinetic head + pressure head $= 2 + 0.0827 + 20 = 22.0827$ m |
| Note that the value of kinetic head is very low compared to pressure head and datum head. Therefore often the kinetic head is neglected in calculations. |
| Power of water flow $= \gamma \, Q_x H_x = 9810 \times 0.01 \times 22.0827 = 2166.313$ N-m/s. |

# Assessment

## Part A
## Fill in the blanks

1. The pressure acting on a point in a water mass is _____ in all directions. (**equal**)
2. Pressure of water acts _____ to the inner surface of the container. (**perpendicular**)
3. Mass density of water is _____. (**1000 kg/m³**)
4. Weight density of water is _____. (**9810 N/m³**)
5. Total head is the total energy possessed/unit _____ of water. (**weight**)

## True or False

1. Datum head is due to velocity of water flow. (**False**)
2. Pascal's law deals with pressure inside water. (**True**)
3. Work is the product of force and distance moved. (**True**)
4. Energy and work done have different units.(**False**)
5. Power is rate of transfer of energy. (**True**)

## Multiple-choice questions

1. The unit for pressure head is

    (a)  $N/m^2$
    (b)  $N/m$
    (c)  $m/N$
    (d)  **m**

2. Product of mass and acceleration is

    (a)  **Force**
    (b)  Velocity
    (c)  Power
    (d)  Energy

3. 1 kg(f)/cm$^2$ is approximately equal to

    (a)  **10 m of water**
    (b)  1 m of water
    (c)  5 m of water
    (d)  11 m of water

4. Power of water flow at any section for 1 m3/s discharge rate and 1 m total head is

    (a)  **9810 N m s$^{-1}$**
    (b)  918 N-m
    (c)  10 N m
    (d)  1 N-m

5. The datum head of an unit weight of water at a height of 4 m from a datum is

    (a)  2 m
    (b)  **4 m**
    (c)  2 N
    (d)  1 N-m

**Part B**

**Short answer questions**

1. What is Pascal's law?
2. What are different forms of energy possessed by water?
3. What is the formula for datum head?
4. What is the formula for pressure head?
5. What is the formula for velocity head?
6. What is the formula for power of fluid flow?

# References and Further Reading

1. Cengel, Y. A., & Cimbala, J. M. (2014). *Fluid mechanics*. McGraw Hill.
2. https://en.wikipedia.org/wiki/Pascal%27s_law

# Bernoulli's Law

## Introduction

Bernoulli's law is the most applied law in water engineering. Bernoulli's law is useful in understanding the relationship between transfer of energy between pressure energy, kinetic energy and datum energy. Design of microirrigation pipes and selection of pumps could be done if one has a good grasp of the application of this law. A major reason for the loss of pressure energy in pipe flow is friction. Equations and methods to calculate frictional energy losses are introduced here.

### Learning Outcomes

- To understand and apply Bernoulli's law
- To understand working of Siphon using Bernoulli's law
- Pressure loss due to friction in pipe flow
- Pressure loss and gain in pipe flow in slopes.

## Bernoulli's Equation

Bernoulli's law states that along a streamline say a pipe flow; total head is constant along all the locations of flow if no loss of energy occurs along the path. But individual components of energy such as datum head, kinetic head and also pressure head would vary from one location to the other keeping the total head constant. Bernoulli's equation was derived by Euler and presented in Fig. 1.

When water moves along a pipe, the three components of total head, namely pressure head, kinetic head and datum head change along the way. In Fig. 2, point $A$ is in a higher cross-sectional area of flow and point $B$ is in a lower cross-sectional area of flow. Both points $A$ and $B$ are at a same elevation from the ground. Therefore both the points $A$ and $B$ will have the same datum head. Let us assume that, there is no loss of energy between points $A$ and $B$ as the distance between them is very less.

V. Ravikumar, *Sprinkler and Drip Irrigation*,
https://doi.org/10.1007/978-981-19-2775-1_19

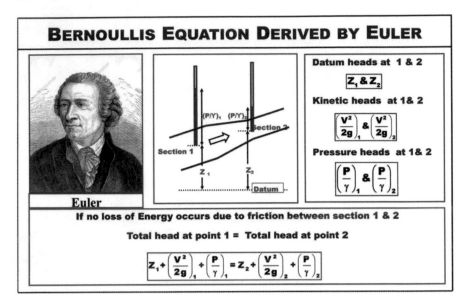

**Fig. 1** Bernoulli's equation

**Fig. 2** Energy change on diameter change

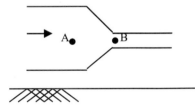

In this situation, when flow is steady, the velocity at point $B$ will be higher than point $A$ because at $A$ cross-sectional area of flow is less. Therefore, at point $B$, pressure head will be less than at point $A$. The decrease in the pressure head at point $B$ is equal to the increase in velocity head at point $B$.

For instance, let us assume the following: At point $A$, pressure head is 4 m, kinetic head is 3 m and datum head is 2 m. Between points $A$ and $B$, there is no loss of energy due to friction. If the kinetic head at point $B$ is 5 m, we can find out the pressure head at the point $B$ by applying Bernoulli's equation:

Total head at point A = Total head at point B

$$z_A + \left(\frac{v^2}{2g}\right)_A + \left(\frac{P}{\gamma}\right)_A = z_B + \left(\frac{v^2}{2g}\right)_B + \left(\frac{P}{\gamma}\right)_B \tag{1}$$

If the sample data discussed above is substituted in the preceding equation, the result is as follows:

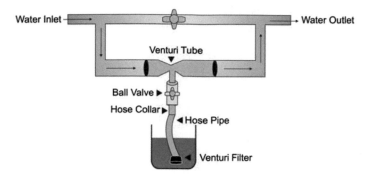

**Fig. 3** Venturi for fertilizer injection (www.ksnmdrip.com)

$$2 + 3 + 4 = 5 + 2 + z_B$$
$$z_B = 2 \text{ m}$$

It can be noted that the pressure at $B$ is less than the pressure at $A$. So now one can understand that when water passes from a higher pipe diameter to lower pipe diameter velocity head increases and pressure head decreases. This principle is used in a microirrigation accessory called as venturi for injecting fertilizer solution in irrigation water. Venturi has a narrow cross-sectional area (called as throat) and when water passes through this section, pressure head drops and the pressure reduces to a level lower than atmospheric pressure. Hence fertilizer solution from the container placed below the venturi is sucked into the venturi.

Normally farmers place the fertilizer solution container tank below the venturi as shown in Fig. 3. But if the fertilizer tank level is raised higher than the elevation of the venturi, we can substantially increase the fertilizer solution suction rate due to siphon effect (Fig. 4).

Working of siphon is a big debated subject among scientists around the world even today. But the normal siphons we use in daily life with water as flowing fluid is fairly explainable with the help of Bernoulli's equation. Water should be filled in the bent siphon tube before the starting of siphon to work (Fig. 5). Assume that the valve in Fig. 5 is closed, and so there will not be flow in the siphon. Under no-flow conditions, if we apply Bernoulli's law between points $A$ and $B$ by assuming $A$ as datum, we get,

$$z_A + \left(\frac{v^2}{2g}\right)_A + \left(\frac{P}{\gamma}\right)_A = z_B + \left(\frac{v^2}{2g}\right)_B + \left(\frac{P}{\gamma}\right)_B$$

$$0 \text{ m} + 0 \text{ m} + 10 \text{ m} = 0.1 \text{ m} + (P/\gamma)_B$$
$$(P/\gamma)_B = 9.9 \text{ m}$$

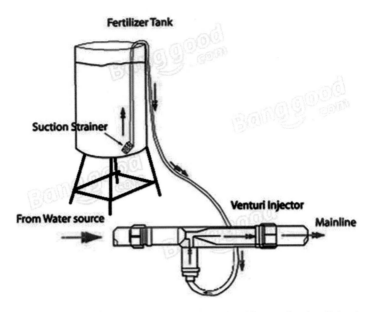

**Fig. 4** Elevated fertilizer solution tank (https://m.made-in-china.com/product/Irrigation-Venturi-Fertilizer-Injector-896109035.html)

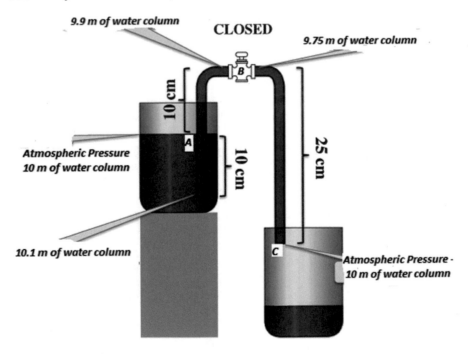

**Fig. 5** Siphon (Wikipedia)

Under no-flow conditions, if we apply Bernoulli's law between points $B$ and $C$ by assuming C as datum, we get,

$$z_B + \left(\frac{v^2}{2g}\right)_B + \left(\frac{P}{\gamma}\right)_B = z_C + \left(\frac{v^2}{2g}\right)_C + \left(\frac{P}{\gamma}\right)_C$$

$$0.25 \text{ m} + 0 \text{ m} + (P/\gamma)_B = 0 \text{ m} + 0 \text{ m} + 10 \text{ m}$$

$$(P/\gamma)_B = 9.75 \text{ m}$$

From the preceding calculations, it can be seen that pressure at $B$ due to right-side limb is less than the left-side limb and its difference (=0.15 m) is equal to the elevation difference between water level in the left limb container and the exit of siphon. So when the valve at $B$ is opened naturally flow occurs towards right side.

Following equation is a slight modification of Bernoulli's law which also accounts for the loss of energy between points $A$ and $B$.

$$z_A + \left(\frac{v^2}{2g}\right)_A + \left(\frac{P}{\gamma}\right)_A = z_B + \left(\frac{v^2}{2g}\right)_B + \left(\frac{P}{\gamma}\right)_B + \Delta h$$

Usually in microirrigation pipe design, the term $(v^2/2g)$ is neglected as the value of the term is insignificant. The term $P/\gamma$ is also always represented with $h$. Therefore, the preceding equation is written as follows:

$$z_A + h_A = z_B + h_B + \Delta h$$

$$h_B = h_A + z_A - z_B - \Delta h$$

Generally it is written as,

$$h_B + h_A \pm \Delta z - \Delta h$$

Where $\Delta z$ is difference in elevation between points $A$ and $B$. Plus (+) is used when elevation of $A$ is greater than $B$ and minus (−) is used when elevation of A is less than B.

## Friction Head Loss

When water passes through a pipe, water loses energy along the direction of flow. This is because of friction between the water and the pipe material and also friction between water molecules (Fig. 6). When flow is steady in Fig. 6, water will have the same velocity at point A and also at B. If we take datum as central longitudinal axis along the pipe as shown in the figure, points A and B have zero datum head and

**Fig. 6** Friction head loss in
a pipe

since velocity is same at points A and B, velocity head will also be equal at both the
points. The height of water standing in the piezometer at point A is greater than at
point B. This indicates that there is a loss of energy head when water travels from
point A to point B and the loss of head is in terms of pressure head in this case.

In a pipe flow, friction loss is directly proportional to discharge rate and length
and inversely proportional to diameter of pipe.

## Friction Loss Formulae

In microirrigation design, finding frictional pressure head loss is an essential job.
Basically, there are two kinds of equations available for frictional loss calculations.
One is Hazen–Williams equation and two is Darcy–Weisbach equation.

If a discharge rate of $Q$ enters a pipe at point $A$ and if the same $Q$ exits after a
length $L$ at point $B$, then the head loss due to friction ($\Delta h$) as per Hazen–Williams
equation is as follows:

$$\Delta h = \frac{1.22 \times 10^{10} Q^{1.852} L}{C^{1.852} D^{4.87}} \tag{2}$$

where

$\Delta h$    is in m
$Q$     is discharge rate in l/s
$D$     is inside diameter of pipe in mm
$L$     is length of pipe in m
$C$     is called as friction coefficient depending on the pipe material (Table 1).

For the same case, the head loss due to friction ($\Delta h$) as per Darcy–Weisbach
equation is derived as below:
The Darcy–Weisbach equation normally used in fluid mechanics is as follows:

$$h_f = \frac{f L v^2}{2g D} \tag{3}$$

where

**Table 1** Values of Hazen–Williams $C$

| S. No. | Pipe material | $C'$ |
|---|---|---|
| 1 | Plastic | 150 |
| 2 | Epoxy coated steel | 145 |
| 3 | Cement asbestos | 140 |
| 4 | Galvanized steel | 135 |
| 5 | Aluminium and steel (new) | 130 |
| 6 | Steel (old) or concrete | 100 |

$h_f$  is head loss (m);
$f$  is called Darcy–Weisbach friction factor;
$L$  is length of pipe (m);
$V$  is velocity of flow (m);
$g$  is acceleration due to gravity (m/s$^2$);
$D$  is diameter of pipe (m).

The flow in microirrigation pipes is generally turbulent flow, and the pipes are smooth. For these conditions, the expression for finding friction factor f was given by Blasius, which is as follows:

$$f = 0.32 \text{Re}^{-0.25} \tag{4}$$

where Re is Reynolds number of flow. This equation is valid for the Reynolds number between 2000 and 100,000. It should be noted that this equation is valid for any pipe material which is smooth.

The expression for Reynolds number (Re) is as follows:

$$\text{Re} = \frac{\rho v d}{\mu} \tag{5}$$

where $\rho$ is mass density of water and $\mu$ is dynamic viscosity of water.

Substitute $\rho = 1000 \text{ kg/m}^3$ and $\mu = 0.001002 \text{ NS/m}^2$ for temperature of 20 °C. If velocity of flow ($v$) is substituted with flow rate ($Q$) divided by cross-sectional area of flow as circle, we get the following equation:

$$\Delta h = \frac{K Q^{1.75}}{D^{4.75}} L \tag{6}$$

In microirrigation, head loss is expressed in metres (m), flow rate is in litres per second (l/s) and length of pipe in metres (m) and diameter of pipe in millimetres (mm). For such situation, the value of $K = 7.89 \times 10^5$.

$$\Delta h = \frac{789,000 Q^{1.75}}{D^{4.75}} L \tag{7}$$

It has been observed that, the Darcy–Weisbach equation is more accurate for pipe diameters less than 125 mm. One more advantage of this equation is that, the frictional loss does not depend on the pipe material. Mostly drip irrigation pipes are less than 125 mm and hence, in this book, all further equations are derived based on Darcy–Weisbach equation. Where ever pipe diameters more than 125 mm are encountered, the Hazen–Williams equation can be used.

## Pressure Change Due to Elevation

When water flows in downslope pipe, there will be pressure head gain due to reduction in datum head from $A$ to $B$ and the pressure head gain is equal to difference in elevation ($\Delta z$). Pressure head is also lost due to friction ($\Delta h$) on the way from point $A$ to $B$ (Fig. 7). So we cannot say whether pressure at $B$ will be more or less compared to point $A$ because one term is positive and another is negative. If $\Delta z$ is more than $\Delta h$, $h_B$ will be more than $h_A$. If $\Delta z$ is less than $\Delta h$, $h_B$ will be less than $h_A$. The pressure head at $B$ can be found out using the following formula:

$$h_B = h_A - \Delta h + \Delta z \qquad (8)$$

$\Delta h$    is pressure head loss due to friction
$\Delta z$    is pressure head gain/loss due to elevation.

If the water flows in an upsloping pipe, pressure head at $B$ will always be less than the pressure at A because both $\Delta h$ and $\Delta z$ are negative (Fig. 8). The formula for finding pressure for upslope case is as follows:

$$h_B = h_A - \Delta h + \Delta z \qquad (9)$$

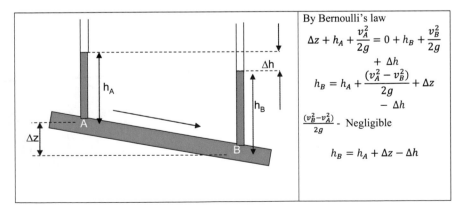

By Bernoulli's law

$$\Delta z + h_A + \frac{v_A^2}{2g} = 0 + h_B + \frac{v_B^2}{2g} + \Delta h$$

$$h_B = h_A + \frac{(v_A^2 - v_B^2)}{2g} + \Delta z - \Delta h$$

$\frac{(v_B^2 - v_A^2)}{2g}$ - Negligible

$$h_B = h_A + \Delta z - \Delta h$$

**Fig. 7**  Pipe flow in downslope

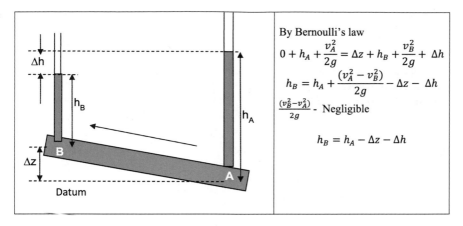

By Bernoulli's law

$$0 + h_A + \frac{v_A^2}{2g} = \Delta z + h_B + \frac{v_B^2}{2g} + \Delta h$$

$$h_B = h_A + \frac{(v_A^2 - v_B^2)}{2g} - \Delta z - \Delta h$$

$\frac{(v_B^2 - v_A^2)}{2g}$ - Negligible

$$h_B = h_A - \Delta z - \Delta h$$

**Fig. 8** Pipe flow in upslope

## Flow in Microirrigation Pipes

Water flow in microirrigation pipes is influenced by changing slopes, frictional head losses and pressure developed by the pumping units. In Fig. 9, pressure head distribution of a typical lateral pipe is shown as a group of vertical lines. To properly design sizes of microirrigation pipes, estimation of pressure distribution in microirrigation system is possible based on the procedures discussed in preceding sections.

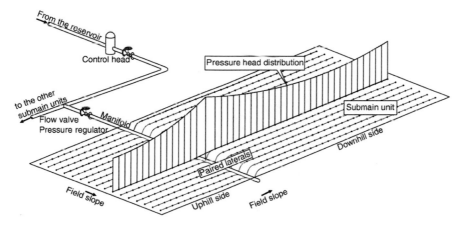

**Fig. 9** Pressure head distribution in a typical drip irrigation lateral [1]

## Summary

The total head of flowing of water is the sum of datum, kinetic and pressure heads. Bernoulli's law provides a framework for understanding the change of energy components such as datum, kinetic and pressure heads in pipe flow. Understanding of working of Siphon is useful in many occasions in microirrigation operation. Energy is lost due to friction in pipe flow. If the pipe diameter is less than 125 mm, pipe friction head loss is calculated with Darcy–Weisbach formula which does not consider the type of pipe material. If pipe diameter is more than 125 mm, pipe friction head loss is calculated with Hazen–Williams equation with specific pipe friction constants for each type of material. The magnitude of velocity head is negligible and hence not considered in design calculations.

## Points to Ponder

1.  Bernoulli's law states that the total head is constant at any two sections in a pipe flow, but individual components of energy such as datum head, kinetic head and pressure head may vary at every location.
2.  Energy is lost in pipe flow due to friction. Darcy–Weisbach equation is used for pipe diameter less than 125 mm and Hazen–Williams equation is used for pipe diameter greater than 125 mm.
3.  Datum head is converted into pressure head for downslope pipes, and pressure head is converted into datum head in upslope pipes. Pressure head will always be less when compared to the upstream location in an upsloping pipe. Pressure head may be higher or lower when compared to the upstream location in downslope pipe.

## Solved Exercises

1.  In a venturi flow (Fig. 10) at location 1, the diameter ($A_1$) is 20 mm and at location 2, diameter ($A_2$) is 10 mm. The discharge rate is 0.5 L per second (l/s). The pressure head at location 1 is 10 m. Find out the pressure head at location 2.

**Fig. 10**  Venturi problem

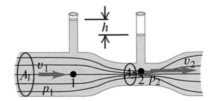

Area of flow at location 1: $A_1 = (\pi/4) \times \text{diameter}^2 = (\pi/4) \times 0.02^2 = 0.000314 \, \text{m}^2$

Velocity of flow at location 1:   $v_1 = \text{Discharge rate}/A_1 = 0.0005 \, \text{m}^3/\text{s}/0.000314 \, \text{m}^2$
$$= 1.592 \text{ m/s}$$

Velocity head $(v^2/2g)_1 = 1.592^2/(2 \times 9.81) = 0.129$ m

Area of flow at location 2: $A_2 = (\pi/4) \times 0.01^2 = 0.0000785 \, \text{m}^2$

Velocity of flow at location 2:   $v_2 = \text{Discharge rate}/A_2 = 0.0005 \, \text{m}^3/\text{s}/0.0000785 \, \text{m}^2$
$$= 6.369 \text{ m/s}$$

Velocity head $(v^2/2g)_2 = 6.369^2/(2 \times 9.81) = 2.068$ m

Applying Bernoulli's law at locations 1 and 2 as follows:

(Assume central axis of pipe as datum)

$$z_1 + \left(\frac{v^2}{2g}\right)_1 + \left(\frac{P}{\gamma}\right)_1 = z_2 + \left(\frac{v^2}{2g}\right)_2 + \left(\frac{P}{\gamma}\right)_2$$

$$0 + 0.129 + 10 = 0 + 2.068 + \left(\frac{P}{\gamma}\right)_2$$

$$\left(\frac{P}{\gamma}\right)_{A2} = 8.061 \text{ m}$$

2.  Find out the frictional loss in a plastic pipe of inside diameter ($D$) 102.6 mm of length ($L$) 100 m, discharging ($Q$) 15 lps using Darcy–Weisbach equation and also Hazen–Williams Equations. Use $C$ value for Hazen–Williams equation as 150.

By Darcy–Weisbach equation
   The frictional head loss = 2.525 m

By Hazen–Williams equation
   The frictional head loss = 2.755 m

## Assessment

### Part A

### Fill in the blanks

1.  In a pipe flow, friction loss is directly proportional to_____ and length and inversely proportional to diameter of pipe. (**discharge rate**)
2.  In horizontal pipe flow, if diameter reduced at a point, pressure head gets _____at that point compared to the pressure head upstream of the point. (**reduced**)
3.  Pressure head gets reduced to a level lower than _____. (**atmospheric pressure**) in a venturi throat section.

4. In an upsloping pipe flow, pressure head at a downstream point is always _____ than the pressure head at any upstream point. (**lower**)

5. Water should be filled in the bent siphon tube before the _____of siphon to work (**starting**)

**True or False**

1. The energy head causing the flow in a siphon is equal to the elevation difference between the water level in the upper container and the level of outlet point of the siphon tube. (**True**)

2. In a downslope pipe pressure head at a downstream point is always higher compared to an upstream point (**False**)

3. When water flows in a pipe, energy loss due to friction is inevitable (**True**)

4. Darcy–Weisbach equation takes into account the type of material for calculation of friction head loss (**False**)

5. Venturi injection rate can be increased if siphon principle is used. (**True**).

**Multiple-choice questions**

1. Darcy–Weisbach equation is used for finding pipe diameter less than

   (a) 25 mm
   (b) 50 mm
   (c) 75 mm
   (d) **125 mm**

2. Usually in microirrigation design of pipes, the _____ is neglected as the value of the term is insignificant.

   (a) **Velocity head**
   (b) Pressure head
   (c) Kinetic head
   (d) Total head

3. Pipe friction head losses are directly proportional to _____ and length of flow

   (a) **Discharge rate**
   (b) Diameter of pipe
   (c) Pressure
   (d) All choices a, b and c

4. Pipe friction head losses are inversely proportional to _____ and length of flow

   (a) Discharge rate
   (b) **Diameter of pipe**
   (c) Pressure
   (d) All choices a, b and c

5. _____equation takes into account the type of pipe material for calculation of friction head loss

   (a)  **Hazen–Williams**
   (b)  Darcy–Weisbach
   (c)  Both a and b
   (d)  None of a and B.

**Part B**

1.  State Bernoulli's law
2.  Briefly explain how Siphon principle can be used to increase injection rate in venturi fertilizer injector.

# References and Further Reading

1.  Baiamonte, G., Provenzano, G., & Rallo, G. (2015). Analytical approach determining the optimal length of paired drip laterals in uniformly sloped fields. *Journal of Irrigation and Drainage Engineering.* https://doi.org/10.1061/(ASCE)IR.1943-4774.0000768,04014042
2.  Cengel, A. Y., & Cimbala, J. M. (2014). *Fluid Mechanics.* McGraw Hill.
3.  https://en.wikipedia.org/wiki/Siphon

# Soil Texture and Soil Moisture

## Introduction

When we travel from one place to the other, we are able to see soils like red soils, black soils, sandy soils and also loamy soils. Each type of soil possesses different water-holding capacities and water releasing capacities. For instance, black soils possess very high water-holding capacity but very low water releasing capacity, whereas sandy soils possess low water-holding capacity and high water releasing capacity.

Loamy soils are mostly found in river deltas and are ideal for crop growth because they possess good water retentive and also water releasing capacities. Soils with organic matter content are also good in water-holding and releasing capacities.

In this chapter, we would study about soil, classification of soil based on the size of soil particles, soil water occurrence and methods of specifying soil moisture. The water content in soil can be expressed in either mass basis called as gravimetric moisture content or in volumetric basis.

### Learning Outcomes

- Classify the soil particles based on size
- Understand the nature of occurrence of water in soil
- Distinguish between volumetric water content and gravimetric water content.

## Soil

Soil is a porous mixture of inorganic particles, organic matter, air and water. One important physical property of soil, namely soil texture is relative proportion of various sizes of soil particles. Soil particles are classified as follows:

Sand    0.05 mm to 2 mm
Silt     0.002 to 0.05 mm
Clay    < 0.002 mm

© The Author(s), under exclusive license to Springer Nature Singapore Pte Ltd. 2023    563
V. Ravikumar, *Sprinkler and Drip Irrigation*,
https://doi.org/10.1007/978-981-19-2775-1_20

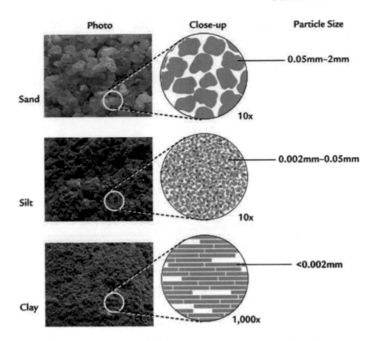

**Fig. 1** Relative sizes of sand, silt and clay (https://madhavuniversity.edu.in/soil-types.html)

In Fig. 1, we can see how sand, silt and clay would appear when they are magnified. It is necessary to note that the clay particles possess platy shape. Soils with more clay fractions will have more water-holding capacity than coarse fractions, but the movement of water is restricted in clay due to very small size of pores and platy shape of clay particles.

Based on the proportion of each type of soil particles, any soil is classified using United States Department of Agriculture (USDA) soil textural classification chart shown in Fig. 2. To classify the soil by this method, advanced soil analysis needs to be done and that kind of results can be obtained by giving soil samples to exclusive soil test laboratories maintained by national and state governments. But practically soil textural classification can be easily found out by a method called feel method and given in Fig. 3.

## Types of Soil Water

The soil water is classified into three forms, based on the suction pressure needed to remove the water from the surface of a soil particle (Fig. 4).

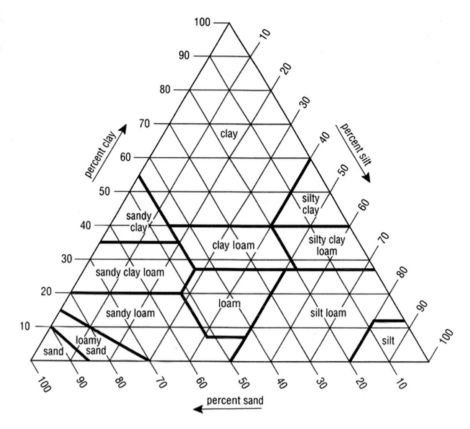

**Fig. 2** USDA soil textural classification chart (Courtesy U.S. Soil Conservation Service.)

## Hygroscopic Water

When water is exposed to solid soil particles, because of adhesion, first a thin layer of water gets attached to the soil. To remove this water, suction pressure needed is very high and hence, this water cannot be used by plants.

## Capillary Water

If more water is added on the surface of hygroscopic water, because of cohesion, some more water gets attached and remains intact because of surface tension. This water is primarily useful for plant uptake.

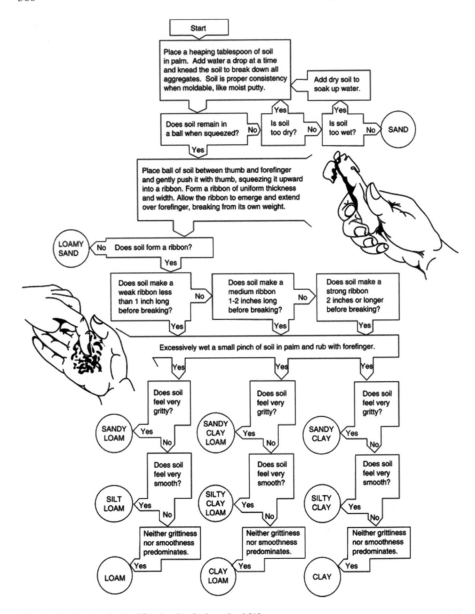

**Fig. 3** Soil textural classification by feel method [1]

## *Gravitational Water*

If more and more water is added, they also fill the macropores. This water cannot be sustained in the macropores, and they get drained down and are called as gravitational

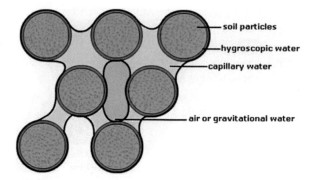

**Fig. 4** Types of soil water (www.site.iugaza.edu.ps)

water. This water can be absorbed by the plant but the opportunity time for the plant to absorb this water is very less as this water would drain down in a short time.

When water is added to dry soil, it displaces the air from the pore spaces and eventually fills the pores when all the macropores and micropores are filled with water, when the soil is said to be saturated, it is at its maximum retentive capacity. The maximum retentive capacity is the sum of the hygroscopic water, capillary water and gravitational water.

After gravitational water is drained down, the capillary water is held by the soil particles due to adhesion between soil particle and water molecules.

## Soil Moisture Specification Methods

Soil contains three components such as air, water and solid. Let us take a 1 m² surface of land and a soil volume of $V_{total}$ m³ (Fig. 5). This $V_{total}$ volume of soil contains

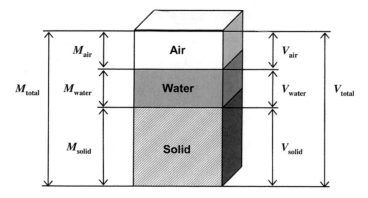

**Fig. 5** Soil as components of solids, water and air

$V_{solid}$ volume of solids, and it represents the solid part of soil alone. Assume that the solid particles are so flexible and is packed as shown in the figure without any water or air in between the solid particles (without any pore spaces). $V_{water}$ is volume of water and $V_{air}$ is volume of air in $V_{total}$ volume of soil. $M_{air}$, $M_{water}$ and $M_{solid}$ are mass of air, water and solid in the same volume, respectively.

$$\text{Mass density of soil } \rho_{soil} = \frac{M_{solids}}{V_{soil}} \tag{1}$$

The water content in soil can be expressed in either mass basis called as gravimetric moisture content or volumetric basis. Gravimetric soil moisture content ($\theta_g$) in the soil is defined as the mass of water per unit mass of solids and is usually expressed as %.

$$\theta_g = \frac{M_{water}}{M_{solids}} \tag{2}$$

Volumetric soil moisture content ($\theta_v$) in the soil is defined as the mass of water per unit volume of soil and is usually expressed as %.

$$\theta_v = \frac{V_{water}}{V_{soil}} \times 100 \tag{3}$$

Relationship between $\theta_g$ and $\theta_v$ is as follows:

$$\theta_v = SG_{soil} \, \theta_g \tag{4}$$

where $SG_{soil}$ is called as apparent specific gravity of soil. It is the ratio of mass density of soil to mass density of water (1 g/cm$^3$).

If $\theta_v$ is divided by area of land (i.e. 1 m$^2$ in this case), we get depth of water ($\theta_d$). It is to be noted that both $\theta_v$ and $\theta_d$ are numerically same and units are different. For instance, if $\theta_d$ is 10, then it should be understood as 10 cm of water per 100 cm of soil.

The gravimetric method of soil moisture estimation is very useful for estimation of moisture content in laboratories, whereas volumetric moisture content is useful in specifying the amount of water to be applied when irrigation scheduling is done. So both the representations are needed. Whenever moisture content is not specified either as volumetric or gravimetric, it can be taken as volumetric moisture content.

## Summary

Soil particles are classified as sand, silt and clay based on the size. Based on the relative proportion of sand, silt and clay in soil mass, there are 12 classes of soil. Soil water is classified into gravitational water, capillary water and hygroscopic water. Capillary water is very much useful for plant uptake. Soil moisture expressed per unit mass of dry soil is called as gravimetric water content and expressed per unit volume of soil is called as volumetric water content.

## Points to Ponder

- Black soil has more water-holding capacity than red soil
- Loamy soils are ideal for plant growth
- Sand size range is 0.05 mm to 2 mm
- Silt size is 0.002 to 0.05 mm
- Clay size range is < 0.002 mm
- Gravitational water is drained after irrigation is done
- After gravitational water is drained, capillary water remains in soil and useful for plant uptake
- Mass density of soil is mass of dry soil per unit volume of soil
- Specific gravity of soil is ratio of mass density of soil to mass density of water
- Gravimetric soil moisture content is mass of water per unit mass of dry soil
- Volumetric soil moisture content is volume of water per unit volume of soil
- Volumetric soil moisture content is the product of apparent specific gravity of soil and gravimetric soil moisture content.

## Exercises

1. Take three funnels as shown in Fig. 6 and place a filter paper over the funnel. Take dry soils of clay, silt and sand. Place equal mass of dry soil in each of the funnels and pour equal amount of water in all the funnels at a rate so that water does not pond over the soil. After the end of the experiment, you will find that the collecting jar below the clay soil would have least amount of water. This indicates that more water is stored inside the clay soil because both capillary pore space and gravitational pore space are more in clay soil. Silt ranks second and sand ranks third in water-holding capacities.
2. Collect different soils around your place and use the feel method for finding the soil texture.
3. Find out mass density of soil in the fields as follows:
   Excavate a small cuboid pit on a level ground. Collect all the soil excavated and oven-dry it. Find out the mass of soil excavated ($M_{solids}$). Find out the volume of soil excavated by multiplying the length, width and height of the pit ($V_{soil}$). Mass density of soil is obtained by dividing $M_{solids}$ by $V_{soil}$ (Fig. 7).
4. Gravimetric moisture content can be found out by taking a moist sample of soil. First find out its mass ($m_1$). Then place the sample in an ordinary oven or

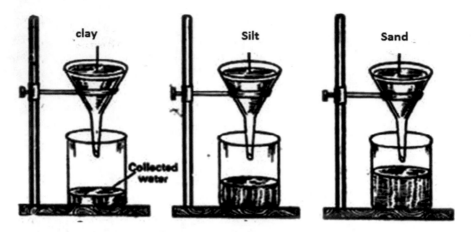

**Fig. 6** Water-holding capacities of different soils

**Fig. 7** Excavation of earth
for finding mass density of
soil

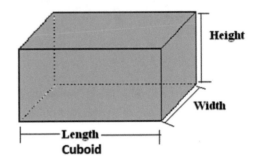

microwave oven and set the temperature to 105 °C. Keep the sample for one day
and take out the sample and find out the mass of the sample ($m_2$). Gravimetric
water content of the soil sample is found out using the following formula:

$$\theta_g = \frac{M_{\text{water}}}{M_{\text{soilds}}} \times 100 = \frac{m_1 - m_2}{m_2} \times 100$$

## Solved Exercises

1.  A small pit was excavated whose dimensions are as follows: Length = 10 cm,
    width = 10 cm and depth = 5 cm. The mass density of the soil is 1.5 g/cm³.
    The mass of the soil before drying is 850 g. The soil from the pit was oven-
    dried at 105 °C for one day. What will be the mass of dried soil? What are the
    gravimetric moisture content, volumetric moisture content and moisture content
    in depth units? Also find out the moisture content in depth units for the 5 cm

depth of soil (Recall: 1 g of water = 1 cm$^3$ of water, mass density of water = 1 g/cm$^3$)

$$\text{Volume of excavated pit} \quad = 10 \times 10 \times 5 = 500 \, \text{cm}^3$$

$$\text{Mass densty of soil: } \rho_{\text{soil}} = \frac{M_{\text{solids}}}{V_{\text{soil}}}$$

$$1.5 \, \text{g/cm}^3 = M_{\text{solids}}/500 \, \text{cm}^3$$

$$\text{Mass of dried soil} = 500 \times 1.5 = 750 \, \text{g}$$

$$\text{Gravimetric moisture content} = \theta_g = \frac{M_{\text{water}}}{M_{\text{solids}}} \times 100$$
$$= (850 - 750) \times 100/750$$
$$= 13.33\%$$

$$\text{Volumetric moisture content} = \theta_v = \frac{V_{\text{water}}}{V_{\text{soil}}} \times 100$$
$$= 100 \, \text{cm}^3 \times 100/500$$
$$= 20\%$$

**By Another Formula**

$$\text{Apparent specific gravity of soil}(SG_{\text{soil}}) \quad = 1.5$$

$$\text{Volumetric moisture content by another formula} = \theta_v = SG_{\text{soil}} \times \theta_g$$
$$= 1.5 * 13.33$$
$$= 20\%$$

$$\text{Moisture content in Depth units}(\theta_d) = 20 \, \text{cm}/100 \, \text{cm depth of soil}$$

$$\text{Moisture content for 5 cm depth of soil} = 20 \times 5/100 = 1 \, \text{cm}$$

## Assessment

### Fill in the blanks

1. _____ soil textural classification chart is useful for soil textural classification. (**USDA**)
2. _____of soil is ratio of mass density of soil to mass density of water. (**Apparent specific gravity**)
3. _____is obtained by multiplying gravimetric moisture content and Apparent specific gravity. (**Volumetric moisture content**)
4. Mass density of soil is mass of dry soil per_____of soil. (**unit volume**)
5. Gravimetric soil moisture content is mass of water per_____of dry soil. (**unit mass**).

### True or False

1. Size of sand varies from 0.05 mm to 2 mm. (**True**)
2. Size of silt varies from 0.002 to 0.05 mm. (**True**)
3. Size of clay is less than 0.002 mm. (**True**)
4. Clay particles are round-shaped. (**False**)
5. Volumetric moisture content and moisture content expressed in depth units are numerically same but units are different. (**True**)

### Multiple-Choice Questions

1. _____ water is not available for plant uptake.

   a. **Hygroscopic**
   b. Capillary
   c. Gravitational
   d. Pressurized

2. Water-holding capacity of _____ soil is maximum among the soil types mentioned below:

   a. Sand.
   b. **Clay**
   c. Silt
   d. Red soil

3. _____ water is very much available for plant uptake.

   a. Hygroscopic
   b. **Capillary**
   c. Gravitational
   d. Pressurized

4.  _____ pores store more gravitational water.

    a.  **Macro**
    b.  Micro
    c.  Nano
    d.  Pico

5.  _____has more water-holding capacity than red soil

    a.  **Black**
    b.  Brown
    c.  Sand
    d.  Silt

**Part B**

1.  Write a short note on gravitational water.
2.  Write a short note on capillary water.
3.  Write a short note on hygroscopic water.

# Reference and Further Reading

1.  Thien, S. J. (1979). A flow diagram for teaching texture by feel analysis. *Journal of Agronomic Education* 8, 54–55.
2.  Hillel, D. (1998). *Environmental soil physics.* Academic Press.

# Soil Water Movement

## Introduction

Soil water movement study is very much essential in microirrigation for design of sand filters, selection of dripper discharge rate and spacing of drippers. Water movement in soils is distinguished into two types of flow. One is saturated flow where in all the pores in the soil are filled with water. Two is unsaturated flow where in pores are filled with water and also with air.

### Learning Outcomes

- Saturated flow and unsaturated flow
- Equations for saturated flow and unsaturated flow
- Applications of saturated and unsaturated flow studies in microirrigation
- Solve problems of filter design and drainage below root zone.

## Saturated Flow

Flow of water in a soil occurs between any two locations when there is total energy difference between the two locations. In a container filled with sand, water flows from upstream (section-1) to downstream (section-2) (Fig. 1). The datum heads ($z_1$ & $z_2$) and pressure heads ($P_1/\gamma$ & $P_2/\gamma$) are shown in the figure. The total energy of water flow in a saturated soil is conventionally the sum of pressure head and datum head which is called as hydraulic head. Conventionally, the velocity head for water flow in soils is treated as negligible.

Similar to energy loss due to friction in pipe flow, here also there is a loss of energy for the flow through the soil ($h_L$). Note that flow of water inside soil pores will be tortuous depending on the nature of connectivity of pores. But the length of flow ($L$) considered here is only measured externally. Hydraulic gradient is change in hydraulic head per unit length of flow ($h_L/L$).

© The Author(s), under exclusive license to Springer Nature Singapore Pte Ltd. 2023
V. Ravikumar, *Sprinkler and Drip Irrigation*,
https://doi.org/10.1007/978-981-19-2775-1_21

**Fig. 1** Saturated flow in soils

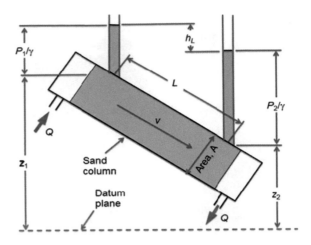

A French scientist, namely Darcy postulated a following law: The discharge rate through a saturated soil column is directly proportional to the cross sectional area of soil through which flow occurs and hydraulic gradient. According to this law following equation is written:

$$Q = K_s A \, h_L / L \tag{1}$$

where $Q$ is discharge rate, $A$ is area of cross section of soil, $h_L/L$ is hydraulic gradient and $K_s$ is called as saturated hydraulic conductivity. The value of $K_s$ is a function of type of soil. Table 1 shows typical values of hydraulic conductivity of different soils.

**Table 1** Saturated hydraulic conductivity and Gardener's flow exponent for different soil texture classes

| S. No. | Soil texture | Sat. Hyd. Cond (m/d) | Gardner's flow exponent $\alpha(m^{-1})$ |
|--------|--------------|----------------------|-------------------------------------------|
| 1 | Sand | 7.120 | 14.5 |
| 2 | Loamy sand | 3.500 | 12.4 |
| 3 | Sandy loam | 1.060 | 7.5 |
| 4 | Loam | 0.250 | 3.6 |
| 5 | Silt | 0.060 | 1.6 |
| 6 | Silty loam | 0.110 | 2.0 |
| 7 | Sandy clay loam | 0.310 | 5.9 |
| 8 | Clay loam | 0.065 | 1.9 |
| 9 | Silty clay loam | 0.017 | 1.0 |
| 10 | Sandy clay | 0.029 | 2.7 |
| 11 | Silty clay | 0.005 | 0.5 |
| 12 | Clay | 0.045 | 0.1 |

**Fig. 2** Sand filter

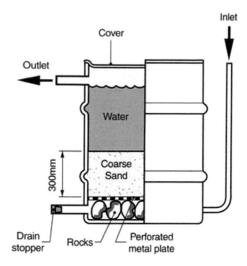

Darcy's law is useful in many situations like sand filter design, leaching salts in greenhouse soils. Let us assume that one wants to design a sand filter for collecting and filtering rainfall-runoff water from rooftop with a setup as shown in Fig. 2. From the figure it can be seen that water goes into the drum from bottom and comes out at the top. Generally a 300 mm depth of sand is recommended for filtering dirt from water coming from rooftops. By collecting data such as peak expected runoff from the rooftop ($Q$), and depth of water flow in sand ($L$) and pressure head loss between inlet and outlet of filter, one can estimate the area of filter medium needed. A worked-out example elsewhere would be very much helpful to understand this design process.

# Unsaturated Flow

Figure 3 shows a wetting pattern of a dripper operated in a vertical face of a trench. A small pond forms around a dripper, when water from a dripper falls on the soil. Flow of water in the soil very near to the dripper will be saturated flow. After some distance from the dripper within the wetted zone, all the pores will not be filled with water, but flow of water will occur. This flow is called as unsaturated flow. In unsaturated flow, the major cause for the flow is capillary force.

Now let us discuss capillary action. Capillary rise occurs when a narrow glass tube is immersed in water. Due to the adsorption of water to glass surface, water rises and due to cohesion between water molecules, a water column is formed. The column is sustained by the surface tension. The weight of the hanging column is supported by the surface of water in the capillary tube. Less diameter pipes have more surface area for adhesion of the water particles. Therefore, when the diameter of the pipe is

**Fig. 3** Wetting pattern of a dripper

**Fig. 4** Capillary rise versus diameter

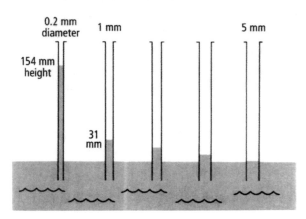

less, capillary rise is more (Fig. 4). The combination of adsorption, cohesion surface tension forces is called as capillary forces.

The capillary forces are in action in the soil water also because the size of the pores is very small. When we put a drop of water on soil and observe the movement closely, we can find the water moving in the horizontal direction as well as in vertical direction (downwards). The horizontal direction movement is due to the capillary forces, and downward direction movement is due to the combination of gravitational force and capillary force.

You may have seen water droplets on leaf surfaces during morning (Fig. 5). These droplets have some height. The water droplet does not get broken down because the droplet possesses surface tension. When the height of water droplet increases, the hydrostatic pressure increases, whereas the surface tension of the water remains constant. Then the water droplet breaks down if the hydrostatic pressure becomes more than the surface tension. The movement of water inside the droplet due to

**Fig. 5** Water droplets on leaf

hydrostatic pressure is in the opposite direction of resistance caused due to surface tension.

When you turn the leaf upside down, large water droplets fall down and small water droplets stick to the surface of leaf. Small droplets do not fall downs due to the adsorptive force existing between water particles and the leaf surface.

Therefore, it is understood that the capillary forces tend to stick the water towards soil matrix. If we want to remove this water from the soil, we need to apply suction. The energy needed to remove a unit weight of water from the soil and bring the water to the normal atmospheric condition when soil water is capillary water or hygroscopic water is called as matric head of water.

When the pressure head is more than atmospheric pressure, we can derive work out of water while bringing the water to the atmospheric pressure. When elevation of water is above a reference datum, we can derive work out of it if we bring the water to the datum. Hence, hydrostatic pressure above atmospheric pressure and datum energy when water is above datum are specified with a positive sign.

When the water is below the reference datum, we need to do work on the water to bring it to datum. Similarly, the matric head also indicates the amount of work to be done on the water to remove from the soil and to bring the water to the standard state of atmospheric pressure. Hence, the datum head below the datum and the matric head in unsaturated flow are specified with the negative sign.

## Energy of Unsaturated Flow

The total head of water in unsaturated flow is the sum of matric head and datum head. Similar to saturated flow, unsaturated flow also occurs from a location of higher total

**Fig. 6** Unsaturated flow in
soil pores

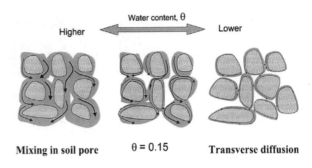

Mixing in soil pore          θ = 0.15          Transverse diffusion

energy to lower total energy. Therefore Darcy's law developed for saturated flow is
suitably modified for unsaturated flow.

From Fig. 6, it can be seen that the area of occurrence of flow in the pore is not
through all the porous area in case of unsaturated flow. Therefore, this difference
in the unsaturated flow compared to the saturated flow is accounted by having a
hydraulic conductivity function instead of using a constant value. This unsaturated
hydraulic conductivity ($K(h)$) is a function of pressure head at any location and
following expression developed by Gardner is used as a first-level approximation:

$$K(h) = K_s \, e^{\alpha h}$$

where $K(h)$ is unsaturated hydraulic conductivity at the suction pressure head of $h$,
e is an exponential constant approximately equal to 2.718, $K_s$ is saturated hydraulic
conductivity, $h$ is pressure head (negative quantity) and $\alpha$ is called as Gardner's
exponent. The use of $K(h)$ in the place of $K$ in Darcy's equation is called as Darcy–
Buckingham equation.

Figure 7 provides an overview of how pressure head varies with water content
for different soils. The pressure head mentioned in the figure means suction pressure
head, and it is termed as negative and it is conventional not to explicitly write the
negative sign. From Fig. 8, you can notice that, the value of hydraulic conductivity
remains constant for some extent of increase of suction pressure head, then after
that when suction pressure increases further the hydraulic conductivity values fall
rapidly. From this discussion, it can be understood that after water is absorbed by the
plant roots at a location, the movement of water towards the roots is not rapid when
unsaturated flow occurs.

For instance for loamy soils, when water content decreases by 50% of its saturated
water content, the suction pressure head is $-100$ cm (Fig. 7). We can apply the
preceding formula to find out how much reduction in hydraulic conductivity occurs if
the soil water content decreases by 50% of the saturated water content. The saturated
hydraulic conductivity value for loam is 25 cm/d, and Gardner's exponent value for
loam is 0.036 cm$^{-1}$.

$$K(h = -100 \, \text{cm}) = 25 \times 2.718^{(0.036 \times (-100))}$$

**Fig. 7** Water content versus
suction head

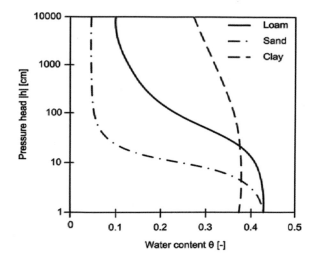

**Fig. 8** Suction head versus
hydraulic Conductivity

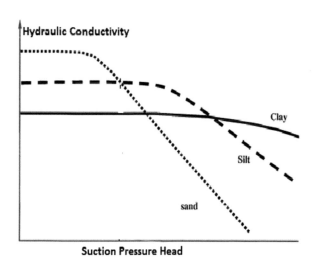

$$= 0.683 \, \text{cm/d}$$

This value of hydraulic conductivity is 36.6 times lower than the saturated
hydraulic conductivity value of Loam. From this result, we can understand that
the extent of reduction in hydraulic conductivity value as water content decreases is
very high.

**Fig. 9** Tensiometer

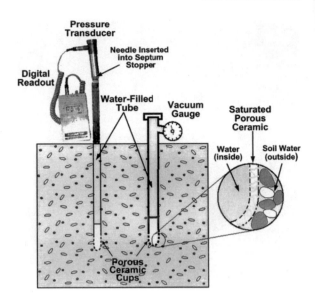

## *Tensiometer*

Tensiometer is one of the very simple and reliable instrument used for measuring the matric head of soil water.

Tensiometer is an instrument used to measure the suction pressure head needed to remove a unit weight of water attached in the soil when water is held in the pores in capillary phase as shown in Fig. 9. Tensiometer consists of a tube with a ceramic porous block with airtight lid and a vacuum gauge. It is filled with water and placed in the soil. The water from the tensiometer may go out or come in through the porous block. The pores in the porous block of tensiometer are very small in order to retain the water in the tensiometer tube. If the pores in the porous block are bigger, air will easily enter into the tube through its bottom.

Water moves either from soil to the porous block or from the porous block to the soil until the suction pressure inside the porous block and the soil around the porous block are the same. When this water movement gets stopped, equilibrium is said to be reached between the soil and the tensiometer. The vacuum gauge measures the suction pressure of the water in the tensiometer tube at the point of connection of vacuum gauge.

Normally tensiometers are designed so that the air would enter into the porous cup when the pressure head inside the porous cup is less than one atmosphere. Therefore when soil moisture suction pressure head falls below one atmosphere, air will enter through the porous cup and the vacuum cannot be maintained inside the tensiometer tube and all the water inside the tube would be sucked out of the tube with in short time. Then the tensiometer will not show any reading.

**Fig. 10** Correction of
Tensiometer observations

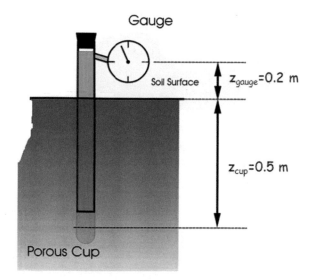

### Correction of Tensiometer Reading

Let us assume that a tensiometer installed in a location as shown in Fig. 10 shows an observation of negative pressure head $h = -1.8$ m. This pressure head is the observation for the location of centre of the dial of the gauge. The pressure head at the porous cup would be more than the value of $-1.8$ m because there is a standing water column above the porous cup. In the figure, we see that the depth of porous cup is 0.7 m (i.e. 0.2 m + 0.5 m). Pressure head at the porous cup should be found out by adding the pressure head due to standing water with the pressure head shown by the gauge (i.e. $h_{cup} = -1.8$ m + 0.7 m = $-1.1$ m).

### Summary

Soil water movement is studied by distinguishing into two types of flow. One is saturated flow and another is unsaturated flow. Hydraulic head of soil water is the sum of pressure head and gravitational head. The pressure head in saturated flow is taken as positive as the pressure is above atmospheric pressure. The pressure head in unsaturated flow is taken as negative as water is held by the soil particles and energy is needed to remove the water from the soil and to bring the water to atmospheric pressure. Hydraulic gradient between any two points is the ratio of change in hydraulic head between the points to the distance between the points. Hydraulic conductivity is the ability of soil to transmit water and it varies from one type of soil to the other. Hydraulic conductivity is expressed as a function of pressure head in unsaturated soils. Darcy's law is used for quantifying flow of water in both saturated flow and unsaturated flow using hydraulic gradient and hydraulic conductivity. Darcy's law is useful for design of filters and also studying the flow of water in soil applied from drippers.

## Points to Ponder

1. Hydraulic conductivity varies from one soil to the other.
2. Unsaturated soil hydraulic conductivity is a function of soil moisture tension.
3. The combination of adsorption, cohesion surface tension forces is called as capillary forces.
4. Tensiometer measures soil moisture suction for unsaturated soils and the observations have to be corrected according to the depth of porous cup.
5. The energy needed to remove a unit weight of water from the soil and bring the water to the normal atmospheric condition when soil water is capillary water or hygroscopic water is called as matric head of water.
6. The discharge rate through a saturated or unsaturated soil is directly proportional to the cross-sectional area of soil through flow occurs and hydraulic gradient.

## Solved Exercises

1. A sand filter has to be designed for filtering water harvested from a house with the following data:

The thickness of the sand layer is 0.3 m, depth of water standing above sand is 0.6 m, inner cross-sectional area of the drum is 0.9 $m^2$, maximum rate of discharge from the roof is 0.00072 $m^3$/s, saturated hydraulic conductivity of sand is $8 \times 10^{-5}$ m/s, pressure head at the bottom of sand is 1.5 m (Fig. 11).

For the preceding data find out the number of drums required for filtering the water.

**Fig. 11** Illustration for
problem 1

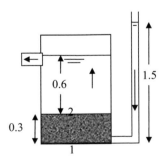

## Solution

We have to find out the head loss ($h_L$) between locations 1 and 2 because only between 1 and 2 saturated flow occurs through sand. Let us take datum at location 1

Total head at the entry of water into sand location 1 in Fig.) $= P_1/\gamma + z_1$
$$= 1.5 + 0 = 1.5\,m$$

Total head at the exit of water from sand location 2 in Fig.) $= P_2/\gamma + z_2$
$$= 0.6 + 0.3 = 0.9\,m$$

Hydraulic gradient ($h_L/L$) $= (1.5 - 0.9)/0.3 = 2$

Discharge rate through the sand column $= Q = Ks\,A\,h_L/L$
$$= 8 \times 10^{-5} \times 0.9 \times 2 = 0.000144\,m^3/s$$

Number of sand columns required
$$= \text{Maximum discharge rate from the roof}/0.000144$$
$$= 0.00072/0.000144 = 5\ \text{Numbers}$$

2. A container of 0.15 cm diameter is used for growing a plant. There are two tensiometers whose cups are below the root zone as shown in Fig. 12. The soil in the container is sandy with a hydraulic conductivity 1 m/d and Gardener's flow exponent of 7.5 $m^{-1}$. The corrected tension value at the centre of porous cup for the upper tensiometer is $-0.5$ m and for the lower tensiometer is also $-0.5$ m. Calculate the rate of drainage below the root zone.

**Fig. 12** Drainage below root zone

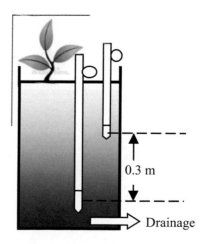

0.3 m

Drainage

Cross sectional area of flow through soil(A)

$$= \pi r^2 = 3.14 \times 0.15^2 = 0.07065 \text{ m}^2$$

Assume the datum is passing through the centre of the lower porous cup.

Hydraulic head at the porous cup of upper tensiometer $= -0.5 + 0.3 = -0.2 \text{ m}$

Hydraulic head at the porous cup of lower tensiometer $= -0.5 + 0.0 = -0.5 \text{ m}$

Hydraulic gradient between the upper and lower porous cups$(h_L/L)$

$$= (-0.2 - (-0.5))/0.3$$

$$= 1$$

Unsaturated hydraulic conductivity K $(h = -0.5 \text{ m}) = K_s e^{\alpha h}$

$$= 1 \times e^{7.5 \times (-0.5)}$$

$$= 0.02352 \text{ m/d}$$

Total Discharge below the root zone $(Q) = K(h)A(h_L/L)$

$$= 0.02352 \times 0.07065 \times 1$$

$$= 0.00166 \text{ m}^3/\text{d}$$

$$= 1.66 \text{ litres per day}$$

## Assessment

### Part A

### Fill in the blanks

1.  Flow of water in a soil occurs between any two locations when there is _____ between the two locations. (**total energy difference**)
2.  According to Darcy's law, the discharge rate through a saturated soil column is directly proportional to the cross-sectional area of soil through flow occurs and _____. (**hydraulic gradient**)
3.  Gardener's flow exponent is useful to get an estimate of _____. (**unsaturated hydraulic conductivity**)
4.  _____is an instrument used to measure the suction pressure head needed to remove a unit weight of water attached in the soil when water is held in the pores in capillary phase. (**Tensiometer**)

5.  Normally tensiometers are designed so that the air would enter into the porous cup when the pressure head inside the porous cup is less than _____. (**one atmosphere**)

**True or False**

1.  Unsaturated hydraulic conductivity values for any soil will be less than the value of saturated hydraulic conductivity for the soil. (**True**)
2.  Hydraulic gradient is the change in hydraulic gradient per unit length. (**True**)
3.  The combination of adsorption, cohesion surface tension forces is called as capillary forces. (**True**)
4.  When we put a drop of water on soil and observe the movement closely, we can find the water moving in the horizontal direction as well as in vertical direction (downwards). The horizontal direction movement is due to the capillary forces. (**True**)
5.  When we put a drop of water on soil and observe the movement closely, we can find the water moving in the horizontal direction as well as in vertical direction (downwards). The downward direction movement is due to the combination of gravitational force and capillary force. (**True**)

**Multiple-Choice Questions**

1.  Flow of water in a soil when 100% of pore spaces are not filled up with water is called as

    a.  **Unsaturated flow**
    b.  Saturated flow
    c.  Saline flow
    d.  None of these

2.  Generally a _____ depth of sand is recommended for filtering dirt from water coming from rooftops.

    a.  **300 mm**
    b.  500 mm
    c.  100 mm
    d.  1000 mm

3.  The energy needed to remove a unit weight of water from the soil and bring the water to the normal atmospheric condition when soil water is capillary water or hygroscopic water is called as _____ of water.

    a.  **Matric head**
    b.  Determinant head
    c.  Datum head
    d.  Gravitational head

4.  Tensiometer is normally fitted with a _____ gauge.

    a.  **Vacuum**

    b.   Pressure
    c.   Length
    d.   Mass

5.   A tensiometer's porous cup is at a depth of 0.7 m from the vacuum gauge. The observation in the vacuum gauge is −1.8 m of water column. What is the corrected tension at the porous cup ?

    a.   **−1.1 m**
    b.   2.5 m
    c.   1.1 m
    d.   3.5 m

**Part B. Short Answer Questions**

1.   What is Darcy's law?
2.   What is matric head of water?
3.   What are the components of a tensiometer?
4.   Draw a typical illustration of a water content versus suction head.
5.   Draw a typical illustration of a suction head versus hydraulic conductivity.

# Reference and Further Reading

1. Hillel, D. (1998). *Environmental soil physics*. Academic Press.

# Salinity

## Introduction

All forms of irrigation to lands would lead to salinity; of course drip irrigation salinizes the lands slowly. When saline lands are irrigated, a fraction of water over and above the consumptive use requirements of the crop is applied and is called as leaching fraction. Therefore a rudimentary understanding of the processes causing the salinity of lands is very helpful. Knowledge of basic soil chemistry is a necessity to comfortably discuss the salinity topic. Also a microirrigation engineer's work often is associated with interpretation of soil and water chemical test results. Hence this section starts with a humble introduction to soil chemistry.

### Learning Outcomes

- Know mechanism of how water brings in electro-neutrality in soil solution
- Understand the parameters of water chemical test
- Able to derive an expression for Leaching Requirement Ratio
- Understand chemistry of leaching process of sodic soils.

## Atom

Atom is a basic element with which materials are made of. Every atom of any element is composed of particles such as proton, electron and neutron. Protons are designated as positive particles, and electrons are designated as negative particles. For any atom, if it has to be in its stable form, number of electrons must be equal to the number of protons. Neutrons are electro-neutral particles.

## Electro-Neutrality

When atoms of two different elements come together, a new level of electro-neutrality is reached by sharing of electrons of the atoms. For instance, when iron (Fe), oxygen ($O_2$) in air and water ($H_2O$) come together, rust ($Fe_2O_3H_2O$) gets formed on the surface of the iron. Rust is a salt, and we see many different salts in real life. Common salt, namely sodium chloride (NaCl), is the most commonly seen salt in real life. When salts are dissolved in water, mostly the salt compounds get dissociated into ions. Ions are in electrical equilibrium when in solution. But ions are either positively charged (cation) or negatively charged (anion). For instance, each sodium ion ($Na^+$) in water solution lacks one electron for its electro-neutrality and chloride ion ($Cl^-$) in the solution has an excess electron. Water molecules behave both as positive charges and negative charges. Therefore for each $Na^+$ ion and $Cl^-$ ion, surrounding water molecules bring in electro-neutrality. The dissociated salts also get precipitated out of water if physical conditions like pressure and temperature change and also if there is change in chemical constituents like dissolved gases.

## Clay Particles

Clay particles are plate like shaped when viewed closely. They have a longer linear length and smaller thickness. They are negatively charged. If the soil particles are larger, they can be treated electro-neutral. As the size of the clay particles becomes smaller and smaller, more and more negative charges can be expected on a clay particle.

For electro-neutrality to exist in soil, cations are needed to be supplied to soil. Let us assume that $H^+$ ions are used to bring electro-neutrality to pure clay soil particles. We know that one $H^+$ ion has one positive charge and one gram of $H^+$ has $6.0221409 \times 10^{23}$ (Avogadro's number) ions. Let us assume that 'x' is the mass of a sample of pure clay particles which gets neutralized with one milligram of hydrogen ions. If we neutralize the same sample of clay particles with $Na^+$ ions, we need 23 g of $Na^+$ ions because $Na^+$ ion is approximately 23 times heavier than $H^+$ ions because the atomic mass of Na is 23. Likewise, if we use $Ca^{2+}$ ions to neutralize the same sample of soil particles, we need 20 g of $Ca^{2+}$ ions. This is because the atomic mass of Ca is 40 and each $Ca^{2+}$ ion contributes 2 positive charges. It is to be recalled that the valency of calcium is 2 (valency of an element is the number of electrons, an atom of the element shares in bonding) and equivalent mass of Ca needed for contributing one positive charge is 20 times hydrogen. Similarly $Mg^{2+}$ has an atomic mass of 24 and equivalent mass of 12 approximately.

# Salinity of Soil

Salinity of soil is the salt concentration in soil water when the soil moisture content is at field capacity. The criterion used to gauge the salinity of the soil is Total Dissolved Salts (TDS), and the unit used is milligram per litre of water (mg/l). There is also another indirect criterion to measure the salinity of soil known as electrical conductivity of soil (EC). The unit for measuring EC is deciSiemens/m (dS/m). One dS/m is approximately equivalent to 640 mg/l.

Finding salinity of soil solution when the soil moisture level is at field capacity is difficult. The reason is that extraction of water adsorbed on soil particles at field capacity is difficult. Hence a simple and a standard conventional method is adopted which is as follows: For a dry soil sample of 100 g, 100 g of water is added and stirred, and then the soil solution is extracted. Then the EC of the soil solution extract is found out and is denoted as $EC_e$. $EC_e$ is found to be approximately related with EC at field capacity ($EC_{fc}$) by an equation which is as follows:

$$EC_{fc} = 2\,EC_e$$

## *Leaching Requirement*

In arid regions, leaching of salts from root zone due to rainfall is insufficient, and in such situations, irrigation must cater to both evapotranspiration and leaching of salts. Here in this section a steady-state salt balance equation mostly used in practice is derived.

If the root zone is to be maintained at a constant concentration, the salts added to the root zone must be leached down the root zone (Fig. 1). For such a condition, following equation can be written:

**Fig. 1** Steady-state salt balance in root zone

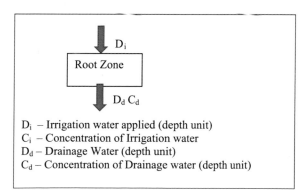

$D_i$ – Irrigation water applied (depth unit)
$C_i$ – Concentration of Irrigation water
$D_d$ – Drainage Water (depth unit)
$C_d$ – Concentration of Drainage water (depth unit)

$$D_i C_i = D_d C_d \tag{1}$$

$$\frac{D_d}{D_i} = \frac{C_i}{C_d} = LR \tag{2}$$

where LR is called as Leaching Requirement Ratio. Electrical conductivity (EC) of water and saline concentration of water are directly proportional. Hence following equation can be written:

$$LR = \frac{EC_i}{EC_d} \tag{3}$$

Let us make an assumption that the drainage water has a concentration of salt equal to the concentration of salt at field capacity. Under such a situation following equation can be written:

$$EC_d = EC_{fc} = 2\,EC_e \tag{4}$$

Since, evapotranspiration (ET) $= D_i - D_d$, following equation can also be written from the preceding equations, which is useful for finding the depth of irrigation for saline lands:

$$D_i = ET \frac{2EC_e}{2EC_e - EC_i} \tag{5}$$

## Cation Exchange Capacity (CEC)

CEC of soil is the total exchangeable cation content in 100 g of soil. CEC of a soil sample is the mass of $H^+$ ions needed to neutralize 100 g of the soil sample. Let us assume that 5 mg of $H^+$ ions are needed to neutralize 100 g of soil. Then the CEC of soil is expressed as 5 milliequivalent (meq)/100 g of soil. In the preceding statement, the meaning of 5 meq is 5 mg of $H^+$. When neutralizing process is done, any cation other than $H^+$ ions may also be used. For instance if we use $Ca^{2+}$ ions to neutralize the same soil sample of 100 g of soil, 100 mg of $Ca^{2+}$ ions are needed (i.e. 5 meq × Equivalent mass (20) = 100). Obviously clayey soil would have more CEC than sandy soil.

## Exchangeable Sodium Per Cent (ESP)

It is the fraction of Na ions in total cations. In order to find the exchangeable sodium per cent, first find exchangeable Na content in meq per 100 g of soil and divide it by total exchangeable cation content in meq per 100 g of soil.

## Sodium Adsorption Ratio (SAR)

$$SAR = \frac{Na^+}{\sqrt{\frac{Ca^{++}+Mg^{++}}{2}}}$$

In the preceding equation, $Na^+$, $Ca^{++}$ and $Mg^{++}$ concentrations are expressed as meq/l of soil solution extract. There is a commonly used relationship between ESP and SAR used for finding ESP from lab analytic results which is as follows:

$$ESP = \frac{100(-0.0126 + 0.01475\ SAR)}{1 + (-0.0126 + 0.01475\ SAR)}$$

## Hydration

Hydration is the process of adsorbing of water molecule on a clay particle or cation or anion. Water molecules are bipolar in nature because a water molecule exhibits positive charge on one side and negative charge on the other side. Therefore, water molecules are capable of hydrating both cations and anions.

On the surface of soil particle, few layers of cations and water molecules get formed and they do not move and behave as solid layer (Fig. 2). This layer is called as stern layer. Beyond the stern layer, hydrated cations and anions exist along with free water molecules. These hydrated cations and anions would have some affinity towards the clay particle. These hydrated ions adjacent to stern layer can move but not very freely. The zone in which such kind of restricted movement of ions occurs is called as diffuse layer. Conventionally both the stern layer and the diffuse layer are together called as diffuse double layer (DDL).

The thickness of DDL would vary depending on the type of cations in the soil solution. $Na^+$ ions have single positive charge, and therefore, clay binding ability of sodium is poor. $Na^+$ ions get more hydrated than $Ca^{++}$. Therefore when $Na^+$ ions are present, the distance between adjacent clay particles gets increased and dispersion of soil occurs in sodic soils. When $Ca^{++}$ ions are present, the distance between adjacent clay particles is less than when $Na^+$ ions are present.

**Fig. 2** Diffuse double layer
in saline soils

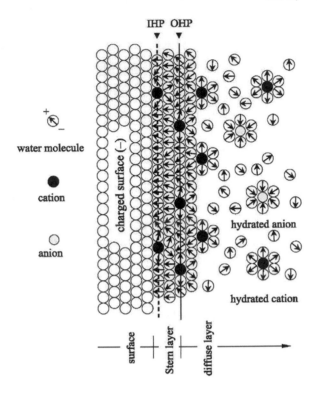

The thickness of DDL is also dependent on pH. When pH is lower, more $H^+$ ions are present and the thickness of DDL is lower. That is why pH is kept at lower level when sodium is leached. It must also be noted that sodic soils can be leached with water by removing $Na^+$ ions. But the clay particles without cations would be opposing each other due to their negative charges. That is the reason why sodic soils are not to be leached with water alone. Instead, if leaching is done by replacing with $Ca^{++}$ in place of $Na^+$, the soil properties like structure and infiltration capacity get improved.

## Problem

Calculate the gypsum requirement for a 1 ha field for which 15 cm depth of soil must be leached. ESP must be reduced to 15%. The bulk density of soil is 1.3 g/cm³. The factor of safety to be taken is 1.25 in order to take care of gypsum purity as well as deficiency of soil chemical reaction.

Soil test results are as follows:

$$Ca^{2+} : 7.5\,meq/100\,g$$
$$Mg^{2+} : 6.6\,meq/100\,g$$
$$Na^+ \ \ : 12.3\,meq/100\,g$$
$$K^+ \ \ \ : 1.2\,meq/100\,g$$

**Solution**

Molecular mass of Gypsum$(CaSO_4, 2H_2O)$
$$= 40 + 32 + (4 \times 16) + (2 \times 2) + (2 \times 16) = 164$$

Valency of Calcium $\qquad = 2$
Equivalent mass of Gypsum $\quad = 164/2 = 82$
Cationic Exchange Capacity (CEC) $\quad = 7.5 + 6.6 + 12.3 + 1.2$
$$= 27.6 \, \text{meq}/100 \, \text{g}$$

$$\text{ESP} = (\text{Exchangeable Na/CEC}) \times 100$$
$$= (12.3/27.6) \times 100$$
$$= 44.6\%$$

$$\text{Reduction of ESP} = 44.6 - 15 = 30\% (\text{app})$$

$$\text{Exchangeable Na} = 0.3 \times 27.6$$
$$= 8.28 \, \text{meq}/100 \, \text{g}$$

Gypsum needed/100 g of soil

$$= 8.28 \times \text{Eq.mass of gypsum}$$
$$= 8.28 \times 82 = 678.96 \text{mg of Gypsum}/100 \, \text{g}$$

$$\text{Bulk density of soil} = 1.3 \, \text{g/cm}^3$$

Volume of soil in 15 cm depth for 1 ha

$$= 1 \times 10,000 \times 0.15 = 1500 \, \text{m}^3$$

$$\text{Mass of } 1500 \, \text{m}^3 \text{of soil} = 1500 \, \text{m}^3 \times 1300 \, \text{kg/m}^3$$
$$= 1.95 \times 106 \, \text{kg}$$

Mass of gypsum needed

$$= 1.95 \times 106 \, \text{kg} \times 678.96 \left(10^{-6}\right) \text{mg of Gypsum}/100 \, \text{g} \left(10^{-3}\right)$$
$$= 13,239 \, \text{kg}$$

These kinds of computations are very much useful while reclaiming sodic lands.

# Summary

Salinity of irrigation water is measured with EC. The measurement unit is dS/m. Approximately 1 dS/m is equal to 640 mg/l. EC of soil water is to be known when soil moisture content is at field capacity. Conventionally EC of soil water is found out from the soil water extract taken by adding water equal to the mass of soil. This is called as EC of soil water extract, and the EC of soil water is taken as twice that of EC of soil water extract. Leaching fraction ratio is a ratio of EC of irrigation water to EC of drainage water below root zone. The depth of irrigation in saline lands is expressed as a function ET, EC of irrigation water and EC of saturation extract.

# Assessment

### Part A. Say True or False

1. Water behaves both as positive charge and negative charge (True).
2. While water dissolves salts into ions, water molecules neutralize both positive ions and also negative ions (True).
3. Clay particles are negatively charged (True).
4. Mass of one mole of hydrogen is approximately one gram (True).
5. Valency of an element is the number of electrons an atom of the element shares in bonding (True).
6. Clayey soil would have more CEC than sandy soil (False).

### Part B. Short Answers

1. Why EC at field capacity of soil is found out by saturating the soil sample?
2. If EC of saturation extract of a soil is 3 dS/m, find out the EC of soil at field capacity?
3. Explain the characteristics of diffuse double layer in sodic soils with reference to leaching process?

# Reference and Further Reading

1. Ritzema, H. P. (1994). *Drainage principles and applications.* ILRI Publications 16, Netherlands.

# Pumps

## Introduction

Pumps are mechanical devices used to lift water from lower elevation to higher elevation or for transporting water from one place to the other. Pumps are driven either by electric motors or by diesel engines. Mostly centrifugal pumps and its variants are used for irrigation. The heart of the most irrigation systems is the pump. To make an irrigation system as efficient as possible, the pump must be selected to match the requirements of the water source, the water piping system and the irrigation equipment. In this chapter an introduction to centrifugal pump and its variants are provided.

### Learning Outcomes

- Know expressions for torque
- Distinguish between centripetal force and centrifugal force
- Know the principle of working of centrifugal pump and know the variants of centrifugal pumps and pump troubles and remedies
- Derive expressions for work done by the pump, total dynamic head of a pump and minimum speed needed to initiate flow in a pump
- Know limitations in suction lift of a centrifugal pump
- Derive expressions for specific speed of pump and classify based on the specific speed
- Know and interpret performance characteristics of centrifugal pumps

## Basic Physics to Understand Pumps

Some basic physics concepts like relative motion, vector addition and vector subtraction are presented in Figs. 1, 2 and 3, respectively.

© The Author(s), under exclusive license to Springer Nature Singapore Pte Ltd. 2023
V. Ravikumar, *Sprinkler and Drip Irrigation*,
https://doi.org/10.1007/978-981-19-2775-1_23

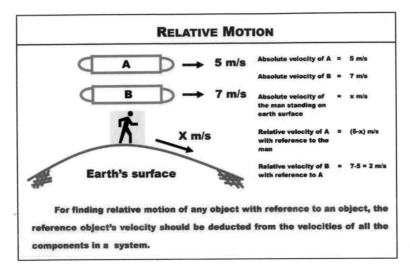

**Fig. 1** Relative motion

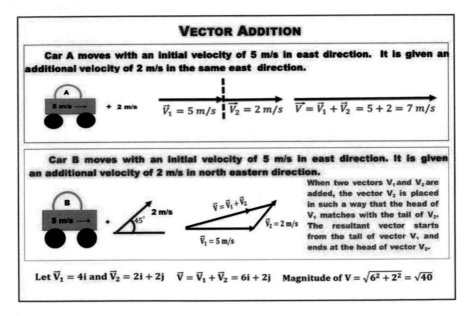

**Fig. 2** Vector addition

## Angular Momentum

Consider a body revolving with a linear tangential velocity '$V$' at a radial distance of '$r$' from the centre. If '$l$' is the circumferential distance travelled in time $t$, then

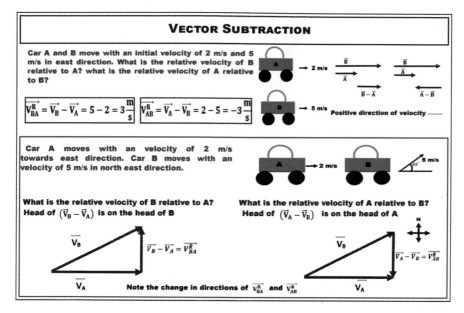

Fig. 3  Vector subtraction

d$l$/d$t$ is its linear tangential velocity '$V$' (Fig. 4).

$$V = \frac{dl}{dt}$$

$$l = r\theta$$

Fig. 4  Linear velocity and angular velocity

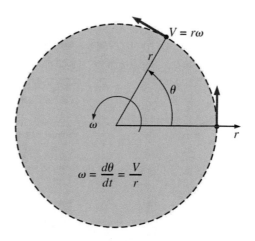

where $\theta$ is the angle in radians.

$$V = \frac{dl}{dt} = r\frac{d\theta}{dt} = r\omega \tag{1}$$

where $\frac{d\theta}{dt}$ is angular velocity denoted by $\omega$.

Angular momentum ($L$) is the moment of a rotating body. If a body of mass '$m$' rotates about a point whose radial distance is '$r$' from the centre of rotation with a tangential velocity '$V$', then the angular momentum possessed by the body is as follows:

$$L = mVr$$

Put $V = r\omega$ in the preceding equation, we get:

$$L = mr^2\omega$$

$$L = I\omega$$

where $I$ is called as moment of inertia of the revolving body.

$$\text{Rate of change of Angular momentum} = \frac{dL}{dt} = I\frac{d\omega}{dt} \tag{2}$$

## *Torque*

Torque acting on a revolving body is defined as the product of tangential force acting on the body and the radial distance of the revolving body.

$$T = m\frac{dV}{dt}r = mr^2\frac{d\omega}{dt} = I\frac{d\omega}{dt} \tag{3}$$

It can be seen that the rate of change of angular momentum is equal to the torque on the revolving body.

## *Rotational Power*

We know that linear work done is the force times the distance and the power ($P$) is rate of doing work. For rotation to occur, tangential force on the object is to happen.

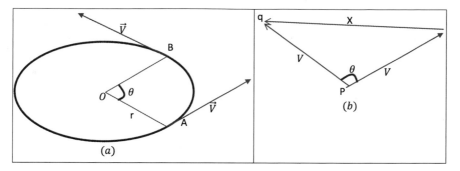

**Fig. 5** Centripetal acceleration

$$P = m\frac{\mathrm{d}V}{\mathrm{d}t}V = m\frac{\mathrm{d}V}{\mathrm{d}t}r\omega = T\omega \tag{4}$$

Hence it is understood that that the rotational shaft power needed is equal to the product of torque and angular velocity.

## *Centripetal Force and Centrifugal Force*

It must be kept in mind that centrifugal force is a reaction force to centripetal force. As long as centripetal force is in action only, centrifugal force would exist. Fig. 5

Let us assume that a static body of mass '$m$' starts to move with a linear velocity '$V$'. If the mass has to remain in circular motion, its direction of velocity must be changed continuously for which we need to apply a force on the body. This force is called as centripetal force [1].

Let '$A$' be the position of the object at the start and '$B$' be the position of the object after a time '$\mathrm{d}t$'. So, within time '$\mathrm{d}t$', the object must be given an additional velocity '$X$' towards the centre. Let the additional velocity be denoted as $\left(\overrightarrow{X}\right)$.

In (b) of Fig. 5, the velocity vector at the point A is drawn first and then the velocity vector $\left(\overrightarrow{X}\right)$ is drawn in the direction towards the centre. The resultant velocity vector $(\overrightarrow{pq})a$ also must have a magnitude of '$V$' and its direction must be the velocity vector at the point B. The triangle OAB in (a) and the triangle in (b) of Fig. 5 are similar triangles, and hence following equation can be written:

$$\frac{x}{v} = \frac{AB}{r}$$

$$x = \frac{AB}{r}v$$

Acceleration of the object $\left(\frac{x-0}{dt}\right) = \frac{v}{r}\frac{AB}{dt} = \frac{v^2}{r}$

$$\text{Centripetal force} = \frac{mv^2}{r}$$                                      (5)

## Centrifugal Pump Operating Principle

Basically, centrifugal pumps add pressure energy to the water. Most of the time, it is misunderstood that the centrifugal pumps create vacuum. But the centrifugal pumps do not operate by creating vacuum. If you take a half-filled bucket of water and put your hand in the bucket and churn the water in the bucket, the water surface in the bucket will assume the shape as shown in Fig. 6. Let us see why water rises like this.

When water is rotated, water moves towards periphery and hits on the wall of the bucket. The wall directs the water towards the centre. This force given by the wall of the bucket is the centripetal force. The reaction force (centrifugal force) given by the water against the centripetal force acting on it causes pressure to get increased between the water molecules. The dissipation of the momentum of the water is the basic reason for the increase of pressure energy. How the dissipation of momentum causes pressure increase in between molecules can be easily understood when you throw a soil clod on a wall, it gets broken down into pieces because the momentum is dissipated. It can also be seen that the pressure is low at the centre and the pressure keeps on increasing towards the periphery.

The process occurring in centrifugal pumps is much similar to this. Instead of hand rotating the fluid, a rotating element rotates the water in the pump. In this case, the bucket is under open condition whereas the pump is closed. At the centre, the

**Fig. 6** Shape of water surface due to churning of water

r – radius from the centre

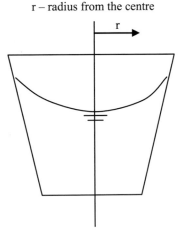

pressure drops down and inflow into the pump is connected at the centre called as suction pipe, and the delivery pipe is connected at the periphery.

## Components of Centrifugal Pump

Figure 7 shows different components inside a centrifugal pump. Figure 8 shows the accessories along with the pump while in operation.

### *Impeller*

It is a wheel provided with a series of curved vanes. It is mounted on a shaft, which is coupled to an external source of energy (electric motor or diesel engine), which imparts required energy to the impeller to rotate.

Figure 9 shows different types of impellers normally used.

1. *Shrouded (or) Closed*: In this type vanes are covered with metal plates on both the sides. The closed impeller provides a better guidance for the flow, and so it is efficient. If the liquid to be pumped is free from debris, this is used.

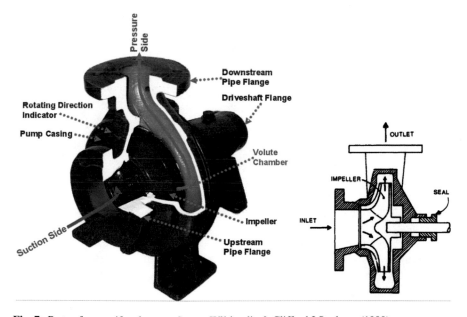

**Fig. 7** Parts of a centrifugal pump. *Source* Wikipedia & Clifford J Studman (1990)

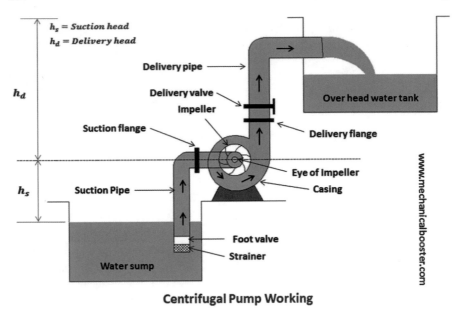

**Fig. 8** Components in a centrifugal pumping system

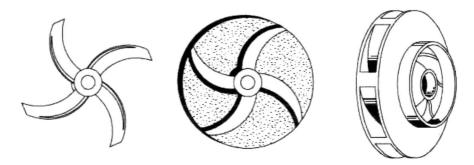

**Fig. 9** Types of impeller (www.joinoilandgas.com)

2.  *Semi-open* type: In this type only base plate is provided, and crown plate is not provided. If the liquid contains debris, semi-open type is used in order to prevent clogging.
3.  *Open type*: In open type, only vanes exist. No base plate and crown plate are provided. This is useful in pumping of liquids containing suspended solid matter such as paper pulp, sewage water. If closed or semi-open other impellers are used for this purpose, the impellers get clogged.

## Casing

It is an airtight chamber, which surrounds the impeller. The air tightness is ensured by the provision of a seal that occupies the space between rotating shaft and casing (Fig. 7—Right-side image). This seal would be continuously rubbing on the rotating shaft and hence would out. Failure of air tightness due to wearing out of the seal is often go unnoticed and the pump performance gradually gets reduced. The radial space between casing and impeller is called spiral volute chamber.

## Suction Pipe

It is a pipe section, which is connected from eye to foot valve.

## Foot Valve

It is a non-return valve. While the pump is in operation, it remains opened. When the pump is stopped, it immediately closes automatically preventing water in the suction line from going out.

## Strainer

Strainer prevents debris and plant leaves from going into the pump.

## Delivery Pipe

It is a pipe section connected from the casing to the point of water delivery.

## Delivery Valve

It is a valve connected in the delivery pipe to regulate the discharge rate of the pump.

## Working of Centrifugal Pump

### *Priming*

It is an operation in which the suction pipe casing and the portion of the delivery pipe up to the delivery valve are completely filled with the liquid to be pumped. There should not be any air packets left inside the water.

The rotation of the impeller in the casing full of liquid produces pressure increase with respect to radial distance. The pressure increase is also a function of mass density of rotating fluid and also the speed of rotation.

Let us assume that the pump impeller is surrounded by air instead of water. If the impeller rotates, the impeller will rotate the air surrounding it. The increase in pressure of the air will not be sufficient for lifting the water. Similarly, if air packets remain in the liquid, density of air plus liquid mixture is lower than water alone. Therefore, with the liquid plus air mixture also the pressure increase is lower compared to the liquid without any air packets. This is the reason why the pump is to be filled without air packets.

After the pump is primed, the delivery valve is kept closed and the electric motor is started to rotate the impeller. The delivery valve is closed because it reduces the starting torque required to rotate the impeller. If it is not closed, the impeller raises the water in the delivery pipe and this water column will increase the pressure on the impeller in a short time. The motor gains full torque only after its full speed is reached. When the delivery valve is closed, the impeller just churns the liquid, and after a few seconds, full torque level of motor is also reached and the pressure is also developed in the pump and the delivery valve is opened. Because of the high-pressure water leaving the impeller, the liquid is lifted in the delivery pipe and the water is delivered out. As the water starts to go out, at the eye of the impeller, the pressure becomes low. So, water from the suction pipe is sucked in and the flow becomes continuous.

## Volute Chamber

When the water comes out of the impeller, it has a very high kinetic energy. This kinetic energy should also be converted into useful pressure energy. This is achieved in volute chamber of casing. As soon as the water comes out of the impeller the cross-sectional area of the flow in the volute chamber of casing is enlarged. So, the velocity decreases, which causes increase in pressure. From the throat, the water travels along the periphery in anti-clockwise direction until delivery pipe inside the volute chamber. Along the periphery more and more water gets added. If we have a constant cross-sectional area of flow through the volute chamber, the velocity will get increased along the path of the volute chamber. The objective is to decrease the velocity by increasing the pressure head. Therefore, the cross-sectional area of flow is increased gradually so that the velocity gets decreased gradually.

## Volute Pump with Vortex Chamber

An improvement in the volute pump called vortex chamber or whirlpool chamber improves the efficiency of the pump (Fig. 10). This is introduced as the part of the casing. In ordinary volute pumps, the water leaving the impeller enters directly into the volute chamber. This would cause eddy losses significantly. But in the vortex chamber, the water leaving the impeller just forms a free vortex. The water also has a radial component of velocity. So it enters into the volute chamber. Free vortex is also a rotating mass of fluid, but while the fluid is under rotation, no external force acts on it. The water when it is rotating inside the impeller, it is called as forced vortex. After water leaves the impeller, the water will still have rotation, but during this rotation no external force is acting on the fluid.

Coming back to our rotation of water in a bucket with hand, forced vortex happens during the churning of water by hand. Even after you stop the churning by hand, the water would continue to rotate. This is called as free vortex.

In the free vortex, it has been found that as the radial distance of the water increases, the tangential velocity decreases. In a continuum of flow when velocity decreases, pressure increases.

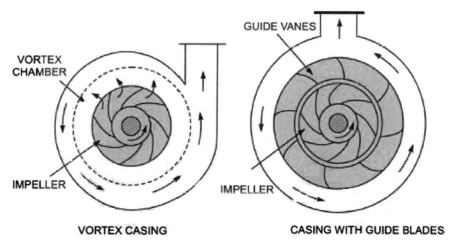

**Fig. 10** Free vortex casing and casing with stationary guide blades. (www.hkdivedi.com)

## Casing with Stationary Guide Blades

In these pumps (Fig. 10), the impeller is surrounded by a series of guide vanes mounted on a ring called diffuser ring. The diffuser ring along with guide vanes is fixed in position. They do not rotate. The water from the impeller enters into the guide vanes in such a way that it does not cause any shock and enters tangentially. The velocity gets decreased and the pressure increases inside the guide vanes, since they provide gradual increase in cross-sectional area for flow. These pumps will work efficiently for only one discharge rate and for the other discharge rates there will be impact of water coming out of impeller on the diffuser ring and so loss of energy will be more.

## Classification of Centrifugal Pumps

The centrifugal pumps are classified based on different criteria. Based on the head developed the centrifugal pumps are classified as follows:

| | |
|---|---|
| Low head | : Up to a head of 15 m |
| Medium head | : 15–40 m |
| High head | : More than 40 m |

Based on the relative direction of flow through the impeller, the centrifugal pumps are classified as follows (Fig. 11).

### *Radial Flow*

In these impellers, liquid flows from the centre to the periphery radially outward. Normally, many centrifugal pumps are radial flow pumps. Radial flow pumps are capable of producing more pressure. Table fans we use in houses have the air flow in radial direction.

### *Mixed Flow*

In these impellers, liquid flows axially as well as radially. These pumps are low head and high discharge pumps.

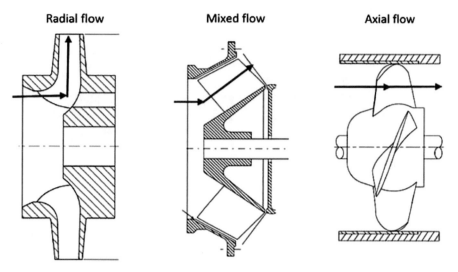

**Fig. 11** Impeller types based on direction of flow (https://vertiflopump.net/tag/horizontal-centri fugal-pump)

## *Axial Flow Pumps*

In the impellers of these pumps, flow occurs axially or along the direction of shaft (like screw conveyor). These pumps produce lower pressure and work with a higher discharge. The ceiling fans, we use in houses, have axial air flow. As far as energy efficiency is concerned, axial flow pumps are the best and followed by the mixed flow, and finally the radial flow pumps.

Based on the number of entrance to the impeller the centrifugal pump is classified as single suction and double suction pumps (Fig. 12). In single suction pumps, liquid is admitted into impeller in one direction only. In double suction pump, liquid is admitted into the impeller in both the directions. Double suction pumps have an advantage of balanced axial force on the bearings, and hence wear and tear are comparatively reduced.

## *Self-priming Pumps*

A self-priming pump has the ability to remove air bubbles and resume the normal pumping without outside attention (Fig. 13). The suction pipe is connected to the pump above the pump rotation axis to collect some water within the pump casing as soon as the pump is stopped. The pump casing of self-priming pumps is also large to store more water and it is called as priming chamber. When pump is restarted, the water in the priming chamber will be churned by the impeller, and at the air

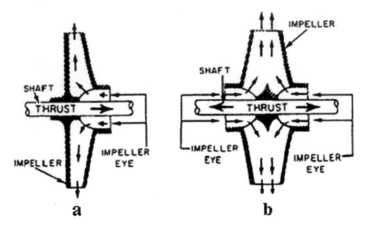

**Fig. 12** Single suction and double suction pumps (Ahmed Farid Saad Ayad Hassan/researchgate)

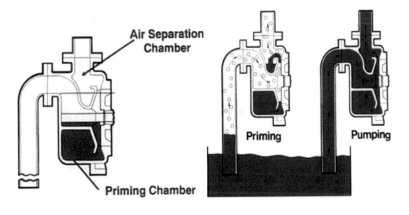

www.gouldspumps

**Fig. 13** Self-priming pumps (www.gouldspumps)

separation chamber, the air bubbles will escape out. The air separation chamber is also a large chamber to facilitate easy air escape. When impeller rotates, a partial vacuum is formed at the eye of the impeller which causes air from suction pipe to be sucked into the casing and the air plus water mixture is churned and the air bubbles continue to escape from the air separation chamber. At the same time, air from delivery pipe will not get back into the casing. After all the air from the suction pipe is gradually pumped out, the pressure head produced in the pump is sufficient to continue pumping normally.

The static lift and suction piping should be minimized to keep priming time to a minimum. Excessive priming time can cause liquid in the priming chamber to vaporize before the priming is achieved. All connections in the suction piping should

**Fig. 14** Priming bypass line
(www.gouldspumps)

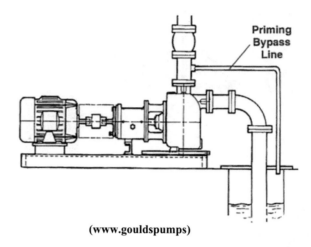

**(www.gouldspumps)**

be leak-free as air could be sucked in, thus extending/compromising the priming of the pump. A priming bypass line should be installed so that back pressure is not created in the discharge piping during priming (Fig. 14).

## Centrifugal Pump Installation

The suction piping size and design is very important. Many centrifugal pump troubles are caused by poor suction conditions. The function of suction piping is to supply an evenly distributed flow of liquid to the pump suction, with sufficient pressure to the pump to avoid excessive cavitations in the pump impeller. The suction pipe should never be smaller than the suction connection of the pump and in most cases should be at least one size larger. Suction pipes should be as short and as straight as possible. Suction pipe velocities should be in the 1.5–2.5 m/s range unless suction conditions are unusually good. Higher velocities will increase the friction loss and can result in air or vapour separation. This is further complicated when elbows or tees are located adjacent to the pump suction nozzle, in that uneven flow patterns or vapour separation prevents the liquid from evenly filling the impeller. This upsets hydraulic balance leading to vibration, possible cavitation, and excessive shaft deflection. Especially on high and very high suction energy pumps, shaft breakage or premature bearing failure may result.

On pump installations involving suction lift, air pockets in the suction line can be a source of trouble. If elbow at the eye is used, it should not be horizontal. If the site condition warrants for horizontal elbows, a straight run pipe of length at least 5–6 diameters should be provided before the elbow is connected. The suction pipe can be horizontal, or with a uniform slope upward from the sump to the pump. But the suction pipe should not slope downwards from the sump to the pump. This may

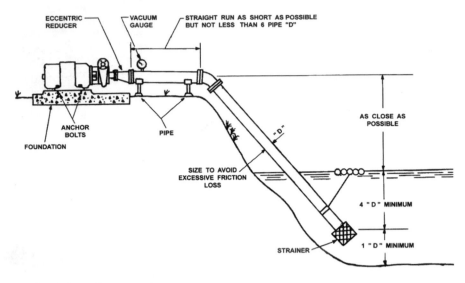

**Fig. 15**  Suction pipe connection

cause air packets to get formed in the suction line. There should be no high spots in suction where air can collect and cause the pump to lose its prime. Eccentric reducers should always be used, on horizontal installations, with the flat side located on top (Fig. 15).

Never fit any control valve in suction line, if there is any suction lift. This may cause air to enter in the suction line.

## Centrifugal Pump Troubles and Remedies

Some of the troubles commonly experienced during the operation of the centrifugal pumps and the remedial measures to be undertaken for the same are as listed below.

### Pump Fails to Start Pumping

- Pump is probably not properly primed. Re-prime the pump, opening the air vent until a steady and unbroken stream of liquid is obtained.
- Total static head is probably much higher than that for which the pump is designed. Check the same with accurate vacuum and pressure gauges, or probably determined by actual measurement, the difference in elevation between pump and suction liquid level, and also between pump and point of discharge. Add to this

static head, the loss of head due to friction in pipes and that in the other fittings used in the installation.

- Wrong direction of rotation of the impeller. Arrow on the pump case shows the proper direction of rotation.
- Impeller may be clogged. Examine carefully for solids or foreign matter lodged in the impeller.
- Suction lift may be too high. Check with vacuum gauge or by actual measurement where ever possible, adding to the static suction lift, the loss of head in the pipe and fittings.
- Strainer or suction line may be clogged. This would cause an excessive suction lift.
- Speed may be too low. Check the speed with a tacheometer and compare it with the one given on the name plate of the pump. When the pump is being driven by electric motor, check up to see whether the voltage may be too low.

## *Pump Working but not Up to Capacity and Pressure*

- Air may be leaking into pump through suction line or through stuffing boxes.
- Speed may be too low.
- Discharge head may be higher than anticipated.
- Suction lift may be too high.
- Foot valve or end of suction pipe may not have sufficient submergence, or it may be partly clogged or entirely too small.
- Impeller may be partly clogged or too small in diameter.
- Rotation may be in the wrong direction.

## Multi-stage Pump

The head produced by a centrifugal pump depends on the speed of the impeller and diameter of the impeller. If pumping is to be done for higher heads, either the speed has to be increased or the size of the impeller has to be increased. Increasing the speed and diameter beyond a certain limit has many practical difficulties. So, for deep well pumping, multi-stage pumps were developed (Fig. 16).

A multi-stage pump consists of two or more identical impellers mounted on the same shaft and enclosed in a same casing. All the impellers are connected in series, i.e. the liquid pumped from one impeller passes through a connecting passage in which guide vanes are placed. During that process the velocity head is converted into pressure head and the velocity is be brought to the same velocity with which it enters into the impeller. Like this, when water passes through successive impellers the pressure head increases step by step.

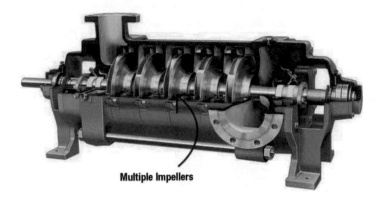

**Multiple Impellers**

**Fig. 16** Multi-stage pump (www.castlepumps.com)

## Turbine and Submersible Pumps

Turbine pump is a multi-stage pump in which a motor is placed at the ground level and a pump is placed inside the water, and they are connected by a shaft. Deep well turbine pumps are adapted for use in cased wells or where the water surface is below the practical suction limits of a centrifugal pump. Turbine pump efficiencies are comparable to or greater than most centrifugal pumps. They are usually more expensive than centrifugal pumps and more difficult to inspect and repair (Fig. 17). Submersible pump is also a multi-stage pump in which both motor and pump are placed inside the water (Fig. 18).

## Propeller Pump

This is also called as axial flow pump (Fig. 19). In this pump, inlet guide blades, propeller vanes and outlet guide blades are so designed that when propeller vanes rotate, the vanes impart only velocity head to the water not pressure head. The water is not given any rotation (no forced vortex). So, no pressure head increases. Water flows axially upward.

It is a high-speed pump with very high discharge and the total head for which it can be used is around 9–12 m. The maximum suction lift is 1 m. The disadvantage is that when suction lift exceeds the limit, cavitations will be more.

**Fig. 17** Turbine pump.
*Source* BookFire Pumps and
Water Supplies, Robert C.
Till and J. Walter Coon,
Springer

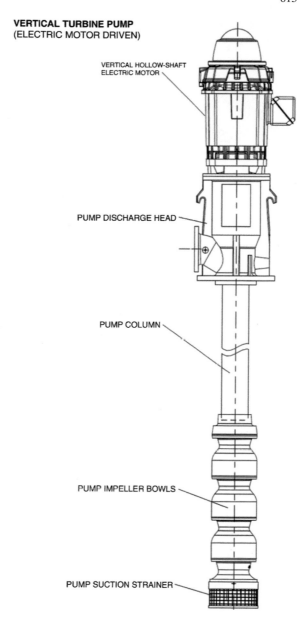

**VERTICAL TURBINE PUMP**
(ELECTRIC MOTOR DRIVEN)

VERTICAL HOLLOW-SHAFT
ELECTRIC MOTOR

PUMP DISCHARGE HEAD

PUMP COLUMN

PUMP IMPELLER BOWLS

PUMP SUCTION STRAINER

# Jet Pump

In this pump, water at high pressure is passed through a nozzle (Fig. 20). Since velocity is high at the throat, pressure drops down and vacuum is produced. The drop in pressure sucks the water from below. Jet pumps use only centrifugal pump.

**Fig. 18** Submersible pump
(Lund. J. W. 2013)

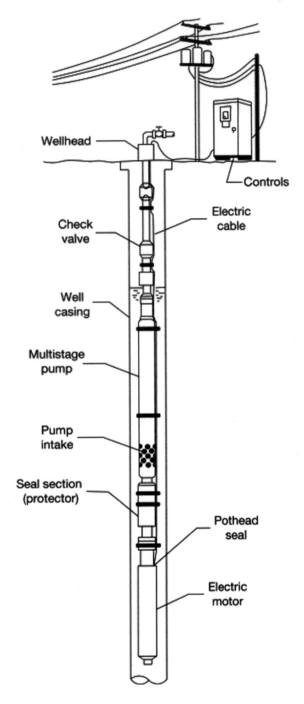

**Fig. 19** Propeller pump
(braincart.com)

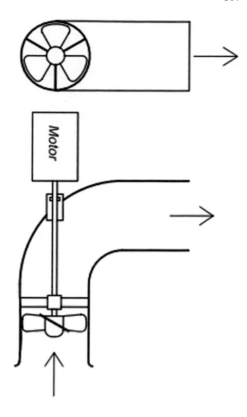

A valve controls the delivery of the pump and a part of the water is diverted down to pass through the jet. A series of jets may also be used for lifting water from large depths. The actual rate of discharge is less than the total quantity of water handled by the impeller as some amount of water is always in circulation. This pump is used for deep well pumping, and its efficiency is less than ordinary centrifugal pumps. These pumps do not have any limitation on suction lift.

## Airlift Pump

In this pump, compressed air is introduced through a nozzle at the foot of rising main (Fig. 21). Since air is introduced, the density of water in the delivery line reduces. Therefore, high density water inside the well pushes up the water plus air mixture through the delivery line. For proper working of the pump, useful lift should be less than the depth of submersion. One important advantage of this pump is that no moving part is inside the water, and therefore, wear and tear are less and repairs are less and easy. The disadvantages include, it is the pump with the lowest efficiency and the depth of well needed is more. Since the use of this pump is increasingly seen

**Fig. 20** Jet pump (fao.org)

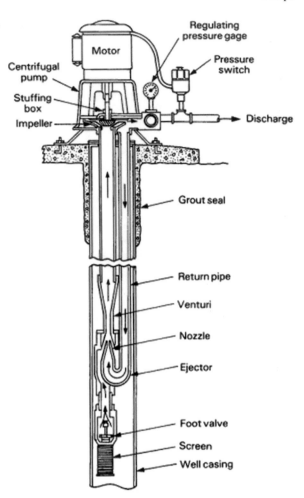

in many places and also in the aquaponics, the design aspects of this pumping system is also discussed later in this chapter.

## Work Done by a Centrifugal Pump Impeller

When water enters into the eye of the pump, it radially spreads (Fig. 22). Let us assume that the water travels with a radial velocity '$v_1$' just before entering a vane. Let '$\omega$' be the angular velocity of the impeller. Apply the principle of relative motion for all the components by deducting '$\omega$' in opposite direction. Then the impeller comes to rest. The process of deducting '$\omega$' is equivalent to an observer revolving

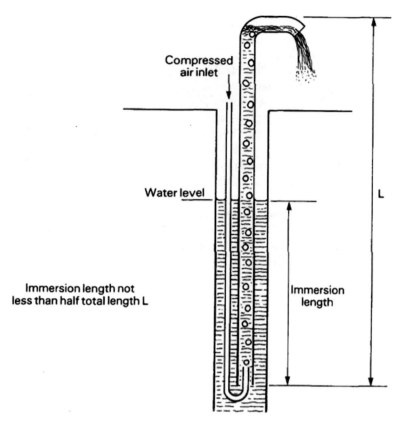

**Fig. 21** Airlift pump (fao.org)

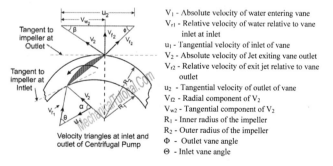

$V_1$ - Absolute velocity of water entering vane

$V_{r1}$ - Relative velocity of water relative to vane inlet at inlet

$u_1$ - Tangential velocity of inlet of vane

$V_2$ - Absolute velocity of Jet exiting vane outlet

$V_{r2}$ - Relative velocity of exit jet relative to vane outlet

$u_2$ - Tangential velocity of outlet of vane

$V_{f2}$ - Radial component of $V_2$

$V_{w2}$ - Tangential component of $V_2$

$R_1$ - Inner radius of the impeller

$R_2$ - Outer radius of the impeller

$\Phi$ - Outlet vane angle

$\Theta$ - Inlet vane angle

**Fig. 22** Velocity triangle for an impeller vane (www.mMechanicalTutorial.com)

along the impeller vane. For the observer rotating along with the impeller, the velocity of impeller is zero.

Water moves with a linear radial velocity '$v_1$' before entering the vane. For the person moving with the vane, the relative velocity of jet ($v_{r1}$) is ($\vec{v}_1 - \vec{u}_1$). The linear velocity '$u_1$' of the inlet of impeller vane is $R_1\omega$. To get ($\vec{v}_1 - \vec{u}_1$), draw the vector "$u_1$" as shown in Fig. 22. The velocity triangle thus formed at the inlet of vane is called as inlet velocity triangle. The angle $\theta$ is called as inlet vane angle. This angle is very important design parameter. This $\theta$ should be such that, entering jet should not have any shock due to impact. Shock does not occur if the water enters tangentially to the vane for the observer rotating along the vane. The term 'tangent' in the preceding sentence must not be confused with the tangents to impeller inlet and outlet drawn in Fig. 22. $v_{r1}$ is the direction and magnitude as seen by observer sitting on the vane.

Then the jet leaves the vane tangentially with velocity $v_{r2}$ with respect to the observer on the vane. For the person revolving along the vane $v_{r2}$ is the relative velocity of jet leaving the vane. We are interested in finding the velocity of leaving jet for an observer standing on the earth ($v_2$). If the person revolving along the vane gets down from the vane, '$\omega$' gets added to $v_{r2}$. Therefore, $v_2$ is obviously ($\vec{v}_{r2} + \vec{u}_2$). $V_{f2}$ is radial component of $v_2$ and $v_{w2}$ is the tangential component of $v_2$.

The triangle at the outlet is called as the outlet velocity triangle. The outlet vane angle $\varphi$ is also an important design parameter of impeller.

Figure 23 shows an impeller. Let us assume that the water in the impeller at the time $t$ occupies the zone A and B. After time '$dt$' the water in zone A and B would occupy the zone B and C when there is a steady flow of water radially outward. Let $Q$ be the discharge rate through the impeller. Let $\rho$ be mass density of water. The mass of stored water in the annular space A and C will be obviously equal to $\rho Q dt$.

Change of angular momentum of the water in the impeller during time dt

$$= (L_B + L_C) - (L_A + L_B)$$

**Fig. 23** Angular momentum of water in impeller

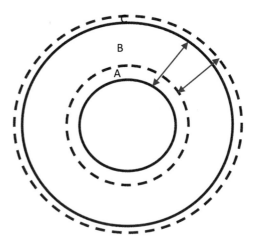

$$= L_C - L_A$$
$$= \rho Q \, dt \, v_{w2} R_2 - \rho Q dt \, v_{w1} R_1$$

where $L$ is the angular momentum.

Rate of change of angular momentum of the water in the impeller

$$= \rho Q (v_{w2} R_2 - v_{w1} R_1)$$

Rate of work done by the impeller $= \rho Q (v_{w2} R_2 - v_{w1} R_1) \omega$     (6)

where $\omega$ is angular velocity in radian/s, and other terms are as defined in Fig. 22. $v_{w1}$ can be assumed as zero since the water enters radially.

Work done by the impeller/second (WD) $= \rho Q v_{w2} R_2 \omega$

$$= \rho Q v_{w2} u_2$$

Work done by the impeller per unit mass of the fluid $- v_{w2} u_2$

It is also possible to convert Eq. 6 to a different form. According to cosine rule, for the outlet velocity triangle,

$$v_{r2}^2 = v_2^2 + u_2^2 - 2u_2 v_2 \cos \beta$$
$$vr_2^2 = v_2^2 + u_2^2 - 2u_2 v_{w2}$$
$$v_{w2} = \left( v_2^2 + u_2^2 - v_{r2}^2 \right)/2u_2 \tag{7}$$

Similarly for inlet velocity triangle,

$$v_{w1} = \left( (u_1)^2 + (v_1)^2 - (v_{r1})^2 \right)/2u_1 \tag{8}$$

Substituting values $v_{w1}$ and $v_{w2}$ in Eq. 6,

$$\text{WD}/s = \rho Q \left[ u_2^2 + v_2^2 - v_{r2}^2 \right) u_2/2u_2 - \left( u_1^2 + v_1^2 - v_{r1}^2 \right) u_1/2u_1 \right] \tag{9}$$

$$= \rho Q \left[ (u_2^2 - u_1^2)/2 + (v_2^2 - v_1^2)/2 + (v_r^2 - v_{r1}^2)/2 \right] \tag{10}$$

$\rho Q$ may also be expressed as $W/g$, where $W$ is weight rate of discharge, and g is acceleration to gravity.

$$= W/g \left[ (u_2^2 - u_1^2)/2 + (v_2^2 - v_1^2)/2 + (v_{r1}^2 - v_{r2}^2)/2 \right] \tag{11}$$

Work done/unit weight of liquid

$$= \left(u_2^2 - u_1^2\right)/2g + \left(v_2^2 - v_1^2\right)/2g + \left(v_r^2 - v_{r1}^2\right)/2g$$

$$= \left(u_2^2 - u_1^2\right)/2g + \left(v_2^2 - v_1^2\right)/2g + \left(v_{r2}^2 - v_{r1}^2\right)/2g \tag{12}$$

is called as Euler's energy transfer equation. $\left(u_2^2 - u_1^2\right)/2g$ is called as centrifugal head. $\left(v_2^2 - v_1^2\right)/2g$ is called as kinetic energy head, and $\left(v_{r1}^2 - v_{r2}^2\right)/2g$ is called as relative velocity head.

From the preceding derivation, some salient inferences can be made. The relative velocity of jets moving on the vanes depends on the discharge rate of the pump. If the discharge rate varies, the relative velocity of jets entering and exiting the vanes would also vary. But while jets enter the vane, the jet should be tangential to the vane. The inlet vane angle and outlet vane angle cannot change, and it is an important design parameter. Therefore, when the discharge rate varies, the relative velocity of jet also varies and the entry and exit of jets will not be tangential. Therefore, one impeller working efficiently for a specific design discharge may not work efficiently for other discharges. The entering jet and exiting jet will unnecessarily impinge on the vanes which will cause wastage of work done. Pumps with a specific horse power say, 5 H.P. will have different models with different discharge rates and pressure development based on variations in impeller design and motor speed.

## Heads of a Pump

### *Static Head*

The static head is the vertical distance between the liquid surface in the sump and the delivery end (Fig. 24):

$$H_s = h_s + h_d \tag{13}$$

$h_s$ is static suction head. This is the vertical distance between the water level in the sump and the centre of the pump. $h_d$ is static delivery head. This is the vertical distance between the centre lines of the pump and the delivery end. Sometimes, instead of free-falling discharge as shown in the figure, the discharge pipe may be submerged in the tank. In such cases, the depth of submergence should also be added.

**Fig. 24** Heads of a pump

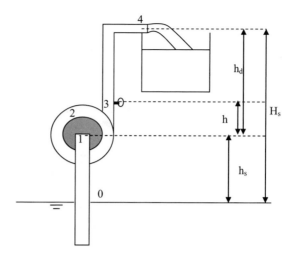

## Suction Dynamic Head

Suction dynamic head is the sum of the static suction head and friction head loss in the suction pipe.

## Delivery Dynamic Head

Delivery dynamic head is the sum of the static delivery head, friction head loss in the delivery pipe and kinetic head at the delivery.

## Total Dynamic Head

Total dynamic head (TDH) is the total pressure head that must be produced by the pump. It is the useful work done by the impeller.

$$\text{TDH} = \text{Work done by the impeller} - \text{Losses of head in pump} \qquad (14)$$

In following sections, an expression for TDH is derived by applying the Bernoulli's law:

Let $p_s$ be pressure at the eye of the pump (point-1), $\gamma$ be the weight density of water, $v_s$ be the velocity at the eye of the pump, $g$ be acceleration due to gravity and $h_s^f$ be head loss due to friction in suction pipe. Let $h_i^L$ be head loss in impeller. Applying Bernoulli's equation between points 0 and 1 for figure yields the following equation:

$$0 = \frac{p_s}{\gamma} + \frac{v_s^2}{2g} + h_s + h_s^f \tag{15}$$

$$\frac{p_s}{\gamma} = -\left(\frac{v_s^2}{2g} + h_s + h_s^f\right) \tag{16}$$

Applying Bernoulli's equation between points 1 and 2 yields the following equation:

$$\frac{p_s}{\gamma} + \frac{v_s^2}{2g} = \frac{p_2}{\gamma} + \frac{v_2^2}{2g} + h_i^L - \frac{v_{w1}u_1}{g} \tag{17}$$

Applying Bernoulli's equation between points 2 and 3 yields the following equation:

$$\frac{p_2}{\gamma} + \frac{v_2^2}{2g} = \frac{p_d}{\gamma} + \frac{v_d^2}{2g} + h + h_c^L \tag{18}$$

In the preceding equation $h_c^L$ is called as head loss in casing. $p_d$ is pressure at point 3, and $v_d$ is velocity in the delivery pipe. It should be noted that friction loss between 2 and 3 is neglected because the length between points 2 and 3 is small.

Putting Eq. 18 in Eq. 17 yields the following equation:

$$\frac{p_s}{\gamma} + \frac{v_s^2}{2g} = \frac{p_d}{\gamma} + \frac{v_d^2}{2g} + h + h_c^L + h_i^L - \frac{v_{w1}u_1}{g} \tag{19}$$

$$\frac{v_{w1}u_1}{g} - h_c^L - h_i^L = \frac{p_d}{\gamma} + \frac{v_d^2}{2g} + h - \frac{p_s}{\gamma} - \frac{v_s^2}{2g} \tag{20}$$

$$\text{TDH} = \frac{p_d}{\gamma} + \frac{v_d^2}{2g} + h - \frac{p_s}{\gamma} - \frac{v_s^2}{2g} \tag{21}$$

Applying Bernoulli's equation between points 3 and 4

$$\frac{p_d}{\gamma} + \frac{v_d^2}{2g} = h_d - h + h_d^f + \frac{v_d^2}{2g} \tag{22}$$

$$\frac{p_d}{\gamma} = h_d - h + h_d^f \tag{23}$$

substitute for $P_d/\gamma$ and $P_s/\gamma$ in Eq. 21 and simplifying, we get

$$\text{TDH} = \left(h_s + h_s^f\right) + \left(h_d + h_d^f + \frac{v_d^2}{2g}\right) \tag{24}$$

$$\text{TDH} = \text{Suction dynamic head} + \text{Delivery dynamic head} \tag{25}$$

Intuitively one will be easily able to guess that the total energy needed from pump is the sum of energy for lifting water from sump to the tank, energy for overcoming frictional losses in pipes and push out of the pipe at the delivery with some kinetic head. On observing each term of Eq. 23, one can easily understand that manometric head is the useful energy delivered by the pump.

## Minimum Speed Needed to Initiate Flow

Before flow is initiated, when the impeller rotates, the liquid inside the impeller rotates with forced vortex. The radial components of velocity of flow over the vanes will be very negligible before initiation of flow. For the forced vortex, the following equation holds good:

$$p_2 - p_1 = \frac{\rho\left(u_2^2 - u_1^2\right)}{2} - \rho g (Z_2 - Z_1) \tag{26}$$

where

$p_1$        is the pressure at the inlet of the vane.
$p_2$        the pressure at the outlet of the vane.
$u_1$        is the tangential velocity of at the inlet of the vane.
$u_2$        is the tangential velocity of at the outlet of the vane.
$(Z_2 - Z_1)$   is the elevation difference between inlet of the vane and at outlet of the vane which is negligible and neglected. Hence following equation is obtained:

$$p_2 - p_1 = \frac{\rho\left(u_2^2 - u_1^2\right)}{2} \tag{27}$$

Dividing Eq. 27 by $\gamma$ on both sides yields following equation:

$$\frac{p_2 - p_1}{\gamma} = \frac{u_2^2 - u_1^2}{2g} \tag{28}$$

$\frac{p_2}{\gamma} - \frac{p_1}{\gamma}$ is the difference in pressure head between the inlet of vane and outlet of vane, and this is the pressure head developed by rotation of the impeller. This pressure head should be sufficient to lift the water. If $H_m$ is the manometric head needed to be developed, then the pressure head developed should be more than $H_m$.
Therefore,

$$\left(\frac{p_2}{\gamma} - \frac{p_1}{\gamma}\right) > \text{TDH} \tag{29}$$

$$\frac{u_2^2 - u_1^2}{2g} > \text{TDH} \tag{30}$$

$$\frac{\left(\frac{\pi D_2 N}{60}\right)^2 - \left(\frac{\pi D_1 N}{60}\right)^2}{2g} > \text{TDH} \tag{31}$$

where $D_1$ and $D_2$ are the diameter of inlet of vanes and outlet of vanes, respectively, and $N$ is number of revolutions of the impeller per minute (rpm). At the limiting case the left-hand side of the equation should be at least equal to right-hand side. Therefore,

$$\frac{\left(\frac{\pi D_2 N}{60}\right)^2 - \left(\frac{\pi D_1 N}{60}\right)^2}{2g} = \text{TDH} \tag{32}$$

where $N$ is the minimum rpm needed to initiate flow.

The minimum expression is useful when pumps are coupled to diesel engine. The speed reduction from diesel engine's flywheel rpm to the impeller rpm can be decided using this equation.

## Impeller Diameter

Usually inside diameter of the impeller $(D_1)$ is taken as half of the outside diameter $(D_2)$.

$$D_2 = 0.5 D_1 \tag{33}$$

Substituting Eq. 33 in Eq. 32 yields the following approximate equation:

$$D_2 = \frac{97.44 \sqrt{H_m}}{N} \tag{34}$$

Equation 34 is the expression for finding the least diameter of the impeller.

## Diameter of Suction Pipe and Delivery Pipe

The velocities of flow in both pipes are limited between 1.5 and 3 m/s. So if discharge rate and velocity of flow in pipes are specified, then the diameter of suction pipe and delivery pipe is found out using following Equations:

$$Q = \frac{\pi d^2 v}{4} \tag{35}$$

$$d = \sqrt{\frac{4Q}{\pi v}} \tag{36}$$

Usually the diameter of suction pipe and delivery pipe is of same size or suction pipe will be slightly larger than the delivery. The advantage of a larger suction pipe in the following sections.


## Limitation of Suction Lift


Applying Bernoulli's equation between water level in sump and eye of impeller and taking into account the atmospheric pressure (absolute pressure system), following equation is obtained:

$$\frac{p_a}{\gamma} = \frac{p_s}{\gamma} + \frac{v_s^2}{2g} + h_s + h_{fs} \tag{37}$$

where $\gamma$ is the specific weight of water. Rearranging the preceding equation yields the following:

$$\frac{p_s}{\gamma} = \frac{p_a}{\gamma} - \left[ \frac{v_s^2}{2g} + h_s + h_{fs} \right] \tag{38}$$

From the preceding equation, it is understood that the pressure at the eye is less than the atmospheric pressure. The extent of reduction depends on the height of suction lift, kinetic head and friction loss in suction.

If the pressure at suction goes below vapour pressure of water, water will get vaporized at eye. If water gets vaporized, steam bubbles will be formed. If there is a slight vaporization, as the water goes along the impeller, the pressure increases and once again the bubbles break on the impeller. Because of this the impeller material gets corroded. This process is called cavitations (Fig. 25). If the pressure at the eye is very much less than the vapour pressure, then pumping itself will be stopped, because the pressure developed depends on the density of liquid pumped. If water gets vaporized, the density would become less and the sufficient pressure would not get developed. The vapour pressure head for water in absolute pressure unit system is 0.24 m of water. Therefore $p_s/\gamma$ should not be less than 0.24 m. Let us assume $\frac{v_s^2}{2g}$ and $h_{fs}$ are negligible and substituting the atmospheric pressure as 10.33 m of water column at the mean sea level in the preceding equation, the value of $h_s$ obtained is 10.1 m. Therefore, 10.1 m is the maximum possible suction head. For mathematical simplicity, $h_{fs}$ and $\frac{v_s^2}{2g}$ are assumed as negligible. If $h_{fs}$ and $\frac{v_s^2}{2g}$ are also considered, the maximum value of $h_s$ will get reduced further. Also, the value of atmospheric

**Fig. 25** Cavitation (www.lig
htmypump.com)

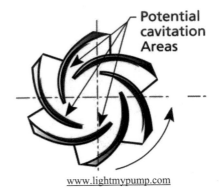

**Potential
cavitation
Areas**

www.lightmypump.com

pressure head has been assumed as 10.33 m of water column and this value also would get reduced when the elevation of location of the pump is above mean sea level. If the actual atmospheric pressure head at the pump site (say 10.0 m) is substituted in the preceding equation, the maximum suction head will get reduced by a value of 0.33 m. So $h_s$ will become still less. This is one big disadvantage of the centrifugal pump. That is the reason why when the water level in the wells goes down the pumps are also brought down.

Following equation can be used for estimating atmospheric pressure head ($p_a/\gamma$) in m of water column, for any elevation (H in m):

$$p_a/\gamma = 10.33 - 0.00119H \tag{39}$$

It can be observed from the Eq. 38 that if diameter of suction is more, $v_s^2/2g$ will get decreased. Then $p_s/\gamma$ will be more. Because of this fact only, sometimes the diameter of suction pipe is more than the suction pipe. Since the suction pipe is working below atmospheric pressure, if there is any leak in the suction pipe, air from atmosphere will get sucked inside and pumping will get affected. But in the delivery side, the pipe is operating at pressures above atmospheric pressure. Therefore, if there are any leaks in delivery, there will be water gushing out of the passages. The pumping will not get affected. The degree of submergence of the suction inlet is also important. If there is no sufficient submergence, air from atmosphere may get into the suction pipe as shown in Fig. 26. At least, the minimum submergence should be

**Fig. 26** Suction
submergence

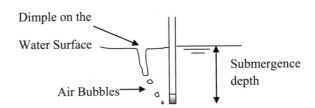

Dimple on the
Water Surface
Air Bubbles
Submergence
depth

more than 5.5 times the suction diameter. In practice, normally 6 m of suction lift is the maximum lift adopted.

## Loss of Head and Efficiencies

Various head losses that would occur during the operation of centrifugal pump are as follows:

1.  Hydraulic losses
2.  Mechanical losses
3.  Leakage losses.

Hydraulic losses are divided into loss with in the pump and other hydraulic losses. The head losses with in the pump are as follows:

1.  Shock losses at the suction and entry and exit of vanes
2.  Friction loss in impeller
3.  Eddy loss in casing.

The other hydraulic losses are as follows:

1.  Friction loss in suction and delivery pipes
2.  Loss at foot valve strainer, bends, valves, etc.

## *Mechanical Losses*

They occur because of friction between impeller and the liquid, which fits the clearance or space between impeller and casing and also due to mechanical friction of bearings.

## *Leakage Losses*

Within the pump casing, both high-pressure zone (volute chamber) and also low-pressure zone (eye) are existing. Because of this pressure difference, some amount of water leaks again to the eye of the pump from volute chamber. Thus, a small amount of total discharge is unnecessarily under circulation. Because of this some energy is lost.

## Performance of Pumps—Characteristic Curves

A pump is usually designed for one set of speed, flow rate and pressure head, but in actual practice, the operation may be at some other condition of head or flow rate. For the changed conditions, the behaviour of the pump may be quite different. As indicated earlier, if the flow through the pump is less than the designed quantity, the value of the velocity of flow of liquid through the impeller will be changed, thereby changing the head developed by the pump, and at the same time the losses will increase, and the efficiency of the pump will be lowered.

Therefore, in order to predict the behaviour and performance of a pump under varying conditions, tests are performed, and the results of the tests are plotted. The curves thus obtained are known as the characteristic curves of the pump. The following three types of characteristic curves are usually prepared for the centrifugal pumps:

(a)    Main and operating characteristics
(b)    Constant efficiency curves

In order to obtain the main characteristic curves of a pump, it is operated at different speeds. For each speed, the rate of flow (Q) is varied by means of a delivery valve and for the different values of Q the corresponding values of TDH, shaft horse power (P), and overall efficiency are measured or calculated: The same operation is repeated for different speeds of the pump. Then TDH versus Q, P versus Q and Efficiency versus Q curves for different speeds are plotted, so that three sets of curves, as shown in Fig. 27 are obtained, which represent the main characteristics of a pump. The main characteristics are useful in indicating the performance of a pump at different speeds.

During the operation, a pump is normally required to run at a constant speed, which is its designed speed (same as the speed of the driving motor). As such that particular set of main characteristics which corresponds to the designed speed is mostly used in the operation of a pump and is, therefore, known as the operating characteristics. A typical set of such characteristics of a pump is shown in Fig. 28. The value of the head and the discharge corresponding to the maximum efficiency are known as the normal (or designed) head and the normal (or designed) discharge of a pump.

The head corresponding to zero or no discharge is known as the shut off head of the pump. From these characteristics, it is possible to determine whether the pump will handle the necessary quantity of liquid against the desired head and one can know what will happen if the head is increased or decreased. The P versus Q curve will show what size motor will be required to operate the pump at the required conditions and whether or not the motor will be overloaded under any other operating conditions.

Figure 29 shows constant efficiency curves. In this graph, it can be seen that for every impeller speed, there are different combinations of pressure head and discharge. For each of these points, an associated overall efficiency exists. Using those efficiencies the curves as shown in the figure can be drawn. When variable speed pumps are used, operation can be done along the line of maximum efficiency. Variable speed pumps can be seen in situations where power is obtained from tractor power take off.

(a) Main and Operating characteristics

(b) Constant efficiency curves

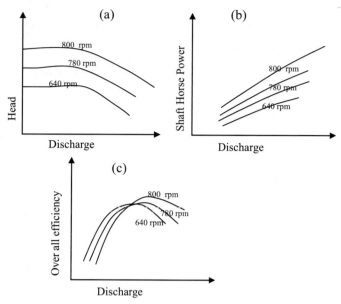

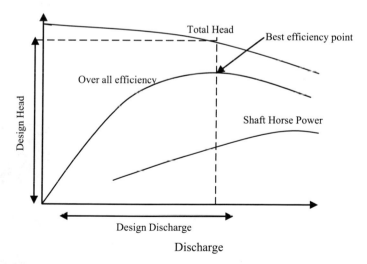

**Fig. 27** Main characteristics

**Fig. 28** Operating characteristics

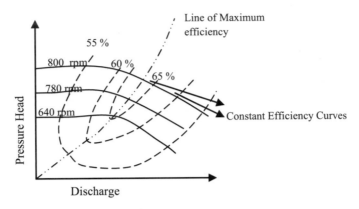

**Fig. 29** Constant efficiency curves

## Specific Speed of a Pump

It is the speed of a geometrically similar pump delivering a unit rate of discharge at unit head. If a pump delivers 10 L per minute at the head of 5 m and rotates at a speed of 1440 rpm, the specific speed of the pump is, at what speed if the same pump rotates, it will deliver 1 m³/s of discharge at the head of 1 m.

Discharge rate (Q) is a product of area of cross section of flow at the outlet of impeller and radial velocity at the outlet of impeller. The cross-sectional area of flow is equal to the product of outer perimeter of the impeller and width of impeller (B) shown in Fig. 30.

$$Q \alpha \pi D_2 B V_{f1}$$

**Fig. 30** Width of impeller
(www.ksbpumps.com)

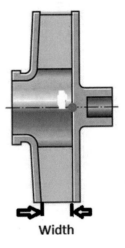

**Width**

## Expression for Specific Speed

Where $D_1$ is outer diameter of the impeller; $V_{f1}$ is radial velocity of water jet at the exit of impeller. When we make geometrically similar pump, the relationship between the width of the impeller and the diameter of the impeller will be maintained. Hence.

$$Q \alpha D_2^2 B V_{f1}$$

According to Toricelli's law, pressure head and velocity are related as follows:

$$V_{f1} \alpha TDH^{0.5}$$

where $H_m$ is the manometric head developed.
Therefore,

$$Q \alpha D_2^2 TDH^{0.5}$$

$$\frac{Q}{D_2^2 \sqrt{TDH}} = k \tag{40}$$

where '$c$' is a constant. Further, the tangential velocity of the impeller at the exit of the impeller ($u_1$) is,

$$u_2 = \frac{\pi D_2 N}{60}$$

$$\Rightarrow D_2 \, \alpha \frac{u_2}{N}$$

It is known that $u_2 \, \alpha \, \sqrt{TDH}$.
Therefore, $D_2 \, \alpha \, \frac{\sqrt{TDH}}{N}$.
Substitute the preceding relationship in Eq. 40

$$\frac{Q \, N^2}{TDH^{3/2}} = k$$

$$\frac{\sqrt{Q} \, N}{TDH^{3/4}} = C$$

For, $Q = 1$ l/s, TDH $= 1$ m, $C$ is equal to $N$. This $N$ is the specific speed ($N_s$), as per the definition of specific speed. Therefore, the specific speed is as follows:

$$N_s = \frac{N \sqrt{Q}}{TDH^{3/4}}$$

where

$N_s$    is specific speed.
$N$     is the speed of the pump in revolutions per minute (rpm).
$Q$     is the flow rate in litres per minute.

TDH is the total dynamic head in metres. For multi-stage pumps, the head developed per single stage should be used.

By calculating the specific speed from any pump data available, it is possible to find out whether the pump is radial or mixed or axial flow pump. Specific speed information can be used to:

- Select the shape of the pump curve
- Determine the efficiency of the pump
- Anticipate motor overloading problems
- Select the lowest cost pump for their application.

## *Specific Speed Versus Operating Characteristics*

Figure 31 shows the different shapes of impeller for different specific speeds. Following points can be noted from Fig. 32.

- The steepness of the head/ capacity curve increases as specific speed increases.
- At low specific speed, power consumption is the lowest at zero discharge and rises as flow increases. This means that the motor could be over loaded at the higher flow rates unless this was considered at the time of purchase.

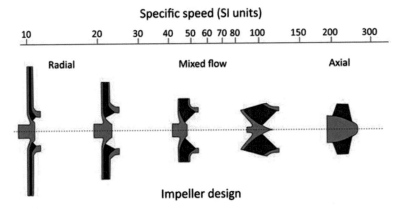

*Typical Suction Speed values for various impeller designs (units used are rpm, m³/s, m)*

**Fig. 31** Specific speed ranges for different impeller types (https://www.michael-smith-engineers. co.uk/resources/useful-info/specific-speed)

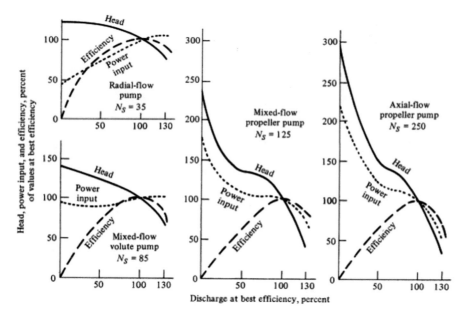

**Fig. 32** Operating characteristics for different impeller types

- At medium specific speed, the power curve peaks at approximately the best efficiency point. This is a non-overloading feature meaning that the pump can work safely over most of the fluid range with a motor speed to meet the best efficiency point requirement.
- High specific speed pumps have a falling power curve with maximum power occurring at minimum flow. These pumps should never be started with the discharge valve completely closed. If throttling is required a motor of greater power will be necessary.
- As a rule of thumb, lower specific speeds produce flatter curves, while higher specific speeds produce steeper ones

If the depth of water level fluctuates appreciably, radial flow pumps are better. If the depth of water level is approximately constant either mixed flow or axial flow pumps can be opted.

## Assessment

### Part A

1. Identify the correct statement among the following:

   a. Electric motor overloading problem occurs in mixed flow pumps
   b. Electric motor overloading problem occurs in radial flow pumps

   c.   **Electric motor overloading problem occurs in axial flow pumps if delivery valve is closed too much**

   d.   Electric motor overloading problem occurs in all kinds of pumps

2.   Identify the correct statement

   a.   Diffuser pumps work well for a constant discharge rate

   b.   Diffuser pumps work well only for sewage pumping

   c.   Diffuser pumps do not work well for sprinkler irrigation

   d.   **Diffuser pumps work well if groundwater level fluctuates significantly**

3.   _____ happens in volute chamber of a centrifugal pump

   a.   **Pressure is converted into kinetic energy**

   b.   Kinetic energy is converted into pressure energy

   c.   No energy conversion occurs

   d.   Forced vortex

4.   What happens when installing a larger diameter pipe in a pumping system ?

   a.   **Reduction in frictional head**

   b.   Reduction in static head

   c.   Both a and b

   d.   None of the above

5.   How does centrifugal pump lift water?

   a.   Creation of vacuum

   b.   **Addition of pressure**

   c.   Addition of gravitational energy

   d.   Addition of air

6.   In a centrifugal pump, the tangential velocity of water at the outlet of vane is 9.81 m/s and the tangential velocity of impeller at the exit is 10 m/s. Find out the theoretical work done by the impeller per unit weight of water?

   a.   10 m

   b.   9.81 W

   c.   98.1 W

   d.   15 J

7.   Find out the specific speed of a pump whose discharge rate is 1 $m^3$/s and it works against a head of 1 m. The speed of the impeller is 1440 rpm?

   a.   **1440 rpm**

   b.   1 rpm

   c.   2880 rpm

   d.   10 rpm

8. Reducing the discharge by using delivery valve of a radial flow pump results in

   a. **Decrease of power**
   b. Decrease of head and increase of power
   c. Increase of power
   d. Decrease of head

9. Shutting off the delivery valve of a propeller pump _____

   a. **is dangerous**
   b. Does not cause any harm
   c. Decreases power consumption
   d. Decreases the manometric head

10. Electric motor overloading problem does not occur in

   a. Radial flow pumps
   b. **Mixed flow**
   c. Axial flow
   d. Jet flow

11. Circulation of a fraction of water from high-pressure zone to low-pressure zone with in casing is called as

   a. **Leakage loss**
   b. Forced vortex
   c. Free vortex
   d. Vortex

12. In a Centrifugal pump, the outer perimeter of the impeller is 0.75 m, the thickness of the impeller is 0.01 m, the radial component of velocity is 2.0 m/s and tangential component velocity is 1 m/s, the discharge rate is

   a. **0.015 cubic metres per second**
   b. 0.25 L per second
   c. 0.0075 cubic metres per second
   d. None of the above

13. A pump has a mechanical efficiency of 90%, motor efficiency of 90%, manometric efficiency of 90. What is the overall efficiency of the pump?

   a. **72.9%**
   b. 90%
   c. 75.2%
   d. None of the above

14. Function of foot valve is to have _____, when pump is started every time.

   a. **Priming**
   b. Cavitation

c.   Addition of kinetic energy

d.   Addition of pressure energy

15. _____ is bursting of steam bubbles over impeller vanes when the pressure at the eye of the pump goes below vapour pressure of water

a.   **Cavitation**

b.   Priming

c.   Trimming

d.   Washing

16. An impeller has a linear velocity at the outer rim of 15 m/s and the tangential component of velocity of water at the outlet of vanes is 15 m/s. Find out the theoretical work done per unit weight of water?

a.   **22.93 m**

b.   22.93 W

c.   22.93 J

d.   None of the above

17 _____ pumps have the highest specific speed

a.   **Axial flow pumps**

b.   Mixed flow pumps

c.   Radial flow pumps

d.   Double suction pumps

18. _____pumps have the highest efficiency

a.   **Axial flow pumps**

b.   Mixed flow pumps

c.   Radial flow pumps

d.   Single suction pumps

## Part B

1.   A centrifugal pump operates against a manometric head of 8.5 m. The inner diameter of impeller is 25 cm and outer diameter of the impeller is 50 cm. Find out the minimum speed of the impeller needed to initiate the flow? (Answer— 570 rpm)

2.   A pump was evaluated for its performance with standard method and accessories. The elevation of the pressure gauge in suction pipe (point 1 in figure) is at the centre of eye of the pump and the value of the vacuum pressure shown by the gauge is—31.4 kPa. The pressure at the delivery gauge is 176.5 kPa. The discharge rate of the pump is 35 L per second. The difference in elevation between suction and delivery gauge is 50 cm (h in the figure). The inside diameter of the suction and delivery pipe is 10 cm. Find out the manometric head developed by the pump and also power delivered by the pump?

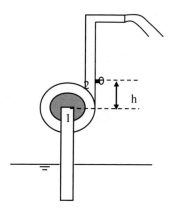

# References and Further Readings

1. Cengel, A. Y., & Cimbala, J. M. (2014). *Fluid mechanics*. McGraw Hill.
2. Modi, P. N., & Seth, S. M. (2018). *Hydraulics and fluid mechanics including hydraulic machines*. Standard Book House. https://en.wikipedia.org/wiki/Centrifugal_pump

Printed in the United States
by Baker & Taylor Publisher Services